Innovations in Engineering and Food Science

Shilpa Mehta
Auckland University of Technology, New Zealand

Fakhar Islam
NUR International University, Pakistan

Ali Imran
Government College University, Pakistan

A volume in the Advances in Environmental Engineering and Green Technologies (AEEGT) Book Series

IGI Global
PUBLISHER of TIMELY KNOWLEDGE

Published in the United States of America by
IGI Global
Engineering Science Reference (an imprint of IGI Global)
701 E. Chocolate Avenue
Hershey PA, USA 17033
Tel: 717-533-8845
Fax: 717-533-8661
E-mail: cust@igi-global.com
Web site: http://www.igi-global.com

Library of Congress Cataloging-in-Publication Data

Names: Mehta, Shilpa, 1987- editor.
Title: Innovations in engineering and food science / edited by Shilpa Mehta, Fakhar Islam, Ali Imran.
Description: Hershey, PA : Information Science Reference, [2024] | Includes bibliographical references and index. | Summary: "This book identifies and promotes emerging technologies and solutions for engineering and food science domains"-- Provided by publisher.
Identifiers: LCCN 2023037446 (print) | LCCN 2023037447 (ebook) | ISBN 9798369308196 (hardcover) | ISBN 9798369308202 (ebook)
Subjects: LCSH: Engineering--Technological innovations. | Food science--Technological innovations.
Classification: LCC TA7 .I634 2024 (print) | LCC TA7 (ebook) | DDC 620--dc23/eng/20240126
LC record available at https://lccn.loc.gov/2023037446
LC ebook record available at https://lccn.loc.gov/2023037447

This book is published in the IGI Global book series Advances in Environmental Engineering and Green Technologies (AEEGT) (ISSN: 2326-9162; eISSN: 2326-9170)

British Cataloguing in Publication Data
A Cataloguing in Publication record for this book is available from the British Library.

Advances in Environmental Engineering and Green Technologies (AEEGT) Book Series

ISSN:2326-9162
EISSN:2326-9170

Editor-in-Chief: Sang-Bing Tsai, Zhongshan Institute, University of Electronic Science and Technology of China, China & Wuyi University, China; Ming-Lang Tseng, Lunghwa University of Science and Technology, Taiwan; Yuchi Wang, University of Electronic Science and Technology of China Zhongshan Institute, China

MISSION

Growing awareness and an increased focus on environmental issues such as climate change, energy use, and loss of non-renewable resources have brought about a greater need for research that provides potential solutions to these problems. Research in environmental science and engineering continues to play a vital role in uncovering new opportunities for a "green" future.

The **Advances in Environmental Engineering and Green Technologies (AEEGT)** book series is a mouthpiece for research in all aspects of environmental science, earth science, and green initiatives. This series supports the ongoing research in this field through publishing books that discuss topics within environmental engineering or that deal with the interdisciplinary field of green technologies.

COVERAGE

- Sustainable Communities
- Green Technology
- Contaminated Site Remediation
- Renewable Energy
- Industrial Waste Management and Minimization
- Electric Vehicles
- Water Supply and Treatment
- Pollution Management
- Policies Involving Green Technologies and Environmental Engineering
- Cleantech

IGI Global is currently accepting manuscripts for publication within this series. To submit a proposal for a volume in this series, please contact our Acquisition Editors at Acquisitions@igi-global.com or visit: http://www.igi-global.com/publish/.

The Advances in Environmental Engineering and Green Technologies (AEEGT) Book Series (ISSN 2326-9162) is published by IGI Global, 701 E. Chocolate Avenue, Hershey, PA 17033-1240, USA, www.igi-global.com. This series is composed of titles available for purchase individually; each title is edited to be contextually exclusive from any other title within the series. For pricing and ordering information please visit http://www.igi-global.com/book-series/advances-environmental-engineering-green-technologies/73679. Postmaster: Send all address changes to above address.

Titles in this Series

For a list of additional titles in this series, please visit:
www.igi-global.com/book-series/advances-environmental-engineering-green-technologies/73679

Modeling and Monitoring Extreme Hydrometeorological Events
Carmen Maftei (Transilvania University of Brasov, Romania) Radu Muntean (Transilvania University of Brașov, Romania) and Ashok Vaseashta (International Clean Water Institute, USA)
Engineering Science Reference • © 2024 • 338pp • H/C (ISBN: 9781668487716) • US $250.00

Convergence Strategies for Green Computing and Sustainable Development
Vishal Jain (Sharda University, Greater Noida, India) Murali Raman (Asia Pacific University of Technology, Malaysia) Akshat Agrawal (Amity University, Guragon, India) Meenu Hans (K R Mangalam University Haryana) and Swati Gupta (Center of Excellence, K. R. Mangalam University, Gurugram, India)
Engineering Science Reference • © 2024 • 330pp • H/C (ISBN: 9798369303382) • US $270.00

Novel AI Applications for Advancing Earth Sciences
Sudesh Yadav (Department of Higher Education, Government PG College, Ateli, India) Satya Prakash Yadav (GL Bajaj Institute of Technology and Management, India) Pethuru Raj (Edge AI Division, Reliance Jio Platforms Ltd., Bangalore, India) Prabhakar Tiwari (Madan Mohan Malaviya University of Technology, India) and Victor Hugo C. de Albuquerque (Department of Teleinformatics Engineering (DETI), Federal University of Ceará, Brazil)
Engineering Science Reference • © 2024 • 405pp • H/C (ISBN: 9798369318508) • US $265.00

Exploring Ethical Dimensions of Environmental Sustainability and Use of AI
Hemachandran Kannan (Woxsen University, India) Raul Villamarin Rodriguez (Woxsen University, India) Zita Zoltay Paprika (Corvinus University of Budapest, Hungary) and Abejide Ade-Ibijola (University of Johannesburg, South Africa)
Engineering Science Reference • © 2024 • 426pp • H/C (ISBN: 9798369308929) • US $265.00

IGI Global
PUBLISHER of TIMELY KNOWLEDGE

701 East Chocolate Avenue, Hershey, PA 17033, USA
Tel: 717-533-8845 x100 • Fax: 717-533-8661
E-Mail: cust@igi-global.com • www.igi-global.com

Table of Contents

Detailed Table of Contents

Mobile networks play a crucial role in facilitating communication through the transmission and reception of radio wave signals. These networks are composed of interconnected cells provided by base stations, enabling wide geographic coverage. The evolution of mobile networks has progressed through several stages. It began with the analog-based first-generation systems that provided basic voice communication. The current phase is the fifth generation, which aims to deliver exceptional performance with faster speeds, low latency, and connectivity density. Ongoing research and development continue to shape the evolution of mobile networks, with technologies like 6G on the horizon, promising even faster speeds and transformative use cases. The network infrastructure for 5G and 6G plays a crucial role in enabling the capabilities and delivering the promised benefits of these advanced wireless communication technologies. The industry applications of 5G and the anticipated applications of 6G are diverse and have the potential to revolutionize various sectors.

The waves that occur in an earthquake affect the behavior of the soil and the building due to the structural condition of the bedrock and the physical properties of the soil near the surface. To predict this interaction, the deep underground structure models of the areas considered as settlements should be determined and how the earthquake waves will be directed (macro-zoning) depending on this should be revealed. Since the near surface features of the ground can increase the effect of earthquake waves, it is necessary to determine the near surface properties and create their maps (micro-zoning). Earthquake waves are formed in thick sediments compared to the rock environment and cause great damage to both the ground and the buildings. In order to be protected from these damages, it is necessary to determine focusing and scattering conditions arising from the deep underground base topography. Therefore, parameters such as maximum horizontal ground acceleration, soil liquefaction, soil dominant vibration period, soil amplification, and near-surface soil properties should be determined.

As urbanization continues to accelerate, the concept of smart cities has emerged as a promising solution to address the complex challenges faced by rapidly growing urban areas. Smart cities leverage advanced technologies to enhance the quality of life for residents, optimize resource utilization, and promote sustainable development. However, the successful implementation of smart cities heavily relies on the engineering of resilient and sustainable infrastructure. This chapter highlights the key aspects of engineering resilient and sustainable infrastructure for smart cities, focusing on the integration of technological innovations, resilience planning, and sustainable practices. The engineering of resilient infrastructure plays a crucial role in smart cities' ability to withstand and recover from various shocks and stresses, including natural disasters, climate change impacts, and cyber-attacks. Resilient infrastructure emphasizes the integration of robust design principles, redundancy, adaptability, and effective risk management strategies.

The advent of Industry 4.0 brought about a transformative wave of automation and digitalization in manufacturing and industrial processes. However, the next phase

In order to ensure a consistent supply of nutrient-rich nutritious food, food safety and security are gaining a lot of attention on a global scale. Numerous technical activities, such as the creation of novel materials and the improvement of food safety and security, need nanotechnology. Nanomaterials are utilized to enhance food protection and aid in the detection of pesticides, toxic chemicals, and microbiological contamination. Nanomaterials are utilized to enhance the detection of pesticides, toxic chemicals, and microbiological contamination. Nanocapsulation, which enables the release of bioactive substances, raises food bioavailability, and lengthens food shelf life, improves food processing even further. The main purpose of food processing is to fortify food by adding nutrients and beneficial ingredients. This chapter covers a variety of subjects, including the use of nanotechnology in food packaging, food safety, food preservation using smart nanocarriers, the detection of allergens and food-borne viruses using nanosensors, and crop growth and yield enhancement.

Food safety ensures reduced food losses and therefore contributes to food security. The factors causing the contamination of the food may threaten the safe consumption and make the foods harmful. For this reason, it is necessary to utilize various resources to prevent the food from being contaminated in all stages of the food chain. Purified high titer phage lysates have been used for the species-specific control of bacteria during the pre and postharvest phases of food production and storage. This chapter presents recent advances in bacteriophage research and their application in the area of food safety. Some of the components of this chapter are general facts on phage biology that are relevant to their application for control and detection of bacterial pathogens in food and environmental samples, application of bacteriophages to control growth of bacterial pathogens and spoilage organisms in food and food processing environment, detection and identification of bacterial pathogens, and advantages of bacteriophage-based methods and their shortcomings.

Edible food wrappers are safe packaging systems that can be consumed with food. Edible packaging developed using natural renewable resources come under the category of bio-based and biodegradable packaging. The main components of edible wrappers include a biopolymer solubilized in a solvent to form film casting solution and certain additives to improve functional characteristics of packaging. Biopolymers are frequently used in food packaging applications due to their flexibility in film formation and biodegradability. The development of green packaging materials utilizing discarded edible materials will be beneficial to confront the challenges associated with traditional packaging systems, attaining sustainability, and fostering recycling in the food industry. Edible food wrappers present a number of benefits over synthetic materials. These packaging systems can replace and possibly strengthen the outside layers of packed items in order to elude moisture loss, flavors, and bioactive components from the foods as well as between them.

The recovery of valuable sources from waste material is an excellent and encouraging approach towards the ecological significances and food security. Agro-industrial surplus produced by agricultural and food processing activities signifies a rich source of beneficial compounds like minerals, vitamins, and other essential components that are helpful in different industries for different purposes. The demands of plant-based protein extraction are very high. Along with proteins, various bioactive components, like carotenoids, lycopene, polyphenols, sugars, and polysaccharide, have also been extracted with the help of eco-friendly technologies. The purpose of this chapter is to study the latest

Preface

Welcome to *Innovations in Engineering and Food Science.* As editors, it is our pleasure to present this comprehensive compilation exploring the dynamic intersection of engineering and food science.

In our rapidly evolving world, engineering stands as a cornerstone for developing products and services that enhance lives. Technologies like Artificial Intelligence, Blockchain, Quantum Computing, and others are seamlessly integrating into modern products, ushering in a new era of convenience, efficiency, and capability. This book delves into these advances, showcasing how they intersect with and transform the realms of engineering and food science.

The challenges we face in feeding a growing global population while mindful of depleting resources demand innovative solutions. This book addresses these concerns by discussing strategies to decrease food loss, enhance production, explore new ingredients, and digitalize the food system. From alternative food sources like microalgae and insects to leveraging technologies such as Artificial Intelligence and Blockchain, this book encapsulates the innovative pathways shaping the future of food production.

Our primary goal is to illuminate emerging technologies and solutions within these domains. We aim to offer a comprehensive guide, not only detailing advancements but also presenting practical insights into their implementation. Researchers, professionals, students, and academicians will find within these pages a rich resource to expand their understanding and apply these insights in their respective fields.

The chapters span an array of topics, from Soft Computing and Computational Intelligence in Engineering to Nanotechnology's applications in improving food safety and sustainability. Each chapter contributes a unique perspective, fostering a deeper comprehension of these rapidly evolving fields.

We extend our gratitude to the contributors whose expertise and dedication have enriched this compilation. It is our hope that this edited reference book serves as a catalyst for innovation, guiding and inspiring future breakthroughs in engineering and food science.

ORGANIZATION OF THE BOOK

Chapter 1: 5G-6G – Infrastructure and Industrial Applications

This chapter illuminates the evolution of mobile networks, detailing the transition from first-generation analog systems to the current fifth generation. It anticipates the promises and possibilities of 6G technology, emphasizing its potential industry applications and transformative impact.

Chapter 2: Local Site Effects Evaluation Using Geophysical Methods

Understanding seismic waves' impact on soil and buildings during earthquakes is vital. This chapter delves into the assessment and prediction of these effects, highlighting the necessity to determine both deep underground structures and near-surface soil properties for comprehensive seismic risk management.

Chapter 3: Crafting Intelligent Urban Environments Innovative Engineering Approaches for City Evolution

As cities evolve into smart cities, this chapter focuses on engineering resilient and sustainable infrastructure critical for their success. It emphasizes technology integration, resilience planning, and sustainable practices for cities to withstand challenges like natural disasters and cyber threats.

Chapter 4: Industry 5.0 From Automation to Autonomy – Engineering the Shift: The Transformation

Moving beyond automation, Industry 5.0 aims for seamless human-machine integration. This chapter explores the paradigm shift, emphasizing how technologies like AI, IoT, and robotics empower human workers to achieve higher productivity and innovation in manufacturing.

Chapter 5: Policies for Advanced Dynamic Load Balancing in Cloud Computing

Examining the complexities of cloud computing, this chapter dives into load balancing mechanisms to improve resource utilization and task response time. It discusses load distribution methods and proposes solutions to optimize throughput and latency in various cloud-based applications.

Chapter 6: Revolutionizing Food – The Latest Frontiers in Food Science and Technology

Highlighting innovative non-thermal food processing technologies, this chapter explores advancements in PEF processing, HPP, and AI in food production. It underscores the challenges and benefits these technologies offer in terms of food safety, quality, and sustainability.

Chapter 7: Applications of Nanotechnology for Improving Food Safety and Security

Focused on ensuring food safety, this chapter delves into nanotechnology's role in detecting contaminants, enhancing food preservation, and improving crop yield. It explores nanomaterials' applications in food packaging, biosensors, and nutrient delivery systems.

Chapter 8: Bacterial Pathogens in Food and Their Control by Bacteriophages

Addressing food safety concerns, this chapter explores bacteriophages' role in controlling bacterial pathogens throughout the food chain. It details phage biology, their application in detecting and controlling pathogens, and the advantages and limitations of phage-based methods.

Chapter 9: Development of Edible Food Wrappers – An Eco-Friendly Approach Towards Sustainability

Edible food wrappers are safe packaging systems that can be consumed with food. Edible packaging developed using natural renewable resources come under the category of bio based and biodegradable packaging. The main components of edible wrappers include a biopolymer solubilized in a solvent to form film casting solution and certain additives to improve functional characteristics of packaging. Biopolymers are frequently used in food packaging applications due to their flexibility in film formation and biodegradability.

Chapter 10: Extraction of Protein and Polyphenols from Agro-Industrial Waste Through Eco-Innovative Technologies

Focusing on eco-friendly methods, this chapter explores the recovery of valuable compounds from agro-industrial waste. It highlights the extraction of proteins and polyphenols and their significance in producing nutrient-rich products while minimizing food wastage.

Chapter 11: Nanoclay Packaging Materials as Biodegradable

Examining nanoclay's role in food packaging, this chapter discusses its potential in enhancing mechanical and barrier properties, antimicrobial activity, and controlled release of active ingredients. It reviews current achievements, challenges, and future prospects in nanoclay technology.

Chapter 12: Recent Advances in the Utilization of Artificial Intelligence in the Food Industry

Detailing AI's transformative role in agriculture and food processing, this chapter explores its applications in crop monitoring, quality control, distribution, and food safety. It emphasizes AI's potential in creating a secure, efficient, and sustainable food ecosystem.

Chapter 13: Effects of Gluten on Health; Pseudocereals as Gluten Substitute

This chapter delves into the impact of gluten on health and explores pseudocereals as gluten substitutes. It addresses gluten-related disorders, dough rheology, and the potential of alternative grains like quinoa and amaranth in the food industry and nutrition routines.

IN SUMMARY

Innovations in Engineering and Food Science brings together a mosaic of cutting-edge insights and pioneering advancements that define the forefront of these dynamic fields. As editors, our journey through the diverse chapters of this book has unveiled a tapestry of knowledge, innovation, and transformative possibilities.

From the evolution of mobile networks in the 5G-6G spectrum to the intricate understanding of seismic effects on urban environments, each chapter unravels a realm of expertise and exploration. The trajectory from Industry 4.0 to the collaborative synergy of humans and machines in Industry 5.0 charts a transformative course in industrial evolution.

Cloud computing's dynamic load balancing mechanisms, nanotechnology's impact on food safety and packaging, and the intersection of AI with food science underscore the profound implications of technology on our future sustenance and progress.

Moreover, our exploration delved into the depths of ecological significance in food production, emphasizing the extraction of valuable components from agro-industrial waste and the quest for biodegradable packaging materials.

We've examined the intricacies of gluten and its substitutes, shedding light on the impact of dietary choices on health and the promising potential of alternative grains.

This compilation is more than a repository of knowledge; it's a testament to collaborative efforts, interdisciplinary research, and the pursuit of solutions that transcend conventional boundaries. The chapters within this volume represent not just individual insights but interconnected facets of a larger mosaic that defines the forefront of innovation.

We extend our deepest gratitude to the contributors whose expertise and dedication have enriched this compilation. Their relentless pursuit of excellence and commitment to pushing the boundaries of knowledge have shaped this book into a beacon of inspiration and knowledge.

As we conclude this journey through the realms of engineering and food science, we envision this book as a catalyst, igniting further discussions, innovations, and breakthroughs. May it serve as a guiding light for researchers, professionals, academicians, and students, inspiring them to push the boundaries of what's possible in shaping a technologically advanced and sustainable future.

Shilpa Mehta
Auckland University of Technology, New Zealand

Fakhar Islam
NUR International University, Pakistan

Ali Imran
Government College University, Pakistan

Chapter 1
5G-6G:
Infrastructure and Industrial Applications

Roopashree Nayak
https://orcid.org/0000-0002-1653-3392
Sahyadri College of Engineering and Management, Mangaluru, India

Pavanalaxmi S.
https://orcid.org/0000-0002-0356-7083
Sahyadri College of Engineering and Management, Mangaluru, India

Praveen Kumar M.
https://orcid.org/0000-0001-9291-3074
Sahyadri College of Engineering and Management, Mangaluru, India

ABSTRACT

Mobile networks play a crucial role in facilitating communication through the transmission and reception of radio wave signals. These networks are composed of interconnected cells provided by base stations, enabling wide geographic coverage. The evolution of mobile networks has progressed through several stages. It began with the analog-based first-generation systems that provided basic voice communication. The current phase is the fifth generation, which aims to deliver exceptional performance with faster speeds, low latency, and connectivity density. Ongoing research and development continue to shape the evolution of mobile networks, with technologies like 6G on the horizon, promising even faster speeds and transformative use cases. The network infrastructure for 5G and 6G plays a crucial role in enabling the capabilities and delivering the promised benefits of these advanced wireless communication technologies. The industry applications of 5G and the anticipated applications of 6G are diverse and have the potential to revolutionize various sectors.

DOI: 10.4018/979-8-3693-0819-6.ch001

1. INTRODUCTION

A wirelessly connected mobile network that covers a large section of the planet's land surface and is made up of transceivers at fixed locations known as base stations or cell sites. The base stations' network coverage facilitates the transmission of speech, data, and various types of information within each cell. To ensure high service quality and avoid interference, neighboring cells often utilize different frequency ranges. These cells are interconnected to provide comprehensive radio coverage across a large geographic area. This permits continuous communication between multiple portable transceivers (such as tablets, pagers, mobile phones, laptops, etc.) as well as fixed telephones and transceivers within the network through the use of base stations. This becomes specifically vital when some users are moving across multiple cells simultaneously while engaging in data transmission. The majority of populous regions on Earth have data and voice cellular networks installed by major telecommunications companies. Now, computers and mobile phones can connect to the public switched telephone network and the public Internet.

2. EVOLUTION OF MOBILE NETWORK (ARSHAD, 2019; DE VRIENDT, 2002)

- **1G:** In the 1980s, analog cell phone networks were released. In order to prevent economies of scale and limit usage to internal boundaries, each country at the time developed its own system. One can make voice-only calls over 1G. The 1G technology was analog and had inferior voice quality, battery life, and security in the phones that used it. They frequently dropped calls as well. The top speed of 1G technology is 2.4 Kbps. In the early to mid-1990s, 2G (second generation) cellular technologies took the role of 1G.
- **2G** (Stuckmann, 2002 & Halonen, 2004)**:** It makes it possible to digitize and compress voice transmissions. As a result, they achieve more efficient utilization of the frequency spectrum compared to 1G. They introduced SMS text messaging-based data services for mobile devices. Both data and audio signals are digitally encrypted, leading to substantial enhancements in security against fraud and eavesdropping. Additionally, digital transmissions consume less battery power, further contributing to improved efficiency. Therefore, mobile sets use substantially less energy than their 1G equivalents.

2G systems Include:

D – AMPS (Digital Advanced Mobile Phone Systems): During its existence, Advanced Mobile Phone Systems (AMPS) operated alongside D-AMPS, a digital variant of AMPS. The system facilitated multiple calls through the use of time division multiplexing (TDM).

Global System for Mobile Communications (GSM): 2G was launched using GSM. The GSM standard pioneered the definition of a digital circuit-switched network designed for full duplex voice transmission. Over time, this evolved to support data communications, initially through circuit-switched transport, then through packet data transport using General Packet Radio Service (GPRS), and finally through Enhanced Data Rates for GSM Evolution (EDGE). The most widely adopted technology in modern times is the GSM network. It utilizes both frequency division multiplexing (FDM) and temporal division multiplexing (TDM) to handle multiple calls simultaneously. The GSM network comprises three primary systems: the Switching System (SS), Base Station System (BSS), and Operation and Support System (OSS).

CDMA (Code Division Multiple Access): It was standardized by International Standards IS-95 and employs a different technology than the one mentioned above.

i. **2.5G:** It is employed to define 2G systems that incorporate both a circuit-switched domain and a packet-switched domain. Because circuit-switched data services also use bundling of timeslots, it may not necessarily result in speedier performance.

ii. **2.75G:** 8PSK encoding enabled the conversion of GPRS networks into EDGE networks. Although each symbol now had three bits as opposed to one, the symbol rate remained constant at 270.833 samples per second. Faster data transfer rates are integrated into traditional GSM through a backward-compatible digital mobile phone technology known as EDGE, Enhanced GPRS (EGPRS), or IMT Single Carrier (IMT-SC). Beginning in 2003, AT&T in the US took the lead in putting EDGE on GSM networks.

- **3G:** The 3rd generation of mobile wireless technology is known as 3G, offers faster information transfer and better speech quality than 2G, 2.5G, GPRS, and 2.75G EDGE networks.

Information transfer rates minimum of 144 kbit/s are supported by 3G telecommunication networks. Later 3G versions, it is sometimes referred to as 3.5G and 3.75G, also offer access to mobile broadband for smartphones and as mobile modems to laptop computers at speeds of several Mbit/s. This enables the usage of this generation in mobile TV, fixed wireless Internet, mobile Internet, video calls, and wireless voice technologies.

Numerous 3G technologies exist, including W-CDMA, GSM EDGE, UMTS, DECT, WiMax, and CDMA 2000. Because it can function with older devices, enhances the data rates for the GSM evolution, or EDGE, which is referred to as a backward digital technology. Compared to the current GSM, EDGE provides faster data transfer. AT&T first offered EDGE in 2003. The GSM coverage has expanded by up to three times with EDGE. Since EDGE is a 3G technology (third-generation technology), packet-switching systems can use it. Mobile communication solutions are globally applicable. UMTS complies with IMT 2000 of the ITU. The USIM (subscriber identity module) and radio access are covered by this sophisticated network. Because it needs new and distinct infrastructure for its coverage, it is a relatively expensive technology for network operators. The foundation of this technology is GSM. IMT-MC is another name for CDMA. Due to its backward compatibility, this technology is similar to 2G GSM. DECT is a type of 3G technology (third-generation technology), which is another wireless technology. Although DETC was created by European Telecommunications Standards Institute, it is now widely used in many other nations as well. It uses a 1900 MHz frequency. WiMAX is a 3G technology (Third Generation Technology). A wireless technology, that is. It sends out a wide range of wireless signals. It can be operated in both point and multi-point modes. Based on wireless internet access, this technology. The WiMAX forum came up with the WiMAX name in 2001. This method eliminates the need for cables and can deliver 10 m bits/second. You can use it to access a hotspot.

1. 3.5G: Building upon GSM standards, 3.5G represented an enhanced version of WCDMA technology. The theoretical data rate on the uplink/upload side was boosted to 14 Mbps, while on the downlink/download side, it reached 5.76 Mbps. Compared to W-CDMA technology, that was 4-5 times more. The theoretical download and upload speeds of 3.5G are 500 kbps and 200 kbps, respectively.3.75G: The top data rates for this generation are 5.74 Mbps and 14 Mbps, respectively.
2. 3.9G: The top data rates for this generation are 11 Mbps for uploads and 28 Mbps for downloads.

- **4G** (Campos, 2017): Mobile communication refers to the 4th generation of wireless communication technology, which is the 3G network's replacement. High-definition video calling, quick downloads and uploads, live streaming, online gaming, and other events are all made possible by it. The capabilities and features outlined by the ITU (International Telecommunication Union) in IMT advanced, including transmission technology and data speed, must be adhered to by a 4G system. Users of 4G networks have access to speeds of up to 100 Mbps, which is faster than 3G networks.

Users may stream high-definition audio and video uninterrupted thanks to 4G's rapid speed. Furthermore, it offers wireless broadband, allowing customers to browse the internet devoid of a permanent cable connection.

Categories of 4G: There are 2 categories.

1. Long-Term Evolution (LTE): A high-speed wireless communication system which is often used in 4G connections is called Long-Term Evolution, or LTE.
2. In technical terms, 4G and 4G LTE are not interchangeable. The utilization of LTE results in faster information transmission and reduced latency compared to standard 4G networks. Anywhere in the world, whether for personal or business use, the LTE network is connectable.

At the time, the new speed and technical demands of 4G were not achievable, leading to the adoption of the term LTE primarily for marketing purposes. LTE offers considerably improved speed and bandwidth compared to 3G, without necessitating any specific additional charges. The speed varies based on the carrier from 20 Mbps to 100 Mbps.

3. WiMAX (Worldwide Interoperability for Microwave Access) serves as an example of 4G wireless internet technology. Similar to WiFi, it provides wireless internet connectivity without the need for wired connections. However, WiMAX surpasses WiFi's coverage range, as it can reach much greater distances, comparable to broadband-like high-speed internet access found in cell phone networks. It's important to note that while a 4G network can be likened to mobile WiMAX, not all 4G networks utilize WiMAX technology. WiMAX offers peak data rates of 128 Mbps on the downlink and 56 Mbps on the uplink, using channels that are 20 MHz wide.

- **5G** (Saha, 2016 & Rost, 2016): In comparison to the 1 Gbps theoretical maximum speed of 4G, the 5G technology boasts a theoretical maximum speed of 20 Gbps. 5G's commitment to lower latency can lead to improved performance for corporate applications and various digital activities, including online gaming, video conferencing, and self-driving cars.

Millimeter-wave technology with low, mid, or high bands can be used to achieve 5G. Low-band 5G, which uses the same frequency band as 4G mobile devices (600-900 MHz), may provide download speeds that are up to 250 megabits per second (Mbit/s) faster than 4G. Similar to 4G towers, low-band cell towers have a similar coverage area and range. Mid-band 5G uses microwaves at frequencies between 1.7 and 4.7 GHz, enabling speeds of 100 and 900 Mbit/s, and each cell tower offers service out to several kilometers. The most prevalent level of service, it was introduced in several major cities in 2020. Due to some places' lack of low-band implementation, mid-band is now the required service level. Although greater frequencies could be used in the future, high-band 5G employs frequencies between 24 to 47 GHz, which are near the bottom of the millimeter wave range. It frequently achieves download speeds that are on par with cable internet, in the gigabit-per-second (Gbit/s) range. Millimeter waves (also known as mmWave or mmW) have a more constrained range and require numerous tiny cells. They may be hindered or prevented by objects in windows, barriers, or by pedestrians. Plans call for exclusively using these cells in crowded urban regions and public gathering places like stadiums and convention centers because of their higher cost. The majority of 5G mmWave licenses and deployments worldwide have been in the range of 24.25-29.5 GHz.

- **6G** (Alraih, 2022): Applications using AI will differentiate 6G from earlier generations. Despite being in its early stages, the autonomous 6G network is expected to serve as the core of 6G technology. In comparison to the existing 5G capacity, data rates and security quality will increase while latency will decrease. Predicted 6G speeds range from 1 to 10 Tbps. The frequency seems to increase as generations move forward. A reliable indicator of high transmission rates is the frequency of THz. With the introduction of 6G, latency will decrease to 10–100 s, connectivity density will increase to 10 devices per km^2, and traffic throughput will increase to 1 $Gb/s/m^2$. Furthermore, 6G will boost those newly stated KPIs' receiver sensitivity to more than -130dBm, reliability to 99.9999%, positioning inaccuracy from the current meter level to the centimeter level, and coverage percentage to 99%.

3. COMPARISON BETWEEN 4G, 5G & 6G MOBILE COMMUNICATION SYSTEMS (HUANG, 2019; BANAFAA, 2022)

- As we progress from 4G to 5G and then to 6G, there is a remarkable reduction in latency. In 4G, latency is around 100 ms, whereas, in 5G, it drops to just 1 ms. With 6G, we can expect an incredibly low latency of 0.1 ms.
- The peak data rate per device sees a significant boost with each generation. In 4G, it offers 1 Gbps, whereas, in 5G, it goes up to 10 Gbps. With the advent of 6G, users can experience a jaw-dropping peak data rate of 1 Tbps.
- The maximum spectral efficacy also improves with the advancements in technology. While 4G offers 15 bps/Hz, 5G doubles it to 30 bps/Hz. With 6G, the efficiency jumps to an impressive 100 bps/Hz.
- Energy efficacy is a crucial factor, and 6G is expected to be significantly more energy-efficient than its predecessors. While 5G is already 1000 times more energy-efficient than 4G, 6G is projected to surpass even that.
- Connection concentration takes a substantial leap with 6G. While 4G supports 2000 devices per km2, 5G pushes it to 1 million devices per km2. With 6G, we can expect it to support more than 10 million devices per km2, opening up vast possibilities for the Internet of Things (IoT).
- Coverage percentage also improves dramatically as we move towards 6G. While 4G has coverage of less than 70%, 5G offers an 80% coverage percentage. With 6G, coverage is expected to surpass 99%, enabling widespread connectivity.
- Locating precision becomes significantly more accurate with each generation. In 4G, it's meters precision (50 m), which improves to meters precision (20 m) in 5G. However, 6G brings it to centimeter precision, allowing for highly accurate location-based services.
- The end-to-end consistency of the network improves with each generation. 4G achieves 99.9% consistency, while 5G elevates it to 99.999%. With 6G, we can expect an even higher level of reliability at 99.9999%.
- Receiver sensitivity plays a vital role in signal reception. 6G is expected to offer an impressive sensitivity of less than -130dBm, surpassing both 4G (around -100dBm) and 5G (around -120dBm).
- Mobility support in 6G reaches new heights, making it suitable for high-speed scenarios. While 4G supports mobility at 350 km/h, 5G goes up to 500 km/h. 6G is projected to support mobility at a whopping speed of at least 1000 km/h, catering to applications like high-speed trains and flying vehicles.

- Satellite integration becomes a full-fledged reality with 6G. Unlike 4G and 5G, which have no satellite integration, 6G fully embraces satellite communication technology.
- Artificial Intelligence (AI) integration also progresses with each generation. While 4G and 5G have limited or partial AI integration, 6G is expected to be fully integrated with AI, enabling advanced smart services and applications.
- Autonomous vehicles, Extended Reality (AR/VR), and Haptic Communication are progressively adopted in each generation. 6G marks the era of fully appreciated potential for these technologies, enabling safe and seamless autonomous vehicles, immersive extended reality experiences, and haptic communication.
- THz communication, which is not available in 4G and has limited presence in 5G, becomes widely available with 6G, opening up new frontiers for high-speed and high-capacity communication.
- Service level offerings evolve with each generation. While 4G primarily focuses on video services, 5G introduces AR and VR services. With 6G, we enter the era of tactile services, enabling haptic feedback and enhanced sensory experiences.
- Architecture undergoes significant changes with each generation. MIMO is a prevalent technology in 4G, followed by Massive MIMO in 5G. 6G introduces the concept of an Intelligent surface, revolutionizing how wireless signals are manipulated and optimized.
- The maximum frequency bands expand as we advance to 6G. While 4G operates in the 6GHz range, 5G extends it to 90 GHz. 6G takes it a step further, reaching frequencies up to 10 THz, enabling extremely high data rates and capacities.

These comparisons showcase the continuous progress and improvements in key performance indicators (KPIs) as we evolute from 4G to 5G and eventually to 6G, promising a future of unprecedented connectivity and technological possibilities.

4. 5G INFRASTRUCTURE

The infrastructure needed for fifth-generation cellular networks is made up of a network of small-cell and macrocell base stations. Large information streams that power applications like IoT devices, semiautonomous vehicles, and AR can be covered by 5G infrastructure with low latency.

4.1 Requirement of New Infrastructure for 5G

Mobile operators must modify their infrastructure to remove bottlenecks and guarantee that users can completely benefit from 5G's performance advantages for the technology to become widely adopted. Operators must upgrade their infrastructure to support a 5G future due to a combination of rising standards and increasing demand. Since 5G is built based on the principles of 4G technology, operators need not build the infrastructure from scratch, instead have to upgrade the technology available to them. Upgrades to the infrastructure don't always call for new equipment or tower climbers. In non-standalone architecture (NSA) 5G provides service over the 4G mobile core. Within the 4G core, users can experience performance similar to the 5G network, but with certain limitations in terms of service support. In an NSA approach, carriers can upgrade 4G towers by using dynamic spectrum sharing and spectrum reframing.

5. WHY IS 5G SO DIFFERENT?

5G provides improved speed, new technologies, and better services. For enterprises, 5G offers a very dependable type of wireless communication that guarantees both speed and capacity. Businesses no longer require a cable connection for their most mission-critical equipment thanks to private 5G. Administrators can modify 5G infrastructure to work in difficult workplaces where existing wireless technologies struggle. Similar to how IT workers create Wi-Fi networks, businesses can configure access points, small cells, and long-range antennas to fit their 5G demands.

The brains of a 5G network are the 5G Core (5GC), which controls data and control plane activities. The 5GC collects data traffic, establishes connections with UE, provides essential network services, and adds extra security layers, among other things. Authentication, authorization, and data and policy management are a few of the functions provided by 5G core which makes communication possible. Non-standalone 5G uses the existing 4G LTE network core to provide services to the users. A standalone configuration uses the 5G new radio access network on a brand new 5G network core. For both end users and operators, 5G SA offers easier network management, enhanced network performance, and a more adaptable environment. The 5G core employs a service-based architecture (SBA) aligned with cloud principles. This architecture facilitates authentication, security, session management, and traffic aggregation from connected devices. This integration of network functions requires a sophisticated approach to ensure seamless operation. The network's configuration should be flexible enough for 5G Core to introduce new services while using a common infrastructure. Time to market for new services should be accelerated from months to days thanks to the 5GC, which should reduce installation times from hours to minutes. The 5GC should provide telco-grade dependability and quick scaling. The 5G Core network must demonstrate the ability to quickly adapt and optimize itself based on operational conditions. Additionally, it ought to enable an affordable switch from 4G to 5G using the access-neutral common core. The 5G core network has been purposefully developed to incorporate features such as SBA, network slicing, mobile edge computing, and others. These enhancements not only enhance flexibility but also facilitate easy integration of third-party software and the provision of services for diverse verticals. Additionally, the improvements contribute to enhanced Quality of Service (QoS) and various other benefits. The modular design known as "Service-Based Architecture" centers around services that have the capability to self-register and provide specific functionalities while also being able to subscribe to other services. By using these services as the foundation, more complicated services can be offered.

6. COMPONENTS OF 5G INFRASTRUCTURE

6.1 Small Cells

The small cell concept will be necessary for 5G technology to employ to extend coverage for more customers and give a greater bandwidth signal. These are low-power, short-range wireless transmission systems specifically designed to cater to indoor and outdoor applications within a limited geographic area. These can handle large data rates for individual users and have all the central behavior of conventional base stations. Small cells are classified into three main categories based on their coverage area and user capacity. Small mobile base stations called femtocells are made to offer greater coverage for commercial and residential applications. Femtocells are typically used to increase indoor customers' data capacity, increase coverage, and offload network congestion. With a wired fiber connection, femtocell provides a service to a coverage area of 10 to 50 meter with support to 8 to 16 users. Pico cell is suitable for small enterprise applications with a coverage area of 100 to 250meter and supports 32 to 64 users. Due to its high transmission power, microcells can cover larger cell sizes and is therefore appropriate for use in smart city and smart metro applications, among others. This cell provides services to 250simultaneous users with a coverage area of about 500 meter to 2.5 kilometers. Small-cell hardware is designed with simpler components, resulting in a speedier and easier implementation process. Small base stations can be put on walls for use within, while small towers or lampposts can be used for usage outside. To offer the greatest service feasible in a specific area, small cells might rebroadcast mid-band transmissions as a high band.

6.2 Macrocells

A macrocell is a cellular base station equipped with large antennas and towers to facilitate the transmission and reception of radio signals. Strong transmission from macrocell enables coverage across thousands of square km. Lower frequencies from macrocells can travel great distances and are not obstructed by walls, windows, or other physical obstacles. In general, macrocells are a little larger than pico and femto cells. They can accommodate an increased number of connected users and a larger geographic area. These networks are predominantly employed in commercial spaces. Leveraging macrocells, 5G networks have the capability to provide services that can cover an entire city efficiently. Macrocells offer broader cellular network coverage, enabling the seamless connection of billions of devices with faster service and reduced system latency or response time.

7. 5G MOBILE CORE NETWORK

The mobile core is in charge of traffic routing and offering users the cellular services they subscribe to. Numerous tasks are completed by the core, including handling authentication services, processing subscriber data, and performing traffic routing duties. The inclusion of network slicing in the new mobile 5G core is a significant change. A network slice refers to a conceptually separate, secure, and self-contained portion of the network, specifically tailored to cater to different services with varying requirements in terms of speed, latency, and reliability. More control over how cellular resources are used is made possible by network slicing. As a result, routing is more effective, and multiple networks may be managed within a single slice. Each slice or segment of the network can be assigned to an individual user depending on the application requirements or use case or the customer himself. Enterprise users expect all business and operational operations, including ordering, activating, delivering, and decommissioning services, to be automated. They anticipate receiving services more quickly and with more security. Communication service providers can satisfy all of their enterprise customers' needs using network slicing. When numerous businesses with strict connectivity requirements are assembled on premises in remote locations, far from well-developed connectivity infrastructure, network slicing solutions offer special advantages. In certain contexts, slicing can offer greater flexibility and cost-efficiency than the fixed connectivity alternatives available today. Broadcasters of all sizes could see a significant change in how live content is transmitted from faraway areas through 5G and network slicing. High speeds, smaller teams, and the ability to cover more ground for less money will be advantageous to them.

7.1 Access Network

A part of a mobile network called a radio access network connects mobile users to the core network. Every mobile system must have a mechanism for effective communication between the base station and the end users. To link devices wirelessly and give the highest network performance, a 5G RAN relies on a fully coordinated, multi-layer network including low-band, mid-band, and high-band. There are numerous RAN architectures available such as fog-RAN (FRAN), heterogeneous cloud-RAN (HCRAN), and cloud-RAN (CRAN). Centralized processing, cooperative radio, and real-time cloud infrastructure form the foundation of CRAN. The CRAN design attempts to diminish the number of Base Stations and the network's energy usage. Heterogeneous networks and cloud computing are combined in the HCRAN to create a cost-effective solution. FRAN architecture seeks to report the matter of increasing traffic needs while simultaneously enhancing QoS for the users. It is built on the advantages of CRAN and fog computing (Pana, 2022).

8. CONCEPTS USED IN 5G TECHNOLOGIES

8.1 Carrier Aggregation

Advanced modulation techniques and network terms let 5G technology to operate with the great efficiency that is expected. To increase system efficiency, carrier aggregation is employed in LTE advanced. Carrier aggregation involves the grouping of two or more carrier signals to support increased bandwidth, reaching up to 100 MHz. This technique utilizes various aggregation algorithms, including intra-band contiguous, intra-band non-contiguous, and inter-band aggregation. With the help of 5G technology, mobile devices can automatically switch to faster internet just when it is required while maintaining reduced data usage when no applications are being used. This technique dramatically prolongs battery life.

8.2 Millimeter Waves

Using millimeter waves (between 24GHz and 86GHz range), 5G will be able to provide greater data throughput. Utilizing these waves permits the distribution of more reliable services with increased bandwidth capacity and faster download speeds. As 5G has to handle more traffic millimetre wave would be the better option to consider for communication. 5G supports many high-bandwidth applications which require higher speed with improved reliability and reduced latency. These waves can support such high bandwidth applications developed over a 5G network.

8.3 Small Cell Concept

To improve the effectiveness of the network, the cell is divided into micro and picocells. Reusing spectrum enables the addition of more users in a constrained geographic area and improves network management.

8.4 MIMO Concept

MIMO technique with Beamforming is one of the primary transmission technologies for supercharged 5G networks and IoT. Many antennas are to be mounted on the communication tower. Using multiple antennas simultaneously, 5G technology enables precise user location and communication at specific places. The user will be controlled and any changes in the user's location can be tracked by the antennas positioned in the user's chosen direction.

8.5 Wi-Fi Offloading

Wi-fi offloading enables the user to connect through a wi-fi network while allowing other users to take up space on the cellular network.

8.6 Device-to-Device Communication

The D2D (Device-to-Device) communication method enables direct connections between two nearby devices within the network. Devices will be controlled by the network, which will also let an operator choose between direct and network paths for traffic routing. One device can connect to another device even while the network is down.

8.7 Cloud-Radio Network Access

A part of a mobile network called a radio access network connects mobile users to the core network. C-RAN is a network technology that allows for efficient connectivity while centrally processing information is done remotely inside of a cloud system. It will speed up deployment overall and provide a safe network ecosystem. The architecture provides for easier network enhancements and scalability.

9. 5G PROTOCOL STACK

Open Wireless Architecture (OWA) Layer functions as the physical layer and data link layer of the OSI stack. Functions of the layer include Error detection, encoding/ decoding, Hybrid ARQ soft-combining, Modulation and demodulation of physical channels, Frequency and time synchronization, MIMO antenna processing, Transmit Diversity, Digital and Analog Beamforming, RF processing, etc. Routing data from the source IP device to the target IP device or system falls under the responsibility of the network layer. Within the network layer, there are lower layers and upper layers. The open transport layer combines the functionalities of the session layer and the transport layer. The data is marked by the application layer in the appropriate format. The data is also encrypted and decrypted by it. For a certain service, it chooses the optimum wireless connection.

9.1 5G Frequency Bands

5G requires a high-frequency spectrum of sub-6 GHz and millimetre waves to support increased bandwidth. Traditional GSM and LTE networks operate in the sub-4 GHz range. It has bandwidth restrictions. To enhance the mobile broadband experience, a significant increase in capacity is necessary. Many underused spectrums at higher frequency ranges, including millimetre waves, can be utilized effectively with more bandwidth. A new standard for super-fast 5G networks is 5G NR or new radio. It will utilize unlicensed airwaves in the millimetre band, both below and above 6 GHz. The goal of 5G NR is to enable a highly expandable network that can accommodate future improvements. According to their frequencies, the radio frequency bands used in the 5G technology are separated into three groups: In 5G, a band of useable frequencies below 1GHz is referred to as the low band. Low-frequency spectrum can better pass through barriers and offers more thorough coverage. The mid-band frequency strikes a compromise between coverage and capacity, making it a desirable option for both urban and suburban locations. Due to their increased capacity and ability to deliver faster data rates than low bands, mid bands were frequently chosen in early 5G deployments. High-frequency bands are used in dense cities because of their restricted coverage and quick response times, but they also offer speedy data transfer. The millimetre wave frequency range is where the high band spectrum works, making it ideal for applications requiring greater bandwidth for rapid communication.

9.2 5G NR (New Radio)

It is the global standard for a unified, more capable 5G wireless air interface. It is delivering significantly faster and more responsive mobile experiences. It continues to evolve to connect and redefine a multitude of new industries. The new standard for 5G wireless technology, known as 5G NR (new radio), allows for much faster, more effective, and scalable networks. With minimal latency and lightning-fast connections, 5G NR technology will let us communicate with a wide range of environmental objects. 5G NR will be using millimeter wave spectrum above the 24 GHz range. 5G utilizes orthogonal frequency-division multiplexing (OFDM) scheme for information transmission. 5G's capability to support various spectrum bands, including mm-Wave, allows for significantly higher available bandwidth.

9.3 Services Provided by 5G

Ambient Assisted Living (AAL) refers to a concept encompassing services, ideas, and products that utilize social environments and technological advancements to enhance the quality of life for individuals. AAL was founded to give elderly people the freedom to live comfortably and safely in their own homes. AAL administrations offer services that include self-governance and remote assistance when a problem arises (Schotten, 2014).

Indirect Emergency Healthcare (IEH) is a vital service that plays a critical role in situations where conventional healthcare services face challenges in deployment due to factors like adverse climate conditions, fire accidents, or in-flight emergencies. IEH focuses on providing essential arrangements and support during these emergencies (Braeken, 2021).

Third-party designers require generic systems with appropriate components to develop context-aware (CA) medical and healthcare applications on an Internet of Things (IoT) network. The Embedded Context Prediction (ECP) service proves valuable in such scenarios. Similar provisions are made for universal healthcare applications (Anwar, 2018).

9.4 Applications of 5G

9.4.1 Healthcare Sector

The delivery of healthcare services could be completely changed by adopting 5G technology in telemedicine and healthcare.

- **Faster and more reliable communication**: Compared to earlier generations of wireless technology, 5G offers significantly higher data speeds and lower latency. This makes it possible for medical professionals to reliably and quickly transmit large amounts of data, such as real-time video and high-resolution medical images. It makes it easier for healthcare professionals, patients, and distant specialists to communicate effectively, which improves diagnosis, treatment, and monitoring (Schotten, 2014).
- **Remote patient monitoring**: It is made possible by 5G, which allows wearables, sensors, and medical equipment to collect and transmit patient data in real time. This enables medical professionals to keep an eye on patients' vital signs, medication compliance, and disease development from a distance. Healthcare professionals can receive immediate alerts and make timely interventions thanks to 5G's low latency and high bandwidth, improving patient care and enabling the early detection of health issues (Braeken, 2021).
- **Teleconsultations and telemedicine**: These are made possible by 5G, which enables real-time, high-definition teleconsultations between patients and medical professionals. Without the need for in-person visits, this enables virtual doctor visits, remote diagnoses, and follow-up appointments. In particular, for people in remote locations or with limited mobility, telemedicine powered by 5G expands access to healthcare and eases the burden on medical facilities (Anwar, 2018).
- **Enhanced surgical procedures:** Remote surgeries and surgical assistance may be supported by 5G's ultra-low latency and dependable connectivity. Remote control of robotic surgical systems by surgeons allows for expert guidance and cross-border collaboration with little delay. This opens up opportunities for gaining access to specialized surgical expertise, enhancing surgical outcomes, and minimizing patient travel requirements (Georgiou, 2021).
- **Internet of Medical Things (IoMT):** 5G technology is essential for the IoMT, which links wearables, medical devices, and healthcare systems to enable seamless data exchange. IoMT devices can transmit data in real-time, supporting applications like remote monitoring, personalized medicine, and proactive healthcare interventions, thanks to 5G's high capacity and low latency (Magsi, 2018).

9.4.2 Agriculture Sector

The deployment of 5G technology in the agriculture field has the potential to revolutionize the industry by enabling a range of advanced applications and improving overall efficiency. Below are several advantages of 5G technology in the field of agriculture:

- **Smart Farming:** By tying together various devices and sensors on the farm, 5G can support the development of smart farming techniques. This makes it possible to track crop health, soil moisture, and weather information in real time, empowering farmers to make informed choices. For instance, farmers can collect data about crop health and apply targeted treatments using drones that are fitted with cameras and sensors (Roopashree, 2022).
- **Agri-robotics:** 5G networks make it easier for agri-robotic systems to communicate and be controlled in real-time. Robots can more accurately and efficiently carry out tasks like planting, weeding, and harvesting. Even in remote agricultural areas, remote operation and control of robots are possible thanks to 5G's low latency and high bandwidth capabilities (Tang, 2021).
- **Precision Agriculture:** 5G networks offer fast, low-latency connectivity, which is necessary for systems that use precision agriculture. Farmers can use autonomous vehicles, such as self-driving tractors and harvesters, to optimize farming operations with precise location tracking and quick data transfer. These vehicles can gather and transmit information about the growth and yield of crops as well as the quality of the soil, enabling more precise and effective resource management (Arrubla-Hoyos, 2022).
- **Livestock Monitoring:** 5G technology makes it possible to track the health and behavior of animals in real time, which can improve the management of livestock. Connected sensors and wearables can gather information on activity levels, feeding habits, and vital signs, offering insights into the health of particular animals or entire herds. Farmers can quickly identify and address health issues thanks to this, which enhances animal welfare and productivity (Razaak, 2019).

9.4.3 Manufacturing and Industrial Automation

Significant advantages and improvements result from the adoption of 5G technology in manufacturing and industrial automation. Here are some significant ways 5G improves this industry.

- **Real-Time Communication:** To support real-time connectivity and data exchange between machines, devices, and control systems, 5G enables ultra-low latency and high-speed communication. Faster decision-making, better coordination, and effective manufacturing process control are made possible as a result (Brown, 2018).

- **Robotics & Automation:** In terms of productivity and accuracy, autonomous machines and guided vehicles are revolutionizing conventional manufacturing and warehousing operations. High-accuracy robotics that can operate nonstop for 24 hours without rest breaks or risk of injury is advantageous for many automated assembly lines. Robotics can help make jobs safer, boost employee productivity, and make the workplace more enjoyable even where humans are required. Consistent and dependable mobile network communications are essential to their efficient operation (Wang, 2020).
- **Smart Factories:** 5G is essential to the development of interconnected machines, devices, and systems in smart factories. It allows for real-time data analytics, process improvement, energy management, and effective supply chain management, which boosts output, reduces costs, and boosts operational effectiveness (Walia, 2019).
- **Quality Control and Product Traceability:** 5G supports manufacturing processes that require real-time quality control and product traceability. It makes it possible to gather and analyze data from sensors and manufacturing processes, ensuring product quality, finding flaws, and enabling quick interventions (Zhong, 2019).

9.4.4 Augment Reality and Virtual Reality

5G technology offers significant advantages that improve user experience and open up new opportunities. Some particular AR and VR uses can benefit from 5G technology.

- **Immersive Entertainment:** Users will be able to enjoy high-quality content with minimal latency due to the ability of 5G to stream and download AR and VR experiences at lightning-fast rates. This makes it possible for users to interact in real-time with other users and virtual environments during immersive gaming, interactive storytelling, and virtual theme park experiences.
- **Real estate and virtual travel:** 5G can improve virtual travel experiences by giving users easy access to high-resolution 360-degree videos and virtual tours. Remote exploration of tourist attractions, lodgings, or real estate properties allows users to get a realistic sense of the area and make decisions.
- **Training and Education:** By providing immersive and interactive learning experiences, 5G-powered AR and VR can revolutionize training and education. For instance, technicians can receive hands-on training in virtual environments to learn complex procedures, and medical students can simulate surgeries in virtual operating rooms. 5G ensures high-quality, low-latency

educational content streaming, enabling in-person communication between instructors and students.

9.5 Area of Use for 6G Networks

6G technology is still in its early stages of research and development, and its precise use cases are not yet fully established. Here are some potential use cases, though, based on the potential advancements anticipated in 6G networks:

- **Robotic Fabric Network**: By the year 2030, there is a projected substantial rise in the deployment of autonomous mobile robots, drones, and automatic guided vehicles (AGVs) to facilitate packet delivery and personal transportation. Key The main goals will be to maintain this traffic's efficiency, safety, and lack of collisions. when coordinating robot movements from a central location and using a data set that has been compiled containing unrelated objects and additional robots. Paths through the stated are negotiated Robotic intentions can guarantee collision avoidance. The robots can effectively see their surroundings thanks to dynamic mapping in high-precision 3D.
- **Conversational Cobots**: Cobots, or collaborative robots, are anticipated to bring humans and robots into much closer contact.

In the future, networked collaborative robots (cobots) are expected to possess the capability to effectively interpret and comprehend human actions and intentions. This will enable them to respond reliably and dependably, assisting humans in productive and secure ways. Utilizing digital twins, these cobots will be adept at performing intricate and challenging tasks in work environments, effectively functioning as coworkers. Moreover, they could serve as valuable care assistants in the residences of elderly or disabled individuals, undertaking laborious human tasks with efficiency and care. A cobot could be a standalone device that responds to commands or changes its behavior based on the environment, or it could be more human-like like an exoskeleton or an adaptive wheelchair.

- **Digital Twins and Real-Time Interactive Mapping**: As precise representations of machines and structures, digital twins are a common tool in a variety of industries. They are anticipated to be commonplace by the year 2030, serving as the cornerstone of smart cities and digitalized factories. A city's worth of digital twins can be linked together to create a real-time interactive map that is essentially a digital representation of the real world. This makes it possible to manage large-scale systems—like waste, heating, and transportation—efficiently. Through a network of wireless sensors and actuators, it also enables real-time environmental representation, surveillance, and navigation. The digital map makes it easier to plan for the future and simulate different scenarios using data from the past, giving important insights. As a network service, this processing power and analysis can be provided along with ongoing synchronization with the physical world.

10. CONCLUSION

With much faster communication rates and lower latency, 5G has been rapidly embraced in many areas. Despite some difficulties, 5G kept increasing its coverage, and ongoing research attempted to improve its efficiency. Regarding 6G, it was still in its infancy and no firm conclusions could be drawn. In speculative forecasts, even faster data speeds, lower latency, and integration of cutting-edge technology like AI and quantum computing were predicted for 6G. Although new applications like holographic communications and intensely immersive experiences were envisaged, 6G's installation and effects had not yet been fully determined. The future of both 5G and 6G networks will depend on international cooperation, research initiatives, and the changing needs of society and businesses as technology advances.

REFERENCES

Alraih, S., Shayea, I., Behjati, M., Nordin, R., Abdullah, N. F., Abu-Samah, A., & Nandi, D. (2022). Revolution or evolution? Technical requirements and considerations towards 6G mobile communications. *Sensors (Basel)*, *22*(3), 762. doi:10.339022030762 PMID:35161509

Anwar, S., & Prasad, R. (2018). Framework for future telemedicine planning and infrastructure using 5G technology. *Wireless Personal Communications*, *100*(1), 193–208. doi:10.100711277-018-5622-8

Arrubla-Hoyos, W., Ojeda-Beltrán, A., Solano-Barliza, A., Rambauth-Ibarra, G., Barrios-Ulloa, A., Cama-Pinto, D., Arrabal-Campos, F. M., Martínez-Lao, J. A., Cama-Pinto, A., & Manzano-Agugliaro, F. (2022). Precision Agriculture and Sensor Systems Applications in Colombia through 5G Networks. *Sensors (Basel)*, *22*(19), 7295. doi:10.339022197295 PMID:36236394

Arshad, Q. K. U. D., Kashif, A. U., & Quershi, I. M. (2019, January). A review on the evolution of cellular technologies. In *2019 16th International Bhurban Conference on Applied Sciences and Technology (IBCAST)* (pp. 989-993). IEEE.

Banafaa, M., Shayea, I., Din, J., Azmi, M. H., Alashbi, A., Daradkeh, Y. I., & Alhammadi, A. (2022). 6G mobile communication technology: Requirements, targets, applications, challenges, advantages, and opportunities. *Alexandria Engineering Journal.*

Braeken, A., & Liyanage, M. (2021). Highly efficient key agreement for remote patient monitoring in MEC-enabled 5G networks. *The Journal of Supercomputing*, *77*(6), 5562–5585. doi:10.100711227-020-03472-y

Brown, G., Analyst, P., & Reading, H. (2018). Ultra-reliable low-latency 5G for industrial automation. *Technol. Rep. Qualcomm*, *2*, 52065394.

De Vriendt, J., Lainé, P., Lerouge, C., & Xu, X. (2002). Mobile network evolution: A revolution on the move. *IEEE Communications Magazine*, *40*(4), 104–111. doi:10.1109/35.995858

Georgiou, K. E., Georgiou, E., & Satava, R. M. (2021). 5G use in healthcare: The future is present. JSLS. *JSLS : Journal of the Society of Laparoendoscopic Surgeons*, *25*(4), e2021.00064. doi:10.4293/JSLS.2021.00064

Halonen, T., Romero, J., & Melero, J. (Eds.). (2004). GSM, GPRS and EDGE performance: evolution towards 3G/UMTS. John Wiley & Sons.

Huang, T., Yang, W., Wu, J., Ma, J., Zhang, X., & Zhang, D. (2019). A survey on green 6G network: Architecture and technologies. *IEEE Access : Practical Innovations, Open Solutions*, *7*, 175758–175768. doi:10.1109/ACCESS.2019.2957648

Magsi, H., Sodhro, A. H., Chachar, F. A., Abro, S. A. K., Sodhro, G. H., & Pirbhulal, S. (2018, March). Evolution of 5G in Internet of medical things. In *2018 international conference on computing, mathematics and engineering technologies (iCoMET)* (pp. 1-7). IEEE.

Pana, V. S., Babalola, O. P., & Balyan, V. (2022). 5G radio access networks: A survey. *Array (New York, N.Y.)*, *14*, 100170. doi:10.1016/j.array.2022.100170

Razaak, M., Kerdegari, H., Davies, E., Abozariba, R., Broadbent, M., Mason, K., ... Remagnino, P. (2019). An integrated precision farming application based on 5G, UAV and deep learning technologies. *Computer Analysis of Images and Patterns: CAIP 2019 International Workshops, ViMaBi and DL-UAV, Salerno, Italy, September 6, 2019 Proceedings*, *18*, 109–119.

Roopashree, K., Kanmani, Babitha, & Pavanalaxmi. (2022). Smart Farming with IoT: A Case Study. *Internet of Things and Analytics for Agriculture*, *3*, 273–286. doi:10.1007/978-981-16-6210-2_13

Rost, P., Banchs, A., Berberana, I., Breitbach, M., Doll, M., Droste, H., Mannweiler, C., Puente, M. A., Samdanis, K., & Sayadi, B. (2016). Mobile network architecture evolution toward 5G. *IEEE Communications Magazine*, *54*(5), 84–91. doi:10.1109/MCOM.2016.7470940

Saha, R. K., Saengudomlert, P., & Aswakul, C. (2016). Evolution toward 5G mobile networks-A survey on enabling technologies. *Engineering Journal (New York)*, *20*(1), 87–119. doi:10.4186/ej.2016.20.1.87

Schotten, H. D., Sattiraju, R., Serrano, D. G., Ren, Z., & Fertl, P. (2014, June). Availability indication as key enabler for ultra-reliable communication in 5G. In *2014 European Conference on Networks and Communications (EuCNC)* (pp. 1-5). IEEE. 10.1109/EuCNC.2014.6882630

Stuckmann, P. (2002). *The GSM evolution: mobile packet data services*. John Wiley & Sons.

Tang, Y., Dananjayan, S., Hou, C., Guo, Q., Luo, S., & He, Y. (2021). A survey on the 5G network and its impact on agriculture: Challenges and opportunities. *Computers and Electronics in Agriculture*, *180*, 105895. doi:10.1016/j.compag.2020.105895

Walia, J. S., Hämmäinen, H., Kilkki, K., & Yrjölä, S. (2019). 5G network slicing strategies for a smart factory. *Computers in Industry*, *111*, 108–120. doi:10.1016/j.compind.2019.07.006

Wang, X., & Gao, L. (2020). *When 5G Meets Industry 4.0*. Springer. doi:10.1007/978-981-15-6732-2

Zhong, M., Yang, Y., Yao, H., Fu, X., Dobre, O. A., & Postolache, O. (2019). 5G and IoT: Towards a new era of communications and measurements. *IEEE Instrumentation & Measurement Magazine*, *22*(6), 18–26. doi:10.1109/MIM.2019.8917899

Chapter 2
Local Site Effects Evaluation Using Geophysical Methods

Osman Uyanık
Suleyman Demirel University, Turkey

Nurten Ayten Uyanık
Isparta University of Applied Sciences, Turkey

Nevbahar Ekin
Suleyman Demirel University, Turkey

ABSTRACT

The waves that occur in an earthquake affect the behavior of the soil and the building due to the structural condition of the bedrock and the physical properties of the soil near the surface. To predict this interaction, the deep underground structure models of the areas considered as settlements should be determined and how the earthquake waves will be directed (macro-zoning) depending on this should be revealed. Since the near surface features of the ground can increase the effect of earthquake waves, it is necessary to determine the near surface properties and create their maps (micro-zoning). Earthquake waves are formed in thick sediments compared to the rock environment and cause great damage to both the ground and the buildings. In order to be protected from these damages, it is necessary to determine focusing and scattering conditions arising from the deep underground base topography. Therefore, parameters such as maximum horizontal ground acceleration, soil liquefaction, soil dominant vibration period, soil amplification, and near-surface soil properties should be determined.

INTRODUCTION

SDG 11 (Sustainable Cities and Communities) emphasizes considering natural events to create safe and sustainable cities (SDG, 2023). To build sustainable, safe,

DOI: 10.4018/979-8-3693-0819-6.ch002

and durable cities against the effects of earthquakes, which are the most important natural events, the structural and physical properties of the underground should be determined with macro and micro-zoning. This book chapter emphasizes the importance and necessity of macro and micro-zoning studies. Different damages are observed in areas affected by earthquakes in many parts of the world (Oluwafemi et al., 2018; Wada et al., 2018; Koren and Rus, 2021; Zarski et al., 2022). The most critical factors in earthquakes that cause damage to structures are the factors such as the magnitude, mechanism, location, and duration of the earthquake, as well as the physical and structural properties of the underground where quake waves radiate and the regional behavior of that environment. Some situations, such as the transmission characteristics of these environments, their dynamic elastic properties, the absorption and expansion of earthquake vibrations, the focusing effect of earthquake waves originating from the base topography, the channeling of earthquake waves into a region as a result of successive reflections in low-velocity layers constitute the regional characteristics of the environment. These features significantly affect the ground structure near the surface and its engineering structures. Therefore, it is essential first to reveal the deep underground system. The structural condition of deep underground structures can be determined by applying many geophysical methods (Aki and Richard, 1980; Alverez, 1990; Ulugergerli et al., 2007; Pamukçu et al., 2014; Chaljub et al., 2021).

The duration and intensity of elastic wave vibration at a particular location depend on many factors, such as the space from the quake epicenter, the earthquake's magnitude, and the soil properties of that location (Bullen and Bolt, 1985; Cilsalar and Temiz, 2020). Although seismic waves emanating from the source travel many kilometers within the crust, forming bedrock until they reach the soil layers, their distance within the soil layers is usually less than 100 m. However, soil layers are indispensable in deciding the features of the observed motion on the earth. Soil layers are like a filter for seismic waves. While seismic waves at some frequencies are damped, others are amplified. This situation is revealed by the geophysical methods realized in the field and the determination of the characteristics of the artificially produced seismic wave frequencies in the ground.

While earthquake waves propagate, a seismic impedance difference occurs between the rock and soil. This status causes the amplitude of earthquake waves to increase in the soil near the surface. Aki and Richards (1980) also stated that seismic impedance refers to the medium's resistance against particle movement. From this point of view, the importance of the seismic velocity values, the dominant periods of the bedrock, and the upper layers in soil amplification are revealed. Loose, porous ground conditions amplify the seismic wave amplitude. Therefore, loose and abundantly porous units are shaken more than tight and hard rocks. Geophysical engineers and international organizations (Regulations such as NEHRP, EuroCode,

TBDY) effectively use shear wave velocities (Vs) to define the soil's tightness or looseness. Soils with small Vs magnify the earthquake shaking and cause greater damage.

Existing settled cities cannot show sufficient resistance against the effects of natural events such as earthquakes. This situation is seen especially in undeveloped and developing countries. Large earthquakes such as the 6-7 February 2023 Kahramanmaraş (Türkiye) with Mw=7.8 and 7.5 resulted in the death of tens of thousands of people in collapsed cities, injuries of tens of thousands of people, and economic losses of billions of dollars. These results arise from poor construction and the wrong selection of areas where cities are settled. Even today, when choosing residential areas, micro zoning is often considered while macro zoning is ignored. This book chapter emphasizes the need for macro zoning before micro zoning.

BACKGROUND

As it is known, the engineering structure has a static effect on the environment in which it is installed and tries to disrupt the balance of the domain. Therefore, it is necessary to know the environment where the building will be built and to establish a balance between the engineering structure and the environment. To create this balance, the propagation and oscillation of earthquake waves on the ground and the strength and dynamic features of the subterraneous system to earthquake waves should be known (Uyanık et al., 2006; Bozcu et al., 2007). When trying to determine the effects of an earthquake that may occur in a residential area, it is vital to predict how the soil properties of this region will affect the constructions and the change of these effects (Türker et al., (1996; 1999); Uyanık et al., 2006; Uyanık and Çekmen, 2009; Çimen et al., 2010; Karabulut et al., 2011; Isik et al., 2019; Uyanık et al., 2019; Yuksel et al., 2020, Varolgüneş, 2021). Therefore, to reduce the damages resulting from the earthquake, it is necessary to determine the regions that will behave differently under dynamic forces. In this context, soils that may be damaged are graded by geophysical studies. In addition, body waves occur in earthquakes because surface waves occur in places close to the surface, depending on the characteristics of the ground.

Geophysical engineers determine where surface waves can occur by making surface wave measurements (Foti et al., 2018). It is known that the degree of damage increases in environments where surface waves are active. In addition, these surface waves significantly increase the effects of earthquake waves due to the internal characteristics of the soil (Uyanık et al., 2006). The best example is soil liquefaction, which developed in water-saturated loose settings in the 1999 Kocaeli earthquake. Soil liquefaction can be determined depending on seismic velocities (Andrus and

Stokoe II, 2000; Uyanık, (2002; 2020); Uyanık and Taktak, 2009; Uyanık et al., 2013a). In addition, the porosity of saturated or dry loose environments is obtained by seismic velocities (Salem, 2000; Uyanık, (2011; 2019)) and this porosity is also considered a factor that prolongs the propagation time of earthquake waves and increases the intensity of the earthquake. One of the fastest and most economical methods to determine the strength of soils is seismic methods that produce earthquake waves artificially. For this, in the regulations prepared by international organizations, the mean shear wave velocity (V_{S30}) of 30 m depth is utilized to classify the soil depending on its strength (BSSC, 1997; CEN, 2004; TBDY, 2018).

Areas with radiological risk should be determined, especially in volcanic regions, and suitability for settlement should be based on radiological risk (Uyanık et al., 2010; Uyanık et al. (2013c; 2013d); Uyanık et al., (2015a; 2015b); Uyanık et al., 2022; Uyanık, 2022). Then, when the suitability for settlement is investigated concerning the effects of quake waves, it is necessary to investigate mainly the deep underground structural condition of the area where the settlement is considered and where it directs the earthquake waves on the ground surface. After determining the areas that are least affected by earthquake waves, the most suitable areas for micro-zoning studies should be chosen, and convenient buildings for these areas should be constructed. Especially when the vibration characteristics of the ground are examined, if the building frequency compatible with the dominant frequency of the earth is created, there will be a resonance situation in the structure. Therefore, determining the soil's dominant frequency will be essential to ensure that the construction is appropriate for that place and that the earthquake waves affect the structure less. Therefore, to minimize the damage caused by earthquake waves, it is crucial to determine how these dynamic forces will affect the place and to classify the area in terms of suitability for settlement. In addition, using the methods of geophysical engineering, the structure parameters required for the analysis of existing structures are also used effectively in determining (Uyanık, (2012; 2014); Uyanık and Tezcan, 2012; Uyanık et al., (2011; 2013b); Öziçer and Uyanık, 2017; Sabbağ and Uyanık (2016; 2017; 2018a; 2018b; 2020); Uyanık et al. 2019; Ekin and Uyanık (2021a; 2021b)).

THE MAIN FOCUS OF THE CHAPTER

This book chapter explains the steps that must be taken according to the regulations regarding suitability for settlement. These steps are generally made by considering the characteristics of the ground for a maximum depth of 30 m. However, it is understood that the causes of heavily damaged areas vary depending on the model of the deeper underground structure. As a result of obtaining these deep underground models with geophysical methods, the causes of heavily damaged areas were tried to be explained. In addition, the characteristics of heavily damaged areas were discussed with the micro-zonation studies carried out regarding suitability for settlement. As a result, micro-zonation studies will be incomplete without determining deep underground structure models.

For this reason, macro-zonation and micro-zonation studies carried out in Türkiye will be mentioned by investigating the macro-zoning research carried out in seriously damaged fields worldwide and will emphasize the importance of defining the severely damaged areas before an earthquake. These studies are essential for cities as they will reduce planning costs, renew (revise) planned cities, and retrofit existing structures. The contributions of geophysical engineering to urban planning and construction should not be ignored.

FACTORS AFFECTING BEHAVIORS OF PLACES AND STRUCTURES IN AN EARTHQUAKE

Knowing earthquake parameters and local ground conditions is vital to predicting heavily damaged areas in an earthquake. Obtaining the parameters such as the earthquake distance, focal depth, magnitude, duration, acceleration, and fault geometry is important for earthquake scenarios. Considering the influence of local soil terms on quake damage, at least the following parameters (soil liquefaction, soil amplification, soil dominant vibration period, PGA value, alluvial thickness effect, soil V_{S30} value, soil class) should be determined.

Today, predicting earthquakes in advance is impossible, but it is possible to predict severe damage locations. There is no need to wait for a big quake for this. It is necessary to carry out four-stage studies to determine the heavily damaged areas in the earthquake. These stages can be listed as macro-zonation, micro-zonation, plot-based studies, and analysis of existing structures. However, in this book chapter, local ground effects will be discussed in macro-micro zoning studies.

Macrozonation

Many effects such as focusing, scattering, and amplification occur because of directing earthquake waves originating from deep underground structural features as well as the impact that may appear on the surface (such as radiological risk, landslide, flood, slope, etc.) to the areas where the settlement will be preferred. Impacts arising from deep underground structural features are not included in the regulations in our country. The importance of deep underground structural properties will be expressed in this part of the study.

Analysis of the deep underground structure for macro-zonation can be done using geophysical methods (Ulugergerli et al., 2007; Pamukçu et al., 2014; Oruç, 2014; Pamuk et al., 2018). If the physical and structural features of the earth's internal structure were homogeneous, the earthquake waves should have a circular intensity distribution on the planet. Many researchers have stated that heavy damage can occur in areas with rocks or soil in many earthquakes worldwide, while much less damage is done in loose soils. The researchers concluded that this damage is due to the heterogeneous structural features of the underground, increasing the earthquake damage intensity. For example, Motosaka and Nagana (1997) stated that in the 1995 Kobe earthquake, earthquake waves focused and created heavily damaged areas. In addition, Alverez (1990) stated that in the 1985 Mexico City earthquake, a heavily damaged area was formed when a size of 2 m width and 1.5 km length collapsed in the city, 400 km away from the epicenter. Geophysical methods explained this situation by determining the formation of sequential reflections at a low-velocity level in the limestone and a deep canyon under the city (Keçeli, 2009).

Depending on the state of the in-ground structure, earthquake waves are directed. In particular, the undulating form of the deep bedrock topography focuses or scatters seismic waves. Therefore, the areas where the earthquake waves are focused can be revealed by analyzing the internal structure of the ground. USGS (2000) also stated that seismic waves change their velocity and orientation as they pass through the deep folded interface and change again at the near-surface folded interface. Hence, the deep and near-surface subsurface structure and the lateral propagation of the layers direct the seismic waves to focus or scatter. Focusing zones create heavily damaged areas, while scattering zones reduce vibration intensity by 75% (Figures 1 and 2).

Therefore, focusing areas should not be preferred as residential areas. In addition, loose and water-saturated thick soils are shaken more because they increase the amplitude of the earthquake waves. The best example of the focus of earthquake waves and the effects of wide and loose sediments is given the heavy damage caused by the 6.9 magnitude earthquake near Samos Island on 30 October 2020 in the Bayraklı district of İzmir.

Figure 2 shows that Bayraklı district is on a soft filling area of approximately 300-350 m deep into the most damaged area. In addition, the Vs value of the filled area is around 250 m s^{-1}, indicating that it has weak strength. Pamuk et al. (2018) stated that there is engineered rock (Vs $\geq$ 760) just below this fill area. The concave mirror-shaped boundary between the rock and soft soil area and the seismic velocity difference caused the earthquake waves to be focused on the middle parts of the filling area, magnifying the earthquake effect and causing destruction. In addition, while seismic waves move toward the surface with P and S waves in the rock, when they come to soft sediments, additionally to these waves, surface waves (R and L) formed by the interference of these waves greatly enlarge the earthquake effect. In

Figure 1. Computer simulation of focusing and scattering effects
Edited from USGS (2000)

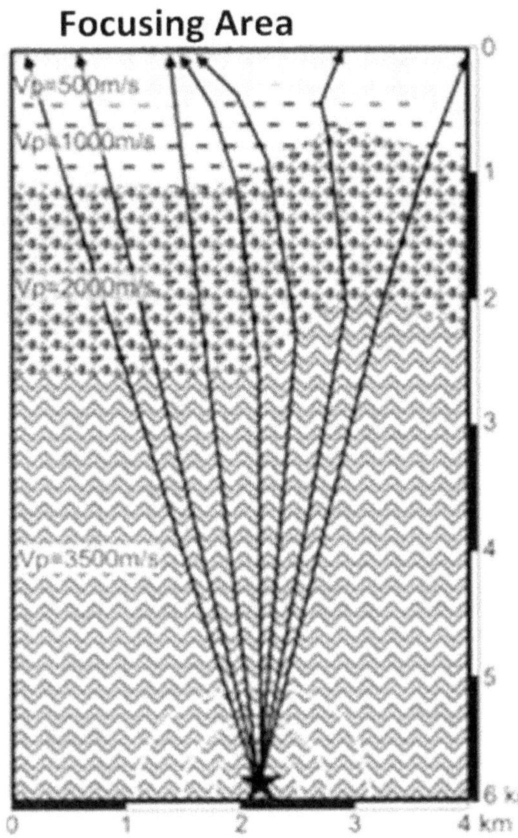

other words, while the impact of P and S waves is observed in the rock, the effect of surface waves is observed in addition to P and S waves in weakly resistant sediments. In this case, the damage effect on normal structures increases. Geophysics engineers determine the impact of surface waves by forming in an environment.

As a result, if the bedrock under the basin is concave, earthquake waves focus and create heavily damaged areas. At the same time, convex topography scatters earthquake waves and creates undamaged or less damaged regions (Figures 1 and 2). In addition, reflected and refracted waves due to seismic velocity differences in vertical and lateral directions of seismic waves can collect in certain areas and create heavily damaged areas. In addition, heavy damages occur in areas close to the ground surface that form surface waves. Therefore, determining the structural status of deep underground structures is important for directing earthquake waves. Therefore, while making city plans, it is necessary to determine the deep underground structure

Figure 2. The 2020 Samos earthquake and the focus of seismic waves

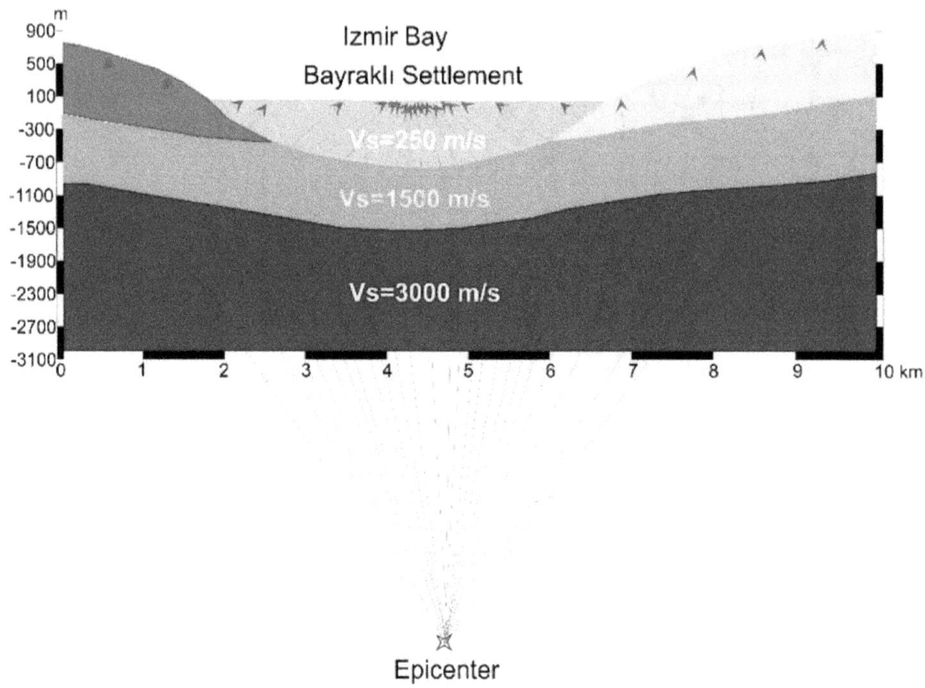

models and the places recommended by using them as suitable areas for settlement. Later, it would be appropriate to conduct micro-zonation studies in these areas.

Microzonation

Microzonation is defined as dividing big and large work areas into smaller zones. However, microzonation today should not be perceived as simply dividing extensive areas into smaller zones based on soil types and dynamic behavior. Microzonation should be perceived as the types of soils in the study area, the dynamic properties of these soils, and the behavior of these soils during distant or near earthquakes and the pre-calculation of the forces that the structures will be exposed to be built on these soils (Uyanık, 2015).

Türkiye is an earthquake country. In today's technology, it is possible to construct buildings suitable for earthquakes. First, it is necessary to analyze the underground structure and establish a balance between the engineering structure and the environment. The propagation and frequency of earthquake waves on the ground and the resistance and dynamic properties of the underground structure to earthquake waves should be known to create this balance. Seismic methods are effectively used to analyze engineering problems. In addition, with the dynamic parameters of the soil obtained from seismic methods, the porous structure of the soils and the fractured, cracked, and altered (decomposed) zones of the rocks can be revealed.

Earthquake Waves and Their Effects

When an earthquake occurs, seismic waves reach the surface as longitudinal (P), shear (S), Rayleigh (R), and Love (L) waves, respectively. The first to reach the surface is the P waves and, a few seconds later, the S waves. The S wave has two components: SH (Horizontal Shear) and SV (Vertical Shear). In areas close to the surface and suitable, the R wave is formed by the interference of the P and SV waves, while the interference of the P and SH waves creates the L wave. P and S are body waves, while R and L are surface waves. These waves depend on the structural and physical properties of the ground. Therefore, knowing the properties of these waves in detail enables the physical properties of the earth to be determined accurately. While P waves propagate in a solid-liquid-gas medium, S waves spread in a solid medium. P waves reflect the structural properties of the ground, and S waves reflect the strength properties of the ground. In other words, if the bond (cohesion) between the grains of the soil is strong, or the frictional resistance between the grains is high, S waves pass through this environment with high speed, and vibration is less. As a result, a high or low S wave velocity value is obtained depending on the strength of the soil. Surface waves occur effectively in water-saturated weak-strength environments and increase damage to structures above these environments. As a result, while the effect of surface waves increases in weak-strength media, their impact is negligible in solid rocks.

S waves are the second to reach the earth from earthquake waves. S waves propagate in the direction perpendicular to the particle motion. Because S wave velocities are a parameter that reflects the ground strength, soil classifications are given at the international level depending on the S wave velocity. According to Eurocode 8 (CEN), NEHRP, and Türkiye earthquake code classifications shown in Tables 1 and 2, if the V_{S30} value is less than 180 m s^{-1}, it is weak soils, and if it is greater than 760 or 800 m s^{-1}, it is rock. These velocities and other velocity ranges are presented in Tables 1 and 2. V_{S30}, shown in Tables 1 and 2, is obtained from in-situ seismic methods. This parameter is essential in soil classification and the determination of building codes. In addition, the area classifications estimated from the S wave velocity models are also crucial in deriving predictive equations for the effects of strong shaking and in constructing earthquake damage mitigation maps.

As a result, the soil class is given in the tables depending on the S wave velocity (V_{S30} value) in the world. S-wave measurement can be done anywhere (Uyanık and Ulugergerli,

Table 1. Soil classification according to V_{S30} in Eurocode 8

Ground Type	Description	Features V_{S30}
A	Rock or other rock-like geological formations, including at most 5 m of weaker material at the surface.	Vs>800
B	Deposits of very dense sand, gravel, or very stiff clay, at least several tens of meters in thickness, characterized by a gradual increase of mechanical properties with depth	360<Vs≤800
C	Deep deposits of dense or medium-dense sand, gravel, or stiff clay with thickness from several tens to many hundreds of meters	180<Vs≤360
D	Deposits of loose-to-medium cohesionless soil (with or without some soft cohesive layers) or of predominantly soft-to-firm cohesive soil	180<Vs

Source: CEN (2004)

Table 2. Soil classification according to NEHRP provisions and uniform building code V_{S30}

Ground Type	Description	Features V_{S30}
A	Solid Rock	Vs>1500
B	Rock	760<Vs≤1500
C	Very Firm-Hard Soil or Soft Rock	360<Vs≤760
D	Hard-Tight Soil	180<Vs≤360
E	Weak Soil	Vs<180
F	Soils Requiring Expert Evaluation	

Source: BSSC (1997)

2008). It has no limits like tests used in other geotechnical applications. For example, the Standard Penetration Test (SPT) is performed only in sandy environments and gives punctual information (Ulugergerli and Uyanık, 2007). In addition, it is not economical in terms of time and money. Moreover, mechanical and dynamic features can be presented by obtaining point and spatial information along a profile with seismic measurements, including 2- or 3-dimensional (2D/3D) underground structural conditions.

Ground Amplification

The seismic waves in any earthquake propagate at different speeds in the ground from the source of the earthquake. When these waves reach the earth's surface, they vibrate for a few seconds to minutes. The duration and intensity of elastic wave vibration at a particular location depend on many parameters, such as the distance from the earthquake epicenter, the magnitude of the earthquake, and the ground properties (Bullen and Bolt, 1985). Seismic waves spend most of their travel time from source to earth in hard bedrock. The final stage of their travel takes place in layers of the earth, which are pretty different from the bedrock, and the physical properties of these layers largely determine the characteristics of the vibration observed on the earth. Earth layers dampen waves of specific frequencies while amplifying others. Since this amplification is amplitude increase, it is called ground amplification (ground effect). Since this change is generally observed as an increase in amplitudes, the term ground effect is also called amplification (Ateş and Uyanık, 2019).

Factors affecting ground amplification can be listed as follows:

1-Resonance effect
2-Focusing due to the bottom of the basin topography
3-Surface waves formed by the basin edges
4-Non-linear ground behavior
5-Topography effect

When it is considered in terms of construction and the factors affecting the ground amplification, it is necessary to determine the ground amplification value in the area where each structure will be built.

To determine ground magnification:

1-Direct earthquake records,
2-Local seismic velocities
3-Vibration (Microtremor) measurements can be used

The vibrometer in the figures below can choose ground amplifications according to the Nakamura and reference station methods.

Ground amplification (n) can be determined by vibrational measurements as well as from seismic velocities using the following Equation 1 (Medvedev, 1965) and Equation 2 (Yeşil and Uyanık, 2012):

$$n = 1.67 Log \frac{\gamma_{rock} Vs_{rock}}{\gamma_{mea} Vs_{mea}} \tag{1}$$

$$n = \left(\frac{Vp_{30}}{Vs_{30}} \frac{\rho_{\max}}{\rho_{mea}} \right)^{0.1} \left(\frac{Vs_{bound}}{Vs_{30}} \right)^{0.5} \tag{2}$$

Figure 3. Ground magnification according to the Nakamura method

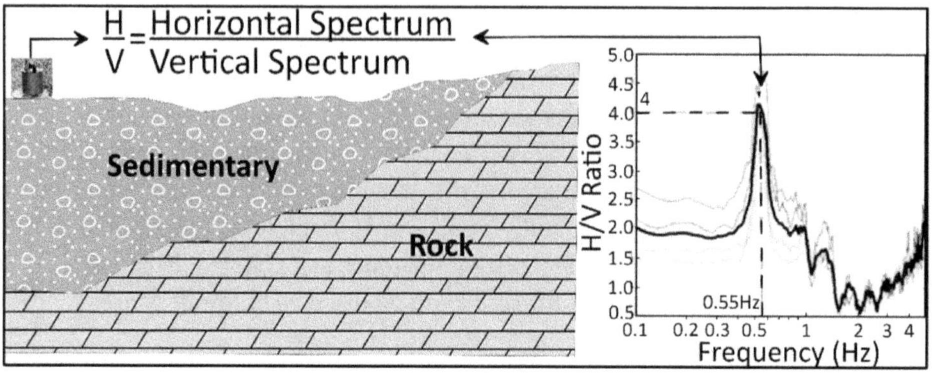

Figure 4. Ground magnification according to the reference station method

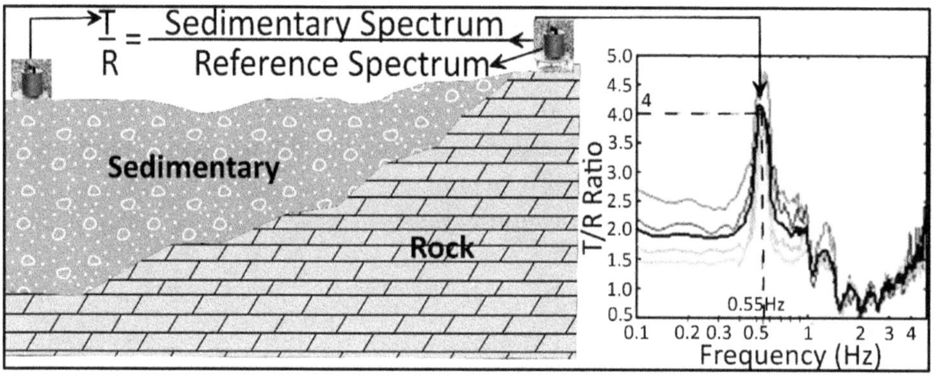

In Equation 2, V_{P30} is the average Vp for 30 m depth, V_{S30} is the average Vs for 30 m depth, ρ_{max} is the highest unit volume weight value that can be obtained in soil and rock (3.5 gr cm^{-3}), ρ_{mea} is the average unit volume weight value for 30 m depth in the study area, and Vs_{bound} is the Vs value (750 m s^{-1}), known as the ground and rock boundary. In cases where the unit volume weight value cannot be measured, it can be calculated from Vp and Vs using the empirical relationship suggested by Uyanık and Çatlıoğlu (2015).

Microtremor measurements and seismic velocities can determine ground amplification (n). As a result of numerous research carried out in the following earthquake regions, it has been determined that some soil types on the surface increase the earthquake intensity. Thus, the damage to the structures and the soils have been separated according to their relative intensity increases (Medvedev, 1965). The intensity increase of the earthquake varies between 1 and 4 according to the location types. While granite-type rocks do not make ground amplification, values up to 4 can be obtained in thick, soft sediments saturated with water. This value can be better understood as follows. For example, when an earthquake with a magnitude of 7 occurs around a region of close granite and water-saturated thick sediments, an earthquake intensity of 7 appears in the granite ground structure (no ground amplification). In contrast, earthquake intensity 11 may occur in saturated, soft, thick sediments (n=4).

It also increases the earthquake intensity depending on the thickness of the fill and the topography of the bedrock expressed in the macro-zonation part. Similarly, the effect of amplification is also seen in the building, depending on where the building is built. Figure 5 shows the outcome of the amplitude and period amplifications of the earthquake signals with the seismometer placed on the rock, soil, and structure.

If the construction were built on the rock, only the effect of the earthquake waves on the rock would be transferred to the structure. However, when the construction is built on the fill soil, as in Figure 5, the effects of earthquake waves on the rock and soil will increase and be transferred to the structure. If specific technologies that reduce the impact of earthquake waves will not be used for designs, public health needs to build high-rise structures on rock and short-story configurations on filled soils. Because it causes more horizontal forces on high-rise structures in loose and short-story constructions in solid, we will continue to face a country that cries and suffers every earthquake. As a result, it should be preferred that the regions to be opened for new settlements should be a sound environment compatible with nature and less affected by earthquake waves.

Ateş and Uyanık (2019), 32 acceleration records of 16 earthquakes of different sizes and distances recorded by accelerometers located on the roof of the Kocaeli

Figure 5. Effects of earthquake waves on rock, soil, and structure

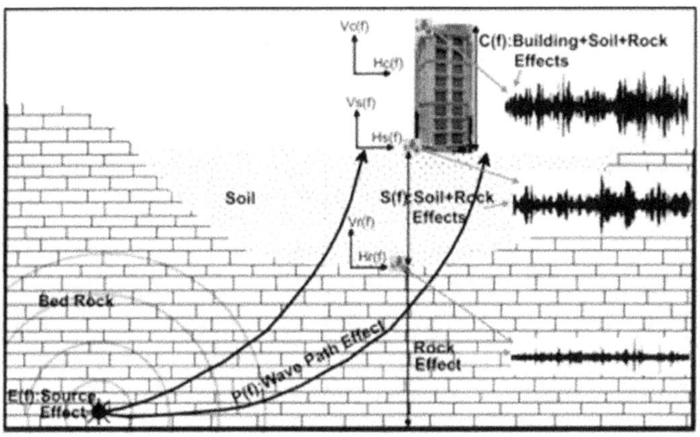

Metropolitan Municipality building and in the garden of the building were used. The records and evaluation results of the 4.5 magnitude earthquake, both on the ground and in the building, are presented in Figure 6.

Figure 6a indicates the S wave window selected in the acceleration records installed in the ground and building. In Figure 6b, the FFT of the selected S wave phase has been taken, the frequency environment has been passed, and two horizontal components have been combined to eliminate the direction dependence. The ground station was considered a reference station for building amplification and finding the dominant frequency. The building record was proportioned to the ground frequency, and the results were graphed separately. Figure 6c presents the two horizontal components of the spectra of the selected S phase of the recordings independently.

In Figures 7a and 7b, building/ground spectral ratio graphs were obtained from the data processing of 16 earthquake data. As can be seen in all recordings, a sudden peak is observed at the frequency of 2 Hz. The dominant oscillation frequency of the building (Mode 1) is thought to be this frequency. Figure 7a shows the average spectral ratio of 16 earthquake records. In the spectral ratio method, the ground effect is eliminated from the building records, and the pure building transfer function is obtained due to the ratio of the horizontal component spectrum of the building to the ground spectrum.

Dominant Vibration Period of Ground

In cases where the ground structure is variable in the lateral and vertical directions, more than one dominant period can be seen. This situation is generally observed

Figure 6. a) 22.10.2014 Mw 4.5 earthquake acceleration records (building and ground) and selected S phase, b) the resultant of horizontal spectra and building/ ground ratio, c) horizontal component spectra of buildings and ground
Source: Ateş and Uyanık (2019)

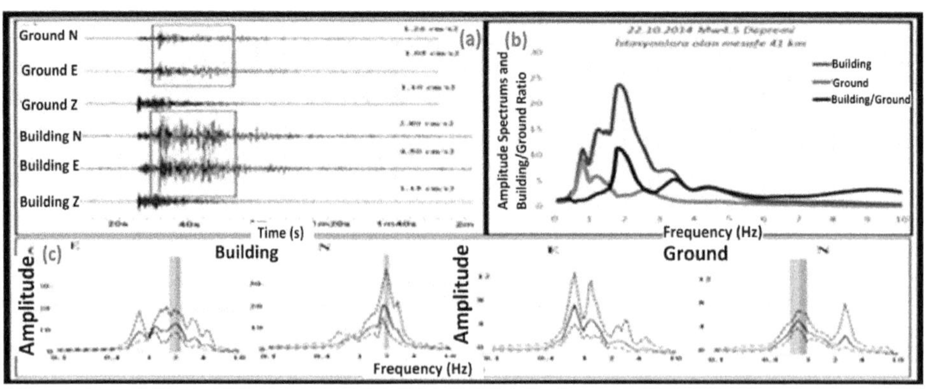

Figure 7. The spectral ratio method (SSR) magnification results from 16 earthquake records recorded in the building and on the ground
Source: Ateş and Uyanık (2019)

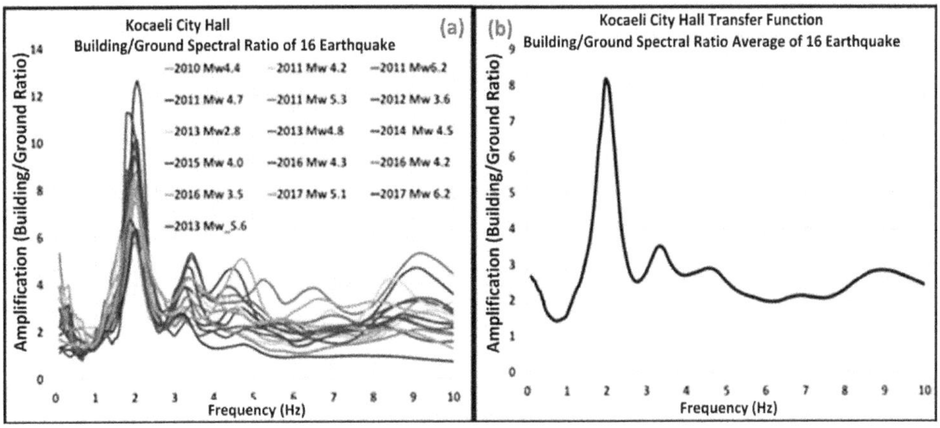

in periods shorter than 0.2 sec and more prolonged than 1 sec. In the case of direct basement rock at the surface, the dominant period of the ground varies between 0.1 and 0.3 sec. In alluvial fillings, the dominant period of the soil is between 0.4 and 0.8 sec. At the same time, values between 1 and 2 sec. are obtained in areas with thick and soft cover layers (for example, İzmir-Bayraklı). The dominant period of the ground is affected by the properties of the uppermost layer, especially in stratified environments. Therefore, since the characteristics of the ground on which each building will sit may vary in residential areas, the dominant vibration period of the ground should be determined by making vibrational measurements and given to civil engineers. Civil engineers design the building period by considering the dominant period of the soil. The frequency value on the horizontal axis of Figures 3 and 4 gives the dominant period of the ground. It is understood that it is an environment with a frequency of about 0.55 Hz, that is, a dominant period of 1.8 sec.

Peak Ground Acceleration (PGA)

Newton's 2nd law is used while calculating the earthquake load in the elastic design calculation in construction. For this, it is necessary to predict the PGA value and the earthquake force concerning the properties of the soil when the earthquake occurs. Tsai (1969), Okamoto (1973), and Ateş and Uyanık (2023) state that the strength of the earth has a significant effect on the PGA value (Figure 8).

Under normal conditions, earthquakes should be recorded by placing acceleration recorders in every area where construction will be made, and acceleration estimation

Figure 8. Acceleration records at different depths and different soil types
Edited from Tsai (1969); Okamoto (1973)

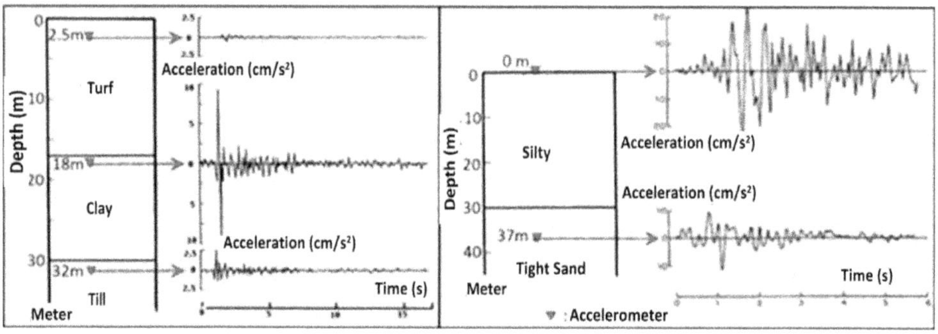

should be made by repeating this situation for earthquakes of different sizes. This status is a costly and time-consuming process. Therefore, acceleration reduction relationships have been developed. However, while these relationships are being developed, acceleration recorders generally produce erroneous results because they are in rock. Therefore, relationships that consider the soil properties come to the fore.

Ground vibrations generated during an earthquake radiate globally from the earthquake focus to the environment, and their effectiveness decreases as they move away from the focus depending on ground conditions. The acceleration reduction relationships show how the ground parameters will change as they move away from the focal point and are inferred by observation. Although the magnitude is a measure of the earthquake's size, it cannot be used directly to design earthquake-resistant constructions. On the other hand, soil liquefaction analysis utilizes quake magnitude and PGA. Because of the importance of PGA in earthquake-related engineering designs, acceleration reduction relationships are improved for diverse areas of the world since horizontal loads affecting the structure during an earthquake arise due to this acceleration. The necessary data to extract the observational attenuation relationships are possible within Türkiye. Acceleration reduction relationships were developed by using the data of our country.

Uyanık et al. (2021) developed an acceleration reduction relationship based on earthquake (M_w, T_D, R_{Hypo}) and soil parameters (soil amplification (b), dominant ground vibration period (T_0), V_{S30}, V_{P30} and intensity (ρ_{30}) (Equations 3-6)). The effect of the ground (ZE) on the acceleration relationship is determined using the parameters given in Equation 7.

$$b = \left(\frac{V_{P30}}{V_{S30}}\frac{3.5}{\rho_{30}}\right)^{0.1}\left(\frac{750}{V_{S30}}\right)^{0.5} \tag{3}$$

$$T_0 = \frac{4h_1}{V_{S1}} + \frac{4h_2}{V_{S2}} + \frac{4(H-(h_1+h_2))}{V_{S3}} \tag{4}$$

$$V_{S30} = \frac{30}{t_d + \frac{30-d}{V_{Send}}} \tag{5}$$

$$V_{P30} = \frac{30}{t_d + \frac{30-d}{V_{Pend}}} \tag{6}$$

$$ZE = 1 + \frac{1}{\sqrt{\left[\frac{1}{b}\left(1+\frac{T_D}{T_0}\right)^2 + \left(1-\frac{T_D}{T_0}\right)^2 \frac{V_{S30}}{V_{P30}}\right]}} \tag{7}$$

After determining the magnitude and epicenter of the quake, the PGA (a) value can be calculated by substituting the ground effect of the place to be built in the following Equation 8:

$$a = \left[10^{\left(0.621\,M_W - 1.179\,\log R_{Hypo} - 0.081\frac{V_{P30}}{V_{S30}}\right)}\right]\left[ZE\right] \tag{8}$$

An example of this calculation for poor soil and solid rock is given in Figure 9.

Figure 10 shows the growth of acceleration signals in different strength environments.

As understood from Figure 10, the earthquake force will be much higher in weakly water-saturated domains than in sound rock. In this case, preferring rock less affected by earthquake forces as settlements would be more accurate. When the spectrum curves are examined, high spectral acceleration occurs in a narrow period range for rock. In contrast, high spectral acceleration forms in a broad period range

Figure 9. Relationship between Moment Magnitude (M_w) and PGA: A) for rock, B) for weak ground (Uyanik et al., 2021)

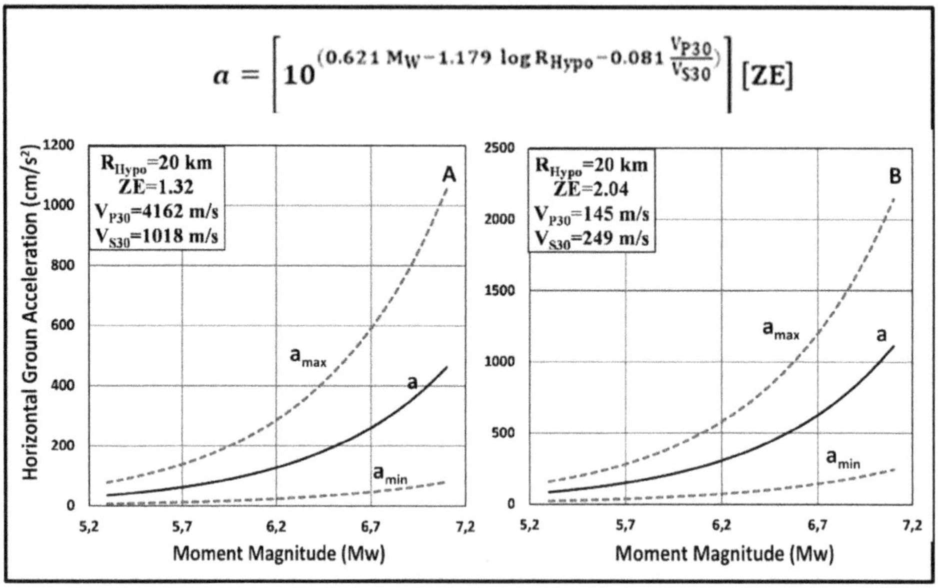

Figure 10. Different acceleration records and spectra obtained in different strength environments
Edited from https://serc.carleton.edu/details/images/185043.html)

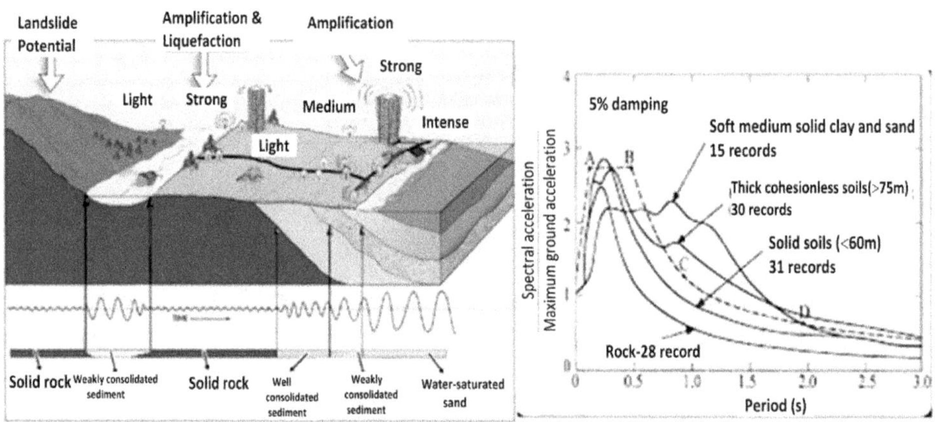

in soft, thick-fill environments. Therefore, structures with different heights in soft, thick-fill environments will be exposed to more earthquake forces.

Soil Liquefaction

Saturated sandy soil layers are forced to become more compact under the repeated loads of the earthquake. At this moment, the pressure of pore water increases and causes the effective vertical stress, which is thought to exist between the soil particles, to decrease or even to zero. Zero effective environmental stress at sandy levels destroys shear strength. In this case, the ground behaves like a liquid and shows large deformations. Depending on the topography and stratification of the land in nature, it can be said that liquefaction in the sand layers during an earthquake generally causes two types of events. The first is the formation of a slope slip due to liquefaction in sloping areas, and the second is the collapse and settlement of foundations in buildings due to liquefaction in areas where the land is flat (Uyanık, 2015).

Liquefaction in Sloping Areas: The 1920 Kansu earthquake is an example that needs to be learned regarding the magnitude of the damage caused by the slope failures caused by the liquefaction. Slope failures in the Kansu earthquake affected an area of 160 km to 480 km and caused the complete disappearance of 10 cities and hundreds of villages within this area. In this region with loess layers, slopes that were thought never to slide have slipped, and significant changes have occurred in the topographic condition of the area. It was understood that the main reason for the slip was liquefaction, as the weather was rainy before the earthquake, and the loess layers became saturated with water (Uyanık, 2015).

During the 1964 Alaska earthquake, the slope in Seward, located on the sediments consisting of medium-density sand and gravel layers, was around 15-20 degrees. Sliding occurred gradually in this region, and it was observed that the ground slipped a little more at each level. Here, the continuous and gradual state of the slip provides a valuable example to explain the mechanics of the event. The shear stress leading to liquefaction increases with the increase of the effective environmental pressure and causes shear of a certain depth. Since there will be a decrease in the environmental stress due to shear, it allows liquefaction to occur in the deeper layers, thus causing a second part to slip (Uyanık, 2015).

Liquefaction in Flat Areas: SPT values near the surface in Adapazarı and Gölcük cities, which are located on a flat area, are below 5, Vs values are below 200 m s^{-1}, SPT value is around 10, Vs value is about 250 m s^{-1} at medium depths. At 20 m and deeper, the SPT value is between 20 and 30, and the Vs value is between 300 and 400 m s^{-1}. In the 1999 Izmit earthquake with a magnitude of 7.4, the PGA was 480 cm s^{-2}. Liquefaction has caused tremendous damage to many buildings. It has been observed that in buildings sitting on shallow foundations, where Vs is less than 250 m s^{-1} under the foundation, the damage is more, and where Vs is more than 300 m s^{-1}, the damage is less (Uyanık, 2015).

As a result, it is known that the liquefaction analysis of soil can be made from seismic velocities (Andrus and Stokoe, 2000; Uyanık, (2002; 2020); Uyanık and Taktak, 2009; Uyanık et al., 2013a). The distinction between liquefied and non-liquefied data using the Vs-dependent liquefaction analysis technique is shown in Figure 11. Liquefaction is not a point but a spatial phenomenon. If geophysical methods are used, the ground can be modeled in the lateral and vertical directions to produce spatial solutions.

The Burdur region is built on both sloping and flat areas. In these areas, groundwater generally varies between 4 and 15 m. The Vs value at saturated sandy and silty levels below this groundwater level is below 300 m s^{-1}. The liquefaction analysis from the corrected Vs obtained by updating the Vs of these levels with effective stretching was done by Uyanık (2002), Uyanık (2006), Uyanık and Taktak (2009), Uyanık et al. (2013) and Uyanık (2020) (liquefaction analysis method from seismic velocities) have drawn the boundaries of the region with liquefaction potential (Figure 12). Earthquake damage will further increase with the effect of liquefaction in the buildings in the regions within the liquefaction limits. The black areas in Figure 12 indicate environments that may liquefy. In these areas, Vs is below 250 m s^{-1}. This map was obtained based on the factor of safety (SF) that SF <1 was expressed in black color, 1≤SF ≤1.2 in gray color, and SF>1.2 in light gray color (Uyanık, 2015).

Landslide Geometry of the Ground

A landslide is the movement of soils, rocks, or weathered rocks in the direction of a slope, triggered by natural or artificial reasons. Landslides, at least as effective as earthquakes, are a natural event that should be considered. When the examples of landslides in the world are examined, it is known how important the issue is

Figure 11. The soil liquefaction analysis depending on the corrected Vs of Uyanık (2002) data (Uyanık, 2020)

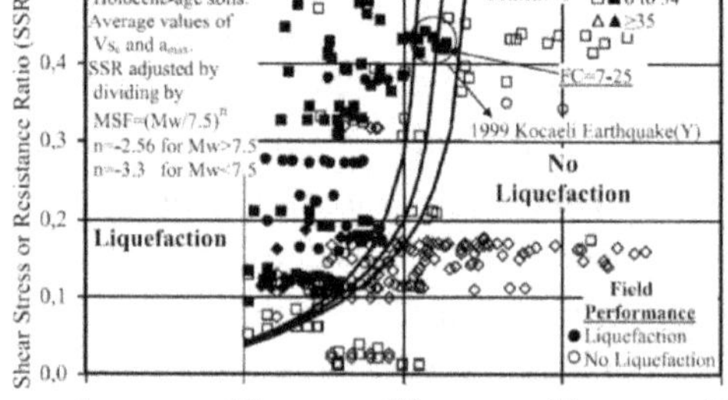

Figure 12. Sand dyke formed because of soil liquefaction and Vs-related liquefaction map of the Burdur Region
Source: Uyanık et al. (2006)

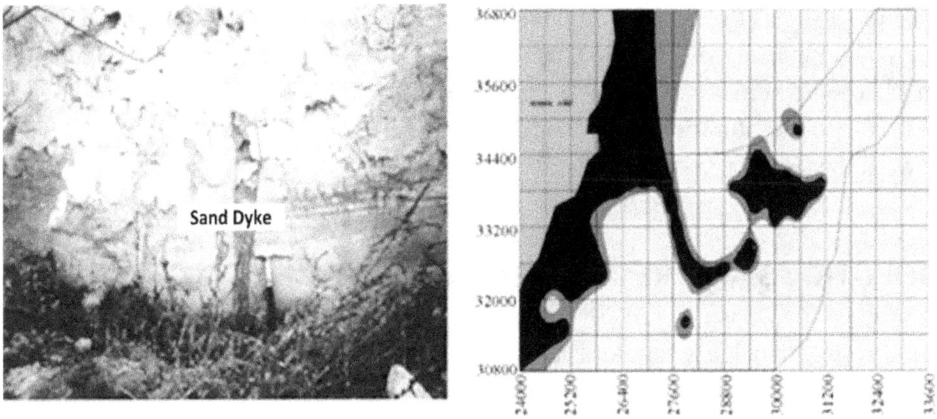

and the necessity of taking precautions. Landslide investigations should be carried out before loss of life, loss of fertile soil, and material losses occur. Due to stress cracks, it is possible to detect landslides before they happen. There is no sudden and surprising slope movement. Even the fastest landslides signal ahead. Despite all this, landslides cause pain because the danger signs are not seen on time (Uyanık and Çatlıoğlu, 2014).

Due to the nature of landslides, it is a complex problem due to the determination of slip surface and depth, movement type, groundwater level, soil type, and dynamic parameters. Geophysical methods are the most widely applied methods for determining the landslide geometry. High-resolution studies are required to determine the slip surface. Inclined and folded zones can be easily determined by collecting data at frequent intervals with geophysical studies. Drilling studies only give information about the type of soil. It is complicated and costly to determine the location and continuity of the structure with a few soundings. Geophysical studies produce fast and economical solutions while describing the underground. Electrical resistivity and seismic methods from geophysical methods (Uyanık and Türker, 2007; Uyanık and Sabbağ, 2013; Senkaya et al., 2020) can determine the slip limit by extracting the section of the underground in the solution of landslide problems. Determining the depth and boundary of the slip surface is the first step necessary to prevent a possible landslide. These limits are determined with geophysical methods (Uyanık and Çatlıoğlu, 2014).

A small-scale landslide occurred because of roadworks in the Alanya region, and the buildings on the upper part of the landslide were damaged. The environment obtained from the geophysical studies is a medium-dense or solid-filled ground with a thickness of between 4 and 25 m and impermeable schists under this ground. The soil on the schist has slipped in the direction of the slope with the saturation, and Figure 13 shows the landslide model.

When Figure 13 is examined, the Vs velocity of mediums 1 and 2 is around 500 m s^{-1}. The Vs velocity of medium 3 is more than 750 m s^{-1} and offers rock velocity. With the water saturation, the 1st and 2nd mediums above the 3rd environment shifted toward the slope.

SOLUTIONS AND RECOMMENDATIONS

As a result, managers should ask the engineers working on geotechnical studies to determine at least the following parameters to create healthier settlement plans.

For macro zoning:

Figure 13. Landslide model with geophysical methods

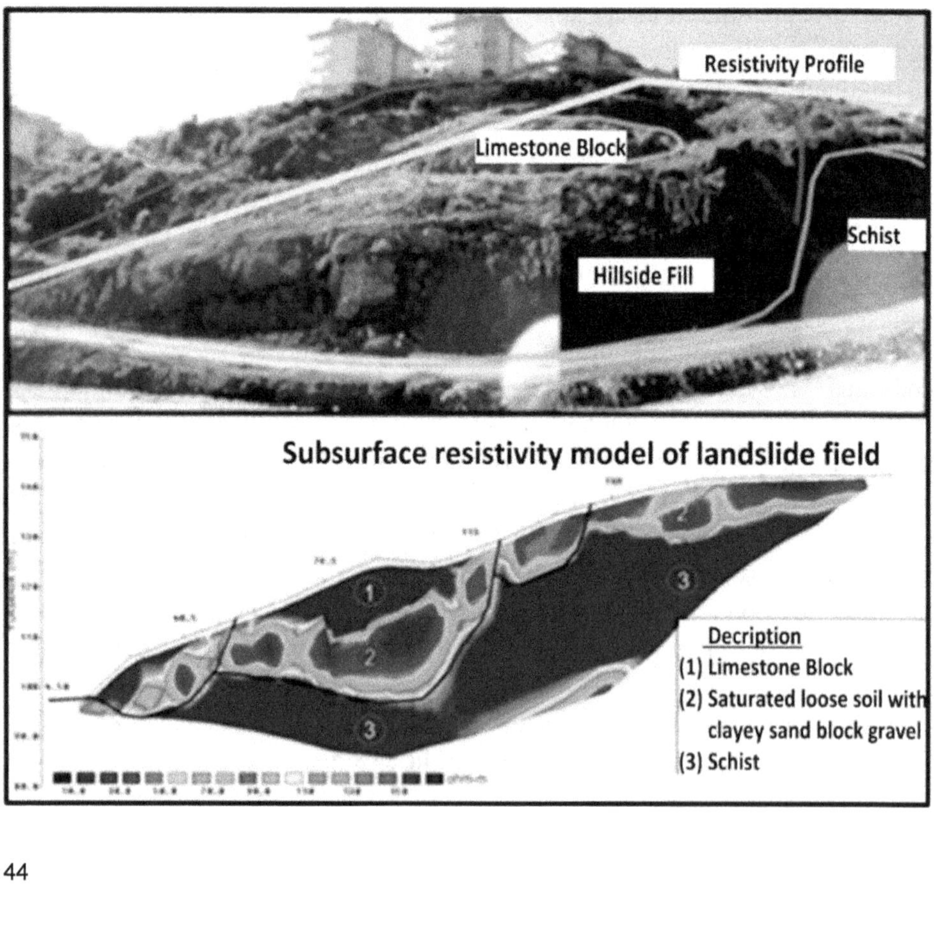

1. Illumination of underground up to a depth of at least 100 m
2. Detection of the lateral changes of the bedrock topography and underground structures
3. Determining the locations of buried valleys and hills
4. Determination of focusing areas of seismic waves
5. Determining the surface and underground locations of possible faults
6. Determination of three-dimensional underground seismic pressure and Vs models
7. Detection of dynamic parameters recommended.

For micro-zoning:

1. Research to a depth of at least 30 m
2. Determining the lateral and vertical spread of the units on the ground
3. Determining the dynamic and mechanical properties of the units on the ground
4. Determination of underground seismic pressure and Vs distribution for 30 m depth
5. Detection of areas that can liquefy
6. It is recommended to determine the maps of soil parameters such as the dominant period, the amplification, and the PGA.

For parcel-based studies:

1. Who will be responsible if severe damage occurs in a possible earthquake in the building in settlements where extreme damage areas have not been determined? For this reason, it should be known that the reports prepared by the engineers working on the parcel basis are not ready for the depths that will determine the possible severe damage areas. Therefore, macro and micro zoning studies should be done by local governments and directed to engineers working based on plots. Otherwise, who will be responsible?
2. The benefit of predetermining the areas that may be heavily damaged in a possible earthquake: It provides economic benefits in reducing earthquake hazard damages by prioritizing strengthening engineering structures in predetermined heavy-damaged areas.
3. Different categories in the current regulation are distinguished depending on the ground class. Geophysical methods are methods that can be carried easily and can be applied anywhere. Therefore, the decision whether the study area is mainly in Category 1 can be made according to the V_{S30} value obtained by seismic study. It is recommended to be given according to the ground class V_{S30} value according to NEHRP or Eurocode 8. In the TBDY regulation, it is decided according to V_{S30}, but for some reason, it is not considered in Category 1.

4. Seismic V_P and V_S velocities of the soil units, V_{S30}, dynamic elastic parameters, Poisson ratio, soil bearing amounts, settlement, bearing coefficient, and other soil static and dynamic parameters should be determined in parcel studies.
5. The ground on which a building will sit is spatial, not point. So, the building sits on an area. Different soil-bearing values are encountered in this area. Therefore, due to seismic refraction inversion studies in this area, images of 2 and 3-dimensional soil-bearing capacity results are obtained more economically and with more information. With the drilling made at five different points on each building floor, the ground is destroyed excessively, and less information is obtained as it is costly. Instead, the earth is not destroyed by seismic refraction studies in the area where the building will sit, and much more information is obtained than drilling. Therefore, more economic, scientific, and technological work is done. Thus, the number and depth of drilling should be determined because of geophysical studies.
6. The bearing capacity relation in the current regulation is impossible for every environment. For example, what will be done in case of not taking samples in water-saturated soils, that is, in case of being unable to obtain the cohesion and internal friction parameters? In addition, the building sits on an area, and as a result of the current regulation drilling and point study, the result is obtained in the laboratory by taking samples. This status ultimately represents how much the building will sit on. Therefore, it is meant to say that only one engineering branch can be determined, and no other engineering branch can interfere with the carrying capacity relation given in the current regulation. However, the engineer evaluates the results of many methods and recommends the most suitable one. In addition, in the current code, Security number 1.4 is accepted. Instead, in geophysical studies, the safety number (Vp/Vs), depending on the nature of the ground, is obtained by measurement. The number of securities should be determined as a result of personal preferences or depending on the characteristics of the place. The seismic velocity ratio is generally around 1.5-2.5 in the rock and about 3-5 in the soil. As a result, it is necessary to determine the security number depending on the property of the place.

FUTURE RESEARCH DIRECTIONS

Is there a focusing problem caused by the underground base topography that increases the ground amplification in terms of ground-structure interaction in an earthquake? What is the type and condition of the ground? What is the frequency of the soil and the PGA? Do surface waves occur on the ground? Can liquefaction occur in saturated sand? To answer these questions, it is necessary to research macro-zonation, micro-zonation, and the parcel base, respectively. If suitable settlement areas are selected based on these suggestions and the appropriate constructions are made by the construction rules and the ground, undamaged structures in earthquakes can be formed.

CONCLUSION

Drilling operations are one-dimensional (i.e., only information is produced in the vertical direction) and point. On the other hand, geophysical studies are done in 2 or 3 dimensions. It is areal, as both depth and lateral changes can be determined. Therefore, geophysical methods should be applied first in the study areas. Then, after the underground's 2- and 3-dimensional models are revealed, drilling studies should be carried out due to the guidance of geophysical investigations. Today, intensive geophysical studies are carried out in our seas, and then drilling sites are determined. Random drilling is not done. When it comes to construction, things turn around and become meaningless. First, drillings are made on the building ground, and then geophysical studies are carried out. This status is against science and logic. This ranking should be applied in construction studies such as oil, natural gas, and geothermal.

The ground is destroyed by random drilling and drilling in water-saturated fill areas close to the sea and the piles created by the high-rise buildings deep into the ground, and the way for seawater to enter deeper into the ground is opened. This situation will lead to the salinization of fertile lands and desert formation in the following years. For this reason, random drilling and construction of high-rise buildings in filling areas are harmful in terms of cost and environment. In addition, more robust constructions are required as earthquake waves exert more force on the structures in thick sediments saturated with water, resulting in high costs.

Environments with $V_{S30} < 360$ m s^{-1} smaller than specified in the NEHRP regulation should not be opened to settlement. In particular, the place where high-rise buildings will be built should be rock. Engineering structures are safer on a rock.

REFERENCES

Aki, K., & Richards, P. G. (1980). Quantitative seismology. W.H. Freeman and Co.

Alverez, R. (1990). Structure of the basin of Mexico City and its relation to destruction in the earthquake of 1985. *Geotechnical and Environmental Geophysics*, *3*, 263–279. doi:10.1190/1.9781560802785.ch21

Andrus, R. D., & Stokoe, K. H. II. (2000). Liquefaction resistance of soils from shear-wave velocity. *Journal of Geotechnical and Geoenvironmental Engineering*, *126*(11), 1015–1025. doi:10.1061/(ASCE)1090-0241(2000)126:11(1015)

Ateş, E., & Uyanık, O. (2019). Jeofizik yöntemler ile yer ve yapı etkileşimi. *SDÜ Fen Bilimleri Enstitüsü Dergisi*, *23*, 46–60.

Ateş, E., & Uyanık, O. (2023). (in press). Comparison of acceleration records measured at different depths. *Turkish Journal of Earthquake Research.*

Bozcu, M., Uyanık, O., Çakmak, O., & Türker, A. E. (2007). Geotechnical properties of Esen I HEPP project field. *SDU Journal of Natural and Applied Sciences*, *11*(1), 75–83.

BSSC (Building Seismic Safety Council). (1997). *NEHRP Recommended Provisions for Seismic Regulations for New Buildings and Other Structures. Part I, Provisions (FEMA 302), 334*. Author.

Bullen, K. E., & Bolt, B. A. (1985). An Introduction to the Theory of Seismology. Cambridge Univ. Press.

CEN. (2004). Eurocode 8-Design of Structure for Earthquake Resistance, Part I: General Rules. Seismic Actions and Rules for Buildings. European Standard EN 1998-1, European Committee for Standardization, Brussels.

Chaljub, E., Celorio, M., Cornou, C., Martin, F. D., Haber, E. E., Margerin, L., Marti, J., & Zentner, I. (2021). Numerical simulation of wave propagation in heterogeneous and random media for site effects assessment in the Grenoble Valley. *The 6th IASPEI / IAEE International Symposium: Effects of Surface Geology on Seismic Motion, August 2021.*

Cilsalar, H., & Temiz, U. (2020). Probabilistic seismic hazard assessment of city center of Kayseri, Erciyes University. *Journal of Institue of Science and Technology*, *36*(2), 286–298.

Çimen, Ö., Uyanık, O., Elmasdere, E., Korkmaz, K. A., & Keskin, S. N. (2010). Mavikent-Isparta örneğinde mikrobölgeleme çalışmaları için yerel zemin koşullarının belirlenmesi. *SDÜ Fen Bilimleri Enstitüsü Dergisi*, *14*(1), 46–54.

Ekin, N., & Uyanık, O. (2021a). Comparison of static and dynamic elastic moduli in concrete: Effects of compressive strength, curing conditions and reinforcement. *Civil Engineering (Shiraz)*, *45*(4), 2327–2343. doi:10.100740996-020-00513-7

Ekin, N., & Uyanık, O. (2021b). Anisotropy of reinforced concrete from geophysical methods. *Science of Sintering*, *53*(3), 323–345. doi:10.2298/SOS2103323E

Foti, S., Hollender, F., Garofalo, F., Albarello, D., Asten, M., Bard, P. Y., Comina, C., Cornou, C., Cox, B., Giulio, G. D., Forbriger, T., Hayashi, K., Lunedei, E., Martin, A., Mercerat, D., Ohrnberger, M., Poggi, V., Renalier, F., & Sicilia, D. (2018). Guidelines for the good practice of surface wave analysis: A product of the InterPACIFIC project. *Bulletin of Earthquake Engineering*, *16*(6), 2367–2420. doi:10.100710518-017-0206-7

Isik, E., Sagir, C., Tozlu, Z., & Ustaoglu, U. S. (2019). Determination of urban earthquake risk for Kırşehir, Türkiye. *Earth Sciences Research Journal*, *23*(3), 237–247. doi:10.15446/esrj.v23n3.60255

Kanai, K., & Tanaka, T. (1954). Measurement of the microtremor. *Bulletin of the Earthquake Research Institute, University of Tokyo*, *32*, 192–209.

Karabulut, S., Ozcep, F., Korkmaz, B., & Zarif, H. (2011). Urban seismic microzonation studies at small scales: Sisli/Istanbul (Turkey). *Arabian Journal for Science and Engineering*, *36*(1), 75–87. doi:10.100713369-010-0012-x

Keçeli, A. (2009). Uygulamalı jeofizik. *TMMOB Jeofizik Mühendisleri Odası Eğitim Yayınları, Ekim Ajans Matbaacılık Hizmetleri*, *9*, 479.

Kıraç. Altınşehir). Kentleşme ve Jeoloji Sempozyum Kitabı, 19-20 Kasım 1998, 189-202, İstanbul.

Koren, D., & Rus, K. (2021). Assessment of a city's performance under different earthquake scenarios. *1st Croatian Conference on Earthquake Engineering 1CroCEE*, 109-120. 10.5592/CO/1CroCEE.2021.155

Motosaka, M., & Nagana, M. (1997). Analysis of amplification characteristics of ground motions in the heavily damaged belt zone during the 1995 Hyogo-Ken Nanbu (Kobe) Earthquake. *Earthquake Engineering & Structural Dynamics*, *26*(3), 377–393. doi:10.1002/(SICI)1096-9845(199703)26:3<377::AID-EQE649>3.0.CO;2-0

Okamoto, S. (1973). Introduction to earthquake engineering. University of Tokyo Press.

Oluwafemi, J., Ofuyatan, O., Sadiq, O., Oyebisi, S., Abolarin, J., & Babaremu, K. (2018). Review of world earthquakes. *International Journal of Civil Engineering and Technology*, *9*, 440–464.

Oruç, B. (2014). Structural interpretation of southern part of western Anatolian using analytic signal of the second order gravity gradients and discrete wavelet transform analysis. *Journal of Applied Geophysics*, *103*, 82–98. doi:10.1016/j.jappgeo.2014.01.008

Öziçer, S., & Uyanık, O. (2017). Beton dayanımının yerinde p dalga hızından belirlenmesi ve İzmir örneği. *SDU International Journal of Technological Sciences, 9*(1), 1–16.

Pamuk, E., Gönenç, T., Özdağ, Ö. C., & Akgün, M. (2018). 3D bedrock structure of Bornova Plain and its surroundings (İzmir/Western Turkey). *Pure and Applied Geophysics, 175*(3), 1–16.

Pamukçu, O., Gönenç, T., Uyanık, O., Sözbilir, H., & Çakmak, O. (2014). A microgravity model for the city of Izmir (western Anatolia) and its tectonic implementations. *Acta Geophysica, 62*(4), 849–871. doi:10.247811600-014-0203-z

Sabbağ, N., & Uyanık, O. (2016). Time-dependent change of seismic velocities on low strength concrete. *The Online Journal of Science and Technology, 6*(4), 49–57.

Sabbağ, N., & Uyanık, O. (2017). Prediction of reinforced concrete strength by ultrasonic velocities. *Journal of Applied Geophysics, 170*(103832), 13–23. doi:10.1016/j.jappgeo.2017.04.005

Sabbağ, N., & Uyanık, O. (2018a). Determination of the reinforced concrete strength by apparent resistivity depending on the curing conditions. *Journal of Applied Geophysics, 155*, 13–25. doi:10.1016/j.jappgeo.2018.03.007

Sabbağ, N., & Uyanık, O. (2018b). Determination of reinforced dry concrete strength by electrical resistivity method. *Pamukkale Univ Journal of Engineering Sciences, 24*(6), 1230–1236. doi:10.5505/pajes.2017.76702

Sabbağ, N., & Uyanık, O. (2020). Determination of anisotropy of reinforced concretes by using resistivity method. *Pamukkale Univ Journal of Engineering Sciences, 26*(3), 570–578. doi:10.5505/pajes.2019.00008

Salem, H. S. (2000). Poisson's ratio and the porosity of surface soils and shallow sediments, determined from seismic compressional and shear wave velocities. *Geotechnique, 50*(4), 461–463. doi:10.1680/geot.2000.50.4.461

SDG. (2023). *The Sustainable Development Goals Report 2023: Special edition.* https://unstats.un.org/sdgs/report/2023/The-Sustainable-Development-Goals-Report-2023.pdf

Senkaya, G. V., Şenkaya, M., Karslı, H., & Gueney, R. (2020). Integrated shallow seismic imaging of a settlement located in a historical landslide area. *Bulletin of Engineering Geology and the Environment, 79*(4), 1781–1796. doi:10.100710064-019-01612-0

TBDY (Türk Bina Deprem Yönetmeliği). (2018). Türkiye Bina Deprem Yönetmeliği. Afet ve Acil Durum Yönetim Başkanlığı, 30364, 416.

Türker, E., Kaya, M. A., Kamacı, Z., Uyanık, O., Büyükköse, N., Mutlutürk, M., Yalcın, A., & Özkan, F. (1996). Dinar Afet Bölgesi Geoteknik Raporu. *Afet İşleri Genel Müdürlüğü-SDÜ.* (Unpublished).

Türker, E., Uyanık, O., Işık, H., & Özalpaydın, C. (1999). *Gölcük Donanma Üssü Yerleşim Alanının Jeolojik ve Jeoteknik Etüdü Projesi.* MTA-SDU (Unpublished).

Ulugergerli, E. U., Seyitoğlu, G., Başokur, A. T., Kaya, C., Dikmen, Ü., & Candansayar, M. E. (2007). The geoelectrical structure of northwestern Anatolia, Turkey. *Pure and Applied Geophysics, 164*(5), 999–1026. doi:10.100700024-007-0200-0

Ulugergerli, E. U., & Uyanık, O. (2007). Statistical correlations between seismic wave velocities and SPT blow counts and the relative density of soils. *Journal of Testing and Evaluation, 35*(2), 187–191. doi:10.1520/JTE100159

USGS. (2000). The local effects of strong ground motion shaking: *USGS Responds Report to an Earthquake.*

Uyanık, N. A. (2022). Determination of the radiological risk and the cancer effect caused by geological units and samples from Afyon, Turkey. *Pure and Applied Geophysics, 179*(4), 1295–1308. doi:10.100700024-022-02978-1

Uyanik, N. A., Akkurt, İ., & Uyanik, O. (2010). A ground radiometric study of uranium, thorium and potassium in Isparta, Turkey. *Annals of Geophysics, 53*(5-6), 25–30.

Uyanik, N. A., Öncü, Z., Uyanik, O., & Akkurt, İ. (2015a). Determination of natural radioactivity from 232th with gamma-ray spectrometer in Dereköy-Yazır (Southwestern Anatolia). *Acta Physica Polonica A, 128*(2B, No 2-B), B441–B442. doi:10.12693/APhysPolA.128.B-441

Uyanik, N. A., Oncu, Z., Uyanik, O., & Bozcu, M. (2022). Determination of alteration zones and geological unit limits using natural radioactivity properties of Sandikli-Suhut areas. *Journal of Applied Geophysics, 196*, 104525. doi:10.1016/j.jappgeo.2021.104525

Uyanik, N. A., Öncü, Z., Uyanik, O., Bozcu, M., Akkurt, I., Günoglu, K., & Yagmurlu, F. (2015b). Distribution of natural radioactivity from 40k radioelement in volcanics of Sandıklı-Suhut (Afyon) Area. *Acta Physica Polonica A, 128*(2B, No 2-B), B438–B440. doi:10.12693/APhysPolA.128.B-438

Uyanık, N. A., Uyanık, O., & Akkurt, İ. (2013c). Micro-zoning of natural radioactivity and seismic velocities of potential residential areas in volcanic fields: The case of Isparta (Turkey). *Journal of Applied Geophysics*, *98*, 191–204. doi:10.1016/j.jappgeo.2013.08.020

Uyanık, N. A., Uyanık, O., Gür, F., & Aydın, İ. (2013d). Natural radioactivity of brick sand brick material in the Salihli-Turgutlu area of Turkey. *Environmental Earth Sciences*, *68*(2), 499–506. doi:10.100712665-012-1754-4

Uyanık, O. (2002). *Kayma dalga hızına bağlı potansiyel sıvılaşma analiz yöntemi*. Dokuz Eylül University Natural and Applied Sciences, Doktora Tezi.

Uyanık, O. (2006). Sıvılaşır ya da sıvılaşmaz zeminlerin yinelemeli gerilme oranına bir seçenek. *DEU. Engineering Faculty Journal of Science and Engineering*, *8*(2), 79–91.

Uyanık, O. (2011). The porosity of saturated shallow sediments from seismic compressional and shear wave velocities. *Journal of Applied Geophysics*, *73*(1), 16–24. doi:10.1016/j.jappgeo.2010.11.001

Uyanık, O. (2012). Sismik hızlardan beton dayanımının belirlenmesi. *Jeofizik Bülteni*, *23*(70), 25–30.

Uyanık, O. (2014). Klasik ve jeofizik yöntemlerle yapı incelemeleri (Yapı Jeofiziği). TMMOB Jeofizik Mühendisleri Odası, JFMO Eğitim Yayınları.

Uyanık, O. (2015). Deprem ağır hasar alanlarının önceden belirlenmesi ve şehir planlaması için makro ve mikro bölgelendirmelerin önemi. *SDU Fen Bilimleri Enstitüsü Dergisi*, *19*(2), 24–38.

Uyanık, O. (2019). Estimation of the porosity of clay soils using seismic P- and S-wave velocities. *Journal of Applied Geophysics*, *170*(103832), 103832. doi:10.1016/j.jappgeo.2019.103832

Uyanık, O. (2020). Soil liquefaction analysis based on soil and earthquake parameters. *Journal of Applied Geophysics*, *176*, 104004. doi:10.1016/j.jappgeo.2020.104004

Uyanık, O., & Çatlıoğlu, B. (2014). Elektrik özdirenç ve sismik kırılma yöntemlerinden heyelan geometrisinin belirlenmesi. *SDU Fen Bilimleri Enst. Dergisi*, *18*(3), 22–29.

Uyanık, O., & Çatlıoğlu, B. (2015). Sismik hızlardan yoğunluğun belirlenmesi. *Jeofizik*, *17*, 3–15.

Uyanık, O., Çekmen V. (2009). Zeminlerin basınç ve kesme mukavemetlerinin sismik hızlardan tahmini. *İstanbul Üniversitesi Yerbilimleri Dergisi*, *22*(2), 95-104.

Uyanık, O., Ekin, N., & Çoşkun, O. (2021). Zemin ve deprem parametrelerinden ivme azalım ilişkisi. *DEÜ Fen ve Mühendislik Dergisi, 23*(68), 575–593. doi:10.21205/deufmd.2021236820

Uyanık, O., Ekinci, B., & Uyanık, N. A. (2013a). Liquefaction analysis from seismic velocities and determination of lagoon limits Kumluca/Antalya example. *Journal of Applied Geophysics, 95*, 90–103. doi:10.1016/j.jappgeo.2013.05.008

Uyanık, O., Kaptan, K., Gülay, F. G., & Tezcan, S. (2011). Beton dayanımının tahribatsız ultrasonik yöntemle tayini. *Yapı Dünyası, 184*, 55–58.

Uyanık, O., & Sabbağ, N. (2013). Determination of landslide geometry by geophysical methods. *4th International Geosciences Student Conference Proceeding Book.*

Uyanık, O., Sabbağ, N., Uyanık, N. A., & Öncü, Z. (2019). Prediction of mechanical and physical properties of some sedimentary rocks from ultrasonic velocities. *Bulletin of Engineering Geology and the Environment, 78*(8), 6003–6016. doi:10.100710064-019-01501-6

Uyanık, O., Şenli, G., & Çatlıoğlu, B. (2013b). Determination from non-destructive geophysics methods of concrete quality of buildings. *SDU Int. Technologic Science, 5*(2), 156–165.

Uyanık, O., & Taktak, A. G. (2009). Kayma dalga hızı ve etkin titreşim periyodundan sıvılaşma çözümlemesi için yeni bir yöntem, *SDU Fen Bilimleri Enst. Dergisi, 13*(1), 74–81.

Uyanık, O., & Türker, E. (2007). Fethiye Eşen II HES şalt ve santral sahasındaki potansiyel heyelanının yerteknik özellikleri ve yorumu. *SDU Fen Bilimleri Enst. Dergisi, 11*(1), 84–90.

Uyanık, O., Türker, E., & İsmailov, T. (2006). *Sığ sismik mikro-bölgeleme ve Burdur/Türkiye örneği. Ekologiya və Su Təsərrüfatı Elmi-Texniki və istehsalat Jurnalı.*

Uyanık, O., & Ulugergerli, E. U. (2008). Quality control of compacted grounds using seismic velocities. *Near Surface Geophysics, 6*(5), 299–306. doi:10.3997/1873-0604.2008004

Varolgüneş, F. K. (2021). Success factors for post-disaster permanent housing: Example of Turkish earthquakes. *The Turkish Online Journal of Design, Art and Communication, 11*(1), 115-130.

Wada, A., Towhata, I., Tamura, K., & Qu, Z. (2018). Moving toward cities where earthquakes will not cause a grievous disaster. *Japan Architectural Review, 1*(4), 10. doi:10.1002/2475-8876.12061

Yeşil, Y., & Uyanık, O. (2012). Zemin büyütmesinin sismik hızlardan belirlenmesi. SDÜ Fen Bilimleri Enstitüsü, Yüksek Lisans Tezi.

Yuksel, A. S., Uyanik, O., & Er, K. (2020). Development of a fuzzy logic based online visualization application for 2D geotechnical cross-section modeling. *Earth Science Informatics*, *13*(4), 1523–1538. doi:10.100712145-020-00500-6

Zarski, M., Wojcik, B., Miszczak, J., Blachowski, B., & Ostrowski, M. (2022). Computer vision based inspection on post-earthquake with UAV synthetic dataset. *IEEE Access : Practical Innovations, Open Solutions*, *10*, 108134–108144. doi:10.1109/ACCESS.2022.3212918

ADDITIONAL READING

Moscatelli, M., Albarello, D., Scarascia Mugnozza, G., & Dolce, M. (2020). The Italian approach to seismic microzonation. *Bulletin of Earthquake Engineering*, *18*(12), 5425–5440. doi:10.100710518-020-00856-6

Uyanık, N. A., Uyanık, O., & Akkurt, İ. (2013c). Micro-zoning of natural radioactivity and seismic velocities of potential residential areas in volcanic fields: The case of Isparta (Turkey). *Journal of Applied Geophysics*, *98*, 191–204. doi:10.1016/j.jappgeo.2013.08.020

Uyanık, O. (2015). Predetermination of Earthquake Heavy Damage Areas and Importance of Macro and Micro Zoning for Urban Planning. *SDU Journal of Natural and Applied Science*, *19*(2), 24–38.

KEY TERMS AND DEFINITIONS

Compressional Wave: The first wave formed during an earthquake spreads in the same direction as the wave propagation direction.

Earthquake: The vibrations occur because of the fractures in the earth's crust and reach the earth by spreading in waves.

Ground Amplification: It is the feature of a wave that occurs on the ground that allows the effect of the earthquake to be felt at different degrees depending on whether the ground is loose or tight.

Landslide: It is the downward slide of a slope consisting of rock or filling material with the effect of gravity.

Liquefaction: It is the behavior of the ground as a liquid by losing its strength.

Macrozonation: To determine the anticline or syncline structures by revealing the deep underground structure. Thus, the settlement selection is ensured by determining the places where the earthquake waves will focus or scatter.

Microzonation: It is the stage of evaluating the features of a region in terms of suitability for settlement.

Shear Wave: It is the second wave formed during an earthquake and propagates perpendicularly to the direction of wave propagation.

APPENDIX: ABBREVIATIONS

NEHRP: National Earthquake Hazards Reduction Program
BSSC: Building Seismic Safety Council
TBDY: Türkiye Building Earthquake Regulation
PGA: Peak Ground Acceleration
USGS: United States Geological Service
SPT: Standard Penetration Test
SF: Safety Factor
V_P: P wave velocity
V_S: S wave velocity
V_{S30}: Average shear wave velocity up to 30 m depth

Chapter 3
Crafting Intelligent Urban Environments:
Innovative Engineering Approaches for City Evolution

S. C. Vetrivel
https://orcid.org/0000-0003-3050-8211
Kongu Engineering College, India

T. Mohanasundaram
https://orcid.org/0000-0003-4511-7539
M.S. Ramaiah Institute of Technology, India

ABSTRACT

As urbanization continues to accelerate, the concept of smart cities has emerged as a promising solution to address the complex challenges faced by rapidly growing urban areas. Smart cities leverage advanced technologies to enhance the quality of life for residents, optimize resource utilization, and promote sustainable development. However, the successful implementation of smart cities heavily relies on the engineering of resilient and sustainable infrastructure. This chapter highlights the key aspects of engineering resilient and sustainable infrastructure for smart cities, focusing on the integration of technological innovations, resilience planning, and sustainable practices. The engineering of resilient infrastructure plays a crucial role in smart cities' ability to withstand and recover from various shocks and stresses, including natural disasters, climate change impacts, and cyber-attacks. Resilient infrastructure emphasizes the integration of robust design principles, redundancy, adaptability, and effective risk management strategies.

DOI: 10.4018/979-8-3693-0819-6.ch003

1. INTRODUCTION

In recent times, the idea of smart cities has gained significant traction, spurred by the compelling objective of utilizing state-of-the-art technologies to enhance the quality of life for urban residents. As our world becomes increasingly urbanized, cities grapple with multifaceted challenges spanning infrastructure, transportation, energy consumption, and ecological sustainability. Within this context, this chapter endeavors to delve into the pivotal role that engineering solutions play in shaping smart cities and how they facilitate profound urban metamorphosis. In response to the rapidly evolving economy and our ever-changing world, especially the increasing demand for sustainability (Fagerberg et al., 2013). Smart cities epitomize urban areas that ingeniously employ advanced technologies and data-driven innovations to elevate the well-being of their residents while advancing sustainability and optimizing resource allocation. These urban landscapes are intricately woven tapestries of interconnected systems and devices, including an array of sensors, IoT (Internet of Things) devices, and intelligent infrastructure. These technological marvels work in harmony to gather and analyze data, automate processes, and empower data-driven decision-making (Crossan and Apaydin, 2010). When it comes to managing the transition to smart cities as processes of innovation, it is essential to consider the dimensions that shape the way these transitions occur.In the rapidly changing global landscape, characterized by a burgeoning need for sustainability, Fagerberg et al. aptly note that smart cities hold the promise of aligning various sectors—transportation, energy, healthcare, infrastructure, public safety, and governance—into a symphony of efficiency, sustainability, and livability. Powered by technology and connectivity, smart cities embark on a mission to confront urban challenges head-on, ultimately aiming to enhance the welfare of their citizens. Smart buildings equipped with automated systems for lighting, heating, and cooling contribute to energy conservation and cost savings (Zhang et al., 2022, 2023). A hallmark of smart cities is the pervasive deployment of sensors and IoT devices throughout the urban fabric. These omnipresent devices collect up-to-the-minute data on essential aspects of urban existence, such as traffic flow, air quality, energy usage, waste handling, and public security. This wealth of data is subsequently closely examined to reveal valuable insights that guide well-informed decision-making.

Transportation systems in smart cities epitomize interconnected networks optimized to alleviate congestion, improve mobility, and promote sustainable transit modes. Examples abound, from intelligent traffic management systems to adaptive traffic signals and real-time public transportation updates. As Crossan and Apaydin aptly emphasize, orchestrating smart city transitions necessitates a careful consideration of the dimensions that define these transformative processes. Energy efficiency and sustainability occupy pivotal positions within the smart city paradigm. Smart grids emerge as powerful tools to optimize energy distribution, integrate renewable energy sources, and effectively manage energy consumption in buildings. Automated systems governing lighting, heating, and cooling in smart buildings contribute to energy conservation and financial savings. Moreover, in the realm of urban and regional studies, scholarly discourse revolves around the technocratic underpinnings of smart city transitions and their ramifications on governance, as highlighted by Lim, Sheikh, and Zhang.

Focusing on the welfare of their inhabitants, smart cities utilize technology to improve healthcare services and public safety. The use of remote health monitoring systems, data-driven healthcare management, and advanced emergency response systems enhances the accessibility and quality of healthcare. In a similar vein, smart surveillance systems, integrated emergency services, and predictive analytics strengthen the safety and security of the population.

A critical facet of smart cities lies in fostering citizen engagement and participation. Digital platforms and mobile applications empower residents to access information, provide feedback, and actively partake in decision-making processes. This engagement cultivates a sense of community and engenders more responsive and inclusive governance. However, as smart cities evolve, it becomes imperative to address a panoply of challenges, including those related to data privacy, cyber security, and equitable technology access. Ethical considerations and safeguarding personal information should remain at the heart of smart city initiatives' design and implementation.

1.1 Characteristics of Smart Cities

Smart cities employ information and communication technologies (ICT) and a range of data sources to enhance the well-being of residents, promote sustainability, and optimize resource utilization. Here are some common characteristics of smart cities:

- Advanced Infrastructure: Smart cities incorporate modern infrastructure that enables efficient and sustainable services. This includes robust communication networks, high-speed internet connectivity, smart grids for electricity distribution, and integrated transportation systems.
- Data-driven Decision Making: Smart cities utilize data gathered from diverse sources like sensors, social media, and public services for the purpose of informed decision-making. To gain insights and enhance resource allocation and service delivery, they harness data analytics and artificial intelligence (AI) algorithms.
- Sustainable Development: Smart cities prioritize sustainability by adopting eco-friendly practices. They focus on reducing energy consumption,

promoting renewable energy sources, managing waste effectively, and implementing green building standards. Smart grids and energy management systems play a vital role in optimizing energy distribution and mitigating emissions.

- Smart Mobility: Smart cities prioritize intelligent transportation systems to improve mobility and alleviate traffic congestion. This includes implementing smart traffic management, intelligent parking systems, real-time public transportation information, and encouraging the use of electric vehicles and bike-sharing programs.
- Citizen Engagement: Smart cities aim to actively involve citizens in decision-making processes. They provide digital platforms and mobile applications for residents to access government services, provide feedback, report issues, and participate in community activities. Open data initiatives ensure transparency and enable public access to information.
- Enhanced Safety and Security: Smart cities employ advanced technologies for public safety and security. This includes video surveillance systems, emergency management systems, predictive policing, and early warning systems for natural disasters. Connected devices and sensors enable real-time monitoring of public spaces.
- Efficient Resource Management: The emphasis in smart cities is on enhancing the efficiency of resource utilization, including energy, water, and waste management. Automated systems and sensors are employed to monitor and manage resource consumption, detect leaks or inefficiencies, and encourage sustainable practices among residents and businesses.
- Quality of Life: Smart cities aim to enhance the well-being of their inhabitants by offering smart healthcare services, digital educational resources, automated and energy-efficient smart homes, and access to cultural and recreational facilities. Intelligent street lighting and noise pollution monitoring contribute to a better living environment.
- Innovation and Collaboration: Smart cities foster innovation and collaboration between the public sector, private companies, research institutions, and citizens. They encourage entrepreneurship, support startups, and provide testbeds for new technologies and services. Partnerships between different stakeholders are formed to drive smart city initiatives.
- Resilience and Adaptability: Smart cities focus on building resilience to cope with challenges like climate change, natural disasters, and cybersecurity threats. They implement robust infrastructure, disaster response plans, and data backup systems to ensure continuity of services during emergencies.

1.2 Importance of Smart Cities in Addressing Urban Challenges

Smart cities play a vital role in addressing the numerous challenges that urban areas face today. With the rapid growth of cities and increasing population, issues such as traffic congestion, environmental deterioration, insufficient infrastructure, and resource management become prominent. Through the utilization of technology and data-driven solutions, smart cities provide innovative approaches to effectively tackle these challenges. For example, smart transportation systems optimize traffic flow, reduce commuting time, and minimize carbon emissions. The integration of advanced sensor networks and IoT devices facilitates efficient waste management, energy conservation, and water usage, thereby promoting sustainability (Van Mierlo et al., 2010). Additionally, smart cities empower citizens by providing them with real-time information, which enhances public safety and enables participatory governance. Through the application of technology to create intelligent and interconnected urban environments, smart cities actively enhance the quality of life, foster economic growth, and support environmental sustainability, ultimately paving the path towards a more resilient and brighter future.

1.3 Overview of the Role of Engineering in Smart City Design

Engineering plays a crucial part in shaping smart cities, bringing about a revolution in urban landscapes and elevating the standard of living. The design of smart cities encompasses various elements such as infrastructure, energy, transportation, communication, and sustainability. It is through the application of engineering principles that these ideas are transformed into reality. Engineers bring their expertise to the table by designing intelligent systems, implementing advanced technologies, and integrating different components to establish efficient and interconnected urban environments. They are instrumental in the deployment of smart grids, intelligent transportation systems, optimized waste management, and innovative energy solutions, among other things. By utilizing their knowledge in fields like civil, electrical, mechanical, and environmental engineering, engineers contribute to the transformation of cities into interconnected ecosystems that enhance resource management, improve citizen services, and foster sustainable growth. Engineers, through their interdisciplinary approach, are instrumental in shaping the future of smart cities, thereby leading the path to improved connectivity, heightened efficiency, and an enhanced quality of life for urban inhabitants.

2. URBAN INFRASTRUCTURE AND CONNECTIVITY

The critical role of urban infrastructure and connectivity in shaping the livability and efficiency of modern cities cannot be overstated. With the ongoing urbanization and population growth, cities are confronted with the task of establishing sustainable and interconnected systems that elevate the quality of life for their inhabitants. Urban infrastructure encompasses a diverse array of physical components, encompassing transportation networks, energy systems, water supply and sanitation, communication technologies, and public spaces (Cugurullo, 2018; Johnson et al., 2020). These elements are intricately interconnected and require careful planning and integration to ensure smooth operation and seamless connectivity. Well-designed transportation systems, such as road networks, public transit systems, and cycling infrastructure, efficiently facilitate the movement of people and goods, alleviating congestion and improving air quality. Likewise, dependable energy systems, including smart grids and renewable energy sources, enhance sustainability and resilience. The provision of clean water, adequate sanitation facilities, and effective waste management contributes to the overall well-being of urban residents. Furthermore, high-speed internet connectivity and digital infrastructure accessibility drive the development of smart cities, fostering innovation and improving access to information and services. By placing a priority on the development of robust and interconnected urban infrastructure, cities can establish dynamic, inclusive, and sustainable environments that enrich the lives of their residents.

2.1 Intelligent Transportation Systems and Smart Mobility Solutions

Intelligent transportation systems (ITS) and smart mobility solutions have fundamentally transformed the management and navigation of transportation networks. These innovations incorporate state-of-the-art developments in information and communication technologies to create transportation systems that are efficient, safe, and sustainable. ITS encompasses a wide range of applications, including real-time traffic monitoring, congestion management, intelligent control of traffic signals, and advanced traveler information systems (Chen & Chan, 2022). These solutions utilize data from various sources, such as sensors, cameras, and GPS devices, to deliver precise and current information to both commuters and transportation authorities. By analyzing this data, ITS can optimize the flow of traffic, reduce congestion, and enhance road safety. Furthermore, smart mobility solutions merge ITS with emerging trends like shared mobility, electric vehicles, and connected infrastructure. These solutions facilitate seamless integration among different transportation modes, promoting options for multimodal travel and decreasing the dependence on private vehicles. Through the integration of smart technologies, transportation systems can become more efficient, environmentally conscious, and user-focused, ultimately reshaping the way we navigate within our cities and beyond.

2.2 Efficient Management of Energy, Water, and Waste Infrastructure

Effective management of energy, water, and waste infrastructure is essential for sustainable urban development and enhancing the overall quality of life in urban areas. Here are some primary approaches to efficiently handle these facets of urban infrastructure and connectivity:

1. Energy Management:
    - Utilize smart grid technologies to enhance energy distribution, minimize transmission losses, and facilitate efficient energy consumption.
    - Encourage the adoption of renewable energy sources like solar, wind, and geothermal in order to diminish dependence on fossil fuels and lower greenhouse gas emissions.
    - Encourage energy-efficient building designs and retrofitting existing buildings with energy-saving technologies.
    - Implement smart metering systems for the purpose of tracking energy consumption and offering feedback to consumers, empowering them to make informed choices regarding their energy usage.
2. Promote sustainable transportation by fostering the adoption of electric vehicles and establishing charging infrastructure.
3. Water Management:
    - Implement water conservation measures such as rainwater harvesting, greywater reuse, and water-efficient irrigation systems.
    - Invest in leak detection and monitoring systems to identify and address water losses in distribution networks.
    - Develop water recycling and desalination facilities to augment water supply and reduce reliance on freshwater sources.
    - Promote public awareness campaigns to encourage responsible water usage and educate residents about water-saving practices.
    - Implement smart water metering systems to monitor and manage water consumption and identify areas of high usage or potential leaks.
4. Waste Management:
    - Establish comprehensive waste management systems that include waste reduction, recycling, and proper disposal.

    - Encourage source separation of waste at the household and commercial levels to facilitate recycling and composting.
    - Establish waste-to-energy facilities that transform organic waste into either biogas or electricity.
    - Develop efficient collection and transportation systems to minimize waste handling and reduce environmental impacts.
    - Promote public awareness and education programs to encourage responsible waste disposal and recycling practices.
5. Integrated Infrastructure Planning:
    - Promote cooperation among various sectors and stakeholders engaged in urban infrastructure planning and development to guarantee the implementation of integrated and effective solutions.
    - Harness data analytics and modeling tools to enhance infrastructure design, operation, and maintenance efficiency.
    - Deploy smart infrastructure systems that employ sensors, automation, and real-time monitoring to enhance resource allocation efficiency and minimize inefficiencies..
    - Plan for compact and mixed-use urban development, reducing the need for long-distance transportation and optimizing infrastructure utilization.
6. Policy and Governance:
    - Develop and enforce regulations and policies that support sustainable infrastructure development and incentivize energy, water, and waste efficiency.
    - Establish public-private partnerships to leverage expertise and resources for infrastructure management and innovation.
    - Promote community involvement and active participation in decision-making processes concerning urban infrastructure and connectivity.
    - Provide financial incentives, subsidies, and grants to support the adoption of sustainable technologies and practices.
    - Conduct regular audits and evaluations to assess the performance and effectiveness of infrastructure management strategies.

Efficient management of energy, water, and waste infrastructure requires a holistic and integrated approach that combines technological advancements, policy frameworks, and public engagement (Tan, 2021). By implementing these strategies, cities can reduce resource consumption, lower carbon emissions, enhance resilience, and create more livable and sustainable urban environments.

2.3 Connectivity and Communication Infrastructure for Smart Cities

The progression and operation of smart cities rely heavily on connectivity and communication infrastructure. As urbanization advances and technology takes center stage, the demand for robust and reliable networks becomes an absolute necessity. The foundation of smart cities relies on the seamless sharing of data among different devices, sensors, and systems to improve services, increase efficiency, and elevate residents' overall quality of life. At the core of this connectivity is a high-speed broadband network, guaranteeing fast and reliable internet access for all residents. This network serves as the cornerstone for the multitude of devices and applications comprising the smart city ecosystem. These interconnected devices, ranging from smart traffic management systems and intelligent street lighting to environmental monitoring and public safety initiatives, require constant communication to operate effectively.

To establish extensive connectivity, smart cities often integrate a combination of wired and wireless technologies. The foundational framework of the wired infrastructure comprises fiber-optic cables, offering high-speed internet access to businesses, residences, and public areas. Furthermore, wireless networks like Wi-Fi, cellular networks (including 4G and 5G), and low-power wide-area networks (LPWAN) extend connectivity beyond stationary locations, supporting the mobility and adaptability needed in urban settings. Communication infrastructure also includes the deployment of sensor networks and Internet of Things (IoT) devices. These interconnected sensors gather data from various sources, including environmental sensors, smart meters, and surveillance cameras. This data is then transmitted and analyzed in real-time to facilitate informed decision-making and optimize resource allocation. For instance, traffic data can be utilized to dynamically adjust signal timings, reducing congestion and improving traffic flow.

Furthermore, a robust communication infrastructure heightens public safety by enabling swift responses and efficient emergency management systems. Real-time data from surveillance cameras and sensors can be analyzed to identify potential threats, while interconnected emergency services can coordinate their actions effectively. Additionally, citizen engagement and participation are enhanced through digital platforms and applications that facilitate feedback, service requests, and information dissemination. As smart cities evolve, the significance of connectivity and communication infrastructure will continue to grow. The expansion of 5G networks, progress in edge computing, and the integration of artificial intelligence (AI) will further augment the capabilities and potential of smart city initiatives. The uninterrupted flow of data and communication will remain pivotal, enabling these cities to become more efficient, sustainable, and responsive to the needs of their residents.

2.4 Integration of Internet of Things (IoT) Devices and Sensors for Data-Driven Decision-Making

The incorporation of Internet of Things (IoT) equipments and sensors has transformed the idea of smart cities, enabling data-driven decision-making like never before. With the extensive deployment of interconnected devices and sensors throughout urban areas, cities are becoming more intelligent and responsive, optimizing resource utilization and humanizing the value of life for their inhabitants. IoT devices and sensors serve as the sensory network of the smart city infrastructure, collecting copious real-time data on various aspects of urban life. These devices can monitor a wide range of factors, including air quality, traffic flow, energy consumption, waste management, public safety, and weather conditions. By harnessing this wealth of data, city administrators, planners, and policymakers gain invaluable insights into the city's operations, enabling them to make well-informed decisions and implement effective strategies. Advanced analytics and machine learning techniques process, analyze, and visualize the data collected by IoT devices and sensors, facilitating the extraction of meaningful patterns, correlations, and trends, which present a complete considerate of the city's dynamics. Decision-makers can utilize these details to improve supply allocation, improve infrastructure planning, enhance transportation systems, and address environmental challenges. Furthermore, the integration of IoT devices and sensors enables proactive and predictive problem-solving approaches. By continuously monitoring and analyzing real-time data, potential issues can be identified and addressed before they escalate. For instance, IoT-based traffic management systems can detect congestion patterns and dynamically adjust signal timings to optimize traffic flow, reducing commute times and minimizing pollution levels. Data-driven decision-making, powered by IoT devices and sensors, is revolutionizing smart cities, making them more sustainable, efficient, and livable. City authorities can tailor their policies and services to meet the evolving needs of their residents by leveraging the wealth of information at their disposal.

3. SUSTAINABLE ENERGY SOLUTIONS

3.1 Renewable Energy Integration and Microgrids in Smart Cities

The successful establishment of smart cities heavily depends on the effective integration of renewable energy and microgrids. These components are critical for creating sustainable and resilient energy systems, reducing greenhouse gas emissions, and promoting localized energy production and consumption. The information below provides a comprehensive overview of how renewable energy and microgrids are integrated into smart cities.

1. Integration of Renewable Energy: The integration of renewable energy involves the incorporation of clean and sustainable energy sources into the existing power grid, enabling the efficient utilization of resources such as solar, wind, hydro, geothermal, and biomass energy. Incorporating renewables into smart cities offers several advantages:
   a. Greenhouse gas reduction: Renewable energy sources produce minimal or no emissions during operation, leading to a significant decrease in pollutants, including carbon dioxide.
   b. Energy source diversification: The integration of renewable energy diversifies the energy mix, reducing reliance on fossil fuels and enhancing energy security.
   c. Economic benefits: As the costs of renewable energy decline, it becomes an economically viable alternative to traditional sources in many regions, resulting in long-term cost savings for both the city and its residents.
   d. Job opportunities: The renewable energy sector creates employment opportunities in manufacturing, installation, maintenance, and operation of renewable energy systems.
   e. Community involvement: Localized renewable energy projects encourage community participation and ownership, fostering a sense of environmental responsibility and social cohesion.
2. Microgrid Implementation: Microgrids are localized energy distribution systems that can operate independently or in conjunction with the primary power grid. They consist of distributed energy resources (DERs), including renewable energy generators, energy storage systems, and demand response technologies. Microgrids in smart cities offer various benefits:
   a. Enhanced reliability: Microgrids improve the resilience of the power supply by providing backup power during grid outages or natural disasters, and they can function independently, supplying electricity to essential loads.

b. Efficient energy management: Microgrids optimize energy generation, storage, and consumption at a local level, facilitating the integration of intermittent renewable energy sources and reducing transmission losses.
c. Grid support and stability: Microgrids can provide ancillary services to the main grid, such as frequency regulation and voltage support, contributing to grid stability.
d. Energy self-sufficiency: Microgrids enable local energy production, reducing dependence on distant power plants and enhancing energy security while decreasing transmission congestion.
e. Electric vehicle integration: Microgrids can include EV charging infrastructure and manage charging patterns, promoting electric transportation and efficient renewable energy use.
f. Peer-to-peer energy trading: Microgrids can facilitate direct energy trading among participants, promoting local economic development and energy sharing.

3.2 Energy-Efficient Buildings and Smart Grid Systems

The establishment of sustainable and eco-friendly smart cities relies heavily on energy-efficient buildings and smart grid systems. These innovations are dedicated to enhancing energy usage, decreasing carbon emissions, and advancing the overall energy efficiency of both buildings and the power grid. Now, let's delve deeper into each of these ideas.

1. Energy-Efficient Buildings: Energy-efficient buildings are designed and constructed with the aim of minimizing energy consumption while maintaining comfort and functionality. Here are some key features and technologies used in energy-efficient buildings:
   a. Insulation and Weatherization: Proper insulation, including high-quality windows and insulation materials, prevents heat transfer between the inside and outside of a building, reducing the need for heating and cooling.
   b. Efficient Lighting: LED lights, occupancy sensors, and daylight harvesting systems help reduce electricity consumption for lighting purposes.
   c. HVAC Systems: By utilizing cutting-edge technology like heat pumps, geothermal systems, and energy recovery ventilation, high-efficiency heating, ventilation, and air conditioning (HVAC) systems lower energy usage.
   d. Renewable Energy Integration: Renewable energy sources, like solar or wind turbines, are frequently incorporated into energy-efficient buildings

to produce clean energy locally and lessen reliance on conventional grid electricity.

e. Smart Energy Management: Intelligent energy management systems enable efficient control and monitoring of energy consumption within a building. These systems use sensors, automation, and data analytics to optimize energy usage and make adjustments based on occupancy patterns and environmental conditions.

f. Building Automation: Automation systems control and optimize various building functions, including lighting, HVAC, security, and occupancy, to minimize energy waste and improve overall efficiency.

2. Smart Grid Systems: To maximize the production, distribution, and consumption of power, a smart grid is a sophisticated electrical infrastructure that integrates automation and digital communication technology. It increases overall grid dependability, permits effective energy management, and makes the incorporation of renewable energy sources easier. Important parts of a smart grid system consist of:

   a. Advanced Metering Infrastructure (AMI): Smart meters collect real-time data on electricity usage, allowing consumers and utility companies to monitor and manage energy consumption more effectively.

   b. Demand Response (DR) programs allow utilities to provide incentives to customers to modify their electricity consumption during periods of peak demand. This lessens the load on the system by balancing the supply and demand for electricity.

   c. Energy Storage: To provide a more consistent and effective power supply, smart grids frequently include energy storage devices, such as batteries, to store extra electricity during times of low demand and release it during times of peak demand.

   d. Microgrids: Microgrids are small-scale energy distribution networks that can function both on their own and in tandem with the larger grid. To provide dependable and resilient power supply to certain locations inside a smart city, they blend renewable energy sources, storage, and cutting-edge controls.

   e. Grid Monitoring and Control: Real-time monitoring and control systems enable utilities to optimize grid operations, detect faults or outages, and respond quickly to maintain grid stability and efficiency.

   f. Integration of Electric Vehicles (EVs): Smart grids facilitate the integration of EV charging infrastructure and manage the charging process to avoid overloading the grid while maximizing renewable energy utilization.

The incorporation of energy-efficient structures and intelligent grid systems within smart urban areas encourages sustainable progress, diminishes the release of greenhouse gases, enhances energy durability, and elevates residents' overall well-being. These technologies play a crucial role in attaining enduring energy and environmental objectives, all the while stimulating economic growth and fostering innovation.

3.3 Demand-Side Management and Energy Optimization Strategies

In the development and operation of smart cities, the significance of demand-side management (DSM) and energy optimization strategies cannot be underestimated. These strategies are essential for effectively managing and improving energy consumption patterns at the individual or consumer level, aiming to enhance energy efficiency, decrease peak demand, and foster sustainability. Van Mierlo et al as a catalyst for change towards a sustainable development". In the subsequent section, a comprehensive elucidation of DSM and energy optimization strategies within the realm of smart cities will be presented.

1. **Demand-Side Management (DSM)**: Demand-side management involves various measures aimed at influencing consumer behavior, load patterns, and energy consumption to achieve a more balanced and efficient use of energy resources. The key objectives of DSM include reducing peak demand, enhancing grid reliability, minimizing energy costs, and promoting environmental sustainability. Here are some common DSM techniques used in smart cities:
    a. Time-of-Use (TOU) Pricing: TOU pricing involves charging consumers different rates based on the time of day and season. By incentivizing off-peak electricity consumption through lower rates, TOU pricing encourages users to shift their energy-intensive activities to non-peak hours, thereby reducing overall demand during peak periods.
    b. Real-Time Pricing (RTP): RTP dynamically adjusts electricity prices based on real-time market conditions and grid demand. Consumers are provided with price signals that inform them about the current electricity costs, enabling them to make well-informed choices regarding their energy consumption. RTP encourages users to reduce consumption during high-price periods, leading to load shifting and demand reduction.
    c. Demand Response (DR): DR programs encourage consumers to voluntarily reduce their electricity usage during peak demand periods or in response to grid emergencies. This can be achieved through financial incentives, such as lower electricity rates or direct payments, or through non-monetary

rewards like energy credits. Automated demand response systems leverage smart meters and IoT devices to facilitate load curtailment in real-time.

d. Energy Efficiency Programs: Energy efficiency initiatives aim to improve the energy performance of buildings, appliances, and equipment. These programs include incentives for adopting energy-efficient technologies, promoting energy audits, and implementing energy-saving practices. By reducing energy consumption, energy efficiency measures help lower overall demand and minimize the need for additional generation capacity.

2. **Energy Optimization Strategies:** Energy optimization strategies concentrate on enhancing the utilization of energy resources within a smart city framework. These strategies harness advanced technologies, data analysis, and automation to enhance energy efficiency and reduce unnecessary energy consumption. Here are some essential energy optimization strategies:

a. Smart Grid Technology: Smart grid technology incorporates digital communication and control systems into the conventional power grid infrastructure, facilitating two-way communication between utility providers and consumers. This technology provides real-time information on energy usage, grid conditions, and pricing to enable consumers to monitor and make informed decisions. Smart grids support load monitoring, demand forecasting, and efficient distribution, leading to improved load balancing and reduced transmission losses.

b. Energy Storage Solutions (ESS): Energy storage solutions, such as batteries, offer the capability to capture and store surplus electricity generated during off-peak periods or from renewable sources. These stored energy reserves can be deployed during peak demand or when intermittent renewable sources are unavailable. ESS plays a crucial role in matching supply with demand, stabilizing the grid, and enhancing the integration of renewable energy sources.

c. Decentralized Energy Resources (DERs): Decentralized energy resources encompass localized power generation and storage technologies, including solar panels, wind turbines, and small-scale co-generation units. By generating electricity in close proximity to the point of use, DERs reduce transmission losses and promote localized energy production. Integrating DERs with smart grid technology enables more effective management and optimization of energy resources.

d. Building Energy Management Systems (BEMS): BEMS use automation, sensor networks, and data analytics to optimize energy consumption within buildings. These systems monitor and control various energy-consuming components, such as lighting, HVAC systems, and appliances, based on real-time conditions and user preferences. BEMS enable energy-efficient operation, demand response capabilities, and occupant comfort optimization.

e. Intelligent Transportation Systems (ITS): ITS optimize energy consumption in transportation systems by employing smart traffic management, route optimization, and vehicle-to-grid (V2G) technology. V2G enables electric vehicles (EVs) to feed surplus energy back to the grid when parked, thus supporting grid stabilization and load balancing. ITS helps reduce congestion, fuel consumption, and emissions associated with transportation.

4. RESILIENCE AND DISASTER MANAGEMENT

Resilience and effective disaster management are integral aspects of building smart cities that can withstand and recover from various challenges. Smart cities are designed to incorporate innovative technologies, advanced infrastructure, and proactive planning to enhance their ability to respond to and rebound from disasters (Wirtz et al., 2021). By implementing robust early warning systems, developing resilient infrastructure, and fostering community engagement, smart cities can strengthen their capacity to handle natural and human-induced emergencies. Integrating data analytics and artificial intelligence enables the identification of vulnerable areas, facilitates real-time monitoring, and enhances emergency response coordination. By prioritizing resilience and disaster management, smart cities can ensure the safety and well-being of their residents while minimizing the impact of adversities on their overall functioning and development.

4.1 Engineering Solutions for Urban Resilience and Disaster Preparedness

Engineering solutions play a significant part in enhancing urban resilience and disaster preparedness within smart cities. These solutions employ innovative approaches to bolster the city's ability to withstand and recover from various challenges. Through the integration of advanced technologies, such as sensor networks, data analytics, and intelligent infrastructure systems, engineers can develop robust early warning systems, efficient evacuation routes, and resilient infrastructure. By leveraging these engineering solutions, cities can enhance their ability to respond swiftly to disasters, mitigate their impacts, and foster a safer and more sustainable urban environment.

4.2 Early Warning Systems and Real-Time Monitoring for Natural Disasters

The safety of residents in smart cities during natural disasters heavily relies on early warning systems and real-time monitoring. These systems are essential in delivering timely alerts and maintaining continuous surveillance, thereby reducing the impact of these events. Early warning systems are designed to detect and anticipate potential natural disasters, including earthquakes, floods, hurricanes, and tsunamis. By employing various sensors, such as seismic sensors, weather stations, and water level gauges, these systems can gather critical data in real-time. The collected information is then processed and analyzed to identify patterns and anomalies that may indicate an impending disaster. Real-time monitoring complements early warning systems by continuously tracking environmental conditions and monitoring changes. Smart cities leverage advanced technologies, such as IoT devices and satellite imagery, to capture and transmit data in real-time. This allows authorities to closely monitor factors like weather patterns, ground movements, and water levels, providing valuable insights for disaster preparedness and response.

The integration of early warning systems and real-time monitoring enables authorities to swiftly issue alerts and notifications to residents in affected areas. These warnings can be disseminated through various channels, including mobile applications, text messages, sirens, and public address systems (Liu et al., 2023). By promptly notifying individuals, communities can take immediate action, such as evacuating to safer locations or preparing emergency supplies. Moreover, these systems facilitate coordination among different stakeholders involved in disaster management. Emergency responders, local authorities, and relief organizations can access the same real-time information, allowing them to align their efforts and resources effectively. This collaboration enhances the overall response and recovery process, minimizing casualties and property damage. In smart cities, early warning systems and real-time monitoring are an integral part of disaster resilience strategies. By leveraging technology and data-driven insights, these systems enhance preparedness, response, and recovery capabilities, ultimately ensuring the safety and well-being of residents during natural disasters.

4.3 Infrastructure Design to Withstand Environmental Challenges

Creating resilient infrastructure capable of withstanding environmental challenges is paramount for the development of smart cities. The design should focus on incorporating robust features that enhance sustainability and adaptability in the face of adverse conditions. Several key considerations play a crucial role in achieving this objective:

1. Climate-responsive design: Incorporate climate-specific features that mitigate the impact of extreme weather events such as hurricanes, floods, or heatwaves. This includes implementing stormwater management systems, elevated foundations, and reinforced structures.
2. Green infrastructure: Emphasize the integration of nature-based solutions like green roofs, vertical gardens, and urban forests. These elements enhance biodiversity, regulate temperature, reduce stormwater runoff, and improve air quality.
3. Resilient energy systems: Deploy distributed renewable energy generation, such as solar panels or wind turbines, to ensure a diversified and reliable power supply. Additionally, integrate energy storage technologies to enhance resilience and provide backup during power outages.
4. Water management: Implement efficient water management strategies to reduce strain on resources and combat water scarcity. This involves incorporating rainwater harvesting systems, wastewater treatment plants, and smart irrigation technologies to optimize water usage.
5. Smart mobility: Develop intelligent transportation systems that prioritize sustainable modes of transportation such as walking, cycling, and public transit. Integrate smart traffic management systems and promote the use of electric vehicles to reduce emissions and enhance mobility efficiency.
6. Digital infrastructure: Establish a robust digital backbone to support smart city operations and services. This includes implementing high-speed broadband networks, data centers, and sensor networks to enable real-time monitoring and data-driven decision-making.
7. Community engagement: Involve the local community in the design and planning process to ensure infrastructure aligns with their needs and priorities. Encourage citizen participation and education programs that promote sustainable practices and raise awareness about environmental challenges.
8. Adaptive land use: Implement flexible land-use planning strategies that can adapt to changing environmental conditions. This involves designing multi-functional spaces that can serve as floodplains, urban parks, or temporary evacuation areas, depending on the situation.
9. Risk assessment and management: Conduct comprehensive risk assessments to identify vulnerabilities and develop mitigation strategies. Implement early warning systems, emergency response plans, and resilient infrastructure standards to minimize the impact of disasters.

10. Continuous monitoring and evaluation: Regularly monitor and evaluate the performance of the infrastructure to identify areas for improvement. Utilize data analytics and predictive modeling to enhance the resilience of the smart city infrastructure over time.

Smart cities can enhance their infrastructure to effectively confront environmental challenges and promote the well-being and sustainability of their residents by embracing these design principles.Through the implementation of these design principles, smart cities can establish resilient infrastructure that can effectively withstand environmental obstacles while safeguarding the welfare and sustainability of their inhabitants.

4.4 Collaborative Approaches for Urban Resilience Planning

Urban resilience planning in smart cities heavily relies on collaborative approaches. These strategies revolve around the active participation of diverse stakeholders, who contribute their valuable expertise, knowledge, and resources to bolster the city's resilience (Jamshed, 2018). By promoting collaboration, cities can successfully confront obstacles and proactively prepare for potential disruptions, eliminating dependence on any singular entity. Below, we outline essential collaborative approaches for urban resilience planning in smart cities:

1. Partnership: Encouraging partnerships between local government, private sector organizations, community groups, and academia promotes collective action and knowledge sharing. This collaborative approach allows diverse perspectives to be considered, leading to more comprehensive resilience planning.
2. Participatory Planning: Engaging citizens and involving them in the decision-making process fosters a sense of ownership and enables a bottom-up approach to resilience planning. Citizens can contribute local knowledge and insights, helping to identify vulnerabilities and prioritize actions that address their specific needs.
3. Multi-Sector Collaboration: Collaboration across various sectors, such as transportation, energy, water, and public health, allows for a holistic approach to resilience planning. This collaborative effort ensures that interdependencies and potential cascading effects are considered, leading to more robust and integrated strategies.
4. Data Sharing: Sharing data among different stakeholders, including government agencies, research institutions, and private companies, facilitates evidence-based decision-making. Collaborative data platforms can help identify risks, monitor performance, and support informed planning and response actions.

5. Capacity Building: Collaborative approaches focus on building the capacity of different stakeholders, empowering them to actively contribute to resilience planning. Training programs, workshops, and knowledge exchange platforms enhance understanding, promote skill development, and enable effective collaboration.
6. Networked Governance: Creating networks and fostering communication channels between different stakeholders enables effective coordination and collaboration. These networks can facilitate the exchange of information, best practices, and lessons learned, fostering innovation and collective problem-solving.
7. Co-Creation: Involving stakeholders in the co-creation of resilience solutions promotes a sense of shared responsibility and fosters innovative approaches. Collaborative workshops, design charrettes, and hackathons encourage stakeholders to work together to develop solutions tailored to the city's specific challenges.

5. DATA ANALYTICS AND URBAN PLANNING

Data analytics plays a fundamental role in shaping urban planning strategies within smart cities. By harnessing the power of data, city planners gain valuable insights into various aspects of urban life, facilitating informed decision-making. Data analytics enables planners to analyze and interpret vast amounts of information related to transportation, energy consumption, waste management, and more (Ranchordás, 2019). These insights provide a deeper understanding of urban dynamics, aiding in the identification of patterns and trends. Such knowledge empowers city planners to optimize resource allocation, improve infrastructure, and enhance the quality of life for residents. By leveraging data analytics, urban planning in smart cities becomes a dynamic and adaptive process, fostering sustainable development and efficient utilization of resources.

5.1 Use of Data Analytics for Urban Planning and Decision-Making

The utilization of data analytics is essential in enabling efficient urban planning and decision-making procedures in smart cities (Cugurullo, 2018). Sustainability of so-called smart cities and eco-cities is done by investigating the extent to which they are developed in a controlled and systematic manner as their developers claim. Through the utilization of data, urban planners can acquire valuable insights that contribute to improving different aspects of city development. Here are a few tangible examples showcasing how data analytics is applied in smart city planning and decision-making:

1. Traffic management: Data analytics enables the collection and analysis of real-time traffic data, including vehicle movement, congestion patterns, and peak hours. This information helps optimize traffic signal timings, identify bottlenecks, and plan efficient road networks and public transportation systems.
2. Waste management: By analyzing data on waste generation, collection routes, and recycling rates, urban planners can optimize waste management systems. This includes determining the most suitable locations for recycling facilities, identifying areas with high waste generation, and optimizing collection schedules to minimize costs and environmental impact.
3. Energy consumption: Data analytics allows for the monitoring and analysis of energy usage patterns within smart cities. Planners can identify energy-intensive areas, detect inefficiencies, and develop strategies to reduce consumption. This may involve implementing smart grids, promoting renewable energy sources, and encouraging energy-saving practices.
4. Public safety: Data analytics can enhance public safety by analyzing crime data, identifying high-crime areas, and predicting crime patterns. This information helps law enforcement agencies allocate resources effectively, implement targeted interventions, and develop strategies to prevent criminal activities.
5. Resource allocation: Data analytics enables the analysis of demographic data, infrastructure usage, and public service demand. Planners can identify areas with high population density, determine resource allocation needs, and make informed decisions about the distribution of healthcare facilities, schools, parks, and other public amenities.
6. Environmental sustainability: Data analytics can contribute to environmental sustainability by monitoring and analyzing air quality, water usage, and waste generation. Planners can identify pollution sources, develop strategies for reducing environmental impact, and promote sustainable practices.
7. Emergency management: Data analytics plays a crucial role in emergency management by analyzing data from sensors, social media, and other sources to detect early warning signs of natural disasters or emergencies. This allows for timely evacuation plans, resource allocation, and effective response strategies.
8. Citizen engagement: Data analytics can enable citizen engagement by collecting and analyzing feedback, sentiments, and preferences from residents. This information helps planners understand community needs, prioritize projects, and involve citizens in decision-making processes.

5.2 Geographic Information Systems (GIS) and Spatial Analysis

Geographic Information Systems (GIS) and spatial analysis enable the integration and analysis of geospatial data to derive valuable insights for urban planning, resource management, and decision-making (Phillips & Jiao, 2023). GIS technology facilitates the collection, storage, and visualization of geospatial information, such as maps, satellite imagery, and sensor data. By leveraging spatial analysis techniques, cities can identify patterns, trends, and correlations within the data, enabling them to optimize transportation routes, identify suitable locations for infrastructure, and enhance emergency response systems. GIS and spatial analysis provide cities with the tools necessary to harness the power of location-based data, enabling them to make informed and efficient decisions towards creating sustainable and livable urban environments.

5.3 Simulation Models and Predictive Analytics for Urban Design

- Modeling and Analytics for Urban Mobility: Utilizing simulation models and predictive analytics to analyze and optimize transportation networks, including traffic flow, public transit, and shared mobility options. This helps in understanding travel patterns, predicting congestion, and improving overall mobility efficiency.
- Energy Consumption Forecasting: Developing simulation models and predictive analytics to estimate energy consumption in smart cities. This enables urban planners to identify energy-saving opportunities, optimize infrastructure, and enhance sustainability efforts.
- Environmental Impact Assessment: Implementing simulation models and predictive analytics to assess the environmental impact of urban design projects. This helps in predicting air and noise pollution levels, identifying areas of concern, and designing strategies for mitigation and sustainable development.
- Urban Growth and Land Use Planning: Employing simulation models and predictive analytics to forecast urban growth and guide land use planning. This enables city planners to make informed decisions regarding zoning, infrastructure development, and allocation of resources.
- Smart Waste Management: Using simulation models and predictive analytics to optimize waste management systems. This involves predicting waste generation patterns, optimizing collection routes, and identifying opportunities for recycling and waste reduction.

- Public Safety and Emergency Response Planning: Utilizing simulation models and predictive analytics to improve public safety and emergency response planning. This includes predicting potential risks, simulating emergency scenarios, and optimizing resource allocation for faster and more effective responses.
- Social Infrastructure Planning: Implementing simulation models and predictive analytics to plan social infrastructure, such as schools, healthcare facilities, and recreational spaces. This helps in predicting population growth, identifying underserved areas, and optimizing the distribution of social services.
- Smart Water Management: Utilizing simulation models and predictive analytics to optimize water management systems in smart cities. This involves predicting water demand, optimizing distribution networks, and identifying strategies for water conservation.
- Urban Heat Island Mitigation: Developing simulation models and predictive analytics to mitigate urban heat islands, which are areas with significantly higher temperatures in cities. This includes predicting heat island effects, optimizing green spaces, and implementing urban design strategies for heat reduction.
- Real-time Monitoring and Decision Support: Implementing simulation models and predictive analytics to enable real-time monitoring of urban systems and provide decision support for urban planners. This involves integrating data from various sources, analyzing trends, and generating actionable insights for efficient and sustainable urban design.

By utilizing simulation models and predictive analytics, these strategies enhance urban design and planning in smart cities. They facilitate data-driven decision-making and promote the development of sustainable and enjoyable urban environments. Simulation models and predictive analytics are employed in these methods to improve urban design and planning in smart cities. They enable decision-making based on data and contribute to the creation of sustainable and livable urban environments.

6. SMART GOVERNANCE AND CITIZEN ENGAGEMENT

Smart cities rely on smart governance and citizen engagement as crucial elements for their modernization and advancement. Through the utilization of technology and innovative solutions, cities can improve their governance systems and encourage active involvement from their inhabitants. In smart cities, governments make use of intelligent tools and data-driven methods to streamline administrative procedures and provide efficient public services. Moreover, citizen engagement is stimulated through digital platforms and interactive mechanisms, enabling residents to express their opinions, offer feedback, and actively participate in decision-making. This inclusive approach empowers individuals and communities to collaborate with local authorities, resulting in a governance model that is more transparent, accountable, and responsive. Ultimately, smart governance and citizen engagement cultivate a sense of ownership and shared responsibility, fostering vibrant and sustainable communities within smart cities.

6.1 E-Governance and Digital Platforms for Citizen Participation

Citizen participation in smart cities is greatly facilitated by e-governance and digital platforms. Such platforms empower residents to actively participate in decision-making processes and make valuable contributions towards the advancement and enhancement of their communities (Zook, 2017). Here are a few instances of e-governance initiatives and digital platforms that encourage citizen participation in smart cities:

- Online Forums: Smart cities often provide online forums where citizens can discuss various issues, share ideas, and provide feedback on urban development projects.
- Mobile Applications: Many smart cities have developed mobile apps that allow citizens to report issues such as potholes, broken streetlights, or garbage accumulation. These apps provide a convenient way for residents to communicate problems directly to the relevant authorities.
- Social Media Platforms: Smart cities leverage social media platforms to disseminate information, gather feedback, and interact with citizens. Municipalities can use platforms like Twitter, Facebook, and Instagram to share updates, conduct surveys, and address citizens' concerns.
- Open Data Portals: Smart cities promote transparency and citizen engagement by providing open data portals. These portals offer access to various datasets, allowing citizens, researchers, and developers to analyze and utilize the data for innovative solutions.
- Online Polls and Surveys: E-governance platforms facilitate online polls and surveys to gauge public opinion on specific issues or policies. This approach ensures that decision-making processes are more inclusive and reflective of citizens' preferences.
- Virtual Town Hall Meetings: Smart cities employ virtual town hall meetings, which allow citizens to participate in discussions and debates remotely.

These platforms enable residents to voice their opinions, raise concerns, and contribute ideas without the need for physical attendance.

- Crowdsourcing Platforms: Digital platforms that enable crowdsourcing can be utilized to gather ideas and solutions from citizens. By involving the community in problem-solving, smart cities can tap into collective intelligence and foster a sense of ownership among residents.
- Civic Engagement Apps: Some smart cities have developed dedicated apps that serve as one-stop platforms for citizen engagement. These apps provide features such as event notifications, service requests, and interactive maps to facilitate participation.
- Online Complaint Resolution Systems: E-governance initiatives incorporate online complaint resolution systems that streamline the process of reporting and resolving issues faced by citizens. These systems enhance efficiency and ensure prompt attention to citizen grievances.
- Virtual Feedback Mechanisms: Smart cities leverage digital platforms to collect feedback from citizens regarding various initiatives, policies, and services. This helps authorities assess the impact of their actions and make data-driven decisions.

6.2 Open Data Initiatives and Transparency in Governance

Open data initiatives involve the release of government-held data sets to the public in machine-readable formats. These datasets can encompass a wide range of information, such as transportation data, environmental data, health data, and more. The availability of such data enables citizens, businesses, and researchers to analyze and utilize the information for various purposes Transparency in governance refers to the visibility and openness of government operations, policies, and decision-making processes. It involves making government information easily accessible to the public and promoting accountability in the actions of public officials. In smart cities, open data initiatives and transparency in governance can lead to several benefits. First, they enhance citizen engagement by enabling individuals to access and understand public information. This empowers citizens to participate actively in decision-making processes, provide feedback, and hold public officials accountable for their actions.

Second, open data initiatives promote innovation and economic growth. By making data available to the public, entrepreneurs, startups, and businesses can use it to develop innovative solutions, products, and services that address urban challenges. This fosters economic development and creates new opportunities for job creation. Third, transparency in governance builds trust between citizens and their government. transactional citizen becomes counted by smart city sensors and integrated into smart city decision-making through the use of certain preferred platforms.[J] When citizens have access to information about government activities, budgets, and policies, they can better understand how decisions are made and how resources are allocated. This transparency helps in reducing corruption, enhancing public trust, and improving overall governance. Furthermore, open data initiatives facilitate evidence-based decision-making. Policymakers can use the insights derived from data analysis to design and implement effective policies, programs, and services. This data-driven approach ensures that decisions are based on objective information, leading to more efficient resource allocation and better outcomes for citizens.To implement open data initiatives and transparency in governance, smart cities need to establish data sharing protocols, develop user-friendly data portals, and ensure data privacy and security. They should also engage with citizens through public consultations, feedback mechanisms, and participatory platforms to promote transparency and accountability.

6.3 Privacy and Security Concerns in Smart Cities

In the realm of smart cities, it is crucial to give due consideration to the issues of privacy and security. As technology and interconnected systems become more prevalent, numerous risks emerge. Here is an outline of the concerns surrounding privacy and security in smart cities:

- Data Breaches: Smart cities generate and collect vast amounts of data from various sources, such as sensors, cameras, and connected devices. If this data is not adequately protected, it can be vulnerable to unauthorized access, leading to data breaches and potential misuse of personal information.
- Surveillance and Tracking: Smart city technologies often involve the use of surveillance systems, including cameras and sensors, to monitor public spaces. While this can enhance safety and security, it also raises concerns about invasion of privacy and the potential for constant tracking and monitoring of individuals' activities.
- Data Misuse: The extensive data collected in smart cities can be misused if not properly regulated. Personal information, such as location data, usage patterns, and behavior, can be exploited for commercial purposes or used to manipulate individuals or communities.
- Lack of Anonymity: In smart cities, the constant collection and analysis of data can potentially compromise individuals' anonymity. The integration of different data sources can lead to the identification of individuals, which may infringe on their privacy rights.
- Inadequate Consent and Transparency: Citizens may not always be fully aware of the data being collected and how it is being used in smart cities.

Lack of transparency and inadequate consent mechanisms can erode trust and raise concerns about the misuse of personal information.

- Infrastructure Vulnerabilities: Smart city infrastructure, including communication networks, sensors, and control systems, can be susceptible to cyberattacks. Hackers may exploit vulnerabilities in the system to gain unauthorized access, disrupt services, or even cause physical harm.
- Public Safety Risks: Smart city technologies rely on interconnected systems, making them potentially vulnerable to cyber threats and attacks. A breach or manipulation of these systems could have significant consequences for public safety, including traffic management, emergency response systems, and critical infrastructure.
- Social Inequity: The implementation of smart city technologies may inadvertently exacerbate existing social inequalities. If certain communities have limited access to technology or are disproportionately targeted by surveillance measures, it can lead to discrimination and privacy violations.
- Lack of Standardization: The absence of uniform standards across different smart city deployments can create vulnerabilities. Inconsistent security protocols and interoperability issues between various systems can increase the risk of unauthorized access and data breaches.
- Long-Term Data Storage: The storage and retention of data in smart cities raise concerns about the long-term security and privacy implications. As data accumulates over time, it becomes increasingly challenging to ensure its integrity, protection, and secure disposal when necessary.

These concerns highlight the need for robust privacy and security measures in smart cities to safeguard individuals' rights and maintain public trust in the technology-driven urban environments.

6.4 The Role of Engineering in Fostering Inclusive and Participatory Governance

Engineering plays a fundamental and instrumental part in promoting inclusive and participatory governance within smart cities. By leveraging technological advancements and applying engineering principles, various aspects of governance can be enhanced to ensure inclusivity and active participation of all citizens. The following are key ways in which engineering contributes to fostering inclusive and participatory governance in smart cities:

1. Infrastructure Development: Engineering plays a crucial role in designing and developing the physical infrastructure of smart cities. By incorporating

inclusive design principles, engineers can ensure that infrastructure is accessible and usable by individuals of diverse backgrounds and abilities. This includes designing transportation systems, public spaces, and buildings that are easily navigable for everyone, including people with disabilities, elderly individuals, and children.

2. Digital Accessibility: Engineering enables the creation of digital platforms and applications that facilitate citizen engagement and participation in governance. Engineers can develop user-friendly interfaces and applications that are accessible to individuals with varying levels of digital literacy and disabilities. This ensures that all citizens can access and engage with government services, participate in decision-making processes, and provide feedback.
3. Data-driven Decision Making: Smart cities generate vast amounts of data through sensors, IoT devices, and various technological systems. Engineers can design data collection mechanisms and develop algorithms that analyze and interpret this data to generate meaningful insights. By leveraging engineering expertise, cities can use data-driven decision-making processes that consider diverse perspectives and inputs from citizens, leading to more inclusive governance practices.
4. Open Data and Transparency: Engineering plays a critical role in implementing open data initiatives within smart cities. Engineers can develop systems and protocols that collect, manage, and share data in a transparent and accessible manner. Open data platforms enable citizens, community organizations, and businesses to access and analyze information about city operations, budgets, and services. This fosters transparency, enables public scrutiny, and empowers citizens to participate actively in governance.
5. Digital Participation Platforms: Engineering enables the creation of digital platforms and tools that facilitate citizen participation in decision-making processes. Engineers can design online platforms for public consultations, surveys, and voting systems that are user-friendly, accessible, and secure. These platforms empower citizens to voice their opinions, contribute to policy discussions, and influence governance outcomes, regardless of their location or physical presence.
6. Cybersecurity and Privacy: Engineering expertise is vital in ensuring the security and privacy of digital systems in smart cities. Engineers play a crucial role in developing robust cybersecurity measures and protocols that protect citizens' personal information and secure government systems. By addressing concerns related to data privacy and security, engineering helps build trust among citizens and encourages their active participation in governance initiatives.

7. CONCLUSION

In this chapter, we explored the main findings and contributions regarding the importance of innovative engineering solutions in driving urban transformation and the future prospects and challenges in designing smart cities. We discovered that innovative engineering solutions play a crucial role in shaping the future of urban areas, as they enable the creation of sustainable, efficient, and livable cities. One of the key findings was the recognition that smart cities heavily rely on cutting-edge technologies, such as the Internet of Things (IoT), artificial intelligence (AI), and data analytics. These technologies enable the collection and analysis of vast amounts of data, allowing for informed decision-making and the implementation of intelligent systems to enhance various aspects of urban life. Moreover, we discussed the significant contributions of innovative engineering solutions in addressing urban challenges, including efficient energy management, optimized transportation systems, intelligent infrastructure, and sustainable resource utilization. These solutions have the potential to revolutionize urban living, improving the quality of life for residents while minimizing environmental impact. Looking ahead, the future prospects of smart cities are promising. The integration of emerging technologies, such as 5G networks, edge computing, and autonomous vehicles, will further enhance the capabilities of smart cities. These advancements will enable real-time data processing, advanced communication networks, and seamless connectivity, paving the way for more efficient urban services and improved citizen experiences. However, despite the exciting prospects, designing smart cities also presents numerous challenges. Privacy and security concerns surrounding the collection and utilization of personal data need to be addressed to ensure citizens' trust. Additionally, the equitable distribution of smart city benefits and the digital divide between different communities must be considered to avoid exacerbating existing social inequalities. In conclusion, this chapter highlighted the crucial role of innovative engineering solutions in driving urban transformation and the potential of smart cities to create sustainable and livable urban environments. While future prospects are promising, it is essential to address challenges related to privacy, security, equity, and inclusivity to ensure that the benefits of smart cities are accessible to all.

REFERENCES

Chen, Z., & Chan, I. C. (2022). Smart cities and quality of life: A quantitative analysis of citizens' support for smart city development. *Information Technology & People*, *36*(1), 263–285. doi:10.1108/ITP-07-2021-0577

Crossan, M. M., & Apaydin, M. (2010). A multi-dimensional framework of organizational innovation: A systematic review of the literature. *Journal of Management Studies*, *47*(6), 1154–1191. doi:10.1111/j.1467-6486.2009.00880.x

Cugurullo, F. (2018). Exposing smart cities and eco-cities: Frankenstein urbanism and the sustainability challenges of the experimental city. *Environment & Planning A*, *50*(1), 73–92. doi:10.1177/0308518X17738535

Fagerberg, J., Martin, B. R., & Andersen, E. S. (Eds.). (2013). *Innovation studies: Evolution and future challenges*. Oxford University Press. doi:10.1093/acprof:oso/9780199686346.001.0001

Jamshed, A. (2018). An Analysis of Trends, Developments and Transparent Issues in the making of Indian Smart Cities. *EAI Endorsed Transactions on Smart Cities.*

Johnson, P. A., Robinson, P. J., & Philpot, S. (2020). Type, tweet, tap, and pass: How smart city technology is creating a transactional citizen. *Government Information Quarterly*, *37*(1), 101414. Advance online publication. doi:10.1016/j.giq.2019.101414

Lim, Y., Edelenbos, J., & Gianoli, A. (2022). Dynamics in the governance of smart cities: Insights from South Korean smart cities. *Int. J. Unity Sci.*, 1–23. doi:10.1080/12265934.2022.2063158

Liu, H. K., Guo, Y., & Chen, L. (2023). Crowdsourcing Smart City: SmartTaipei Project. *Proceedings of the 24th Annual International Conference on Digital Government Research.* 10.1145/3598469.3598474

Phillips, C., & Jiao, J. (2023). Artificial Intelligence & Smart City Ethics: A Systematic Review. *2023 IEEE International Symposium on Ethics in Engineering, Science, and Technology (ETHICS)*, 1-5. 10.1109/ETHICS57328.2023.10154961

Ranchordás, S. (2019). Nudging citizens through technology in smart cities. *International Review of Law Computers & Technology*, *34*(3), 254–276. doi:10.1080/13600869.2019.1590928

Sheikh, H., Mitchell, P., & Foth, M. (2023). Foth More-than-human smart urban governance: A research agenda Digit. *Geogr. Soc.*, *4*, 100045. Advance online publication. doi:10.1016/j.diggeo.2022.100045

Tan, G. K. (2021). Citizens go digital: A discursive examination of digital payments in Singapore's Smart Nation project. *Urban Studies (Edinburgh, Scotland)*, *59*(12), 2582–2598. doi:10.1177/00420980211039407

Van Mierlo, B., Leeuwis, C., Smits, R., & Woolthuis, R. K. (2010). Learning towards system innovation: Evaluating a systemic instrument. *Technological Forecasting and Social Change*, *77*(2), 318–334. doi:10.1016/j.techfore.2009.08.004

Wirtz, B. W., Becker, M., & Schmidt, F. W. (2021). Smart city services: An empirical analysis of citizen preferences. *Public Organization Review*, *22*(4), 1063–1080. doi:10.100711115-021-00562-0

Zhang, J., Bates, J., & Abbott, P. (2022). State-steered smartmentality in Chinese smart urbanism. *Urban Studies (Edinburgh, Scotland)*, *59*(14), 2933–2950. doi:10.1177/00420980211062888

Zook, M. (2017). Crowd-sourcing the smart city: Using big geosocial media metrics in urban governance. *Big Data & Society*, *4*(1), 4. doi:10.1177/2053951717694384

Chapter 4
Industry 5.0 From Automation to Autonomy:
Engineering the Shift

S. C. Vetrivel
https://orcid.org/0000-0003-3050-8211
Kongu Engineering College, India

T. Mohanasundaram
https://orcid.org/0000-0003-4511-7539
M.S. Ramaiah Institute of Technology, India

ABSTRACT

The advent of Industry 4.0 brought about a transformative wave of automation and digitalization in manufacturing and industrial processes. However, the next phase of industrial evolution, Industry 5.0, focuses on achieving a seamless integration of humans and machines, shifting the emphasis from automation to autonomy. This explores the key concepts, challenges, and opportunities associated with this paradigm shift. Industry 5.0 represents a paradigmatic transition that seeks to harness the power of advanced technologies such as artificial intelligence (AI), robotics, the internet of things (IoT), and cyber-physical systems (CPS) to empower human workers and enhance their capabilities. The goal is to create a collaborative and symbiotic environment where humans and machines work together to achieve higher levels of productivity, efficiency, and innovation. This chapter delves into the primary drivers behind the transition to Industry 5.0. One key factor is the increasing complexity and variability of manufacturing processes, which demand greater flexibility and adaptability.

DOI: 10.4018/979-8-3693-0819-6.ch004

1. INTRODUCTION

1.1 Overview of Industry 5.0

Industry 5.0 focuses on achieving a seamless integration of humans and machines, shifting the emphasis from automation to autonomy (Breque et al., 2021). Industry 5.0 is an evolution that leverages the progress achieved in Industry 4.0 with the goal of fostering a manufacturing and production approach that is centered around human involvement (Longo, Padovano, & Umbrello, 2020). Industry 5.0 aims to combine human and machine skills to achieve increased productivity and wider societal benefits, while Industry 4.0 focused on the integration of automation, data sharing, and digital technology.

1.2 Important Aspects of Industry 5.0

- Human-Machine Collaboration: Industry 5.0 places a strong emphasis on human-machine cooperation, recognizing that each possesses unique strengths. While machines handle repetitive tasks and data processing, humans contribute their creativity, problem-solving skills, and emotional intelligence.
- Augmented Workforce: Industry 5.0 envisions a workforce augmented with advanced technology like virtual reality, augmented reality, robotics, and artificial intelligence. These technologies enhance human capabilities, enabling workforces to achieve complex jobs more securely and effectively.
- Customization and Personalization: Industry 5.0 has led to a greater emphasis on product customization. Advanced technologies enable mass customization, where products can be tailored to individual customer needs and preferences without compromising efficiency or cost-effectiveness.
- Localized and Sustainable Production: Industry 5.0 promotes localized production, enabling manufacturing to be closer to the end consumers. This reduces the need for extensive global supply chains and transportation, leading to lower carbon emissions and increased sustainability.
- Ethical and Social Responsibility: Industry 5.0 emphasizes the ethical and social responsibilities of organizations. It encourages companies to consider the well-being of workers, the community, and the environment. This includes fair wages, safe working conditions, and environmentally friendly practices.
- Reskilling and Upskilling: As Industry 5.0 introduces new technologies and shifts job requirements, there is a growing emphasis on reskilling and upskilling the workforce. Continuous learning and development programs

help workers adjust to the varying needs of the business and acquire new skills.

- Cybersecurity and Data Privacy: With increased connectivity and data exchange, Industry 5.0 acknowledges the significance of robust cybersecurity measures and data privacy protection. Safeguarding sensitive data and securing networks and systems are critical to maintaining trust and preventing cyber threats.

The notion of Industry 5.0 envisages a future wherein technology and mankind creativity collaborate closely to establish a manufacturing setting that is further effective, environmentally conducive, and socially accountable (Nahavandi, 2019). By leveraging the advantages of both people and technologies, Industry 5.0 strives to attain economic progress while giving utmost importance to the welfare of individuals and the planet.

1.3 Historical Context of Industrial Revolutions

To comprehend the evolution that culminated in "Industry 5.0: From Automation to Autonomy – Engineering the Shift," it is crucial to consider the historical backdrop of the Industrial Revolution. Let's delve into the preceding industrial revolutions to establish the required context for Industry 5.0.

1. First Industrial Revolution (18th to early 19th century): In Britain, the 1st Industrial Revolution got underway and marked a significant shift from agrarian, manual labor-based economies to machine-driven manufacturing. Key innovations included the steam engine, mechanized textile production, and the advent of factories. This period transformed society, bringing about urbanization, mass production, and economic growth.
2. Second Industrial Revolution (late 19th to early 20th century): The 2nd Industrial Revolution was characterized by advancements in industry, primarily driven by electricity, steel, and the expansion of the railroad network. It witnessed the emergence of the assembly line, the development of telegraphs and telephones, and the widespread use of new materials and chemicals. This era fueled rapid industrialization, increased productivity, and the rise of large corporations.
3. Third Industrial Revolution (1960s to late 20th century): The 3rd Industrial Revolution, sometimes referred to as the Information Age or the Digital Revolution, was driven by the rise of computers, electronics, and the internet. It brought forth automation, computerization, and the integration of information technology into various industries. This revolution revolutionized communication, global connectivity, and the way businesses operated.

4. Fourth Industrial Revolution (late 20th century to early 21st century): The 4th Industrial Revolution built upon the digital revolution and included innovations in robots, big data, the Internet of Things, artificial intelligence, and 3D printing. This era focused on the fusion of digital, physical, and biological systems. It brought about advancements in autonomous systems, smart factories, and interconnected devices.

Now, with the historical context established, we can delve into Industry 5.0 is the next stage of industrial growth, emphasizing the combination of autonomous systems with human workers (Aslam et al., 2020). It intent to harmonize the strengths of computerization and autonomy with human creativeness, problem-solving abilities, and emotional intelligence. This paradigm envisions collaborative robots working alongside humans in shared spaces, leveraging their complementary capabilities. Industry 5.0 identifies the significance of human skills, decision-making, and adaptability in complex tasks, while leveraging cutting-edge technologies like Artificial Intelligence, Machine Learning, and Robotics. The emerging industrial paradigm unites the realms of digital and physical through the integration of Cyber-Physical Systems powered by the Internet of Things. This transformative shift is anticipated to have profound implications for industries, markets, and the economy. It will enhance production processes and boost productivity, impacting the entire product lifecycle. Moreover, it will give rise to novel business models, alter the work environment, and lead to a restructuring of the labor market. The ultimate goal is to create a more flexible, efficient, and sustainable industrial environment that maximizes the potential of both mankind and machines. The rapid shift towards Industry 5.0 represents a response to the limitations and challenges posed by previous industrial revolutions, including the displacement of workers, social inequalities, and environmental concerns.

2. INDUSTRY 5.0: AUTOMATION AND CONNECTIVITY

Industry 5.0, an emerging idea, builds on the foundations of earlier industrial revolutions, specifically Industry 4.0, with its main emphasis on automation and connectivity. It underscores the significance of cooperation between humans and machines and the integration of societal requirements. This idea signifies a shift toward a more human-centered approach to industrial production, aiming to tackle the workforce's challenges and the consequences of automation. Automation: Automation continues to play a pivotal role in Industry 5.0, much like it did in Industry 4.0. However, in Industry 5.0, automation is not solely focused on replacing human workers but rather on augmenting their capabilities (Aruldoss et al., 2014;

Javaid & Haleem, 2020). Cutting-edge robotics, artificial intelligence, and machine learning technologies are employed automating repetitive and routine tasks, thereby liberating human workers to concentrate on more advanced skills. that require creativity, problem-solving, and emotional intelligence.

Connectivity: Connectivity is another key facet of Industry 5.0. It involves the continuous integration and communication between various systems, machines, and devices. Internet of Things (IoT) technologies, cloud computing, and big data analytics empower instantaneous data exchange and analysis, resulting in production processes that are both further effective and responsive. This emerging industrial paradigm can be technically defined as the heightened digitization and automation of the manufacturing landscape, coupled with enhanced communication facilitated by the establishment of a digital value chain. This connectivity extends beyond the factory floor, integrating the entire supply chain and enabling collaboration among different stakeholders, including suppliers, manufacturers, and customers.

Human-Machine Collaboration: Industry 5.0 underscores the partnership between humans and machines. Instead of replacing humans, machines and technologies are designed to work alongside them, enhancing their capabilities and productivity. This cooperation harnesses the advantages of both humans and machines, merging human intuition, creativity, and adaptability with the precision, speed, and scalability of machines. Human workers are involved in decision-making processes, problem-solving, and complex tasks, while machines handle repetitive and physically demanding activities. this novel industrial paradigm welcomes the creation of smart environments capable of seamlessly merging the physical and virtual realms using Cyber-Physical Systems (CPS). It involves the integration of devices, machinery, production modules, and products, enabling them to initiate actions and autonomously control one another.

Societal Considerations: Industry 5.0 also considers societal needs and broader ethical considerations. It seeks to address the possible influence of automation on the employment, skills, and social well-being (Aziez et al., 2019). The vital characteristic of this upheaval, which occurred in the late 1900s in numerous industrialized nations worldwide is the utilization of electronic devices and information technologies to advance automation in production. This may involve redefining job roles, providing training and upskilling opportunities for workers, and ensuring a just transition for affected individuals. Furthermore, Industry 5.0 intents to develop sustainable and environmentally friendly production systems, minimizing waste and optimizing resource utilization.

2.1 Impact of Industry 5.0 on Various Sectors: A Technological Overview

Industry 5.0 is a notion that builds upon the previous industrial revolutions, with a focus on integrating advanced technologies and human-centric approaches in manufacturing processes. It highlights the association between human-beings and machines, aiming to create more sustainable, inclusive, and socially responsible industrial systems. While the term Industry 5.0 is still evolving, several key technologies are commonly associated with it. Here are some of them:

- Internet of Things: IoT, denotes a network of physical devices equipped with sensors, software, and connectivity, facilitating data exchange and communication among these devices. In Industry 5.0, IoT plays a crucial part in collecting instantaneous data from machines, products, and processes, allowing for enhanced monitoring, control, and optimization of manufacturing operations.
- Artificial Intelligence and Machine Learning: These technologies empower machines to study patterns from data and make intelligent decisions with no need of explicit programming. In Industry 5.0, AI and ML are utilized for various purposes, such as predictive maintenance, quality control, optimization of production processes, and human-machine collaboration. These advancements may lead to enhanced efficiency, productivity, and decision-making in manufacturing.
- Robotics and Automation: Robotics and automation technologies have been evolving rapidly, and they have substantial impact in Industry 5.0. The collaborative robots, commonly referred to as cobots, are specifically engineered to operate alongside humans, thereby improving productivity, safety, and adaptability within manufacturing settings. Automation systems, including robotic arms, autonomous vehicles, and smart factories, are transforming traditional manufacturing processes by reducing human error, increasing precision, and enabling the customization of outputs.
- Augmented Reality and Virtual Reality: AR and VR provide immersive experiences and enable the visualization of virtual objects in the real world or entirely simulated environments. In Industry 5.0, AR and VR are applied to support training, remote collaboration, maintenance, and troubleshooting tasks. They provide intuitive interfaces for workers, allowing them to access real-time information, visualize complex systems, and interact with virtual models, leading to improved productivity and reduced downtime.
- Additive Manufacturing: Additive manufacturing, often referred to as 3D printing, enables the fabrication of three-dimensional objects by depositing

materials layer by layer, guided by digital models. In Industry 5.0, additive manufacturing revolutionizes production processes by enabling rapid prototyping, on-demand manufacturing, and customization of products. It reduces waste, enhances design flexibility, and facilitates distributed manufacturing.

- Sustainable Technologies: Industry 5.0 emphasizes sustainable practices and technologies that minimize environmental impact. These include renewable energy sources, energy-efficient manufacturing processes, circular economy principles, waste reduction and recycling methods, and the use of eco-friendly materials. Sustainable technologies enable industries to reduce carbon emissions, conserve resources, and facilitate to a more environmentally friendly and better future.

2.2 The Impact of Automation on Manufacturing Processes

The influence of automation on manufacturing processes has been significant and extensive. It has brought about a revolution in the production of goods, resulting in heightened efficiency, productivity, and cost-effectiveness. Through the substitution of manual labor with machines and advanced technologies, automation has considerably reduced human errors and enhanced the precision and consistency of manufacturing operations. IoT represents a realm in which physical objects seamlessly merge with the digital information network, enabling these objects to actively engage in business processes. It has also allowed manufacturers to streamline their production lines, optimize workflows, and attain higher levels of output. Furthermore, automation has enabled the performance of tasks that are repetitive, hazardous, or physically demanding with minimal human intervention, thereby promoting worker safety and well-being (Ali et al., 2019). Additionally, automation has facilitated the incorporation of data analytics, artificial intelligence, and robotics into manufacturing processes, facilitating instantaneous monitoring, predictive maintenance, and robust decisions. While concerns regarding job displacement have arisen, automation has also generated new opportunities for workers to acquire new skills, adapt existing ones, and engage in more intricate and fulfilling roles within the manufacturing industry. Overall, the impact of automation on manufacturing processes has been transformative, leading to increased efficiency, quality, and innovation, ultimately propelling the advancement of modern manufacturing.

2.3 Challenges and Opportunities of Connectivity in Industry

Connectivity holds immense significance in Industry 5.0 as it enables the fusion of physical and digital systems, highlighting the cooperation between humans and machines. The following are the challenges and opportunities linked to connectivity within the context of Industry 5.0.

2.3.1 Challenges

- Data Security: With increased connectivity, the risk of cyber threats and data breaches becomes more significant. Protecting sensitive information and ensuring the security of connected systems is a major challenge.
- Interoperability: Different machines, devices, and systems may use various procedures and standards, making it hard to establish unified connectivity and interoperability between them. Achieving standardized communication and data exchange can be a challenge.
- Scalability: With the proliferation of connected devices and systems, managing and scaling the connectivity infrastructure becomes complex. Ensuring efficient connectivity across a large-scale network poses a challenge.
- Legacy Systems Integration: Numerous industries continue to depend on outdated legacy systems that were not originally designed with connectivity as a priority. Integrating these older systems with modern, connected technologies can be challenging and may require additional investments.
- Reliability and Downtime: Industry 5.0 heavily relies on connectivity for real-time data exchange and decision-making. Ensuring a reliable and robust network infrastructure becomes crucial to minimize downtime and prevent disruptions.

2.3.2 Opportunities

- Real-Time Monitoring and Control: Connectivity enables real-time monitoring of industrial processes and equipment, allowing for proactive maintenance and optimization. This can result in increased efficiency, reduced downtime, and cost savings.
- Predictive Analytics: Connected systems can collect and analyze large volumes of data, facilitating the implementation of predictive analytics. This enables the identification of patterns, trends, and potential issues, leading to improved decision-making and proactive interventions.
- Collaborative Robotics: The collaboration between humans and machines are fostered by Industry 5.0., where robots and humans work side by side. Connectivity enables seamless communication between robots and humans, fostering safer and more efficient cooperation in various tasks.

- Remote Operations and Maintenance: Connectivity allows for remote monitoring and control of industrial processes and equipment. This opens up opportunities for remote operations and maintenance, reducing the need for physical presence on-site and enabling cost-effective operations in remote locations.
- Supply Chain Optimization: Connected systems can enhance visibility and traceability throughout the supply chain, enabling better inventory management, logistics optimization, and improved coordination among different stakeholders.

2.4 Literature Review

Case Study 1: Waymo's Self-Driving Cars: Waymo, a subsidiary of Alphabet Inc. has been at the forefront of autonomous vehicle development. They have successfully implemented autonomous technology in their fleet of self-driving cars. Waymo conducted extensive testing and refinement of their autonomous system, accumulating millions of miles on public roads and in simulation. Their vehicles use a combination of sensors, including lidar, radar, and cameras, along with advanced machine learning algorithms to perceive their surroundings and make driving decisions. In October 2020, Waymo launched its fully autonomous ride-hailing service, Waymo One, in the Phoenix metropolitan area. The service allows members of the public to hail a self-driving car using a Smartphone app. Waymo's autonomous vehicles have demonstrated impressive safety records and have provided over 20 million miles of rides to date.

Case Study 2: Tesla Autopilot Tesla, the electric vehicle manufacturer led by Elon Musk, has made significant progress in implementing autonomous technology. Tesla's Autopilot system combines various sensors, such as cameras, radar, and ultrasonic sensors, with advanced artificial intelligence algorithms. Tesla's Autopilot is an advanced driver-assistance structure that allows features like adaptive cruise control, lane-centering, and self-parking. While the system is not fully autonomous, it showcases successful implementation of autonomy in certain driving scenarios. Tesla collects vast amounts of real-world driving data from its fleet of vehicles and continuously updates and improves its autonomous algorithms through over-the-air software updates. This iterative approach has allowed Tesla to enhance the capabilities of its Autopilot system over time.

Case Study 3: Autonomous Trucks by TuSimple. TuSimple, a self-driving truck company, has developed and successfully implemented autonomous technology for long-haul freight transportation. Their autonomous trucks are designed to operate on highways and have completed extensive testing in multiple states in the United States. TuSimple's trucks use an array of sensors, including cameras and lidar, to perceive the environment and make decisions. The company has developed sophisticated computer vision algorithms and artificial intelligence systems to handle complex driving scenarios, such as merging, lane changes, and navigating through traffic. In 2019, TuSimple conducted a successful two-week pilot program in partnership with the United States Postal Service (USPS), transporting mail on a 1,000-mile route between Arizona and Texas. The pilot demonstrated the capabilities of TuSimple's autonomous trucks in real-world logistics operations, showcasing the potential benefits of autonomy in the freight industry.

3. THE NEED FOR AUTONOMY

3.1 Understanding the Limitations of Automation

While automation offers numerous benefits, it also has certain limitations that should be understood. Here are some key limitations of automation in Industry 5.0:

1. Complexity of tasks: Automation is most effective when applied to repetitive and well-defined tasks. However, certain tasks require complex decision-making, creativity, and adaptability, which are currently better suited for human workers. Automation may struggle to handle unstructured and unpredictable situations, limiting its effectiveness in certain domains.
2. Human-machine interaction: Collaborative automation in Industry 5.0 emphasizes the interface and cooperation amongst human-beings and machines. However, designing effective human-machine interfaces and ensuring seamless communication between humans and machines can be challenging. The success of automation depends on how well humans can understand and interact with the automated systems.
3. Cost and implementation: Implementing automation technologies can involve significant upfront costs, including the acquisition of hardware, software, and the necessary infrastructure (Leng et al., 2022). Integrating automation systems into existing workflows and processes may require substantial modifications and investments, making it a complex and costly endeavor for many industries.
4. Job displacement concerns: While automation can increase productivity and efficiency, there are apprehensions about work movement. Certain jobs that were previously done by humans may be taken over by machines, potentially leading to unemployment or the need for workers to acquire new skills to

remain employable. Balancing the benefits of automation with the impact on the workforce is a crucial consideration in Industry 5.0.

5. Ethical considerations: Automation raises ethical concerns related to privacy, security, and the impact on society. Automated systems may handle sensitive data, and the potential for misuse or unauthorized access can pose significant risks. Additionally, the societal impact of automation on employment, inequality, and fairness should be carefully evaluated to ensure that the benefits are distributed equitably.
6. Adaptability and flexibility: Automation systems are typically designed for specific tasks or processes, and they may struggle to adapt to new or evolving requirements. Humans, on the other hand, have inherent flexibility and adaptability, allowing them to learn and adjust to changing circumstances. Ensuring that automation systems can effectively handle dynamic environments and diverse tasks is an ongoing challenge.
7. Technical limitations: Automation know-hows like AI and robotics, have made significant advancements. However, they still have limitations. AI models may lack the ability to understand context, make nuanced decisions, or demonstrate common sense reasoning. Similarly, robotic systems may be limited by their physical capabilities, dexterity, and sensory perception.

To fully leverage the advantages of automation in Industry 5.0, it is essential to grasp these limitations. By acknowledging the confines of automation, sectors can formulate approaches to enhance the synergy between humans and machines, leading to optimum outcomes in the form of efficiency and human welfare.

3.2 Benefits of Autonomy in Industry

The incorporation of autonomy in industry, particularly within the framework of Industry 5.0, gives multiple benefits that completely transform business operations. One notable benefit is the heightened efficiency and productivity achieved. Through the integration of autonomous technologies like advanced robotics and AI-driven systems, industries can streamline and optimize their processes, resulting in faster production cycles, decreased errors, and enhanced overall output. Moreover, autonomy contributes to workplace safety by assuming responsibility for perilous tasks that may endanger human workers. By delegating hazardous operations to autonomous machines, employees can be reassigned to more strategic roles that necessitate human expertise, thus ensuring a safer and healthier work environment. Furthermore, autonomy promotes cost-effectiveness by reducing labor expenses and minimizing downtime. Automated systems can function continuously without the need for breaks or shift changes, leading to uninterrupted production and improved utilization of resources. This subsequently boosts profitability for businesses and offers potential cost savings for consumers. Finally, autonomy empowers innovation and creativity by liberating human workers from monotonous and repetitive tasks, enabling them to concentrate on problem-solving, critical thinking, and the exploration of new ideas (Longo, Padovano, & Umbrello, 2020). Digitization involves the convergence of the physical and virtual worlds, with far-reaching implications across all the economic sectors. The combination of human intelligence and autonomous technology opens up avenues for groundbreaking advancements and fresh approaches to industry challenges. In conclusion, the adoption of autonomy in Industry 5.0 bestows remarkable benefits, including heightened efficiency, improved safety, cost-effectiveness, and increased innovation, propelling businesses towards a more productive and prosperous future.

4. AUTONOMOUS SYSTEMS AND TECHNOLOGIES

Autonomous systems and technologies are at the forefront of Industry 5.0, the newest segment in the development of industrial manufacturing. Industry 5.0 builds upon the advancements from its previous setting, integrating emerging areas like artificial intelligence, machine learning, robotics, and IoT to create highly efficient and self-governing systems. Autonomous systems are characterized by their ability to operate independently, making decisions and taking actions without constant human intervention (Yavari & Pilevari, 2020). They are found in numerous sectors, consisting manufacturing, transportation, agriculture, healthcare, and logistics. In Industry 5.0, autonomous technologies are transforming the way businesses operate. Manufacturing processes are becoming more streamlined and adaptive, with robots and machines working collaboratively with humans to increase productivity and ensure high-quality outputs. Autonomous vehicles are revolutionizing transportation and logistics by enabling efficient and safe operations through self-driving cars, drones, and delivery bots. In agriculture, autonomous systems are improving crop yield and reducing resource waste by utilizing precision farming techniques, including automated planting, monitoring, and harvesting. The primary benefit of autonomous systems lies in their capacity to gather and analyze extensive volumes of data instantaneously, allowing for predictive maintenance, optimization of processes, and better decision-making. They can detect anomalies, make adjustments, and learn from their experiences, continuously improving their performance and efficiency. However, ethical considerations and safety measures are crucial in the development and deployment of autonomous technologies to ensure that they operate reliably and responsibly.

While Industry 5.0 presents tremendous opportunities for increased productivity, reduced costs, and improved quality, it also raises apprehensions regarding the displacement of human workforces and the necessity for reskilling and upskilling. The integration of autonomous systems requires a careful balance between automation and human involvement, with a focus on collaboration and utilizing human skills in inventiveness, problem-solving, and critical thinking. (Tiwari et al., 2022) Autonomous systems and technologies are driving the evolution of Industry 5.0, ushering in a new era of efficiency, productivity, and innovation across various sectors. Through nurturing the supremacy of artificial intelligence and automation, businesses can unlock new possibilities, improve their competitive edge, and create a future in which humans and machines collaborate closely to achieve sustainable growth and development.

4.1 Overview of Autonomous Systems Components

Industry 5.0 signifies the subsequent stage of industrial advancement, characterized by the incorporation of self-governing systems into diverse sectors. These systems are imperative in this transformative shift, altering the operational landscape of industries. Comprising multiple essential elements, these systems collaboratively facilitate automation and intelligent decision-making. One pivotal component is sensors, responsible for gathering real-time environmental data for the system. This data undergoes processing by robust computing units, such as artificial intelligence algorithms, which analyze and interpret the information. Actuators, another vital component, act upon physical tasks based on the decisions made by the AI algorithms. These tasks can encompass controlling robotic arms in assembly lines or adjusting machinery operations. In addition, communication networks play a crucial role in guaranteeing smooth connectivity and data interchange between the system's various components. Consequently, Industry 5.0 embraces the integration of these autonomous system components, elevating efficiency, productivity, and safety across multiple sectors and introducing a new era of intelligent industrial automation.

4.2 Artificial Intelligence and Machine Learning in Autonomy

Artificial intelligence (AI) and machine learning play pivotal roles in autonomy, revolutionizing industries such as robotics, transportation, and manufacturing. These technologies empower autonomous systems to perceive, reason, and make informed decisions without human intervention. In the realm of self-driving cars, for instance, machine learning algorithms process vast datasets from sensors, cameras, and lidar to navigate, anticipate traffic, and react to dynamic road conditions. AI models also enable predictive maintenance, ensuring machines and vehicles are serviced before breakdowns occur, increasing efficiency and safety. In manufacturing, autonomous robots with AI capabilities can adapt to changing assembly line demands, making them versatile and efficient. The fusion of AI and machine learning in autonomy is not only reshaping how we live and work but also opening up new frontiers in space exploration, healthcare, and environmental monitoring, offering a glimpse of a future where machines and systems can operate independently with unprecedented precision and reliability.

4.3 Robotics and Autonomous Vehicles in Industrial Settings

In industrial settings, the introduction of robotics and autonomous vehicles has brought about a transformative impact, bringing forth enhanced efficiency, productivity, and safety. Below, we present several essential applications and advantages of robotics and autonomous vehicles within industrial environments.

1. Automated Manufacturing: Robots are extensively used in manufacturing processes, performing repetitive tasks with precision and speed. They can assemble products, handle materials, and perform quality control checks. Autonomous vehicles, such as automated guided vehicles (AGVs), can transport materials and components between workstations, optimizing logistics and reducing human labor.
2. Warehousing and Logistics: Robotics and autonomous vehicles play a vital role in warehouses and distribution centers. Autonomous robots can navigate the warehouse environment, locate and retrieve items, and transport them to the desired location. This automation improves order fulfillment, reduces errors, and enhances inventory management.
3. Material Handling: In industrial settings, robots and autonomous vehicles are employed for efficient material handling. They can load and unload trucks, move heavy objects, and transport goods within a facility. These machines reduce the risk of injuries and allow for the handling of large and bulky items with precision.
4. Inspection and Quality Control: Robots equipped with advanced sensors and cameras are used for inspection and quality control processes. They can perform tasks such as product inspection, defect detection, and measurement accuracy assessment. These automated systems provide consistent and accurate results, minimizing human error.
5. Hazardous Environments: Robotics and autonomous vehicles are well-suited for hazardous environments where human presence poses risks. For example, robots can be deployed in nuclear power plants, chemical factories, or oil

refineries to perform tasks like maintenance, inspections, and repairs. This ensures the safety of human workers and reduces the potential for accidents.

6. Data Collection and Analysis: Autonomous vehicles equipped with sensors and cameras can collect data in real-time, capturing information about environmental conditions, inventory levels, or machine performance. This data can be analyzed to optimize processes, identify bottlenecks, and make informed decisions for continuous improvement.
7. Collaborative Robots (Cobots): Cobots team up with human employees, helping them with various tasks. These robots are designed to be safe, easy to program, and capable of performing delicate operations. Cobots can enhance worker productivity and ergonomics by handling repetitive or physically demanding tasks, while humans focus on more complex activities.
8. Predictive Maintenance: Robotics and autonomous vehicles facilitate predictive maintenance through continuous monitoring of equipment, identifying early signs of potential failures. Through the analysis of sensor and machine-generated data, it becomes possible to schedule proactive maintenance interventions, ultimately reducing downtime and enhancing operational efficiency.

5. HUMAN-MACHINE COLLABORATION

5.1 Importance of Human Involvement in Autonomous Systems

Human involvement in industry autonomous systems is of paramount importance for several reasons:

1. Decision-making and Ethical Judgment: Humans possess the ability to exercise moral judgment and make complex decisions based on ethical considerations. In autonomous systems, human involvement ensures that critical decisions are made with accountability, taking into account societal values, legal requirements, and potential ethical dilemmas. This involvement helps prevent or mitigate harmful consequences arising from machine-based decisions.
2. Adaptability and Flexibility: Humans are highly adaptable and can quickly respond to unforeseen situations or changing circumstances. In industrial settings, human involvement ensures that autonomous systems can be modified, updated, or reprogrammed to suit evolving needs, technological advancements, or unforeseen challenges (Pizoń & Gola, 2023). Human beings have the capacity to analyze system performance, pinpoint areas for improvement, and enact essential changes, thereby augmenting the overall efficiency and effectiveness of autonomous systems.

3. Contextual Understanding and Complex Problem-Solving: Human involvement brings contextual understanding and domain expertise that can significantly improve the performance of autonomous systems. Humans possess the ability to interpret complex situations, understand subtle nuances, and apply reasoning and problem-solving skills that machines may not possess. In industries where real-time decision-making and problem-solving are crucial, human involvement ensures that autonomous systems can leverage the expertise of human operators to navigate challenging scenarios effectively.
4. Safety and Risk Management: Autonomous systems in industries often operate in environments where safety is critical, such as manufacturing plants, energy facilities, or transportation systems. Human involvement is essential for monitoring, assessing risks, and implementing safety protocols to safeguard human lives, property, and the environment. Humans can anticipate potential hazards, evaluate system outputs for anomalies, and intervene when necessary to prevent accidents or mitigate risks.
5. Trust and User Acceptance: Human involvement in autonomous systems helps build trust and user acceptance. People tend to have greater confidence in systems when they know that humans are actively monitoring and overseeing their operations. Human involvement assures users and stakeholders that there is an accountable entity responsible for the system's actions, making it more likely for them to embrace and adopt autonomous technologies in industrial settings.
6. Error Handling and Learning from Mistakes: Autonomous systems may encounter situations or scenarios that fall outside their programmed capabilities or encounter unforeseen errors. Human involvement enables effective error handling and learning from mistakes (Rajumesh, 2023). Humans can analyze system failures or errors, identify root causes, and implement corrective actions or updates to improve system performance. This iterative process of learning from mistakes helps enhance the reliability and robustness of autonomous systems in industries.

5.2 Designing Interfaces for Effective Collaboration

In today's rapidly evolving industries, the importance of designing interfaces for effective collaboration cannot be overstated. The ever-increasing complexity of projects and the necessity for seamless teamwork make intuitive and user-friendly interfaces crucial for fostering productivity and innovation. Professionals in various industries heavily rely on well-designed interfaces that facilitate communication, information sharing, and task coordination. These interfaces should prioritize accessibility, ensuring that users from diverse backgrounds can effortlessly engage with them. It is of utmost importance to integrate instantaneous collaboration techniques like instant messaging, video conferencing, and document sharing, directly into the interface to facilitate effective communication and decision-making. Moreover, interfaces should incorporate clear visual cues and intuitive navigation, enabling users to quickly locate pertinent information, access shared resources, and contribute to collective efforts. To continuously refine and enhance these interfaces, usability testing and user feedback play integral roles, ensuring alignment with the unique requirements and workflows of different industries. By giving collaboration paramount importance in interface design, industries can harness the collective intelligence of their teams and propel innovation to unprecedented levels.

5.3 Ensuring Safety and Trust in Human-Machine Interactions

The utmost importance lies in guaranteeing safety and trust in human-machine interactions across various industries today. The incorporation of progressive technologies like AI, robotics, and automation has brought humans and machines closer than ever before. Industries across manufacturing, healthcare, transportation, and finance heavily depend on human-machine interactions to boost productivity, efficiency, and accuracy. Industry leaders focus on developing and implementing robust safety protocols to ensure safety. This includes conducting thorough risk assessments and establishing strict safety standards that govern the design, operation, and maintenance of machines. Human operators and technicians receive comprehensive training programs to understand potential risks and operate machines safely. Regular maintenance and inspection routines are in place to promptly address any emerging safety concerns.

Trust is another crucial element in human-machine interactions. Humans must have confidence in the reliability, predictability, and ethical behavior of machines. Industries invest in designing machines with transparent and explainable algorithms to build trust among human operators. Rigorous testing and quality control processes are implemented to ensure machines consistently perform as expected and deliver reliable outcomes. Moreover, industries prioritize the ethical use of technology to foster trust. Guidelines and regulations are established to ensure machines operate within ethical boundaries and uphold human values. Privacy and data protection receive significant attention to safeguard sensitive information and maintain confidentiality. Continuous monitoring and feedback mechanisms are established to gather insights from human operators regarding their experiences with machines. This feedback is utilized to improve the user interface, streamline processes, and address any concerns or issues raised by users. By actively involving humans in the design and improvement of machines, industries can enhance user satisfaction and build trust in the technology. Overall, ensuring safety and trust in human-machine interactions requires collaboration among industry stakeholders, regulatory bodies, and end-users. By prioritizing safety protocols, transparency, ethical guidelines, and continuous improvement, industries can create an environment where humans and machines interact seamlessly, leading to enhanced productivity, innovation, and overall success.

6. AUTONOMOUS DECISION-MAKING

6.1 Algorithms and Models for Autonomous Decision-Making

Autonomous decision-making has become a game-changing force in various industries, completely transforming the way organizations function and make critical choices. With the progress made in AI and ML, industries spanning from engineering to finance, healthcare to transportation, are adopting autonomous decision-making systems to enhance efficiency, accuracy, and productivity. These intelligent systems are specifically designed to investigate massive data, detect patterns, and take well-informed decisions without human intervention. This reduces the potential for human error and enables real-time responses to dynamic environments. In manufacturing, autonomous decision-making allows for predictive maintenance, optimization of production schedules, and improved quality control. In finance, it enables automated trading, risk assessment, and fraud detection. The healthcare industry benefits from autonomous systems in areas such as medical diagnosis, patient monitoring, and drug discovery. Autonomous decision-making in transportation facilitates self-driving vehicles, route optimization, and traffic management. Despite the numerous advantages these advancements bring, ethical considerations, transparency, and accountability remain vital to ensure responsible deployment of autonomous decision-making systems across industries.

6.2 Handling Complex and Dynamic Environments

The emergence of Industry 5.0 has brought about a substantial challenge for numerous industries, as they grapple with the task of navigating complex and dynamic environments. Industry 5.0 signifies the upcoming stage of manufacturing revolt, wherein latest technologies like AI, automation, IoT, and big data analytics are seamlessly integrated into manufacturing processes. Within this framework, the effective management of complex and dynamic environments assumes paramount importance in order to maximize productivity, ensure safety, and attain operational excellence. Here are some key considerations for handling complex and dynamic environments in the Industry 5.0 era:

- Instantaneous data collection and analysis: With the spread of IoT tools and advanced sensors, industries can gather vast amounts of real-time data from various sources. This data can provide precise inputs to the existing state of operations, identify potential bottlenecks or issues, and enable timely decision-making. Employing advanced analytics and machine learning programs can aid in analyzing this data and generate actionable intelligence.
- Adaptive and autonomous systems: Industry 5.0 highlights the application of adaptive and autonomous systems that can respond to changing conditions in real-time. These systems can automatically adjust their parameters, workflows and configurations on the basis of existing data and predefined rules. For instance, in a manufacturing environment, autonomous robots can adapt their movements and tasks based on the changing layout of the production floor or dynamically allocate resources to optimize efficiency.
- Collaborative robots and human-robot interaction: Industry 5.0 promotes human-robot collaboration, where robots work alongside humans to leverage their respective strengths. Cobots are designed to operate safely in close proximity to human workers. These robots can assist with complex tasks, handle heavy lifting, and perform repetitive actions, whereas humans can concentrate on higher-level managerial tasks, problem-solving, and creativity.
- Predictive maintenance and condition monitoring: Dynamic environments often involve complex machinery and equipment. Implementing predictive maintenance strategies can help anticipate potential failures or malfunctions by continuously monitoring the condition of machines and predicting maintenance requirements. This approach minimizes the interruption, improves security, and optimize the maintenance schedules.
- Agile and flexible production processes: Complex and dynamic environments often require flexibility in production processes to accommodate changing customer demands, varying product specifications, and market dynamics. Agile manufacturing principles, such as modular production systems, flexible assembly lines, and rapid reconfigurability, enable quick adaptation to changing requirements and improve responsiveness to customer needs.
- Continuous learning and upskilling: In Industry 5.0, technologies and processes evolve rapidly. It is crucial for organizations to foster a culture of continuous learning and upskilling among their workforce. Training programs, reskilling initiatives, and knowledge sharing platforms can help employees stay updated with the latest technologies and develop the necessary skills to navigate complex and dynamic environme

6.3 Ethical Considerations in Autonomous Decision-Making

The future of technology and its impact on society greatly depend on ethical considerations in autonomous decision-making. As artificial intelligence (AI) and autonomous systems advance, their ability to make independent decisions raises ethical questions. Biased decision-making is a significant concern as these systems rely on algorithms and data that can reflect societal biases. To tackle this, it is crucial to carefully design and develop autonomous systems, incorporating diverse perspectives and conducting thorough testing to minimize discriminatory outcomes. Transparency and accountability are also key. Understanding and justifying autonomous decisions become essential as they can have significant consequences (Golovianko et al., 2023). This necessitates the development of interpretable AI models and the establishment of auditing mechanisms for challenging decisions. Moreover, it is important to carefully balance considerations like privacy, consent, and potential harm to protect individuals' rights. Developing and deploying autonomous systems should involve interdisciplinary collaborations with technologists, ethicists, social scientists, policymakers, and other stakeholders to ensure diverse perspectives and values are taken into account. By addressing these ethical considerations, we can promote the responsible and beneficial use of autonomous decision-making systems, thereby contributing to a fair and equitable future.

7. OVERCOMING CHALLENGES

7.1 Technical Challenges in Achieving Autonomy

Achieving autonomy, particularly in the field of robotics and artificial intelligence, involves overcoming several technical challenges. Here are some of the key technical challenges:

1. Perception and Sensing: Autonomous systems need to accurately perceive and interpret the environment around them. This requires robust and reliable sensors, such as cameras, LIDAR, radar, or other specialized sensors, to gather data about the surroundings. Challenges include dealing with sensor noise, handling occlusions, recognizing and tracking objects, and understanding complex scenes.
2. Decision Making and Planning: Autonomous systems must make intelligent decisions and plan actions based on the information they perceive. This involves reasoning, learning, and integrating various inputs to ascertain the most suitable course of action. Challenges include handling uncertainty, optimizing

decisions in real-time, dealing with dynamic environments, and considering safety constraints.

3. Navigation and Localization: Autonomous systems need to navigate through their environment and accurately determine their own position. This requires robust localization algorithms, mapping techniques, and efficient path planning strategies. Challenges include dealing with limited or unreliable sensor data, mapping complex environments, handling dynamic obstacles, and adapting to changes in the environment.
4. Safety and Risk Assessment: Autonomous systems must operate safely and assess potential risks in their actions. This involves developing sophisticated algorithms to predict and mitigate potential hazards, avoiding collisions, and ensuring human safety in collaborative settings. Challenges include determining safe actions in complex scenarios, handling unpredictable human behavior, and integrating safety measures across the system.
5. Learning and Adaptation: Autonomous systems ought to perpetually acquire knowledge from their experiences and adjust their behavior as time progresses. This necessitates the development of algorithms for machine learning, reinforcement learning, and deep learning, enhancing performance and accommodating shifts in dynamic environments. Challenges include data scarcity, dealing with concept drift, avoiding catastrophic forgetting, and achieving efficient and robust learning.
6. Communication and Coordination: In scenarios where multiple autonomous systems need to work together, effective communication and coordination are essential. This involves developing protocols, algorithms, and architectures for decentralized decision-making, task allocation, and cooperation. Challenges include achieving efficient communication, handling asynchronous information, resolving conflicts, and ensuring global system performance.
7. Energy Efficiency and Resource Management: Autonomous systems often operate in resource-constrained environments, such as limited battery capacity or computational power. Optimizing energy efficiency, managing resources effectively, and developing power-aware algorithms are crucial for achieving autonomy in such scenarios. Challenges include designing energy-efficient hardware, optimizing algorithms for low power consumption, and balancing resource allocation across multiple tasks.
8. Ethical and Legal Considerations: Attaining autonomy gives rise to significant ethical and legal inquiries that require consideration and resolution. These challenges include determining liability and responsibility, ensuring privacy and security, addressing biases and fairness, and establishing regulations and standards for autonomous systems.

7.2 Legal and Regulatory Aspects of Autonomous Systems

The advancement of autonomous systems has made the legal and regulatory aspects surrounding them a critical area of focus. As technologies like self-driving cars, drones, and robotic devices continue to progress, they raise complex legal and regulatory questions. One important concern is determining liability in case of harm or damage caused by an autonomous system. Should the responsibility lie with the manufacturer, the operator, or the autonomous system itself? It is critical to create strong rules and standards for allocating liability to ensure accountability and protect the rights of those affected by incidents involving autonomous systems (Raja Santhi & Muthuswamy, 2023). Privacy and data protection present substantial challenges within the realm of autonomous systems. These technologies frequently gather and process extensive quantities of personal data, triggering concerns related to surveillance, consent, and security. Regulations must address these issues by specifying how autonomous systems handle, store, and share data, while also ensuring individuals have control over their personal information. The moral thoughts also act as a major part in regulating autonomous systems. The decisions made by these systems can have far-reaching consequences, and determining their behavior in morally ambiguous situations is a complex task. Guidelines and frameworks for ethical behavior need to be established, taking into account societal values, human rights, and the potential impact on various stakeholders.

Safety and reliability are additional areas that regulatory bodies must address. Autonomous systems need to meet specific standards and undergo rigorous testing to ensure they can operate safely and reliably in different environments. It is crucial to establish comprehensive certification processes and standards that cover the development, deployment, and operation of autonomous systems to guarantee public safety. Furthermore, intellectual property rights and cybersecurity are crucial considerations. Autonomous systems often rely on innovative technologies and algorithms, which raises questions about patent protection, licensing, and preventing unauthorized use or manipulation. Implementing robust cybersecurity measures is necessary to safeguard autonomous systems from potential hacking or malicious activities that could lead to severe consequences.

7.3 Addressing Workforce Concerns and Up Skilling Needs

Addressing workforce concerns and up skilling needs in various industries requires a comprehensive approach that considers the specific challenges and requirements of each sector. Here are some examples of how different industries can address these concerns:

1. Information Technology (IT):
   - Promoting continuous learning: Encourage employees to participate in professional development programs, online courses, and certifications to keep up with the rapidly evolving IT landscape.
   - Cross-training opportunities: Offer opportunities for employees to learn different programming languages, frameworks, or domains within IT to broaden their skill sets.
   - Cooperation and exchange of knowledge: Cultivate a culture of cooperation and knowledge exchange within the organization, whether through mentorship initiatives or regular technology-focused discussions.
2. Manufacturing:
   - Automation and robotics training: As automation technologies advance, provide training programs to help employees learn how to operate and maintain automated machinery, robots, and computerized systems.
   - Advanced manufacturing techniques: Invest in upskilling programs that teach employees about advanced manufacturing techniques like additive manufacturing (3D printing), smart factory technologies, and industrial IoT (Internet of Things).
   - Soft skills development: Recognize the importance of soft skills like problem-solving, teamwork, and adaptability. Offer training programs to enhance these skills alongside technical training.
3. Healthcare:
   - Continuing education opportunities: Provide opportunities for healthcare professionals to pursue advanced degrees, join workshops, discussions, and conferences to remain up-to-date with the newest medical developments and best practices.
   - Specialized training programs: Develop specialized training programs for emerging healthcare fields such as telemedicine, personalized medicine, and health informatics.
   - Interdisciplinary collaboration: Encourage interdisciplinary collaboration between healthcare professionals, researchers, and technology experts to foster innovation and address the evolving needs of the industry.
4. Financial Services:
   - Regulatory and compliance training: Offer regular training sessions on regulatory requirements and compliance practices to ensure employees stay up to date with changing financial regulations.
   - Development of digital skills: To meet the demands of a financial landscape that is becoming more and more digital, invest in training

programs to help staff develop their digital skills in areas like data analysis, cybersecurity, fintech, and digital marketing.

- Customer service and communication training: Provide training to improve customer service skills, effective communication, and relationship management, as these are crucial for client interactions in financial services.

8. THE FUTURE OF INDUSTRY 5.0

The next stage of industrial development Industry 5.0, builds on the progress made during Industry 4.0. Whereas Industry 4.0 concentrated on merging digital technologies and automation in manufacturing, Industry 5.0 goes beyond that by giving greater importance to the collaboration between humans and machines and the societal consequences of technology. Below are several essential elements and forecasts for the forthcoming era of Industry 5.0:

1. Human-Machine Collaboration: Industry 5.0 envisions a future where humans and machines work together in a collaborative and integrated manner. While automation and AI continue to play a significant role, humans bring their creativity, problem-solving abilities, and emotional intelligence to complement the capabilities of machines. This collaboration aims to improve efficiency, productivity, and innovation across industries.
2. Personalized and Customized Production: With Industry 5.0, there will be a shift towards more personalized and customized production. Advanced technologies like additive manufacturing (3D printing) will enable the production of highly individualized products at scale. The integration of data analytics and customer insights will drive the creation of tailored products to meet the unique needs and preferences of consumers.
3. Focus on Sustainability: Industry 5.0 will prioritize sustainability and environmental consciousness (Cimino et al., 2023). The emphasis will be on developing eco-friendly manufacturing processes, reducing waste and emissions, and adopting renewable energy sources. Technologies like Internet of Things (IoT) sensors and artificial intelligence will be used to optimize energy consumption, monitor environmental impact, and promote circular economy practices.
4. Augmented Reality and Virtual Reality: Augmented reality (AR) and virtual reality (VR) will have a significant impact on Industry 5.0. These immersive technologies will revolutionize training, maintenance, and remote collaboration. AR and VR will enable workers to receive real-time guidance, visualize complex

processes, and work on tasks remotely, leading to improved productivity, reduced errors, and enhanced safety.

5. Reskilling and Upskilling: Industry 5.0 will require a workforce with a new set of skills. As machines handle repetitive and mundane tasks, humans will need to develop skills such as critical thinking, problem-solving, creativity, and adaptability. Organizations and educational institutions will need to invest in reskilling and upskilling programs to equip workers with the necessary competencies for the evolving job market.
6. Ethical Considerations and Social Impact: With the advancement of technology, Industry 5.0 will raise ethical considerations and concerns about the social impact of automation. There will be a need to address issues related to job displacement, data privacy, algorithmic bias, and the equitable distribution of benefits. Industry 5.0 will require a balance between technological progress and social responsibility.
7. Connectivity and Interoperability: Industry 5.0 will witness increased connectivity and interoperability between various systems, machines, and devices. The Industrial Internet of Things (IIoT) will facilitate seamless communication and data exchange, enabling real-time monitoring, predictive maintenance, and efficient supply chain management. This connectivity will create a networked ecosystem where different stakeholders collaborate and share information to drive innovation and improve overall performance.
8. Agile and Flexible Manufacturing: Industry 5.0 will promote agile and flexible manufacturing systems that can quickly adapt to changing market demands. Modular production lines and reconfigurable factories will enable efficient customization, faster time to market, and reduced production costs. Automation technologies like collaborative robots (cobots) will assist workers in rapidly switching between tasks and processes.
9. Cross-Sector Collaboration: Industry 5.0 will encourage cross-sector collaboration and knowledge sharing. Industries will come together to address complex challenges and explore synergies between different domains. Collaborations between technology companies, manufacturers, academia, and government bodies will lead to the development of innovative solutions and foster a culture of continuous learning.

8.1 Predictions and Trends for Autonomous Industry

The autonomous industry, as we enter the era of Industry 5.0, is poised to undergo remarkable transformations and witness significant trends in the coming years. Autonomous technologies have already revolutionized various sectors, such as transportation, manufacturing, healthcare, and agriculture, and this trend is expected to continue and expand further. One of the key predictions for the autonomous industry is the widespread adoption of autonomous vehicles, including self-driving cars, trucks, and drones. These vehicles will not only enhance safety and efficiency on the roads but also lead to the development of smart cities with optimized traffic management systems. Additionally, the manufacturing sector will see increased automation through the deployment of autonomous robots and machines that can perform complex tasks with precision and adaptability.

Another significant trend in the autonomous industry will be the integration of artificial intelligence (AI) and machine learning (ML) algorithms. These technologies will empower autonomous systems to continuously learn and improve their performance, enabling them to make more accurate decisions and operate with greater autonomy (Kasinathan et al., 2022). This advancement will lead to the emergence of sophisticated autonomous systems capable of complex problem-solving, decision-making, and adaptation to dynamic environments. In the field of healthcare, autonomous robots and AI-powered systems will play a vital role in assisting medical professionals, performing tasks such as patient monitoring, drug dispensation, and surgical procedures. These advancements will not only enhance the quality of healthcare but also alleviate the burden on healthcare providers, particularly in times of emergencies and shortages.

8.2 Implications for Various Sectors and Job Roles

The next phase of the industrial revolution, known as Industry 5.0, will have significant implications for various sectors and job roles. This era of advanced automation and collaboration between humans and machines will bring transformative changes to industries across the board. In the manufacturing sector, there will be a shift towards hyper-personalized and highly flexible production processes. The incorporation of artificial intelligence (AI) and robotics will result in manufacturing job roles emphasizing programming, maintenance, and supervision, necessitating a combination of technical and analytical skills in workers. Furthermore, the emergence of cyber-physical systems and the Internet of Things (IoT) will create opportunities in data analytics, cyber security, and connectivity management. Service sectors, such as healthcare, will also undergo significant transformations. Industry 5.0 will enable the development of smart hospitals, where advanced technologies like wearable devices, telemedicine, and predictive analytics will enhance patient care and diagnosis. This shift will create a demand for professionals who can leverage data-driven insights and effectively utilize healthcare technologies, thus leading to the emergence of job roles in health informatics, tele health, and personalized medicine. The transportation and logistics sectors will experience increased automation, fueled by technologies like autonomous vehicles and drones. As a result, job roles in these sectors will evolve to focus more on overseeing and managing these advanced systems, optimizing supply chain operations, and leveraging data analytics to improve efficiency and customer experience. In the field of agriculture, Industry 5.0 will bring about the era of smart farming, with the integration of sensors, drones, and AI. Farmers and agricultural workers will need to acquire skills in data analysis, precision farming, and remote monitoring to effectively utilize these technologies and optimize crop yields. Moreover, the development of sustainable practices and eco-friendly solutions will create job opportunities in areas such as agricultural engineering and environmental conservation. Industry 5.0 will reshape the employment landscape, demanding a workforce equipped with a diverse range of skills, including technological proficiency, data analysis, creativity, adaptability, and problem-solving abilities. The key to success in this new era will lie in continuous learning, upskilling, and embracing the collaborative potential of human-machine partnerships.

8.3 Opportunities for Innovation and Growth

The industry today is ripe with opportunities for innovation and growth, driven by the advancements in technology, changing consumer demands, and emerging market trends. One of the key areas for innovation lies in the realm of digital transformation (Elangovan, 2021). As businesses increasingly embrace automation, artificial intelligence, and machine learning, there is immense potential to revolutionize processes, improve efficiency, and enhance customer experiences. Additionally, the growing focus on sustainability and eco-friendly practices opens up avenues for innovation in renewable energy, waste management, and green manufacturing. The healthcare industry presents another significant opportunity, with the potential for breakthroughs in personalized medicine, telemedicine, and healthcare data analytics. Furthermore, the rise of the sharing economy, collaborative platforms, and disruptive business models creates opportunities for startups and entrepreneurs to enter and disrupt traditional industries. As global markets continue to expand, particularly in developing countries, there is vast potential for growth in areas such as e-commerce, financial technology, and infrastructure development. Ultimately, the industry's landscape is brimming with possibilities for those who can seize the opportunities, adapt to change, and leverage innovation to drive growth and success.

9. CASE STUDIES

9.1 Real-World Examples of Successful Industry 5.0 Implementations

- Siemens' Electronic Works Amberg: Siemens' factory in Amberg, Germany, is a prime example of a highly automated and flexible production facility. It combines advanced robotics, data analytics, and human-machine collaboration to achieve high levels of efficiency and customization. The factory seamlessly integrates human workers with intelligent machines, enabling rapid production changes and highly precise manufacturing.
- Tesla's Gigafactory: Tesla's Gigafactories, such as the one in Nevada, demonstrate the implementation of advanced manufacturing techniques in the electric vehicle industry. These factories employ robotics, automation, and data-driven processes to streamline production and increase output. While the focus is on automation, human workers are still essential for quality control, maintenance, and complex problem-solving.
- Airbus' Factory of the Future: Airbus has been working on transforming its production facilities into "Factories of the Future" by incorporating digital technologies and human collaboration. These factories leverage augmented reality, Internet of Things (IoT) devices, and data analytics to optimize production processes, enhance worker safety, and improve overall efficiency.
- BMW's Smart Maintenance Glasses: BMW has implemented smart glasses for maintenance technicians to enhance their work processes. The glasses provide real-time access to technical documents, assembly instructions, and remote expert support. This technology allows workers to collaborate effectively and efficiently, reducing maintenance times and increasing productivity.
- BAE Systems' Shipyard Digitization: BAE Systems, a British defense company, implemented a digital shipbuilding initiative to improve productivity and collaboration in their shipyards. The project involved the integration of digital technologies like virtual reality, robotics, and advanced visualization tools. This digitization effort led to significant improvements in quality control, reduced rework, and enhanced communication among workers.
- Bosch's Connected Industry: Bosch, a multinational engineering and electronics company, has implemented Industry 4.0 principles across its various production facilities. By connecting machines, sensors, and data analytics, Bosch has achieved improved process efficiency, predictive

maintenance, and optimized production flows (Yin & Yu, 2022). These implementations have resulted in cost savings, enhanced product quality, and increased competitiveness.

- General Electric's Brilliant Factory: General Electric (GE) has transformed its manufacturing facilities into "Brilliant Factories" by integrating advanced technologies. By leveraging data analytics, machine learning, and IoT devices, GE has achieved real-time monitoring and optimization of production processes. This has led to improved operational efficiency, reduced downtime, and enhanced product quality.
- Daimler AG's Factory 56: Daimler AG, the parent company of Mercedes-Benz, has established "Factory 56" in Germany as a showcase for Industry 4.0. This production facility incorporates robotics, automation, and data-driven processes to create a flexible and sustainable manufacturing environment. It focuses on customizable production, efficient supply chains, and intelligent logistics systems.

10. CONCLUSION

Industry 5.0 marks a significant paradigm shift in the world of manufacturing and engineering. Moving beyond mere automation, this new era emphasizes the integration of humans and machines to foster autonomy in production processes. The key focus is on leveraging advanced technologies such as artificial intelligence, robotics, and Internet of Things to empower workers, enhance creativity, and drive innovation. Industry 5.0 recognizes the importance of human ingenuity, problem-solving skills, and decision-making capabilities in conjunction with the efficiency and precision offered by machines. By embracing this transformative shift, we can revolutionize industries, create more fulfilling work environments, and unlock unprecedented levels of productivity and progress for a brighter future.

REFERENCES

Ali, M. H., Batai, S., & Sarbassov, D. (2019). 3D printing: A critical review of current development and future prospects. *Rapid Prototyping Journal, 25*(6), 1108–1126. doi:10.1108/RPJ-11-2018-0293

Aruldoss, M., Lakshmi Travis, M., & Prasanna Venkatesan, V. (2014). A survey on recent research in business intelligence. *Journal of Enterprise Information Management, 27*(6), 831–866. doi:10.1108/JEIM-06-2013-0029

Aslam, F., Aimin, W., Li, M., Ur Rehman, K. (2020). Innovation in the Era of IoT and Industry 5.0: Absolute Innovation Management (AIM) Framework. *Information, 11*(2), 124. doi:10.3390/info11020124

Aziez, M., Benharzallah, S., & Bennoui, H. (2019). A full comparison study of service discovery approaches for internet of things. *International Journal of Pervasive Computing and Communications, 15*(1), 30–56. doi:10.1108/IJPCC-04-2019-0038

Breque, M., De Nul, L., & Petridis, A. (2021). *Industry 5.0: towards a sustainable, human-centric and resilient European industry*. European Commission, Directorate-General for Research and Innovation.

Cimino, A., Elbasheer, M., Longo, F., Nicoletti, L., & Padovano, A. (2023). Empowering Field Operators in Manufacturing: A Prospective Towards Industry 5.0. *Procedia Computer Science, 217*, 1948–1953. doi:10.1016/j.procs.2022.12.395

Elangovan, U. (2021). *Industry 5.0: The future of the industrial economy*. CRC Press. doi:10.1201/9781003190677

Golovianko, M., Terziyan, V., Branytskyi, V., & Malyk, D. (2023). Industry 4.0 vs. Industry 5.0: Co-existence, Transition, or a Hybrid. *Procedia Computer Science, 217*, 102–113. doi:10.1016/j.procs.2022.12.206

Javaid, M., & Haleem, A. (2020). Critical components of Industry 5.0 towards a successful adoption in the field of manufacturing. *Journal of Industrial Integration and Management, 5*(03), 327–348. doi:10.1142/S2424862220500141

Kasinathan, P., Pugazhendhi, S., & Alsharif, M. H. (2022). Realization of Sustainable Development Goals with Disruptive Technologies by Integrating Industry 5.0, Society 5.0, Smart Cities and Villages. *Sustainability (Basel), 14*(22), 15258. doi:10.3390u142215258

Leng, J., Sha, W., Wang, B., Zheng, P., Zhuang, C., Liu, Q., Wuest, T., Mourtzis, D., & Wang, L. (2022). Industry 5.0: Prospect and retrospect. *Journal of Manufacturing Systems, 65*, 279–295. doi:10.1016/j.jmsy.2022.09.017

Longo, F., Padovano, A., & Umbrello, S. (2020). Value-oriented and ethical technology engineering in industry 5.0: A human-centric perspective for the design of the factory of the future. *Applied Sciences (Basel, Switzerland), 10*(12), 4182. doi:10.3390/app10124182

Nahavandi, S. (2019). Industry 5.0—A human-centric solution. *Sustainability (Basel), 11*(16), 4371. doi:10.3390u11164371

Pizoń, J., & Gola, A. (2023). Human–Machine Relationship—Perspective and Future Roadmap for Industry 5.0 Solutions. *Machines*, *11*(2), 203. doi:10.3390/machines11020203

Raja Santhi, A., & Muthuswamy, P. (2023). Industry 5.0 or industry 4.0 S? Introduction to industry 4.0 and a peek into the prospective industry 5.0 technologies. *International Journal on Interactive Design and Manufacturing*, *17*(2), 947–979. doi:10.100712008-023-01217-8

Rajumesh, S. (2023). Promoting sustainable and human-centric industry 5.0: a thematic analysis of emerging research topics and opportunities. *Journal of Business and Socio-Economic Development*.

Tiwari, S., Bahuguna, P. C., & Walker, J. (2022). Industry 5.0: A macroperspective approach. In Handbook of Research on Innovative Management Using AI in Industry 5.0 (pp. 59-73). IGI Global.

Yavari, F., & Pilevari, N. (2020). Industry revolutions development from Industry 1.0 to Industry 5.0 in manufacturing. *Journal of Industrial Strategic Management*, *5*(2), 44–63.

Yin, S., & Yu, Y. (2022). An adoption-implementation framework of digital green knowledge to improve the performance of digital green innovation practices for industry 5.0. *Journal of Cleaner Production*, *363*, 132608. doi:10.1016/j.jclepro.2022.132608

Chapter 5
Policies for Advanced Dynamic Load Balancing in Cloud Computing

Dharmesh Dhabliya
https://orcid.org/0000-0002-6340-2993
Department of Information Technology, Vishwakarma Institute of Information Technology, India

Sukhvinder Singh Dari
https://orcid.org/0000-0002-6218-6600
Symbiosis Law School, Symbiosis International University, India

Nitin N. Sakhare
https://orcid.org/0000-0002-1748-799X
Department of Computer Engineering, BRACT'S Vishwakarma Institute of Information Technology, India

Anish Kumar Dhablia
Altimetrik India Pvt. Ltd., India

Digvijay Pandey
https://orcid.org/0000-0003-0353-174X
Department of Technical Education, Government of Uttar Pradesh, India

A Shaji George
https://orcid.org/0000-0002-8677-3682
Almarai Company, Riyadh, Saudi Arabia

A. Shahul Hameed
Department of Telecommunication, Consolidated Techniques Co. Ltd., Riyadh, Saudi Arabia

Pankaj Dadheech
https://orcid.org/0000-0001-5783-1989
Swami Keshvaand Institute of Technology, Management, and Gramothan, India

ABSTRACT

Cloud computing involves virtualization, distributed computers, networking, software, and web services. Clouds have servers, datacenters, and customers. It has fault tolerance, high availability, scalability, flexibility, little user overhead,

DOI: 10.4018/979-8-3693-0819-6.ch005

low ownership cost, on-demand services, etc. These issues demand a robust load balancing mechanism. Balanced load distribution improves resource use and task response time by preventing some nodes from being completely loaded and others idle. Load balancers match processor and network node performance. A load balancing solution that improves throughput and latency for application-based virtual topologies with variable cloud sizes will apply the divisible load scheduling theorem.

INTRODUCTION

By dispersing the entire load throughout the numerous nodes that make up the collective system, this approach hopes to cut down on the amount of time needed to complete a task while simultaneously improving the effectiveness with which resources are put to use. According to Yu, Shucheng, et al. (2010), the goal of this method is to cut down on the amount of time needed to complete the work. Because of this, there is no longer a chance that some nodes may be overloaded as well as underloaded at the same time. This predicament may arise if a few of the nodes are carrying more weight than they should be at any given time (Goyal, S., et al. (2022)). These objectives can be accomplished by enhancing not only the responsiveness of the system but also the effectiveness with which it makes use of the resources that are already available (Boopathi, S., et al. (2023)). A dynamic load balancing technique is one that is defined by the fact that it is based on the behaviour the system is displaying at the moment the approach is being implemented, rather than on the state of the system or its behaviour at any previous time. This is in contrast to a static load balancing strategy, which is defined as one that is based on the behaviour the system exhibited at any earlier time. This indicates that the system's past state or behaviour is not taken into consideration in any way, and that this is the case regardless of the situation (Pramanik, S. et al. (2023)). When developing such an algorithm, some of the most important factors to take into account include the following: load estimation, load comparison, system performance, stability of various systems, node interaction, the kind of work that is to be transferred, node selection, and a whole host of other variables (Shang, Ning, et al. (2010)). This load that is being investigated may be assessed in terms of the quantity of memory that is being consumed, the latency, or the stress that is being placed on the network. All three of these metrics may be taken into consideration. These three strategies are all workable options. There is a possibility that each of these factors will be taken into consideration.

GOALS OF LOAD BALANCING

According to what is said in Nabeel et al. (2012) the objectives of load balancing are as follows:

- Consequently as to significantly raise the level of performance.
- To have a backup plan in place in the event that the system fails in any way.
- In order to preserve the reliability of the system.
- To make room for any changes in the system in the future.

Types of Load Balancing Algorithms

There are three different categories that load balancing algorithms can fall into, depending on who started the process:

- Sender Initiated: In the event that the load-balancing algorithm is set up and initialized by the sender
- Receiver Initiated: In the event when the receiver is the one to commence the load balancing algorithm:
- Symmetric: It is a collection of actions taken by the sender as well as those taken by the receiver.
- Depending on the configuration of the system at the moment, load balancing methods can be separated into one of two groups, as shown in (Wan, Zhiguo, et al. (2012))
- Static: It is not dependent on the way the system is operating at the moment. It is required that you have prior knowledge of the system.
- Dynamic: The current state of the system is taken into consideration while making decisions regarding load balancing. There is no requirement for prior knowledge. Therefore, it is preferable than a static method.
- In this article, we will talk about a variety of potential algorithms for dynamic load balancing for clouds of varying sizes.

Dynamic Load Balancing Algorithm

There are two distinct approaches that can be utilised in the context of a decentralised system in order to carry out dynamic load balancing. Both of these approaches are described in the following sentence. The following text provides an explanation of both of these approaches. Non-distributed load balancing and distributed load balancing are the names given to these two distinct approaches, respectively. The word that is used most frequently is distributed load balancing. A dynamic load balancing process is carried out in a distributed system by each and every node that is a component of the system, and the responsibility of load balancing is distributed among these nodes. This type of system is referred to as a distributed system. The term "distributed load balancing," which is also the name of the strategy itself, is used to refer to this particular kind of load balancing. According to the findings of research that was carried out by Chow, Sherman SM, and others (2006), it is possible for nodes to engage with one another in order to achieve load balancing in either a cooperative or non-cooperative manner. This was found to be the case. These are the two different means by which nodes can communicate with one another. There are two distinct channels through which individual nodes can communicate with one another. The first one involves the nodes working closely together in order to achieve a common goal, which might be anything like reducing the amount of time it takes for the system as a whole to respond, for example. The second one involves the nodes working closely together in order to achieve two separate goals. The second one requires strong collaboration amongst the nodes in order to accomplish two distinct objectives at the same time. The nodes are required to work closely together in order to accomplish a common goal as part of the second method. The second kind can be distinguished from the first by the fact that each node functions independently in order to fulfil a goal that is unique to just itself. This allows the second type to be distinguished from the first. To give you an example of this, a node might make an effort to decrease the amount of time necessary to complete a local task, for example. This is one illustration of an effort that has been put forward in this regard. In general, the non-distributed versions of distributed algorithms for dynamic load balancing create a lesser number of messages than do the distributed versions of the same algorithms. This is because distributed versions of distributed algorithms are more efficient than non-distributed versions. This is due to the fact that non-distributed algorithms do not mandate that each node in the system establish a connection with every other node in the system. This is the rationale behind why things are the way they are. This is because distributed approaches necessitate that each and every node in the system be able to communicate with each and every other node in the system. This is the reason why this is the case. The existence of this factor accounts for the current state of affairs. Because of this, the process of load balancing as a whole will proceed unabated even in the event that one or more of the system's nodes become dysfunctional. This is because the situation is as it is. Instead, there will be a decline in the overall level of performance because the performance of the system will be negatively effected to some degree, which will result in a lower overall level of performance. This will be the case because there will be a decrease in the overall level of performance. One of the many advantages of the arrangement is that it makes things like this feasible, which is one of the many benefits that it offers, which is one of its many perks. Distributed dynamic load balancing has the potential to put an extraordinary amount of strain on a system when it requires every node in the system to communicate its current condition with every other node in the system. This is due to the fact that it is necessary for each individual node in the system to communicate its present status to each and every other node in the system. As a consequence of this, there is a chance that the system will become completely inoperable. It is to everyone's benefit if the bulk of the nodes carry out their tasks in an independent manner and have just a minimal degree of connectivity with the other nodes. This will allow the network to function more efficiently. This will result in the overall efficiency of the network being increased to its maximum potential.

The process of load balancing can be carried out in a non-distributed manner by either a single node or a number of nodes working together as a group. Both options are possible. Either of these choices is an option. Both of these options are available for evaluation and could be used. Dynamic load balancing methods can be implemented for non-distributed systems in either a centralised or a semi-distributed form, depending on the requirements of the system. Semi-distributed models are more common. When the first variation of the algorithm is used, the method of load balancing is only carried out by a single node over the entirety of the system. This is because the algorithm only requires a single input. This is because, as stated by Pandey, B. K. et al. (2011), the algorithm in question only has a single processing core available to it. Because this particular node is the only one that brings the process to a successful conclusion, it is referred to as the central node due to the fact that it is the only one that satisfies the requirements of its role. This particular node is completely responsible for ensuring that the system as a whole keeps an adequate load distribution, and this is the only need that falls inside its scope in order for it to be able to achieve this goal. Because there is no other way for the nodes to communicate with one another, it is necessary to have a central node that acts as a hub in the network. Only then will the other nodes be able to share information with one another. cited in Zadi, A. (2019). This is A. Zadi, A. A semi-distributed system has its nodes arranged into clusters, and the process of load balancing within each cluster is managed by a centralised approach. The semi-distributed system's name comes from the combination of the words "semi-" and "distributed." In some contexts, this kind of system may also be referred to as a hybrid distributed or centralised system. This specific kind of system is referred to as a hybrid distributed and centralised system. This is the term given to the system. In order to choose a central node for each cluster, elections are performed inside each cluster, and the voting mechanism that is determined to be the most appropriate is employed for the elections. It is the responsibility of this central node to ensure that the workload in that cluster is divided in a manner that is both fair and equitable among all of the other nodes that are contained within that cluster. This duty falls under the remit of this node's primary function. According to the findings of research that was conducted by Chow, Sherman SM, and others (2006), the process of load balancing the complete system is accomplished by using the core nodes that are present in each cluster. This was found to be the case. This is done so that the load can be spread equally across the system, and it is done to assure that this can happen.

When compared to the situation of semi-distributed load balancing, the process of arriving at a conclusion with centralised dynamic load balancing requires a substantially lower number of messages. This is because the number of nodes participating in the load balancing process is significantly reduced. This is because centralised load balancing is more effective than other methods in its field. This is the result of the centralised nature of the number of nodes that take part in the load balancing process, as stated by the research that was conducted by Pandey, Jay Kumar, and others (2022). This is the consequence of a considerable reduction in the overall number of interactions that are carried out within the system. If, on the other hand, centralised methods are utilised, this may result in a bottleneck in the system at the central node. In addition, the method of load balancing would be rendered ineffective in the event that the central node experienced a breakdown. When used to networks that have a limited number of nodes, this tactic achieves the highest potential level of efficiency, which is a direct consequence of the fact that it possesses this characteristic.

Policies or Strategies in Dynamic Load Balancing

There are 4 policies:

- **Transfer Policy**: Transfer policy, also known as transfer strategy, is the part of the algorithm for dynamic load balancing that determines which workloads should be moved from a local node to a remote node. It can also be called transfer strategy. This component decides which jobs should be transferred to the new location.
- **Selection Policy**: It provides information on the processors that are participating in the load swap, which is sometimes referred to as processor matching.
- **Location Policy**: The portion of the load-balancing algorithm that is in charge of picking a destination node for a transferred job is referred to as the "location policy" or "location strategy," depending on which name you prefer.
- **Information Policy**: In the context of the algorithm for dynamic load balancing, the word "information policy" or "information strategy" refers to

the component of the algorithm that is in charge of gathering information about the nodes that are present in the system. This component may also be referred to as the "information policy" or the "information strategy."

DIFFERENT ADVANCED DISTRIBUTED LOAD BALANCING FOR THE CLOUDS

Load balancing is an extremely important component that should be kept in mind throughout the entire process of system development when creating systems that are both intricate and enormous in scope. One of the things that can be done to facilitate the process of load balancing on a global scale (for example, in a cloud), is to make use of techniques that would act on the components of the clouds in such a manner that the load of the whole cloud would be balanced. This is one of the things that can be done to facilitate the process of load balancing. In order to help make the process of load balancing go more smoothly, this is one of the things that may be done. According to Babu, S. Z. D. (2022), this is one of the things that could be done that has been suggested. One of the things that can be done to make load balancing simpler on a worldwide scale is to put into practise this particular strategy. There are also a number of other things that can be done. Honeybee foraging, biassed random sampling on a random walk method, and active clustering are the three types of solutions that we are going to go over as part of this topic because

Figure 1. Interaction among components of a dynamic load balancing algorithm

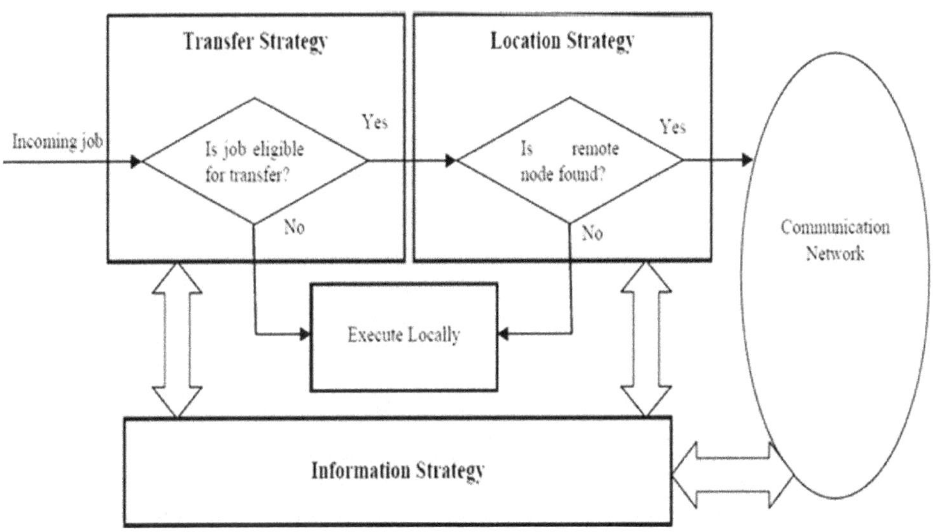

they are all capable of being implemented in a distributed system (Liang, Kaitai, et al (2014)). Honeybee foraging is a solution that uses honeybees to collect data, while biassed random sampling on a random walk approach uses a random walk to collect the data. In addition, honeybee foraging, biassed random sampling based on a random walk approach, and active clustering are the three sorts of solutions that we are going to go over as part of this topic. Honeybee foraging is an approach that uses honeybees to gather nectar and pollen from flowers. The approach of "honeybee foraging" is going to be the first type of solution that we talk about in this article. Honeybee foraging is a method that collects data by employing a honeybee, whereas the approach of biassed random sampling on a random walk collects data by using a random walk. Honeybee foraging and biassed random sampling on a random walk both collect data. Honeybee foraging and the approach of biassed random sampling on a random walk are both ways that data might be collected. Honeybees can only forage for food if they are provided with honeybees.

HONEYBEE FORAGING ALGORITHM

The theory behind this computer programme is based on the way honey bees look for and gather food; to put it another way, bee behaviour served as the primary source of motivation for this programme. Forager bees are the type of bees that travel long distances in order to find new and different kinds of food sources. Forager bees are the ones that are tasked with the responsibility of scouting out new possible food sources. The bees will return to the hive and engage in a type of dance known as the waggle dance when they have been successful in discovering a new source of food. This dance is performed in order to convey their discovery to the other bees in the colony. In addition to providing information about how far away the food is from the hive, the performance of this dance also provides information on the quantity and quality of the food. After that, the scout bees will follow the forager bees to the location of the food, and after they have arrived at the location, they will begin to harvest the food. Afterwards, the forager bees will continue to follow the scout bees. After that, they will go back to the hive and perform a waggle dance, which will provide an assessment of how much food is still available and, as a consequence, will either lead to further consumption of the food source or will lead to abandoning it altogether (Aouafi, R., Zaidi, A., et al., (2023)). [Citation needed] (Aouafi, R., Zaidi, A., et al.

In the case of load balancing, the services are dynamically assigned as a response to changes in the total amount of demand that is positioned on the web servers. This is done in order to keep the load evenly distributed throughout the servers. This is done in order to respond to the ever-evolving needs of the end client, who has requirements that are getting more specific by the day. This is done in order to meet the needs of the end customer. Each of these groups of servers is referred to as a virtual server (VS), and each VS has its very own set of virtual service queues. The servers are arranged into these VS groups using virtual servers. Because of this, each VS is able to operate independently from the others. When a server processes a request from its queue, it will check to see if it has earned a reward or produced a profit for itself. This happens regardless of whether the request was successful or not. In the event that it has, the server will move on to carrying out the request. During the time that they spend doing their waggle dance, the bees will express their degree of expertise in a manner that is analogous to this one. One of the measures that may be used to determine how valuable this incentive is is the amount of time that it takes for the central processing unit (CPU) to deal with a request. The quantity of memory that is necessary to store the information is yet another parameter that may be employed. It is possible to find an idea that is identical to this one in this site; it is presented in the form of an advertisement board, which is similar to the "dance floor" that honey bees use. In addition, the total amount of profit that was made by the colony as a whole is published on this board so that everyone who is a member of the colony can view it (Xue, Kaiping, (2014)).

On each and every one of the servers, players have the choice to play the game as either a scout or a forager. Players can also switch between the two roles at any time throughout the game. The growth of the game depends on both of these roles being played well. After the request has been fulfilled, the server has the opportunity to display their profit on the advertisement boards, which also gives them a chance at positive public relations. A Either the server can check for advertising (see dancing) or serve it in order to exhibit scouting behaviour, or it can select a queue of a virtual server (VS) based on a likelihood of px in order to demonstrate foraging and exploring behaviour. Both of these options are available in order for the server to demonstrate its behaviour. Either the server must check for advertising (which is covered in the section on dancing) or they must serve it in order to display scouting conduct. Another method in which the server can exhibit scouting conduct is by actually serving it (Figure 1). When processing a request, a server will first compute its own profit, then compare it to the profit of the colony, and then set its px value (Figure 3). In the event that a large profit was made, the server will almost certainly continue to be hosted on the same virtual server as it is at the present time, and there is a good probability that PR will be used to publish an advertisement for the server. If it was low, the server would immediately switch its activity to either scouting or foraging, depending on what it had been doing before (Figure 2).

Figure 2. Algorithm used in honey bee technique

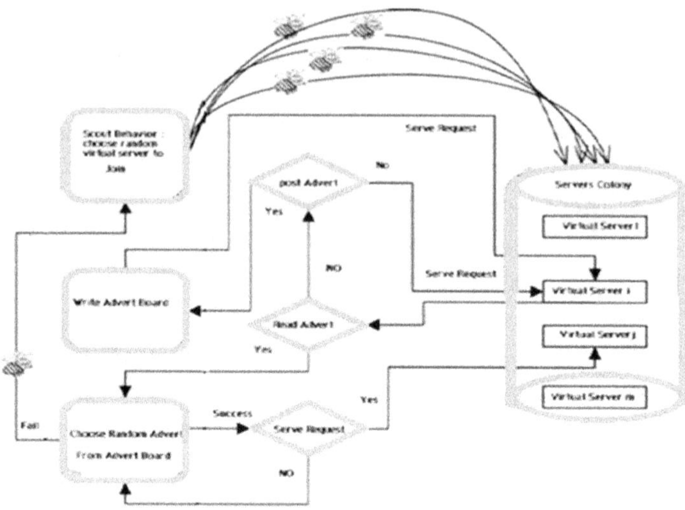

Figure 3. Server allocations by foraging in honey bee technique

```
[A] Initialisation– s_i in V_i serving Q_i, Revenue rate interval T_pr,
                    Advert: posting prob p, reading prob r_i, read interval T_r
  [B] forever
    [C] while Q_i Not Empty do // service queue
          serve request;
          if T_pr expired then
            compute revenue rate;
            adjust r_i from lookup table;
          if Flip(p) == TRUE then Post Advert;
          if T_r expired && Read(r_i) == TRUE then
             if forager then Select/Read advert id V_k // randomly select
               else virtual server id V_k // randomly select
             if V_k Not.Eq V_i then Switch( V_k) // migrate to virtual server V_k
      endwhile
    endforever
```

BIASED RANDOM SAMPLING

During this stage of the process, a virtual graph is being constructed, and the connection of each node (a server is considered a node) is being used to represent the load that is being imposed on the server. Each server is denoted in the graph as a node, and the degree of each node is tied to the total quantity of unrestricted resources that each server provides (Kumar Pandey, et al. (2021)).

When it comes to actually carrying out the work and completing it (Pandey, D. et al. (2022)):

- When a node completes or executes a job, it removes an incoming edge, which results in a reduction in the number of free resources that are available. This may be interpreted as a sign that there are less resources available to be used in the situation (David, S., et al. (2023)).
- When a piece of work is completed, the node will produce an incoming edge, which indicates that there will be more free resources available in the future.

In order to figure out which procedures ought to be retained and which ought to be ditched, a random sampling is carried out as a strategy. The walk can start at any one of the nodes, and at each stage, a different neighbour will be chosen at random to be the next stage's destination. The node that is the last to arrive will be the one to do the load allocation. The selection of a node for load distribution on the basis of particular features, such as the efficiency with which computations are carried out, is an additional technique that may be employed as an alternative. This method can be used to choose a node that will be responsible for load distribution. A additional possibility is to select, for the purpose of load distribution, the node that is not at its capacity, that is, the node with the highest degree. This would be the node with the most degrees. If b is taken to represent the duration of the walk, then an increase in b will lead to an improvement in the efficiency with which load is distributed. We have specified a threshold value for b, and we have found that its value, when obtained empirically, is often equivalent to log n.

A node will only carry out the execution of a job that has been sent to it if the node's current walk length is either equal to or greater than the threshold value. In the event that this requirement is not satisfied, the work that is being examined will have its walk length extended, and a neighbour node will be chosen at random. When a task in the graph is finished being worked on by a node, an incoming edge that connects that node to the rest of the graph is severed. After the task has been completed, an edge will be built that establishes a connection between the node that initiated the process of load allocation and the node that was accountable for the actual work being done.

At the end of the day, all we have is a graph with some directions. The load balancing method that is being used here is very well suited for use in vast network environments, such as those that are found in the cloud, due to the fact that it is not centralised in any way.

ACTIVE CLUSTERING

The concept of active clustering is predicated on the premise that nodes that are connected to one another ought to be grouped together, and after that, the procedure ought to be applied to these groupings. In order to complete the process, the following actions must be taken: • The procedure is kicked off by a node, which then selects another node from its immediate neighbours to serve as the matchmaker node. The matchmaker node will then form a connection with a neighbour of its that is of the same type as the initial node. The matchmaker node will then detach the connection between itself and the initial node. This second node must fulfil the requirement that it must be of a different type than the initial node.

In an iterative manner, the preceding steps in the sequence are carried out again and again.

CONCLUSION

In this chapter, the fundamentals of load balancing are broken down and studied, and issues such the numerous complex load balancing algorithms, the fundamental concept behind dynamic load balancing algorithms, and the policies that may be implemented within such algorithms are brought up for discussion. The many different load balancing strategies that are currently accessible are discussed as well. In the following chapter, we will conduct an in-depth study of the myriad of different distributed load balancing algorithms that are suitable for implementation in cloud-based systems. Research is being done into a variety of alternative algorithmic architectures within the confines of the cloud computing environment.

REFERENCES

Aouafi, R., Zaidi, A., Kouachi, S., & Parshad, R. D. (2023). A remark on “Dynamical behavior of a fractional three-species food chain model”. *Nonlinear Dynamics*, *111*(14), 13641–13651. doi:10.100711071-023-08546-z

Babu, S. Z. D., Pandey, D., Naidu, G. T., Sumathi, S., Gupta, A., Bader Alazzam, M., & Pandey, B. K. (2022, July). Analysation of Big Data in Smart Healthcare. In *Artificial Intelligence on Medical Data: Proceedings of International Symposium, ISCMM 2021* (pp. 243-251). Singapore: Springer Nature Singapore.

Boopathi, S., Pandey, B. K., & Pandey, D. (2023). Advances in Artificial Intelligence for Image Processing: Techniques, Applications, and Optimization. In Handbook of Research on Thrust Technologies' Effect on Image Processing (pp. 73-95). IGI Global.

David, S., Duraipandian, K., Chandrasekaran, D., Pandey, D., Sindhwani, N., & Pandey, B. K. (2023). Impact of blockchain in healthcare system. In *Unleashing the Potentials of Blockchain Technology for Healthcare Industries* (pp. 37–57). Academic Press. doi:10.1016/B978-0-323-99481-1.00004-3

Goyal, S., Pandey, D., Singh, H., Singh, J., Kakkar, R., & Srinivasu, P. N. (2022). Mathematical modelling for prediction of spread of corona virus and artificial intelligence/machine learning-based technique to detect COVID-19 via smartphone sensors. *International Journal of Modelling Identification and Control, 41*(1-2), 43–52. doi:10.1504/IJMIC.2022.127096

Kumar Pandey, B., Pandey, D., Nassa, V. K., Ahmad, T., Singh, C., George, A. S., & Wakchaure, M. A. (2021). Encryption and steganography-based text extraction in IoT using the EWCTS optimizer. *Imaging Science Journal, 69*(1-4), 38–56. doi :10.1080/13682199.2022.2146885

Liang, K. (2014). A DFA-based functional proxy re-encryption scheme for secure public cloud data sharing. *Information Forensics and Security, IEEE Transactions on, 9*(10), 1667-1680.

Nabeel, M., & Bertino, E. (2012). Privacy-Preserving Fine-Grained Access Control in Public Clouds. *IEEE Data Eng. Bull., 35*(4), 21–30.

Pandey, B. K., Pandey, S. K., & Pandey, D. (2011). A survey of bioinformatics applications on parallel architectures. *International Journal of Computer Applications, 23*(4), 21–25. doi:10.5120/2877-3744

Pandey, J. K. (2022). Investigating Role of IoT in the Development of Smart Application for Security Enhancement. In *IoT Based Smart Applications* (pp. 219–243). Springer International Publishing.

Pramanik, S., Pandey, D., Joardar, S., Niranjanamurthy, M., Pandey, B. K., & Kaur, J. (2023, October). An overview of IoT privacy and security in smart cities. In AIP Conference Proceedings (Vol. 2495, No. 1). AIP Publishing. doi:10.1063/5.0123511

Wan, Liu, & Deng. (2012). HASBE: a hierarchical attributebased solution for flexible and scalable access control in cloud computing. *Information Forensics and Security, IEEE Transactions on, 7*(2), 743-754.

Xue & Hong. (2014). A Dynamic Secure Group Sharing Framework in Public Cloud Computing. *Cloud Computing, IEEE Transactions on, 2*(4), 459-470.

Yu, Wang, Ren, & Lou. (2010). Attribute based data sharing with attribute revocation. *Proceedings of the 5th ACM Symposium on Information, Computer and Communications Security*, 261-270.

Zaïdi, A. (2019). Necessary and Sufficient Conditions for the Existence of Robust Whitening Matrices. *IEEE Signal Processing Letters, 26*(6), 863–867. doi:10.1109/LSP.2019.2909651

Chapter 6
Revolutionizing Food:
The Latest Frontiers in Food Science and Technology

Wisha Waheed
Nur International University, Pakistan

Samreen Latif
Nur International University, Pakistan

Alghayia Tahir
Nur International University, Pakistan

Arisha Ahmed
Nur International University, Pakistan

Ali Imran
Government College University, Faisalabad, Pakistan

Fakhar Islam
https://orcid.org/0000-0001-6935-5924
Government College University, Faisalabad, Pakistan

ABSTRACT

The food processing industry has witnessed a shift towards innovative non-thermal technologies to meet the growing demand for high-quality, safe, and minimally processed foods while preserving nutritional value and sensory attributes. This chapter explores recent advances in food science, including pulsed electric field (PEF) processing, sonication, high-pressure processing (HPP), nanotechnology, artificial intelligence (AI), and biodegradable packaging. These advances offer sustainable solutions to improve food processing, safety, and environmental impact, but challenges and potential risks need to be addressed for their effective implementation in the food industry. The purpose of the chapter is to provide some basic knowledge about new thinking in comprehensive food discipline and offer a summary of the latest developments in employing digital technologies in food production systems.

DOI: 10.4018/979-8-3693-0819-6.ch006

1. INTRODUCTION

In the food processing sector, conventional heat treatment techniques like pasteurization and commercial sterilization are commonly used because they may successfully protect food by getting rid of dangerous bacteria, spoilage germs, and enzymes. However, the high temperatures generated by these methods frequently have negative effects on the food's composition, resulting in a degraded quality and diminished nutritional value (Fellows, 2017). A rapid rise has been observed in customer preference for wholesome, natural foods devoid of additives and preservatives. Because of this, customers' demands for high-quality, secure, nutritional, and minimally processed foods are being met through the development of novel, non-thermal technologies (Knorr et al., 2011). Cutting-edge non-thermal or minimal food processing techniques possess the capacity to safeguard food items without significant heat exposure, all the while preserving their nutritional advantages and sensory attributes intact. These innovative methods can prolong the shelf life of food products, maintaining their inherent flavors and colors, resulting in superior quality and heightened consumer appeal.

2. RECENT ADVANCES IN FOOD SCIENCE

Numerous food businesses looking for alternative ways to food processing have shown interest in cutting-edge non-thermal technology (Brennan & Grandison, 2012). The development of these cutting-edge technologies provides the perfect balance between security and minimum processing, thereby resolving financial restrictions while assuring top-notch quality (Tokusoglu, 2015). They also aid in lowering energy and water use, which reduces the carbon and water footprint related to food preparation. These revolutionary innovations are essential for advancing food security and environmental sustainability (Knoerzer, Juliano, Roupas, & Versteeg, 2011; Knoerzer, Buckow, Trujillo, & Juliano, 2015).

A few technologies are now in use, including waste valorization, sonication, HPP nanotechnology, artificial intelligence, and biodegradable packaging. Few, however, are still under development, such as mapping packaging, ohmic heating (thermal food processing technology), and e-beam.

3. PULSE ELECTRIC FIELD

Pulsed electric field (PEF) processing represents an innovative method for pasteurizing food, employing brief, high-voltage electric field bursts to achieve targeted microbial inactivation or modify food structure. In PEF processing, liquid foods are subjected to high-voltage pulses, typically ranging from 20 to 80 kV/cm, with durations spanning milliseconds to microseconds, and this treatment occurs between two electrodes (Zhang et al. 2010; Toepfl, Heinz, and Knorr 2006). For solid foods, where a larger gap exists in the treatment chamber and due to limitations of the pulse generator's power, voltages ranging from 1 to 8 kV/cm are applied. The high-frequency repetition of pulses, reaching rates as high as 3,000 pulses per second, ensures that the entire food sample volume is effectively treated."

PEF processing is utilized for food pasteurization, with commercial applications including organic fruit juices in Oregon, USA. It's employed for liquid foods (e.g., juice, milk, liquid egg) and semi-solids (yogurt, applesauce, salsa, pudding). PEF creates pores in cell membranes via an electric field. It's extensively studied for sterilizing and enzyme inactivation in liquid foods and gaining popularity Recently, PEF has been used for food extraction and dehydration with lower field strength (<10 kV/cm), enhancing nutrient release, liquid diffusion, and reducing dehydration time (Yu, Jin, Fan, and Wu 2018; Yu, Jin, Fan, and Xu 2017).

Caution should be exercised when treating liquid or solid foods with air bubbles. Air, being a poor conductor of electricity, can lead to dielectric breakdown when exposed to PEF, potentially causing equipment failure and fire hazards. It's crucial to remove air from these foods using a deaerator before PEF treatment. PEF treatment may not be suitable for sparkling liquids or foamy liquids for the same reason. Additionally, improperly designed electrodes may experience electro-corrosion, leading to the release of electrode materials like Fe, Cr, Ni, and Mn into the food. Food processors should ensure electrodes have appropriate coatings or materials to minimize electro-corrosion risks.

PEF technology is highlighted for its advantages over methods like heat treatments, as it effectively eliminates microorganisms while preserving the natural color, flavor, texture, and nutritional value of the unprocessed food to a greater extent (Konfo et al., 2023).

The level of microbial inactivation achieved with PEF depends on various factors, including electric field strength, treatment time, pulse frequency, pulse width, and treatment temperature, as well as the specific PEF system used (e.g., batch/static or continuous chamber; coaxial or co-field; square wave, exponential decay, or oscillatory pulses). Additionally, the food product's characteristics, such as electrical conductivity, density, viscosity, pH, and temperature, and the microbial attributes (e.g., bacteria, molds/yeasts, gram-positive/negative, vegetable cells, or spores) play a crucial role. Food processors need to carefully select the appropriate PEF system and determine the specific parameters to apply for their target food products.

4. SONICATION

Ultrasound, as an innovative method for process optimization in the food industry, offers several advantages without compromising food quality. It operates based on acoustic energy, making it a safe, non-ionizing, non-invasive, and environmentally friendly energy source (Alarcon-Rojo et al., 2019).

The core components of ultrasonic equipment are transducers, responsible for converting either mechanical or electrical energy into acoustic waves. These transducers play a pivotal role in Ultrasound-Assisted Extraction (UAE). During UAE, sound waves propagate within a vessel filled with the medium until transducers generate acoustic resonance, leading to the formation of compression and rarefaction regions (Fu et al., 2020). This phenomenon, combined with cavitation and implosion induced by sonication, results in the rupture of cell walls and an increased number of disrupted cells. This disruption facilitates the entry of solvents into cells, enabling the incorporation of intracellular plant material into the solvent (Gallo et al., 2018).

Several factors come into play during UAE, including temperature, ultrasound frequency, extraction time, and the nature of the solvent or medium. These factors not only influence the extraction yield but also impact the composition of the extract.

In several sectors, including medicines, nutraceuticals, and food manufacturing, extraction is a crucial unit function. For the creation of natural functional foods within the food business, extraction in these areas tries to acquire either complete plant extracts or specialized components.

Particularly in the food business, ultrasound has been used for microbial inactivation, curing, and meat tenderization (Liu et al., 2017; Pinton et al., 2021). Additionally, it enhances the content of bioactive elements, Beta-glucans, phenolics in cereals, fruits, and vegetables, and facilitates the extraction of starch from cereals (Ranjha et al., 2020; Manzoor et al., 2019; Li et al., 2020; Alonso-Riaño et al., 2020; Medina-Torres et al., 2017; Jovanovic-Malinovska et al., 2015).

High-intensity ultrasound extraction methods have proven effective in improving the quality and efficiency of extracting various food components, such as oils, flavorings, pigments, as well as bioactive substances like antioxidants and essential oils from aromatic plants like basil, artemisia, and lavender (Žlabur et al., 2018).

Given the current challenges posed by rising energy costs and the imperative to reduce greenhouse gas emissions, industries in the food and plant-based chemical sectors are actively seeking innovative solutions to minimize energy consumption. In this context, sonication is emerging as a promising technology for the extraction of intracellular plant materials, aligning with industrial requirements and sustainability objectives (Chemat & Strube, 2015).

5. HIGH-PRESSURE PROCESSING

High-pressure processing (HPP) is a method for preserving food that utilizes pressure instead of heat to eliminate harmful pathogens and spoilage microorganisms. This technique applies intense pressure, typically in the range of 400-600 MPa, at low temperatures, such as refrigeration levels, to achieve pasteurization. HPP is also referred to as high-hydrostatic pressure processing (HHP) or ultra-high-pressure processing (UHP).

HPP offers several benefits, including the ability to deactivate bacteria and spores at higher temperatures without introducing toxicity. It also helps preserve the nutritional content, colors, and flavors of foods, reduces processing times, ensures consistent treatment, and has the potential to decrease the reliance on chemical preservatives. From a microbial standpoint, HPP enhances food safety and extends shelf life (Penchalaraju, 2013).

However, HPP has some drawbacks, including high equipment costs, the requirement that foods contain at least 40% free water for effective antimicrobial action, batch processing, a lack of flexible packaging options, little effect on food enzyme activity, and the survival of some microorganisms.

The pasteurization and sterilization of fruits, sauces, yoghurt, salad dressings, meats, and vegetables, as well as the decontamination of heat-sensitive substances including flavorings and vitamins, are all commercial uses for HPP. The chemical makeup of the food items and the kinds of bacteria present have an impact on how well HPP treatment works (Alpas et al., 1999; Benito et al., 1999). Depending on the pressure and the makeup of the food, HPP (High-Pressure Processing) inherently results in some temperature rises brought on by adiabatic heating. Low molecular weight molecules are little impacted, keeping vitamins, flavors, and colors while upholding the nutritional value and quality of food. HPP, however, has the ability to permanently change some substances. Carbohydrates may gelatinize under pressure, but proteins can denature under extreme heat. Ideal HPP meals have a high moisture content, no gas inclusions, and empty areas for the best pressure transfer. Packaging should be flexible to withstand compression without structural damage. Common materials include polypropylene (PP), polyester tubes, polyethylene (PE) pouches, and nylon cast polypropylene pouches.

The impact of high pressure on microorganisms parallels its effect on food components. It leads to protein denaturation, disrupting essential functions within bacterial cells, potentially causing cell death when pressure levels are adequate. However, it's important to note that HPP's effectiveness varies among microorganisms. When in their spore condition, spore-forming microorganisms have excellent resistance to HPP, although yeasts and molds are more susceptible. At room temperature, pressures between 300 and 400 MPa can inactivate the majority of vegetative species in a matter of minutes; however, treatment at even higher pressures may be necessary for yeast and mold ascospores.

6. NANOTECHNOLOGY

Food nanotechnology involves the utilization of nano-carrier techniques to enhance the effectiveness of bioactive ingredients, making them more resistant to environmental factors and chemical changes (Mozafari et al., 2008). This technology enhances sensory attributes such as color, flavor, and texture, contributing to improved overall quality (Kalita and Baruah, 2019). Additionally, it enhances the absorption and availability of nutraceuticals (Jafari and McClements, 2017) and drug delivery systems (Safari and Zarnegar, 2014).

In food processing, nanomaterials serve various purposes, including acting as food additives, anti-caking agents, carriers for efficient nutrient delivery, antimicrobial agents, and strengthening agents for packaging materials. Food nano sensing is employed to assess and enhance food quality and safety (Anandharamakrishnan et al., 2013). This technology is especially valuable in the production of nutraceuticals, ensuring improved stability and bioavailability (Duncan, 2011).

Nanotechnology finds multiple applications in food processing, including nano-filtration, nanoscale enzyme-based reactions, nano-encapsulation for ingredient absorption and modification, as well as heat and mass transfer through nanofabrication. Nano-filtration is essential in industries for purifying pharmaceuticals and treating drinking water and dairy products to improve quality (Dasgupta et al., 2015). Nanofabrication enhances the heat resistance of packaging materials. Nanoscale enzyme reactors are employed to modify food processing methods, improving flavor, nutritional value, and health benefits. Nanoencapsulation is widely used to enhance food products, particularly for flavor improvement, preservation, and cooking stability.

The increased application of nanotechnology in agriculture and food products has raised significant concerns among a wide segment of society due to potential adverse effects associated with various nanoparticles (Baruah and Dutta, 2009). The risks associated with nanotechnology primarily stem from the small size of nanoparticles, which have large surface areas and can disperse easily, potentially leading to toxicity as they can penetrate cells and reach distant areas within the body (Dasgupta et al., 2015). Furthermore, the similarity in size between nanomaterials and DNA raises the possibility of interactions with biological specimens (Baruah and Dutta, 2009).

7. 3D/4D PRINTING OF FOOD

The application potential of three-dimensional (3D) and four-dimensional (4D) printing technologies in the food industry is extensive and promising, as evidenced by the design of printing model (Chen et al., 2022). 3D laser printing finds widespread applications across diverse divisions, including textiles, pharmaceuticals, automotive, the military, and the food industry. Integrating 3D printing technology with digital cooking techniques enables personalized customization of food, allowing individuals to modify the shape, structure, and flavor of their meals according to their unique preferences, separating from the conventional approach of mass-produced commercial food. For example, Animal shapes can be crafted by using sugar, while cakes can be decorated with a variety of colorful syrups when printed. Four different 3D printing technologies are employed within the food industry. These are shown in Figure 1 (Zhang et al., 2022).

Currently, the majority of studies focusing on 3D food printing technology predominantly utilize ink extrusion as their primary method. This involves the propulsion of food inks through printing nozzles, which are then dispersed onto designated surfaces.

In food industry, 3D printing technology finds applications in diverse settings, ranging from households and restaurants to space exploration. One notable application involves helping astronauts in crafting personalized, customized food for extended space missions. The electrostatic properties, printing attributes, thermal properties, encompassing rheological properties, nutritional properties, and sensory qualities, of natural food gels are enhanced through the incorporation of additives in 3D/4D food printing. These additives, specifically lipids, hydrocolloids, and carbohydrates,

Figure 1. 3D printing technologies

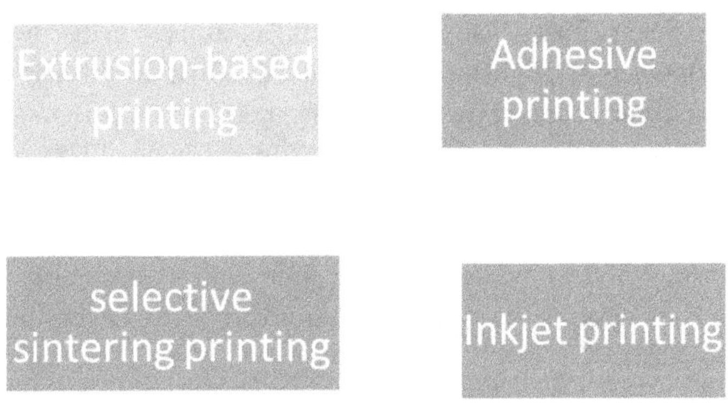

play a crucial role in improving the overall printing process and the quality of the printed food (Terfansky et al., 2013).

8. ARTIFICIAL INTELLIGENCE

AI plays a pivotal role in the food industry by addressing both food security and food quality management. In terms of food security, AI applications encompass image processing and recognition technologies to fortify security measures, AI-driven fertilizer management to ensure the production of safe food, food inspection and grading using AI and image processing, as well as the deployment of robots to bolster food safety in production and storage. Moreover, AI facilitates food security through the implementation of smart technologies and information sharing mechanisms. In the realm of food quality management, AI is instrumental in enhancing food quality, employing mathematical modeling techniques for quality control, utilizing machine learning to optimize the production of high-quality food, and efficiently managing pesticide usage. Collectively, AI technologies play a crucial role in safeguarding and enhancing the safety and quality of food products within the food industry.

In the food processing sector, the task of effectively sorting and packaging food products is a labor-intensive and time-consuming process for manufacturing facilities. AI-powered intelligent decision-making systems incorporate a variety of tools and techniques, including high-resolution cameras, laser-based technology, X-ray systems, and IR spectroscopy. These technologies are employed to scrutinize every aspect of food products like fruits and vegetables as they enter the processing pipeline. Traditional systems can only classify products as good or bad based on their appearance. By implementing these systems, industries have experienced several benefits, including increased production speed, improved product quality, and reduced labor costs.

Additionally, AI-based systems have also addressed personal health and sanitation guidelines. These systems are used to monitor individuals who do not adhere to the guidelines (Rary et al., 2020). AI is also assisting customers in discovering unique flavors and preferences (Dehghan-Dehnavi et al., 2020). It records individual tastes and customer selections, providing valuable insights for the launch of new products in the market (Dehghan-Dehnavi et al., 2020; Yost & Cheng, 2021). Furthermore, in the food processing sector, AI-based systems can easily manage the cleaning and maintenance of processing equipment (Wang et al., 2020). AI is deployed to oversee every stage of the process, including price control, inventory management, forecasting, and monitoring the supply chain from production to customer delivery (Kakani et al., 2020).

AI is poised to revolutionize the food processing industry due to its potential to significantly enhance productivity for both consumers and employees. However, there are limitations and drawbacks to consider, such as the potential for job displacement, technological challenges (machines can only perform tasks they are programmed for), and the expenses associated with AI development and maintenance, which could ultimately raise product prices.

9. BIODEGRADABLE PACKAGING

Plastics have ushered the food industry into a new era, but traditional petroleum-based plastics, being non-degradable, have posed significant environmental challenges such as endangering aquatic life and deteriorating air quality. As a response to these issues, biodegradable polymers, or biopolymers, have emerged as a viable alternative for various industrial applications to mitigate the risks associated with non-biodegradable plastics. As per the American Society for Testing and Materials (ASTM) definition, biodegradable plastics are those that break down due to the activity of naturally occurring microorganisms like bacteria, fungi, and algae (ASTM, 2004). The decomposition of biodegradable plastics results in the production of water, carbon dioxide, inorganic compounds, or biomass.

Modified Atmosphere Packaging (MAP) is characterized as the practice of packaging perishable products in an environment altered from standard air composition. It stands as a widely adopted method for packaging and preserving fruits and vegetables by modifying the gas content within the package's headspace (Coles et al., 2003). This modification of the atmosphere within the packaging extends the shelf life of the products and shields them from potential exposure to diseases and other environmental contaminants (Briano et al., 2015).

A form of packaging called edible packaging allows consumers to consume it together with the food it contains. Films, sheets, coatings, and pouches are the most common forms (Janjarasskul and Krochta, 2010). The same composition is used to create both films and coatings; however, films are deposited as solid sheets, and coatings are applied as liquids (Galus and Kadzinska, 2015).

The three basic components of edible packaging are lipids, proteins, and polysaccharides. Two components that are frequently employed are chitosan and gelatin/collagen. Gelatine coatings limit oil migration by lowering oxygen and moisture exposure. Notably, collagen-based banger casings are an extremely popular and readily accessible edible protein film on the market. These coatings lessen exudation on thawed and chilled beef steaks without affecting colour or lipid oxidation. Collagen-based films are used to improve the juiciness of processed meats, reduce shrinkage, and absorb fluid exudates in a variety of cooked meat products.

Due to their high adhesion to the surfaces of cut fruits and vegetables and robust gas barrier qualities, polysaccharide-based films have demonstrated effectiveness in increasing the shelf life of fruits and vegetables.

The deliberate incorporation of components intended to release or absorb chemicals within the packed food or its immediate environment is known as active packaging. Utilising this novel strategy, food goods' safety and shelf-life are improved. Absorbers, which act as scavenging systems to remove CO2, 02, moisture, odours, or ethylene, and emitters, which act as releasing systems and introduce various compounds like antimicrobial agents, antioxidants, flavours, and more into the headspace, make up the two main parts of the active packaging system (Yildirim et al., 2018).

The efficacy of active ingredients that are added directly to food may occasionally be hampered or reduced as a result of interactions with other elements of the bulk food system. Therefore, regulated release of active ingredients within active packaging proved to be a more effective strategy for bulk food systems. The use of surface coating and direct integration are the two main strategies for accomplishing this regulated release. Notably, a variety of microorganisms, including gram-positive and gram-negative bacteria, yeast, and moulds, have shown the strongest antibacterial activity in response to the surface coating approach (Debiagi et al., 2014).

10. WASTE VALORIZATION

The management of food waste and by-products from food processing has taken on greater significance in recent years as a means of enhancing economic performance and fostering environmental sustainability. The chemicals produced from these food waste products have a great deal of promise for use in different biotechnological and food applications. The use of environmentally friendly extraction techniques like microwave, ultrasound, and pulsed electric field methods represents a significant advancement in the recovery of valuable by-products from food waste in comparison to conventional methods like Soxhlet extraction, liquid-liquid extraction, and mechanical shaking.

One of the main tasks in the food processing industry is the processing of grapes, which generates a lot of waste products such grape shoots, grape pomace, exhausted yeast, and wastewater. The extraction of these useful molecules is a vital stage in the waste valorization process since these waste materials are notably rich in bioactive chemicals. These bioactive substances may be used as additions in both food and cosmetic goods and have been shown to have favourable impacts on health (Barba et al., 2016).

In a similar vein, Pulsed Electric Field (PEF)-based recovery methods have been successfully employed in the extraction of valuable bioactive compounds from waste generated in the cooking oil industry (Boussetta et al., 2014). Waste stemming from dairy operations encompasses a range of materials, including manure, runoff contaminated with waste, milking house waste, bedding, and spilled feed (Ahmad et al., 2019). The utilization of the PEF technique has proven to be an effective means of treating waste whey produced by the dairy industry and offers a viable approach for obtaining protein fractions with potential applications in the food industry.

Similar to this, important bioactive chemicals have been effectively extracted from waste produced by the cooking oil industry using pulsed electric field (PEF)-based recovery techniques (Boussetta et al., 2014). Dairy operations generate a variety of wastes, including manure, runoff that is polluted with trash, waste from the milking house, bedding, and spilt feed (Ahmad et al., 2019). The PEF method has been shown to be a successful method for processing leftover whey generated by the dairy industry and provides a workable method for generating protein fractions with potential uses in the food sector.

Globally, a huge amount of eggshell waste is produced, and these shells are a significant source of many different chemicals. Calcium is the most prevalent of these chemicals, making up 90% of the eggshell's content, mostly in the form of calcium carbonate (Waheed et al., 2019).

Recently, there has been an increase in interest in using fish waste, pushing improved extraction techniques, and using byproducts for their value components in a variety of applications. Fish bones, skin, and fins stand out as particularly good suppliers of collagen and gelatin. Additionally, Yin and He (2008) have successfully extracted soluble calcium from fish bones using the Pulsed Electric Field (PEF) method. Additionally, fish oil made from waste materials serves as a substantial source of the important omega-3 fatty acids.

At present, there is a noteworthy focus on harnessing natural elements sourced from food waste or its secondary products. Phenolic compounds extracted from food waste play a crucial role in safeguarding against free radical damage, cancer, and cardiovascular ailments (Kumar, Yadav, Kumar, Vyas, & Dhaliwal, 2017). The remnants of fruits, known as pomaces, are a valuable source of pectin—a versatile thickening agent employed in a variety of food products like jams and confectioneries (Kumar & Singh, 2018). Additionally, specific tropical fruits contain papain or bromelain, which are natural enzymes capable of breaking down proteins, and these enzymes find practical applications as meat tenderizers or even in laundry detergents.

Figure 2. Shows the outline of green extraction techniques

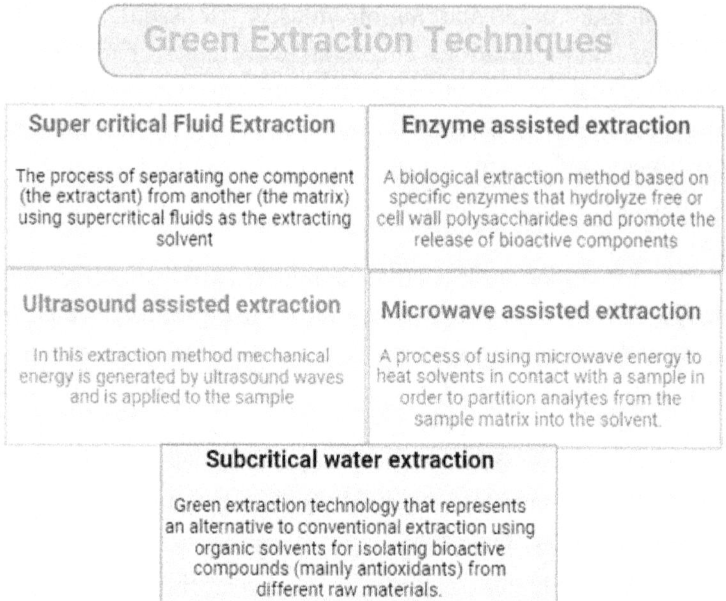

11. CONCLUSION

Innovation serves as the linchpin for ensuring the sustainable growth of the food industry, and one effective avenue to achieve this is through the process intensification approach, which can ultimately result in significant reductions in energy expenditures.

There is a notable association between the nature of food products and the adoption of non-thermal food processing techniques in their production. The primary determinant for selecting non-thermal food processing technology is its ability to enhance both nutrient retention and sensory attributes, with subsequent benefits including prolonged product shelf life and the resolution of safety concerns. However, it's crucial to acknowledge that the substantial initial investment poses a major obstacle to the widespread adoption of non-thermal food processing technologies.

The primary catalysts for driving innovation in this domain are equipment manufacturers and government-sponsored research initiatives. Technological advancements come to fruition when there is substantial research conducted by experts, accompanied by extensive experimental work. It is imperative to prioritize the development of novel and improved innovative non-thermal technologies that strike a harmonious balance between ensuring food safety and minimizing the impact of processing.

12. FUTURE PERSPECTIVES

The future perspectives of revolutionizing food are promising as advancements in technology and innovation continue to play a crucial role in transforming the way we produce, distribute, and consume food. With the increasing global population and urbanization, vertical farming offers a sustainable solution by cultivating crops in vertically stacked layers, often within urban environments, reducing the need for extensive land and transportation. The emphasis on sustainable and eco-friendly food production methods, including circular economy principles, will continue to gain importance, addressing environmental concerns and promoting responsible resource management. The development of lab-grown or cultured meat holds the potential to address environmental concerns, ethical considerations, and resource inefficiencies associated with traditional livestock farming. In summary, the future of revolutionizing food lies in a holistic approach that combines technological innovation, sustainable practices, and a focus on meeting the nutritional needs of a growing global population while minimizing environmental impact.

REFERENCES

Ahmad, T., Aadil, R. M., Ahmed, H., Ur Rahman, U., Soares, B. C., Souza, S. L., ... Esmerino, E. A. (2019). Treatment and utilisation of dairy industrial waste: A review. *Trends in Food Science & Technology*, *88*, 361–372. doi:10.1016/j.tifs.2019.04.003

Alarcon-Rojo, A. D., Carrillo-Lopez, L. M., Reyes-Villagrana, R., Huerta-Jiménez, M., & Garcia-Galicia, I. A. (2019). Ultrasound and meat quality: A Review. *Ultrasonics Sonochemistry*, *55*, 369–382. doi:10.1016/j.ultsonch.2018.09.016 PMID:31027999

Alonso-Riaño, P., Sanz Diez, M. T., Blanco, B., Beltrán, S., Trigueros, E., & Benito-Román, O. (2020). Water ultrasound-assisted extraction of polyphenol compounds from Brewer's spent grain: Kinetic study, extract characterization, and concentration. *Antioxidants*, *9*(3), 265. doi:10.3390/antiox9030265 PMID:32210202

Alpas, H., Kalchayanand, N., Bozoglu, F., Sikes, A., Dunne, C. P., & Ray, B. (1999). Variation in resistance to hydrostatic pressure among strains of food-borne pathogens. *Applied and Environmental Microbiology*, *65*(9), 4248–4251. doi:10.1128/AEM.65.9.4248-4251.1999 PMID:10473446

Anandharamakrishnan, C. (2013). Nanoencapsulation of food bioactive compounds. *Techniques for Nanoencapsulation of Food Ingredients*, 1–6. doi:10.1007/978-1-4614-9387-7_1

Astm, D. (2004). *04 Standard Specification for Compostable Plastics*. J ASTM Int.

Barba, F. J., Zhu, Z., Koubaa, M., Sant'ana, A. S., & Orlien, V. (2016). Green alternative methods for the extraction of antioxidant bioactive compounds from winery wastes and by products: A review. *Trends in Food Science & Technology*, *49*, 96–109. doi:10.1016/j.tifs.2016.01.006

Baruah, S., & Dutta, J. (2009). Nanotechnology applications in pollution sensing and degradation in agriculture: A Review. *Environmental Chemistry Letters*, *7*(3), 191–204. doi:10.100710311-009-0228-8

Benito, A., Ventoura, G., Casadei, M., Robinson, T., & Mackey, B. (1999). Variation in resistance of natural isolates of *escherichia coli* O157 to high hydrostatic pressure, mild heat, and other stresses. *Applied and Environmental Microbiology*, *65*(4), 1564–1569. doi:10.1128/AEM.65.4.1564-1569.1999 PMID:10103251

Boussetta, N., Soichi, E., Lanoisellé, J.-L., & Vorobiev, E. (2014). Valorisation of oilseed residues: Extraction of polyphenols from flaxseed hulls by pulsed electric fields. *Industrial Crops and Products*, *52*, 347–353. doi:10.1016/j.indcrop.2013.10.048

Briano, R., Giuggioli, N. R., Girgenti, V., & Peano, C. (2015). Biodegradable and compostable film and modified atmosphere packaging in postharvest supply chain of Raspberry Fruits (cv. Grandeur). *Journal of Food Processing and Preservation*, *39*(6), 2061–2073. doi:10.1111/jfpp.12449

Chemat, F., & Strube, J. (2015). *Green extraction of natural products theory and Practice*. Wiley-VCH-Verl. doi:10.1002/9783527676828

Chen, X., Zhang, M., Teng, X., & Mujumdar, A. S. (2022). Recent progress in modeling 3D/4D printing of foods. *Food Engineering Reviews*, *14*(1), 1–14. doi:10.100712393-021-09297-6

Chen, Y., Zhang, M., Sun, Y., & Phuhongsung, P. (2022). Improving 3D/4D printing characteristics of natural food gels by novel additives: A review. *Food Hydrocolloids*, *123*, 107160. doi:10.1016/j.foodhyd.2021.107160

Coles, R., McDowell, D., & Kirwan, M. J. (2003). *Food Packaging Technology*. Crc Press.

Dasgupta, N., Ranjan, S., Mundekkad, D., Ramalingam, C., Shanker, R., & Kumar, A. (2015). Nanotechnology in agro-food: From field to plate. *Food Research International*, *69*, 381–400. doi:10.1016/j.foodres.2015.01.005

Debiagi, F., Kobayashi, R. K. T., Nakazato, G., Panagio, L. A., & Mali, S. (2014). Biodegradable active packaging based on cassava bagasse, polyvinyl alcohol and essential oils. *Industrial Crops and Products*, *52*, 664–670. doi:10.1016/j.indcrop.2013.11.032

Dehghan-Dehnavi, S., Fotuhi-Firuzabad, M., Moeini-Aghtaie, M., Dehghanian, P., & Wang, F. (2020). Estimating participation abilities of industrial customers in demand response programs: A two-level decision-making tree analysis. *2020 IEEE/IAS 56th Industrial and Commercial Power Systems Technical Conference*. 10.1109/ICPS48389.2020.9176817

Duncan, T. V. (2011). Applications of nanotechnology in food packaging and Food Safety: Barrier Materials, antimicrobials and sensors. *Journal of Colloid and Interface Science*, *363*(1), 1–24. doi:10.1016/j.jcis.2011.07.017 PMID:21824625

Fellows, P. J. (2017). *Food processing technologies: Princilpes and practices. (4th* (418th ed.). Woodhead Publishing, Elsevier.

Fu, X., Belwal, T., Cravotto, G., & Luo, Z. (2020). Sono-physical and sono-chemical effects of ultrasound: Primary applications in extraction and freezing operations and influence on food components. *Ultrasonics Sonochemistry*, *60*, 104726. doi:10.1016/j.ultsonch.2019.104726 PMID:31541966

Gallo, M., Ferrara, L., & Naviglio, D. (2018). Application of ultrasound in Food Science and Technology: A perspective. *Foods*, *7*(10), 164. doi:10.3390/foods7100164 PMID:30287795

Galus, S., & Kadzińska, J. (2015). Food applications of emulsion-based edible films and coatings. *Trends in Food Science & Technology*, *45*(2), 273–283. doi:10.1016/j.tifs.2015.07.011

Jafari, S. M., & McClements, D. J. (2017). Nanotechnology approaches for increasing nutrient bioavailability. *Advances in Food and Nutrition Research*, *81*, 1–30. doi:10.1016/bs.afnr.2016.12.008 PMID:28317602

Jovanovic-Malinovska, R., Kuzmanova, S., & Winkelhausen, E. (2015). Application of ultrasound for enhanced extraction of prebiotic oligosaccharides from selected fruits and vegetables. *Ultrasonics Sonochemistry*, *22*, 446–453. doi:10.1016/j.ultsonch.2014.07.016 PMID:25116595

Kakani, V., Nguyen, V. H., Kumar, B. P., Kim, H., & Pasupuleti, V. R. (2020). A critical review on Computer Vision and artificial intelligence in food industry. *Journal of Agriculture and Food Research*, *2*, 100033. doi:10.1016/j.jafr.2020.100033

Kalita, D., &Baruah, S. (2019). The impact of nanotechnology on food. *Nanomaterials Applications for Environmental Matrices*, 369–379. doi:10.1016/B978-0-12-814829-7.00011-2

Knoerzer, K., Buckow, R., Trujillo, F. J., & Juliano, P. (2015). Multiphysics Simulation of Innovative Food Processing Technologies. *Food Engineering Reviews*, *7*(2), 64–81. doi:10.100712393-014-9098-3

Knoerzer, K., Juliano, P., Roupas, P., & Versteeg, C. (2011). *Innovative food processing technologies: Advances in multiphysica simulation*. Wiley-Blackwell. doi:10.1002/9780470959435

Knorr, D., Froehling, A., Jaeger, H., Reineke, K., Schlueter, O., & Schoessler, K. (2011). Emerging technologies in food processing. *Annual Review of Food Science and Technology*, *2*(1), 203–235. doi:10.1146/annurev.food.102308.124129 PMID:22129381

Konfo, T. R. C., Djouhou, F. M. C., Hounhouigan, M. H., Dahouenon-Ahoussi, E., Avlessi, F., & Sohounhloue, C. K. D. (2023). Recent advances in the use of digital technologies in agri-food processing: A short review. *Applied Food Research*, 100329.

Kumar, K., Yadav, A. N., Kumar, V., Vyas, P., & Dhaliwal, H. S. (2017). Food waste: A potential bioresource for extraction of nutraceuticals and bioactive compounds. *Bioresources and Bioprocessing*, *4*(1), 18. doi:10.118640643-017-0148-6

Kumar, V., & Singh, L. (2018). Pectin from fruit peels and its uses as pharmaceutical and food grade: A descriptive review. *European Journal of Biomedical and Pharmaceutical Sciences*, *5*, 185–199.

Li, Y., Tao, F., Cui, K., Song, Y., Nan, L., Cui, C., Li, Y., Yang, J., Wang, Y., & Jiang, L. (2020). Optimization of ultrasonic extraction of anthocyanin in mulberry residue by response surface methodology. *IOP Conference Series. Earth and Environmental Science*, *559*(1), 012024. doi:10.1088/1755-1315/559/1/012024

Liu, P., Xu, Y., Gao, X., Zhu, X., Du, M., Wang, Y., Deng, R., & Gao, J. (2017). Optimization of ultrasonic-assisted extraction of oil from the seed kernels and isolation of monoterpene glycosides from the oil residue of Paeonia lactiflora pall. *Industrial Crops and Products*, *107*, 260–270. doi:10.1016/j.indcrop.2017.04.013

Manzoor, M. F., Zeng, X.-A., Rahaman, A., Siddeeg, A., Aadil, R. M., Ahmed, Z., Li, J., & Niu, D. (2019). Combined impact of pulsed electric field and ultrasound on bioactive compounds and FT-ir analysis of Almond extract. *Journal of Food Science and Technology*, *56*(5), 2355–2364. doi:10.100713197-019-03627-7 PMID:31168118

Medina-Torres, N., Ayora-Talavera, T., Espinosa-Andrews, H., Sánchez-Contreras, A., & Pacheco, N. (2017). Ultrasound assisted extraction for the recovery of phenolic compounds from vegetable sources. *Agronomy (Basel)*, *7*(3), 47. doi:10.3390/agronomy7030047

Mozafari, M., Johnson, C., Hatziantoniou, S., & Demetzos, C. (2008). Nanoliposomes and their applications in food nanotechnology. *Journal of Liposome Research*, *18*(4), 309–327. doi:10.1080/08982100802465941 PMID:18951288

Penchalaraju, M., & Shireesha, B. (2013). Preservation of foods by high-pressure processing. *RE:view*.

Pinton, M. B., dos Santos, B. A., Lorenzo, J. M., Cichoski, A. J., Boeira, C. P., & Campagnol, P. C. (2021). Green Technologies as a strategy to reduce NaCl and phosphate in meat products: An overview. *Current Opinion in Food Science*, *40*, 1–5. doi:10.1016/j.cofs.2020.03.011

Ranjha, M. M., Amjad, S., Ashraf, S., Khawar, L., Safdar, M. N., Jabbar, S., Nadeem, M., Mahmood, S., & Murtaza, M. A. (2020). Extraction of polyphenols from Apple and pomegranate peels employing different extraction techniques for the development of functional date bars. *International Journal of Fruit Science, 20*(sup3). doi:10.1080/15538362.2020.1782804

Rary, E., Anderson, S. M., Philbrick, B. D., Suresh, T., & Burton, J. (2020). Smart sanitation—Biosensors as a public health tool in sanitation infrastructure. *International Journal of Environmental Research and Public Health*, *17*(14), 5146. doi:10.3390/ijerph17145146 PMID:32708793

Safari, J., & Zarnegar, Z. (2014). Advanced Drug Delivery Systems: Nanotechnology of health design a review. *Journal of Saudi Chemical Society*, *18*(2), 85–99. doi:10.1016/j.jscs.2012.12.009

Terfansky, M., Thangavelu, B., Fritz, B., & Khoshnevis, B. (2013) 3D printing of food for space missions. *Proceedings of AIAA SPACE 2013 Conference and Exposition*. 10.2514/6.2013-5346

Toepfl, S., Heinz, V., & Knorr, D. (2006). Applications of pulsed electric fields technology for the food industry. *Pulsed Electric Fields Technology for the Food Industry*, 197–221. doi:10.1007/978-0-387-31122-7_7

Tokusoglu, O. (2015). Introduction to innovative food processing. *Natural Science and Discovery*, *1*(4), 85–87. doi:10.20863/nsd.89151

Waheed, M., Butt, M. S., Shehzad, A., Adzahan, N. M., Shabbir, M. A., Suleria, H. A. R., & Aadil, R. M. (2019). Eggshell calcium: A cheap alternative to expensive supplements. *Trends in Food Science & Technology*, *91*, 219–230. doi:10.1016/j.tifs.2019.07.021

Wang, X., Puri, V. M., & Demirci, A. (2020). Equipment Cleaning, sanitation, and maintenance. *Food Engineering Series*, 333–353. doi:10.1007/978-3-030-42660-6_13

Whitley, L. (2008). Food Processing Handbook - edited by Brennan J C, Wiley-VCH. International Journal of Dairy Technology, 61(3), 309–310. doi:10.1111/j.1471-0307.2008.00386.x

Yildirim, S., Röcker, B., Pettersen, M. K., Nilsen-Nygaard, J., Ayhan, Z., Rutkaite, R., Radusin, T., Suminska, P., Marcos, B., & Coma, V. (2018). Active packaging applications for Food. *Comprehensive Reviews in Food Science and Food Safety*, *17*(1), 165–199. doi:10.1111/1541-4337.12322 PMID:33350066

Yin, Y., & He, G. (2008). A fast high-intensity pulsed electric fields (pef)-assisted extraction of dissoluble calcium from bone. *Separation and Purification Technology*, *61*(2), 148–152. doi:10.1016/j.seppur.2007.10.005

Yost, E., & Cheng, Y. (2021). Customers' risk perception and dine-out motivation during a pandemic: Insight for the restaurant industry. *International Journal of Hospitality Management*, *95*, 102889. doi:10.1016/j.ijhm.2021.102889 PMID:36540682

Yu, Y., Jin, T. Z., Fan, X., & Wu, J. (2018). Biochemical degradation and physical migration of polyphenolic compounds in osmotic dehydrated blueberries with pulsed electric field and thermal pretreatments. *Food Chemistry*, *239*, 1219–1225. doi:10.1016/j.foodchem.2017.07.071 PMID:28873543

Yu, Y., Jin, T. Z., Fan, X., & Xu, Y. (2017). Osmotic dehydration of blueberries pretreated with pulsed electric fields: Effects on dehydration kinetics, and microbiological and nutritional qualities. *Drying Technology*, *35*(13), 1543–1551. doi:10.1080/07373937.2016.1260583

Zhang, H. Q., Barbosa-Cãnovas, G. V., Bala, B. V. M., Dunne, C. P., Farkas, D. F., & C., Y. J. T. (2011). Nonthermal Processing Technologies for Food. Wiley.

Žlabur, J., Voća, S., Brnčić, M., & Rimac-Brnčić, S. (2018). New trends in food technology for green recovery of bioactive compounds from plant materials. *Role of Materials Science in Food Bioengineering*, 1–36. doi:10.1016/B978-0-12-811448-3.00001-2

Chapter 7
Applications of Nanotechnology for Improving Food Safety and Security

Aliza Batool
Faculty of Food and Home Sciences, Muhammad Nawaz Shareef University of Agriculture, Multan, Pakistan

Umar Farooq
Faculty of Food and Home Sciences, Muhammad Nawaz Shareef University of Agriculture, Multan, Pakistan

Nida Firdous
Faculty of Food and Home Sciences, Muhammad Nawaz Shareef University of Agriculture, Multan, Pakistan

Afshan Shafi
Faculty of Food and Home Sciences, Muhammad Nawaz Shareef University of Agriculture, Multan, Pakistan

Zulqurnain Khan
Department of Biotechnology, Institute of Plant Breeding and Biotechnology (IPBB),Muhammad Nawaz Shareef University of Agriculture, Multan, Pakistan

Shabbir Ahmad
Faculty of Food and Home Sciences, Muhammad Nawaz Shareef University of Agriculture, Multan, Pakistan

Muhammad Sibt-e-Abbas
Faculty of Food and Home Sciences, Muhammad Nawaz Shareef University of Agriculture, Multan, Pakistan

Muhammad Usman
Faculty of Food and Home Sciences, Muhammad Nawaz Shareef University of Agriculture, Multan, Pakistan

ABSTRACT

In order to ensure a consistent supply of nutrient-rich nutritious food, food safety and security are gaining a lot of attention on a global scale. Numerous technical activities, such as the creation of novel materials and the improvement of food safety and security, need nanotechnology. Nanomaterials are utilized to enhance food protection and aid in the detection of pesticides, toxic chemicals, and microbiological

DOI: 10.4018/979-8-3693-0819-6.ch007

contamination. Nanomaterials are utilized to enhance the detection of pesticides, toxic chemicals, and microbiological contamination. Nanocapsulation, which enables the release of bioactive substances, raises food bioavailability, and lengthens food shelf life, improves food processing even further. The main purpose of food processing is to fortify food by adding nutrients and beneficial ingredients. This chapter covers a variety of subjects, including the use of nanotechnology in food packaging, food safety, food preservation using smart nanocarriers, the detection of allergens and food-borne viruses using nanosensors, and crop growth and yield enhancement.

FOOD SAFETY AND SECURITY

Food security means having both the means and the ability to access a sufficient, safe and nutritious food supply that supports a healthy and active lifestyle. When people don't have enough food security, it can lead to malnutrition and this, in turn, can result in chronic diseases. Food access, food use, food stability, and food availability are the basic pillars of food security. By 2050, the world faces a significant challenge, as projected by the United Nations Food and Agriculture Organization (UNFAO). The demand for food is expected to rise substantially, by 59 to 98%, due to global population growth. This increase will be especially pronounced in developing countries, where the population is expected to reach 9 billion people. To ensure food security, food production needs to increase by 70%. Several factors will influence this increase, including available land, climate change, agricultural productivity and the food supply (Ansari, 2023).

The UNFAO also mentioned that food waste happens for different reasons in various parts of the world. In high-income countries, it's often related to outdated packaging and appearance preferences. In contrast, in underdeveloped countries, it's more common due to issues like spoilage, contamination, and low food quality. It has been demonstrated that abiotic stresses (such as soil salinization, high CO_2 and temperature levels, drought and nutritional imbalances) have a detrimental effect on plant growth, development, production, and quality. These events have sometimes helped in the local extinction of some species. Additionally, it has been noted that most crops experience an average yield loss of more than 50% due to abiotic stressors. Nanotechnology help to enhance the plant development and resistance to abiotic and biotic challenges, in order to improve the quality and agriculture productivity. Food waste negatively impacts the amount of land and water that may be used to grow food as well as the amount of greenhouse emissions that cause climate change (Kalpana et al., 2019).

Food security is expected to become a bigger concern in the near future because of several important changes. These include a growing global population, shifts in dietary habits, increased connections between food and energy resources, especially bioenergy and the effects of climate change (Shafiee-Jood et al., 2016). The challenge of feeding the growing urban population is significant, with an estimated 40% more land required by 2030. Climate change, including rising temperatures and decreased rainfall in certain areas, is expected to harm crop yields for essential crops like corn, wheat and rice. Alongside climate issues, freshwater resources are depleting, land use is changing, and ecosystems on land and in water are being polluted. These challenges, in addition to existing resource strains, pose a threat to food security. Shockingly, over 800 million people worldwide still lack sufficient protein, vitamins and energy in their daily diets (Chaudhary et al., 2018).

Public health is seriously affected by issues with food safety. Its main objective is to protect food from biological, chemical, and physical hazards while it is being handled, prepared, processed and distributed. Changes in food recipes, dietary habits, and the presence of toxins, harmful microorganisms and contaminants can all risk food safety and pose serious health risks to people (Gu et al., 2019).

Food security is a very difficult issue for national control and regulatory agencies in a country like Saudi Arabia because of the country's low agricultural production and total reliance on food imports and adherence to Islamic law. Farmers worldwide are actively seeking new breakthroughs and modern technologies to increase crop yields, improve soil quality, enhance food packaging and distribution, and maintain food safety and security. Traditional approaches to disease detection, contamination analysis, and food security enhancement are frequently time- and money-consuming. Consequently, in recent years, various nanoscale particles and related devices have been developed, offering cost-effective, large-scale applications to boost agricultural productivity, enhance soil quality, improve food processing, extend shelf life, analyse infections and toxins (Rovera et al., 2020). These developments show that fundamental modifications to current methods of food production and distribution are required to assure food security and also decrease the negative environmental effects of food system (Zhou et al., 2021)

NANOTECHNOLOGY

Nanotechnology, a frontier of the 21st century, was born from ambitious dreams. Nanotechnology is all about understanding and controlling matter at sizes between 1 and 100 nanometres, where unique properties open up new possibilities. While humans have been exposed to nanoparticles throughout history, their exposure significantly increased during the industrial revolution. Nanoparticle research is not a relatively new field of study. The 1925 winner of the Chemistry Nobel Prize, Richard Zsigmondy, is credited with coining the term "nanometre." He was the first to use a microscope to measure the size of particles like gold colloids and coined the phrase "particle sizes" (Hulla et al., 2015).

Different disciplines are involved in nanotechnology like engineering, physics, chemistry, and other academic disciplines (Chandra, 2016). These nanoparticles have become more widely used in a variety of industries, such as public health, agriculture, medicine, clothing, cosmetics, and food, because of their exceptional capacity to increase bioavailability and solubility while also safeguarding bioactive components during processing and storage (Hu et al., 2017).

Numerous industries, including healthcare, food safety, water treatment, food preservation, crop protection, use nanomaterials extensively Fig.1. This is because of their unique chemical and physical properties, as well as their ability to combat many harmful bacteria (Baranwal et al., 2018). In the food business, nanostructured materials (NSMs) are also being used as novel packaging materials, food component encapsulants, and nanosensors. This chapter focuses on the use of nanotechnology, in particular nanoparticles and artificial nanostructures, to ensure the safety and security of food. Nanotechnology has found numerous applications in the food industry, where specific nanomaterials are added to foods to impart certain desired qualities. Additionally, nanotechnology has been helpful in the global research and development of large-scale agricultural product production, processed food production, beverage production, and enhanced food packaging. According to several research, these nanoparticles can improve food safety without altering the flavor or physical properties of the food by strengthening the efficiency of food packaging, prolonging the shelf life of products, and enhancing their nutritional value as supplements (Bajpai et al., 2018).

Improvement in Food Safety Through Nanotechnology

Food safety is a significant concern both nationally and internationally, despite advancements in food preservation, sanitation, and regulations. Foodborne illnesses caused by toxins and pathogens pose substantial health risks to people. In the United States, it's estimated that these illnesses lead to around 9.4 million cases annually. The food business must create a strong system to identify and detect foodborne pathogens, like fungus, bacteria, viruses, and other possible pollutants on food products or surfaces, in order to assure food safety. Achieving this goal remains a top priority for the scientific community and researchers. The food sector needs quick, precise contaminant detection techniques that don't demand a lot of labor or technical knowledge. It can be

Figure 1. Nanotechnology in different field of food for improving food safety and security

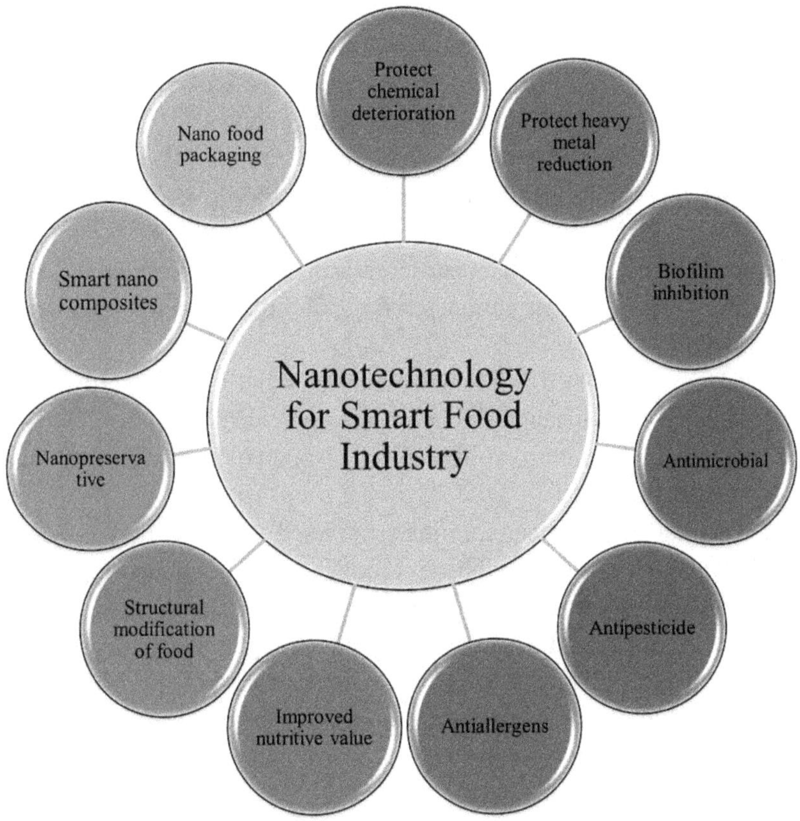

expensive and time-consuming to identify hazardous bacteria or toxins using conventional detection methods and rapid molecular technologies. Moreover, using conventional methods may introduce accuracy issues or self-contamination (Ameta et al., 2020).

In today's food industry, nanotechnology applications have gained widespread use and significantly impact various aspects of the field. Nanotechnology offers solutions to food safety challenges through tools and methods such as microbial and toxin detection, extending shelf life and advancements in food packaging (Kumar et al., 2021). In order to address concerns about food safety, nanotechnology largely focuses on the antibacterial capabilities of nanoparticles and nano sensors for identifying toxins and pathogens that are present in food (Nasr et al., 2015).

Food safety is rapidly being revolutionized by nanotechnology, which offers detection technologies that are quicker, more precise, and more affordable than conventional and molecular ones. Contrary to conventional ways, technologies based on nanotechnology can find dangerous bacteria throughout different stages of food production. These nanotechnology-based methods offer high sensitivity and precision while significantly reducing incubation and measurement times. New nanotechnology techniques are being developed, especially in microbial detection and management, to enhance food quality and safety. These innovations address challenges related to precision and time constraints, enabling the food industry to produce safe, healthy and high quality food products (Eleftheriadou et al., 2017).

Additionally, mixing various nanoparticle materials has demonstrated a synergistic effect, leading to more potent antibacterial capabilities than single nanoparticles (Gomes et al., 2011).

It has been found that when different nanoparticles, like gold, zinc, silver, chitosan, iron, platinum, and carbon nanotubes, are combined with natural essential oils, they create a more powerful antimicrobial effect compared to using them separately (Morsy et al., 2014).

When essential oils and nanoparticles are used together, they enhance their ability to fight against various foodborne pathogens. This combined approach can be especially effective against drug-resistant germs. In the food industry, where many harmful microorganisms can thrive, antimicrobial coatings made with nanotechnology are applied to surfaces. These coatings like nanoscale silver, TiO_2, and ZnO, act as disinfectants and help prevent the formation of biofilms or eliminate existing ones. Biofilms are tough to remove with traditional disinfectants, making nanocoating a valuable solution (Khan et al., 2015).

By preventing microbial adhesion and biofilm formation, antimicrobial nanomaterial coatings have the potential to address surface contamination issues in the food industry. Nano sensor technology offers another promising application in food safety. Nano sensors can quickly detect infections, mycotoxins, and common foodborne pathogens like E. coli, Listeria monocytogenes and Salmonella. They also have potential to detect chemical pollutants in food, making them a valuable tool for safeguarding food quality and safety (Berekaa et al., 2015).

Just like antimicrobial nanoparticles, nano sensors have shown they can be faster, more accurate, and more concise than traditional detection methods when it comes to identifying harmful microbes and their dangerous compounds. Various nano sensors have been created to spot specific foodborne pathogens, their harmful byproducts, and the formation of biofilms. All in all, nanotechnology, with its strong antibacterial capabilities and ability to tackle food safety issues in the food industry, plays a crucial role in ensuring food safety (Adeyeye et al., 2019)

Nanotechnology Role in Food Packaging and Preservation

Freshly harvested fruits and vegetables are quite susceptible to spoiling due to exposure to oxygen, water and ethylene gas (Gaikwad et al., 2020). This is where food packaging comes into play as a crucial solution. Packaging acts as a shield, keeping out oxygen and other harmful gases that can spoil the food. It not only helps control spoilage and contamination but also provides physical protection against temperature changes, bumps, vibrations, and microbial threats. Nanotechnology, specifically the use of nanoparticles and polymer-based composites, has emerged as a promising approach to enhance food packaging (Rashidi et al., 2011). To reduce environmental impact, biodegradable materials are preferred for making food packaging (Ncube et al., 2020). Recently, covering food with a natural polymer or biopolymer has drawn interest and demonstrated success in maintaining freshness (Luo et al., 2020).

Various applications of nanotechnology in food packaging and preservation can be found in Table 1. It's worth highlighting that food industries, as well as scientists and researchers in the field of food packaging, should focus on high barrier plastics, incorporating antimicrobial properties into packaging, and implementing contamination detection measures during the packaging process. Food preservation involves managing and treating food to extend its shelf life and prevent microbial degradation while processing, packaging, and preservation are key stages in the control and handling of food Fig.2 (Pradhan et al., 2015).

Figure 2. Use of nanotechnology for food preservation, safety, security

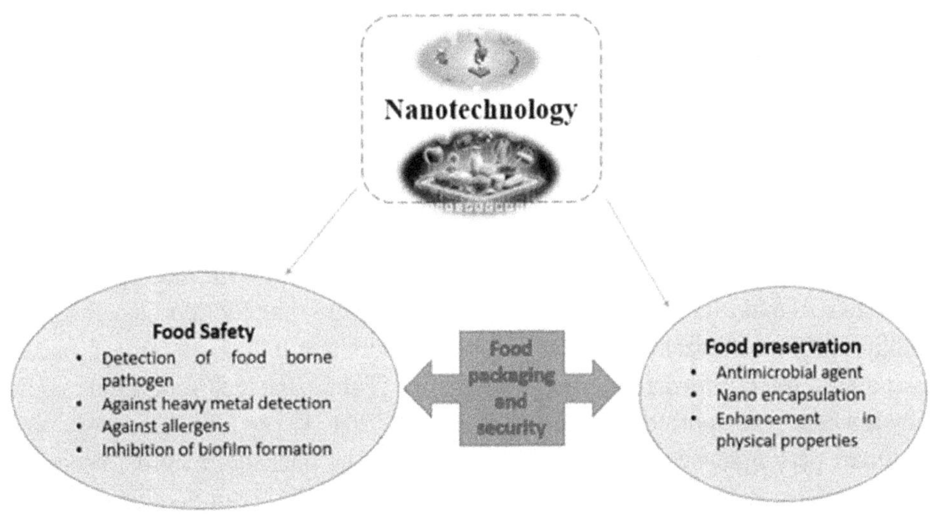

Active Packaging Systems

The main functions of active packaging in food storage are to increase the quality and extend the shelf life of packaged foods. It's a particular sort of packaging with elements designed to release antioxidants or antimicrobials into or from the packaged food, or perhaps to do both. The shelf life and quality of packaged goods are improved when active chemicals including water vapor absorbers, nanoparticles, oxygen absorbers and antimicrobials are included with the packaging material itself (Majid et al., 2018).

In active packaging, various types of nanoparticles, such as silicon oxide, silver, titanium dioxide, gold, and zinc oxide, are widely used (Bikiaris et al., 2013). These nanoparticles can interact with organic substances in the food through gradual migration or direct contact. Among them, silver nanoparticles are commonly used due to their well-known ability to fight harmful bacteria, viruses, and fungi like Monkeypox and others (Sharma et al., 2017).

Interestingly, silver nanoparticles can both encourage and inhibit the production of reactive oxygen species (ROS) and affect respiratory chain enzymes. In one example, films with added silver nanoparticles completely halted the growth of E. coli, while those with silver micro-particles managed to kill about 80% of the bacteria. Silver nanoparticles are also used in sodium alginate films for packaging fresh fruits and vegetables, proving effective. Additionally, by lowering microbial contamination, the addition of nanoparticles like (carbon nanotubes and allyl isothiocyanate) to an anti-microbial film for packaging cooked and chopped chicken meat increased the meat's shelf life by up to 40 days. Additionally, this inclusion assisted in regulating the oxidation process and prevented the meat's color from changing (Fasihnia et al. 2018).

Smart Packaging Systems

Smart packaging systems commonly employ nano sensors, often created using nanoparticles, to detect food contaminants. The ability to monitor biological, physical, and chemical changes that occur during food processing and preservation is greatly enhanced by nanosensors based on nanoparticles. With smart packaging, these sensors make it easier to identify chemicals, toxins, and foodborne pathogens. Additionally, smart packaging that has nanosensors and indicators built in makes it simple to track information about the quality of the packaged food items while they are being transported and stored (Dias et al., 2013).

Smart packaging, with sensors integrated into the package, records responses to internal or external environmental factors. This helps maintain food quality throughout distribution. Some indicators are commonly added to food packaging to measure changes influenced by factors like temperature, time and packaging integrity (freshness). Employing indicators to monitor changes in food products across the supply chain and production process plays a vital role in preserving product quality and extending shelf life (Branton et al., 2010).

NANOSENSORS

Nano sensors, also known as bio-analytical devices, are sophisticated systems composed of various biological receptors and tiny structured materials. Their dependability, rapid detection, and cost-effectiveness have made them very popular in the food processing and packaging industries (Nile et al., 2020). Nano sensors excel in their ability to selectively and sensitively detect substances. They use a range of nanomaterials, like nanoparticles (metal oxide non-metallic and metallic), nanowires, nanofibers, nanorods, and carbon nanotubes, to enhance their performance by optimizing factors like size, optical traits and electrical properties (Ansari et al., 2010).

These nano sensors are adept at identifying changes in food colour and the gases produced during food decomposition. They exhibit remarkable sensitivity to several gases, including ammonia, hydrogen, nitrogen oxides, sulphur dioxide, and hydrogen sulphide. Nano sensors are now frequently employed for detecting chemicals, toxins, pesticides, and pathogens present in food products. Their effectiveness and precision make them superior to traditional sensors.

Nano sensors typically consist of electronic components (data processors) and sensing components, allowing them to convert changes in temperature, light, gas and chemical composition into electrical signals (Kang et al., 2007).

Common materials like gold, platinum and palladium nanoparticles are often used in gas sensors. It employs multiple chemical sensors connected to a data processing system, closely mimicking the functionality of a human nose and earning its nickname, the electronic nose (Coles *et al.,* 2013).

Nanoparticles in Food Packaging

Adding nanoparticles to food packaging enhances its effectiveness, extending the shelf life and improving food quality. Nanoparticles are tiny substances that can encapsulate active components and enhance the functionality, stability, and availability of food packaging. Different nanoparticles have recently been created and employed in the food sector. Many of these nanoparticles possess antibacterial properties, which is highly beneficial for food packaging. For instance, silver nanoparticles are commonly used in food packaging due to their strong antibacterial properties.

Other nanoparticles like zinc oxide, titanium oxide and silicate nanoparticles are also added to plastic films to decrease the oxygen infiltration, prevent moisture loss and extend food shelf life (Chaudhary et al., 2020).

Silicon dioxide, often employed as a food colorant and anti-caking agent, is known for its hygroscopic properties, helping absorb water molecules. These nanoparticles are used for food colouring, as a photocatalytic disinfectant, and to whiten dairy products like milk and cheese. They function in food packaging as a UV barrier as well. As permitted by regulatory agencies, considerable volumes of silicon dioxide and titanium dioxide nanoparticles can be utilized as food additives without causing any harm (Acosta et al., 2009).

Silver nanoparticles, for example, can protect food against microbial contamination and have been deemed safe by the FDA when used within established limits. They are preferred over other antimicrobial nanoparticles because of their ease of integration into packaging materials. Silver nanoparticles hinder ribosome activity in microbial systems, preventing the production of certain enzymes. They have shown promise in combating both gram-positive and gram-negative bacteria, with greater efficacy against the latter (Dakal et al., 2016; Yin et al., 2020). Additionally, silver nanoparticles can absorb and degrade ethylene, extending the shelf life of vegetables and fruits (Jiang et al., 2018).

Nanocomposites

In food packaging, not only nanoparticles but also nanocomposites and nanolaminates play a significant role. These materials have gained popularity in recent years, especially for creating antimicrobial food packaging. Nanocomposites, made of polymers and nanoparticles, are commonly used to extend the shelf life of food products and protect them from physical damage and high temperatures (Honarvar et al., 2016). For example, zinc oxide nanocomposites are well-regarded in the food industry because of their antioxidant properties, making them suitable for active food packaging (Kim et al., 2020).

Nanocomposites enhance the properties of polymers, offering flexibility and chemical resistance, which helps preserve food freshness and delay microbial spoilage. They are often used to create materials with excellent barrier properties, such as gas barriers that prevent carbon dioxide from escaping from bottles and cans of carbonated beverages. Instead of heavy glass bottles or expensive cans, food companies that produce carbonated drinks can apply a nanocomposite layer on their beverage containers to reduce gas leakage (Pandey et al., 2013).

Typically, nanoclay-based nanocomposites are employed for these gas barriers. Commercially available nanoclay-based nanocomposites like Nanocar, Durethan and Imperium, are widely used due to their clearness, low weight, surface properties, biodegradability, and excellent flow characteristics. Polyamide-based Durethan adds rigidity to paperboard fruit juice containers. Imperium, a nanoclay-based polymer, scavenges oxygen to increase the shelf life of food products, and Nano-car is used to create plastic beer bottles, reducing carbon dioxide leakage (He et al., 2010).

Nanoparticle-based nanocomposites are also commonly used in packaging films, and they exhibit a wide range of antibacterial activities. Silver-coated nanocomposites, for instance, have antibacterial properties. When silver comes into contact with bacteria, it damages their cell walls, increasing permeability and causing harm to bacterial DNA. Some pests that can infiltrate packaged food products have been effectively controlled by nanocomposites coated with garlic oil. Another type of nanoencapsulation called nanolaminates is typically used to protect fruits, vegetables, meats and baked goods (Mohammad et al., 2022)

NANOTECHNOLOGY ENSURING FOOD SECURITY

Food safety concerns focus on whether the food we buy will harm our health, including issues like food spoilage caused by microbes or gases and food adulteration, as discussed earlier. On the other hand, food security is a broader issue that impacts human survival and development, addressing fundamental challenges related to the availability of food, water and proper nutrition. Nanotechnology, a modern technology, has the potential to help address these critical food security challenges (White et al., 2018)

Increasing Agricultural Production

The COVID-19 epidemic has disrupted the farming and food supply chains, raising prices, causing famine, and possibly triggering food crises. According to the World Food Program's 2020 Global Report on Food Crises, COVID-19 has put at least 130 million people at danger of being hungry. New methods of agricultural production are required to address the issue of the limited amount of arable land that is currently available worldwide. Nanotechnology presents opportunities for improving agricultural productivity. It can be used to track crop growth, improve soil fertility, reduce fertilizer waste, and enable precision agriculture all of which contribute to increasing agricultural yield and addressing food security concerns (Laborde et al., 2020).

Nanofertilizers

It's widely known that using more fertilizer can boost crop yields, but regular fertilizers can be inefficient and harm the environment. Enter nanofertilizers, a type of tiny material that can make regular fertilizers work better or provide nutrients directly to plants. When used in the right amounts, nanofertilizers can help seeds grow, improve nutrient uptake, and make plant roots and stems longer. Nanoparticles interact with plants, changing how they grow and develop (Usman et al., 2020).

Studies have shown that nanosilica can help plants resist salt stress by improving the growth of squash seeds and their water and nutrient use. It can also boost chlorophyll and proline activity. A nanofertilizer made with zinc oxide nanoparticles using an eco-friendly method was found to provide significant nutritional benefits to tomatoes, especially when used in the right amounts (Jabeen et al., 2018). Nanosilica also proved helpful in tomato seedling germination, showing promise as an alternative fertilizer for sustainable agriculture (Line et al., 2017).

This improves water and nutrient absorption efficiency and boosts the germination of aquaporins, aiding tomato seedling growth. Single-walled carbon nanohorns also have the potential to act as plant growth regulators, promoting the germination and growth of various crops like rice, corn, soybeans, and tomatoes. Using nanofertilizers helps make better use of environmental resources, reduces environmental pollution, lowers input costs and supports sustainable agriculture (Zulfiqar et al., 2019).

Nanocarriers

Crop development can be affected by many factors like pests, soil fertility etc. To increase crop yields, farmers often use agricultural chemicals like pesticides. However, these chemicals can leave behind too many residues in the soil, which can be harmful to both human health and the diversity of soil organisms. Nanocarriers can help reduce the negative effects of these traditional methods compared to regular formulations. Nanocarriers can be used as an additional ingredient to make agricultural chemicals like insecticides and plant growth regulators more stable and reduce the amount needed, ultimately increasing grain yields. There are various types of nanocarriers, like nanoemulsions, nanospheres and nanocapsules, Nanocapsules can contain active chemicals since they have a shell and core structure. Typically, materials that are simple to absorb into plant cells are selected for use as nanocarriers. For instance, chitosan nanoparticles have been created to efficiently protect crops and regulate the release of agricultural pesticides at the proper rate. By combining solutions, one can produce stable colloid systems called nanoemulsions, which have minuscule solute sizes in the nanometer range (Sharma et al., 2020).

Nanocarriers not only transport active chemicals but also benefit the soil directly. Use of pectin calcium nanoparticles, which are crosslinked by pectin and calcium ions as nanocarriers, is one promising strategy for enhancing crop growth, especially in regions with limited water resources. These nanoparticles improve the functionality of water reservoirs in the soil and increase their ability to retain water, which can be beneficial for crop growth (Sharma et al., 2017).

Nanosensors in Agriculture

Agricultural development is uncertain due to various factors like soil quality, water availability, and climate conditions. To ensure sustainable agriculture and food security, it's essential to accurately monitor and manage soil and crop conditions. Nanotechnology can play a role in achieving precision farming and sustainable soil management while also helping in monitoring agricultural processes. This approach can enhance food security. Wireless nano sensors in irrigation systems can provide data on soil moisture and temperature. These sensors are also used for monitoring crop leaf humidity and leaf area in greenhouses to manage conditions effectively. Soil can become contaminated with pollutants like pesticide residues and heavy metals, which can harm agricultural productivity and human health. Around one-third of the world's soil is degraded, often due to poor soil management practices (Kah et al., 2019).

Moreover, smart nano sensors based on nano- and biotechnology can work alongside agricultural electronic equipment to precisely control the distribution of water and pesticides, leading to increased plant yields. Still in its infancy, the application of nanotechnology to increase crop yields has been focused on two areas: utilizing nanoparticles as soil fertilizers and putting insecticides into nanocarriers (Landry et al., 2019).

Nanotechnology Enhance Nutritional Components

Food security also depends critically on nutrition. Malnutrition continues to be a major problem, particularly in poorer nations. More than 790 million people still do not get enough protein, energy, and vitamins in their diets. By increasing the rate at which nutrients are utilized, which might be done by using nanotechnology, the issue of nutritional deficiency may be as much as possible reduced. As a result, it is anticipated that the following objectives can be accomplished: (a) preventing nutritious components from degrading; (b) enhancing the solubility of insoluble components to increase the bioavailability and nutrition value of those components (Rafiee et al., 2019).

Additionally, nutritional or pharmaceutical intervention is required due to the aging population. Similar to this, the use of nanotechnology could boost the effectiveness of food and medicine. Nanocarriers can be employed as an adjunct method to enhance the stability of these active elements such as vitamins. Nanocapsules are one of the most popular types of nanocarriers for nutrient delivery. The majority of its coating ingredients are proteins, lipids, or carbohydrates, which may give sensitive components opportunities to improve stability and retention. Active components can be enclosed in the nanocapsules. Additionally, the units can be inserted into and connected to other structures thanks to the holes on their exterior. Nanocapsules serve a variety of purposes (Souri et al., 2021)

The use of whey protein isolated nanoparticles to encapsulate vitamin D3 at various calcium concentrations, levels of acidity, and levels of alkalinity. The outcome demonstrated that vitamin D3 in nanocapsules had a better level of stability than in the control group. The biological effects of bioactive substances depend on how much of them are exposed to tissues or blood. To achieve the best biological effect, the concentration of bioactive substances should be at a specific level. The intended health benefits can be obtained with low amounts of nutrients delivered using nanocarriers (Zhou et al., 2021).

Drawbacks of Nanotechnology on Human Health

There is growing evidence every day about the health risks associated with inhaling nanoparticles:

- Increased activity, bioaccumulation, and overconsumption of nanotechnology-based products have detrimental effects on health and present risks to safety and health.
- Greater concentrations of these compounds that humans breathe in or absorb through skin may pose serious safety risks that require careful examination through risk assessment, especially with regard to long-term toxicity.
- Despite the paucity of information on their toxicity, certain studies have indicated that food containing silver nanoparticles may seep into the packaging and end up in human food.
- Nanoparticles may accumulate in the spleen, stomach, liver, kidneys, small intestine, and among other organs.
- ZnO nanoparticles can harm the liver, kidneys, and lungs when taken orally in a single dose.
- The utilization of titanium oxide and its eventual disposal could potentially impact both the environment and human health, thereby increasing the risk to both (Pérez-Hernández et al., 2021).

CHALLENGES AND LIMITATION

As was previously mentioned, nanotechnology facilitates the development of materials, devices, or systems at the nanoscale and also has unique physiochemical properties in the agriculture sector. The food business is significantly impacted by nanotechnology since it makes it possible to process, package, and store food more effectively and for a longer period of time. This has increased the food sector significantly by enhancing the taste and texture of meals. The bioavailability and targeted distribution of food bioactive compounds have been improved using nanotechnology-based delivery systems because the bulk of food bioactives that combat various diseases.

It is extremely challenging to produce edible and non-toxic compounds or delivery systems that are hazardous for human use by using cost-effective approaches. According to reports, both humans and food have shown signs of bioaccumulation of nanomaterials like AgNPs produced through nanopackaging. The body's far-flung regions can become harmful from NPs because of their large surface area and nanoscale size, which allows them to enter cells with ease.

In addition, there is a significant risk to farmers' health from the use of nanofertilizers and nanopesticides in agriculture because of the potential for environmental, soil, and water pollution. Despite their great potential to frame novel products and production processes in the food industry, nanotechnologies face significant challenges in developing efficient formulations that are safe for human consumption and in using cost-effective processing operations to develop edible and non-toxic nano-delivery systems. This has prompted concerns about how to convert food-grade materials into safe, non-toxic, and biocompatible nanostructures using useful, practical methods like the layer-by-layer approach.

CONCLUSION

The explanation above makes it clear that nanotechnology holds out a lot of potential for the advancement of improved food security and safety. Food security is a global issue affecting the survival and growth of the global population, while food safety is directly tied to human health. The human body is not directly affected by nanotechnology. Consumers may be more receptive to nanotechnology products. The public's view and comprehension of nanotechnology must be improved. Additionally, regulatory rules and pertinent laws and regulations need to be enhanced in the development of commercial nanotechnology. Effective regulations, rules, and regulatory laws are necessary for a safer deployment of NPs in the food industry. For instance, it's crucial to inform the public on how nanoparticles affect environmental safety, human health, and the environment. Before nanofood may be sold on the market, risk evaluations and safety standards must be met. To guarantee the safety and security of the food, it is suggested that further research and development be done to improve the applications of nanotechnology in the food and agricultural sectors.

Table 1. Nanotechnology in food packaging

Nanoparticles	Food Product	Role	References
AgNPs	Chicken meat	Microbiological studies showed no distinction between chicken meatballs placed in silver nanoparticle polyethylene bags and control bags; the chicken meatballs did not contain any silver.	(Trbojevich et al., 2020)
Silver ion and silver NPs	Meats (turkey, tuna and ham)	With regard to Gram-negative bacteria in Turkey, food stimulants containing silver ions and nanoparticles released from food contact paper possess antibacterial qualities.	(Gallocchio et al., 2016)
Chitosan and ZnO NPs loaded gallic-acid films	Active food packaging	Compared to pure chitosan, films have a significant antibacterial activity and robust antioxidant behavior.	(Yadav et al., 20210
Graphite carbon nitride nanosheets/MoS2 nanodots	Cherry tomatoes	The usage of films is safe and they have broad-spectrum antibacterial activity.	(Oliveira et al., 2019)
Ag NPs were immobilized by carboxymethyl cellulose/cellulose nanocrystals.	Strawberries	Strawberries' shelf life was extended by up to 7 days due to antibacterial properties that inhibited E. coli and S. aureus, as well as improvements to tensile strength and air permeability.	(Ni et al., 2021)
Low-density polyethylene with a nanosilver coating	Active food packaging	Significantly reduced silver migration and antibacterial activity during oven heating.	(Pandey et al., 2020)
Nanocomposite film made of polylactic acid, titanium dioxide, and lycopene	Margarine	Margarine has a longer shelf life thanks to films that serve as visible indicators (red to yellow) of oxidation fluctuations during storage; regulated oxidative variables	(He et al., 2021)
AgNPs mango peel extract film	Strawberries	Increased the shelf life of strawberries, showed good antibacterial performance against E. coli and S. aureus, and improved the mechanical properties of the film as well as its barrier to oxygen and water vapor.	(Pirsa et al., 2021)
Chitosan NPs	Active bioplastic packaging	E.coli-growth-resisting antibacterial activity and strong antioxidant potential	(Cheng et al., 2021)
Composit of chitosan, grape seed extract, and silver nanoparticles	Grapes	Antimicrobial and antioxidant properties; decreased weight loss and decay percentage; prevented the growth of all mold during storage; increased the shelf life of grapes.	(Kongkaoroptham et al., 2021)

REFERENCES

Acosta, E. (2009). Bioavailability of nanoparticles in nutrient and nutraceutical delivery. *Current Opinion in Colloid & Interface Science*, *14*(1), 3–15. doi:10.1016/j.cocis.2008.01.002

Adeyeye, S. A. O., & Fayemi, O. E. (2019). Nanotechnology and food processing: Between innovations and consumer safety. *Journal of Culinary Science & Technology*, *17*(5), 435–452. doi:10.1080/15428052.2018.1476276

Ameta, S. K., Rai, A. K., Hiran, D., Ameta, R., & Ameta, S. C. (2020). Use of nanomaterials in food science. *Biogenic nano-particles and their use in agro-ecosystems*, 457-488.

Ansari, A. A., Alhoshan, M., Alsalhi, M. S., & Aldwayyan, A. S. (2010). Prospects of nanotechnology in clinical immunodiagnostics. *Sensors (Basel)*, *10*(7), 6535–6581. doi:10.3390100706535 PMID:22163566

Ansari, M. A. (2023). Nanotechnology in food and plant science: Challenges and future prospects. *Plants*, *12*(13), 2565. doi:10.3390/plants12132565 PMID:37447126

Bajpai, V. K., Kamle, M., Shukla, S., Mahato, D. K., Chandra, P., Hwang, S. K., ... Han, Y. K. (2018). Prospects of using nanotechnology for food preservation, safety, and security. *Yao Wu Shi Pin Fen Xi*, *26*(4), 1201–1214. PMID:30249319

Baranwal, A., Srivastava, A., Kumar, P., Bajpai, V. K., Maurya, P. K., & Chandra, P. (2018). Prospects of nanostructure materials and their composites as antimicrobial agents. *Frontiers in Microbiology*, *9*, 422. doi:10.3389/fmicb.2018.00422 PMID:29593676

Berekaa, M. M. (2015). Nanotechnology in food industry; advances in food processing, packaging and food safety. *International Journal of Current Microbiology and Applied Sciences*, *4*(5), 345–357.

Bikiaris, D. N., & Triantafyllidis, K. S. (2013). HDPE/Cu-nanofiber nanocomposites with enhanced antibacterial and oxygen barrier properties appropriate for food packaging applications. *Materials Letters*, *93*, 1–4. doi:10.1016/j.matlet.2012.10.128

Branton, D., Deamer, D. W., Marziali, A., Bayley, H., Benner, S. A., Butler, T., Di Ventra, M., Garaj, S., Hibbs, A., Huang, X., Jovanovich, S. B., Krstic, P. S., Lindsay, S., Ling, X. S., Mastrangelo, C. H., Meller, A., Oliver, J. S., Pershin, Y. V., Ramsey, J. M., ... Schloss, J. A. (2008). The potential and challenges of nanopore sequencing. *Nature Biotechnology*, *26*(10), 1146–1153. doi:10.1038/nbt.1495 PMID:18846088

Chaudhary, A., Gustafson, D., & Mathys, A. (2018). Multi-indicator sustainability assessment of global food systems. *Nature Communications*, *9*(1), 848. doi:10.103841467-018-03308-7 PMID:29487286

Chaudhary, P., Fatima, F., & Kumar, A. (2020). Relevance of nanomaterials in food packaging and its advanced future prospects. *Journal of Inorganic and Organometallic Polymers and Materials*, *30*(12), 5180–5192. doi:10.100710904-020-01674-8 PMID:32837459

Cheng, J., Lin, X., Wu, X., Liu, Q., Wan, S., & Zhang, Y. (2021). Preparation of a multifunctional silver nanoparticles polylactic acid food packaging film using mango peel extract. *International Journal of Biological Macromolecules*, *188*, 678–688. doi:10.1016/j.ijbiomac.2021.07.161 PMID:34343590

Coles, D., & Frewer, L. J. (2013). Nanotechnology applied to European food production–A review of ethical and regulatory issues. *Trends in Food Science & Technology*, *34*(1), 32–43. doi:10.1016/j.tifs.2013.08.006

Dakal, T. C., Kumar, A., Majumdar, R. S., & Yadav, V. (2016). Mechanistic basis of antimicrobial actions of silver nanoparticles. *Frontiers in Microbiology*, *7*, 1831. doi:10.3389/fmicb.2016.01831 PMID:27899918

de Oliveira, J. P., Bruni, G. P., El Halal, S. L. M., Bertoldi, F. C., Dias, A. R. G., & da Rosa Zavareze, E. (2019). Cellulose nanocrystals from rice and oat husks and their application in aerogels for food packaging. *International Journal of Biological Macromolecules*, *124*, 175–184. doi:10.1016/j.ijbiomac.2018.11.205 PMID:30471399

Dias, M. V., Nilda de Fátima, F. S., Borges, S. V., de Sousa, M. M., Nunes, C. A., de Oliveira, I. R. N., & Medeiros, E. A. A. (2013). Use of allyl isothiocyanate and carbon nanotubes in an antimicrobial film to package shredded, cooked chicken meat. *Food Chemistry*, *141*(3), 3160–3166. doi:10.1016/j.foodchem.2013.05.148 PMID:23871073

Duncan, T. V. (2011). Applications of nanotechnology in food packaging and food safety: Barrier materials, antimicrobials and sensors. *Journal of Colloid and Interface Science*, *363*(1), 1–24. doi:10.1016/j.jcis.2011.07.017 PMID:21824625

Eleftheriadou, M., Pyrgiotakis, G., & Demokritou, P. (2017). Nanotechnology to the rescue: Using nano-enabled approaches in microbiological food safety and quality. *Current Opinion in Biotechnology*, *44*, 87–93. doi:10.1016/j.copbio.2016.11.012 PMID:27992831

Fasihnia, S. H., Peighambardoust, S. H., & Peighambardoust, S. J. (2018). Nanocomposite films containing organoclay nanoparticles as an antimicrobial (active) packaging for potential food application. *Journal of Food Processing and Preservation*, *42*(2), e13488. doi:10.1111/jfpp.13488

Gaikwad, K. K., Singh, S., & Negi, Y. S. (2020). Ethylene scavengers for active packaging of fresh food produce. *Environmental Chemistry Letters*, *18*(2), 269–284. doi:10.100710311-019-00938-1

Gallocchio, F., Cibin, V., Biancotto, G., Roccato, A., Muzzolon, O., Carmen, L., Simone, B., Manodori, L., Fabrizi, A., Patuzzi, I., & Ricci, A. (2016). Testing nano-silver food packaging to evaluate silver migration and food spoilage bacteria on chicken meat. *Food Additives & Contaminants. Part A, Chemistry, Analysis, Control, Exposure & Risk Assessment*, *33*(6), 1063–1071. doi:10.1080/19440049.2016.1179794 PMID:27147130

Gomes Silva, C., Juárez, R., Marino, T., Molinari, R., & García, H. (2011). Influence of excitation wavelength (UV or visible light) on the photocatalytic activity of titania containing gold nanoparticles for the generation of hydrogen or oxygen from water. *Journal of the American Chemical Society*, *133*(3), 595–602. doi:10.1021/ja1086358 PMID:21142160

Gu, B., Zhang, X., Bai, X., Fu, B., & Chen, D. (2019). Four steps to food security for swelling cities. *Nature*, *566*(7742), 31–33. doi:10.1038/d41586-019-00407-3 PMID:30718889

He, C. X., He, Z. G., & Gao, J. Q. (2010). Microemulsions as drug delivery systems to improve the solubility and the bioavailability of poorly water-soluble drugs. *Expert Opinion on Drug Delivery*, *7*(4), 445–460. doi:10.1517/17425241003596337 PMID:20201713

He, Y., Li, H., Fei, X., & Peng, L. (2021). Carboxymethyl cellulose/cellulose nanocrystals immobilized silver nanoparticles as an effective coating to improve barrier and antibacterial properties of paper for food packaging applications. *Carbohydrate Polymers*, *252*, 117156. doi:10.1016/j.carbpol.2020.117156 PMID:33183607

Honarvar, Z., Hadian, Z., & Mashayekh, M. (2016). Nanocomposites in food packaging applications and their risk assessment for health. *Electronic Physician*, *8*(6), 2531–2538. doi:10.19082/2531 PMID:27504168

Hu, B., Liu, X., Zhang, C., & Zeng, X. (2017). Food macromolecule based nanodelivery systems for enhancing the bioavailability of polyphenols. *Yao Wu Shi Pin Fen Xi*, *25*(1), 3–15. PMID:28911541

Hulla, J. E., Sahu, S. C., & Hayes, A. W. (2015). Nanotechnology: History and future. *Human and Experimental Toxicology, 34*(12), 1318–1321. doi:10.1177/0960327115603588 PMID:26614822

Jabeen, N., Maqbool, Q., Bibi, T., Nazar, M., Hussain, S. Z., Hussain, T., Jan, T., Ahmad, I., Maaza, M., & Anwaar, S. (2018). Optimised synthesis of ZnO-nano-fertiliser through green chemistry: Boosted growth dynamics of economically important L. esculentum. *IET Nanobiotechnology / IET, 12*(4), 405–411. doi:10.1049/iet-nbt.2017.0094 PMID:29768221

Jiang, J., Pi, J., & Cai, J. (2018). The advancing of zinc oxide nanoparticles for biomedical applications. *Bioinorganic Chemistry and Applications*. doi:10.1155/2018/1062562 PMID:30073019

Kah, M., Tufenkji, N., & White, J. C. (2019). Nano-enabled strategies to enhance crop nutrition and protection. *Nature Nanotechnology, 14*(6), 532–540. doi:10.103841565-019-0439-5 PMID:31168071

Kalpana, S., Priyadarshini, S. R., Leena, M. M., Moses, J. A., & Anandharamakrishnan, C. (2019). Intelligent packaging: Trends and applications in food systems. *Trends in Food Science & Technology, 93*, 145–157. doi:10.1016/j.tifs.2019.09.008

Kang, S., Pinault, M., Pfefferle, L. D., & Elimelech, M. (2007). Single-walled carbon nanotubes exhibit strong antimicrobial activity. *Langmuir, 23*(17), 8670–8673. doi:10.1021/la701067r PMID:17658863

Khan, S. T., Al-Khedhairy, A. A., & Musarrat, J. (2015). ZnO and TiO 2 nanoparticles as novel antimicrobial agents for oral hygiene: A review. *Journal of Nanoparticle Research, 17*(6), 1–16. doi:10.100711051-015-3074-6

Kim, I., Viswanathan, K., Kasi, G., Thanakkasaranee, S., Sadeghi, K., & Seo, J. (2022). ZnO nanostructures in active antibacterial food packaging: Preparation methods, antimicrobial mechanisms, safety issues, future prospects, and challenges. *Food Reviews International, 38*(4), 537–565. doi:10.1080/87559129.2020.1737709

Kongkaoroptham, P., Piroonpan, T., & Pasanphan, W. (2021). Chitosan nanoparticles based on their derivatives as antioxidant and antibacterial additives for active bioplastic packaging. *Carbohydrate Polymers, 257*, 117610. doi:10.1016/j.carbpol.2020.117610 PMID:33541641

Kumar, A., Pratush, A., & Bera, S. (2021). Significance of nanoscience in food microbiology: current trend and future prospects. *Nanotechnology for Advances in Medical Microbiology*, 249-267.

Laborde, D., Martin, W., Swinnen, J., & Vos, R. (2020). COVID-19 risks to global food security. *Science*, *369*(6503), 500–502. doi:10.1126cience.abc4765 PMID:32732407

Landry, M. P., & Mitter, N. (2019). How nanocarriers delivering cargos in plants can change the GMO landscape. *Nature Nanotechnology*, *14*(6), 512–514. doi:10.103841565-019-0463-5 PMID:31168070

Liné, C., Larue, C., & Flahaut, E. (2017). Carbon nanotubes: Impacts and behaviour in the terrestrial ecosystem-A review. *Carbon*, *123*, 767–785. doi:10.1016/j.carbon.2017.07.089

Luo, Y., Wang, Q., & Zhang, Y. (2020). Biopolymer-based nanotechnology approaches to deliver bioactive compounds for food applications: A perspective on the past, present, and future. *Journal of Agricultural and Food Chemistry*, *68*(46), 12993–13000. doi:10.1021/acs.jafc.0c00277 PMID:32134655

Mahato, K., Prasad, A., Maurya, P., & Chandra, P. (2016). Nanobiosensors: Next generation point-of-care biomedical devices for personalized diagnosis. *Journal of Analytical & Bioanalytical Techniques*, *7*(2).

Majid, I., Nayik, G. A., Dar, S. M., & Nanda, V. (2018). Novel food packaging technologies: Innovations and future prospective. *Journal of the Saudi Society of Agricultural Sciences*, *17*(4), 454–462. doi:10.1016/j.jssas.2016.11.003

Mohammad, Z. H., Ahmad, F., Ibrahim, S. A., & Zaidi, S. (2022). Application of nanotechnology in different aspects of the food industry. *Discover Food*, *2*(1), 12. doi:10.100744187-022-00013-9

Morsy, M. K., Khalaf, H. H., Sharoba, A. M., El-Tanahi, H. H., & Cutter, C. N. (2014). Incorporation of essential oils and nanoparticles in pullulan films to control foodborne pathogens on meat and poultry products. *Journal of Food Science*, *79*(4), M675–M684. doi:10.1111/1750-3841.12400 PMID:24621108

Nasr, N. F. (2015). Applications of nanotechnology in food microbiology. *International Journal of Current Microbiology and Applied Sciences*, *4*(4), 846–853.

Ncube, L. K., Ude, A. U., Ogunmuyiwa, E. N., Zulkifli, R., & Beas, I. N. (2020). Environmental impact of food packaging materials: A review of contemporary development from conventional plastics to polylactic acid based materials. *Materials (Basel)*, *13*(21), 4994. doi:10.3390/ma13214994 PMID:33171895

Ni, Y., Sun, J., & Wang, J. (2021). Enhanced antimicrobial activity of konjac glucomannan nanocomposite films for food packaging. *Carbohydrate Polymers*, *267*, 118215. doi:10.1016/j.carbpol.2021.118215 PMID:34119169

Nile, S. H., Baskar, V., Selvaraj, D., Nile, A., Xiao, J., & Kai, G. (2020). Nanotechnologies in food science: Applications, recent trends, and future perspectives. *Nano-Micro Letters*, *12*(1), 1–34. doi:10.100740820-020-0383-9 PMID:34138283

Pandey, S., Zaidib, M. G. H., & Gururani, S. K. (2013). Recent developments in clay-polymer nano composites. *Scientific Journal of Review*, *2*(11), 296–328.

Pandey, V. K., Upadhyay, S. N., Niranjan, K., & Mishra, P. K. (2020). Antimicrobial biodegradable chitosan-based composite Nano-layers for food packaging. *International Journal of Biological Macromolecules*, *157*, 212–219. doi:10.1016/j.ijbiomac.2020.04.149 PMID:32339572

Pérez-Hernández, H., Pérez-Moreno, A., Sarabia-Castillo, C. R., García-Mayagoitia, S., Medina-Pérez, G., López-Valdez, F., Campos-Montiel, R. G., Jayanta-Kumar, P., & Fernández-Luqueño, F. (2021). Ecological drawbacks of nanomaterials produced on an industrial scale: Collateral Effect on Human and Environmental Health. *Water, Air, and Soil Pollution*, *232*(10), 1–33. doi:10.100711270-021-05370-2 PMID:34658457

Pirsa, S., & Asadi, S. (2021). Innovative smart and biodegradable packaging for margarine based on a nano composite polylactic acid/lycopene film. *Food Additives & Contaminants. Part A, Chemistry, Analysis, Control, Exposure & Risk Assessment*, *38*(5), 856–869. doi:10.1080/19440049.2021.1891299 PMID:33709879

Pradhan, N., Singh, S., Ojha, N., Shrivastava, A., Barla, A., Rai, V., & Bose, S. (2015). Facets of nanotechnology as seen in food processing, packaging, and preservation industry. *BioMed Research International*, *2015*, 2015. doi:10.1155/2015/365672 PMID:26613082

Rafiee, Z., Nejatian, M., Daeihamed, M., & Jafari, S. M. (2019). Application of different nanocarriers for encapsulation of curcumin. *Critical Reviews in Food Science and Nutrition*, *59*(21), 3468–3497. doi:10.1080/10408398.2018.1495174 PMID:30001150

Rashidi, L., & Khosravi-Darani, K. (2011). The applications of nanotechnology in food industry. *Critical Reviews in Food Science and Nutrition*, *51*(8), 723–730. doi:10.1080/10408391003785417 PMID:21838555

Rovera, C., Ghaani, M., & Farris, S. (2020). Nano-inspired oxygen barrier coatings for food packaging applications: An overview. *Trends in Food Science & Technology*, *97*, 210–220. doi:10.1016/j.tifs.2020.01.024

Shafiee-Jood, M., & Cai, X. (2016). Reducing food loss and waste to enhance food security and environmental sustainability. *Environmental Science & Technology*, *50*(16), 8432–8443. doi:10.1021/acs.est.6b01993 PMID:27428555

Sharma, C., Dhiman, R., Rokana, N., & Panwar, H. (2017). Nanotechnology: An untapped resource for food packaging. *Frontiers in Microbiology*, *8*, 1735. doi:10.3389/fmicb.2017.01735 PMID:28955314

Sharma, S., Cheng, S. F., Bhattacharya, B., & Chakkaravarthi, S. (2019). Efficacy of free and encapsulated natural antioxidants in oxidative stability of edible oil: Special emphasis on nanoemulsion-based encapsulation. *Trends in Food Science & Technology*, *91*, 305–318. doi:10.1016/j.tifs.2019.07.030

Souri, J., Almasi, H., Hamishehkar, H., & Amjadi, S. (2021). Sodium caseinate-coated and β-cyclodextrin/vitamin E inclusion complex-loaded nanoliposomes: A novel stabilized nanocarrier. *Lebensmittel-Wissenschaft + Technologie*, *151*, 112174. doi:10.1016/j.lwt.2021.112174

Trbojevich, R. A., Khare, S., Lim, J. H., Watanabe, F., Gokulan, K., Krohmaly, K., & Williams, K. (2020). Assessment of silver release and biocidal capacity from silver nanocomposite food packaging materials. *Food and Chemical Toxicology*, *145*, 111728. doi:10.1016/j.fct.2020.111728 PMID:32898598

Usman, M., Farooq, M., Wakeel, A., Nawaz, A., Cheema, S. A., Rehman, H., Ashraf, I., & Sanaullah, M. (2020). Nanotechnology in agriculture: Current status, challenges and future opportunities. *The Science of the Total Environment*, *721*, 137778. doi:10.1016/j.scitotenv.2020.137778 PMID:32179352

White, J. C., & Gardea-Torresdey, J. (2018). Achieving food security through the very small. *Nature Nanotechnology*, *13*(8), 627–629. doi:10.103841565-018-0223-y PMID:30082813

Yadav, S., Mehrotra, G. K., & Dutta, P. K. (2021). Chitosan based ZnO nanoparticles loaded gallic-acid films for active food packaging. *Food Chemistry*, *334*, 127605. doi:10.1016/j.foodchem.2020.127605 PMID:32738726

Yin, I. X., Zhang, J., Zhao, I. S., Mei, M. L., Li, Q., & Chu, C. H. (2020). The antibacterial mechanism of silver nanoparticles and its application in dentistry. *International Journal of Nanomedicine*, *15*, 2555–2562. doi:10.2147/IJN.S246764 PMID:32368040

Yotova, L., Yaneva, S., & Marinkova, D. (2013). Biomimetic nanosensors for determination of toxic compounds in food and agricultural products. *Journal of Chemical Technology & Metallurgy*, *48*(3).

Zhao, R., Torley, P., & Halley, P. J. (2008). Emerging biodegradable materials: Starch-and protein-based bio-nanocomposites. *Journal of Materials Science*, *43*(9), 3058–3071. doi:10.100710853-007-2434-8

Zhou, Y. H., Mujumdar, A. S., Vidyarthi, S. K., Zielinska, M., Liu, H., Deng, L. Z., & Xiao, H. W. (2021). Nanotechnology for food safety and security: A Comprehensive Review. *Food Reviews International*, 1–21.

Zulfiqar, F., Navarro, M., Ashraf, M., Akram, N. A., & Munné-Bosch, S. (2019). Nanofertilizer use for sustainable agriculture: Advantages and limitations. *Plant Science*, *289*, 110270. doi:10.1016/j.plantsci.2019.110270 PMID:31623775

Chapter 8
Bacterial Pathogens in Food and Their Control by Bacteriophages

Nida Firdous
MNS-University of Agriculture, Multan, Pakistan

Shabbir Ahmad
MNS-University of Agriculture, Multan, Pakistan

Umar Farooq
MNS-University of Agriculture, Multan, Pakistan

Aliza Batool
MNS-University of Agriculture, Multan, Pakistan

Muhammad Usman
MNS-University of Agriculture, Multan, Pakistan

Muhammad Sibt-e-Abbas
MNS-University of Agriculture, Multan, Pakistan

Zafar Iqbal
Punjab Food Authority, Pakistan

Muhammad Asim Ijaz Sidhu
Punjab Food Authority, Pakistan

Tahira Siddique
Islamabad Food Authority, Pakistan

ABSTRACT

Food safety ensures reduced food losses and therefore contributes to food security. The factors causing the contamination of the food may threaten the safe consumption and make the foods harmful. For this reason, it is necessary to utilize various resources to prevent the food from being contaminated in all stages of the food chain. Purified high titer phage lysates have been used for the species-specific control of bacteria during the pre and postharvest phases of food production and storage. This chapter presents recent advances in bacteriophage research and their application in the area of food safety. Some of the components of this chapter are general facts on phage biology that are relevant to their application for control and detection of bacterial

DOI: 10.4018/979-8-3693-0819-6.ch008

pathogens in food and environmental samples, application of bacteriophages to control growth of bacterial pathogens and spoilage organisms in food and food processing environment, detection and identification of bacterial pathogens, and advantages of bacteriophage-based methods and their shortcomings.

1. INTRODUCTION AND HISTORY

Food Borne diseases are a major problem of these days and increasing day by day. These diseases are caused by various types of pathogens and most of these illnesses are caused due to bacteria (Farooq *et al.*, 2019). These foods borne diseases are affecting health and working efficiency of people and increasing burden on health system along with major economic losses of many countries. World Health Organization has reported that only diarrheal disease caused approximately 1.9 million death of children every year in the world. Pathogens can infect food products at any point in the food supply chain, including during cultivation, harvesting, killing, milking, processing, and, ultimately, eating (Garcia *et al.*, 2008).

Even at the expense of food quality, significant efforts have been taken to battle germs in order to avoid food-borne illnesses. However, these methods are unreliable and alter the organoleptic qualities of food commodities. Examples of these approaches include the use of heat, chemicals, or UV light. Additionally, there are some restrictions on the application of these methods, such as the inability to apply certain chemicals directly on fresh produce or food that is ready to eat. Furthermore, prolonged exposure to antimicrobial substances may cause germs to become resistant (Ryan *et al.*, 2011). Therefore, it is imperative to apply novel techniques or approaches to guarantee food safety and quality and avoid food-borne illnesses. Utilizing natural enemies of bacteria is one strategy to avoid bacterial contamination in food, which can reduce the negative effects of the processing and preservation techniques now in use to guarantee food safety and quality (Ryan *et al.*, 2011).

A class of viruses known as "bacteriophages" (derived from the Greek word "phagein," which means "to eat") infect and multiply within bacteria but do not pose a threat to human health (Chanishvili, 2012). Bacteriophages are found everywhere in the earth in different ecosystem like sahara desert, cold and hot springs (Sharma *et al.*, 2017). It is estimated that there are approximately 10^{31} particles with different size, structure and genome organization. So, in contrast to different antimicrobial approaches bacteriophages is an effective approach to combat bacterial pathogens due to their unique properties. Furthermore, the sensory qualities of food are unaffected by Phage treatments, and issues with bacterial resistance to certain chemicals may be readily resolved (Ishaq *et al.*, 2020; Thung *et al.*, 2019; Li *et al.*, 2022). Additionally, this method is more affordable and safe for humans, which makes it the greatest substitute for conventional antibacterial methods.

Bacteriophages, sometimes known as viruses, target and destroy bacteria. Certain phages are host-specific, whereas other phages may infect several species or genera of bacteria. A phage's host range affects how it interacts with its host, influencing behaviors like attachment, infection, and lysis that are crucial for use against disease targets. For example, when managing a particular disease in a live mammal, a narrow-spectrum phage is preferred to avoid damaging commensal beneficial bacteria. On the other hand, in food processing facilities, a phage that can successfully reduce or eradicate the spectrum of strains within a pathogenic species is preferred. It has been shown that, in the correct conditions, a broad-spectrum phage may transform into a restricted state and vice versa. Consequently, care must be taken when determining the phage host range and when employing phages with a broad or limited host range for specific objectives (Jonge *et al.,* 2019).

1.1 What Are Bacteriophages?

Bacteriophages are inert biological entities with a protein and genome based on either DNA or RNA. Morphologically, these phages are usually consisting of following three dimensional structures Fig 1:

1. **Capsid or Head:** It is often icosahedral, contains protein, DNA, or RNA, and is responsible for carrying genetic information.
2. **Tail:** It consist of spiral or contractile sheet surrounding tail tube.
3. **Tail Fiber:** Protein-based tail fibers are affixed below the base plate and aid in the identification of molecules on the host's surface. This base plate is also in charge of detecting the host cell and coordinating the movement of tail fibers. When the sheet around the tail tube contracts, the genetic material inside the bacteria is released into the host cell.
4. **Base plate:** It is a dome like structure made up of different proteins and helps in the binding of bacteriophage with host bacteria (Leiman and Shneider, 2012).

There are more than 600 phages has been identified in last fifty years and They are categorized by to their morphology, host, life cycle, living environment, and genetic makeup. On the basis of genetic material bacteriophages are classified into four categories:

1. DNA with a single strand
2. DNA with a double strand

Figure 1. Structure of bacteriophages
Sources: Hameed et al. (2022); Wagh et al. (2023)

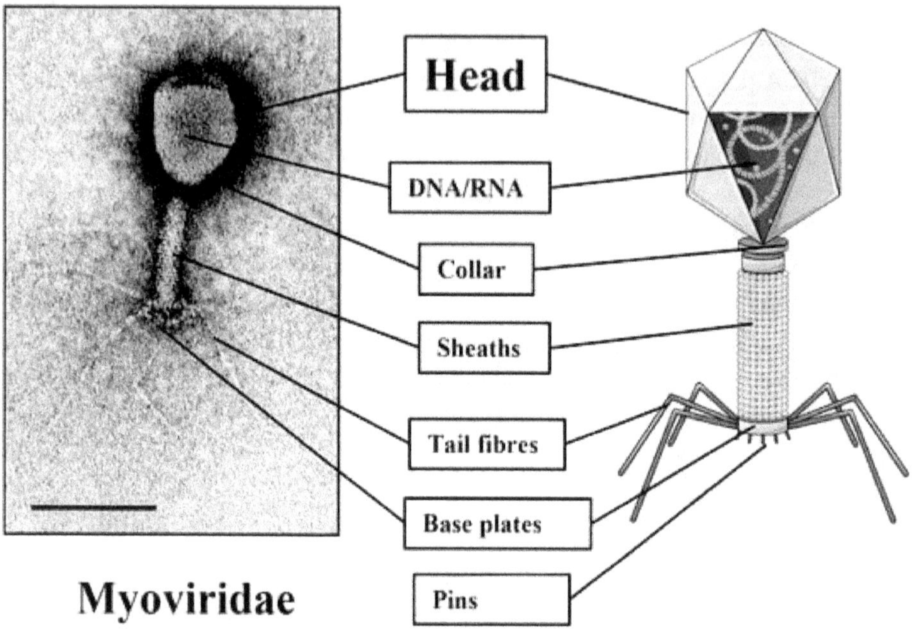

3. RNA with a single strand
4. RNA with a double strand
5. Polyhedral and spiral capsids are distinguished based on their symmetry (Hameed *et al.*, 2022)

1.2 History of Bacteriophages

The first record of bacteriophages dates back to 1896, when British bacteriologist Ernest Hanbury Hankin noticed that filtered water from the Ganga and Jamuna rivers in India had antibacterial action against vibrio cholera. His research was published in French in the Pasteur Institute Annals. Later, in 1915, Fredrick Twort, a fellow British microbiologist, noted that bacteriophages had antibacterial action. Following this, French-Canadian Flix d'Herelle employed them to treat dysentery in 1917 and postulated that viruses may be the origin of translucencies.

But the discovery of antibiotics eliminated the usage of phages, and the use of antibiotics leads to bacterial resistance (Fair and Tor, 2014). The inactivation of E. Coli by phage in rats in 1980 proved the superiority of bacteriophages over antibiotics. Furthermore, compared to antibiotics, bacteriophages are a safer and superior solution since they exclusively infect a certain type of bacteria and do not disrupt the naturally occurring beneficial microflora (Principi *et al.*, 2019; Domingo-Calap and Delgado-Martínez, 2018). Auto dosing is a unique ability of phage which means it is capable to increase number that can affect the host cell (Ryan *et al.*, 2011).

Phage species are the most common in the biosphere and are present in a wide variety of foods and water sources. Phages, which are Greek meaning "bacteria eaters," were initially discovered in 1896 and 1915, respectively, by Ernest Hankin and Edward Twort. The two researchers reported the formation of tiny clean zones on bacterial lawns (plaques) by an isolated filterable entity with antibacterial activity. Nevertheless, in 1917 Felix d'Herelle published the first account of phages being used medicinally to treat diarrhoea. Since then, phage treatment has been used to treat a number of bacterial illnesses, including cholera and typhoid (Garcia *et al.*, 2008). Phage research began to be ignored and marginalized in most of the world after the discovery of penicillin and other antibiotics. However, a few nations in Eastern Europe and the former Soviet Union kept working on phage research. Phage therapy is a common practice in pediatrics, burn care, and surgical hospital settings in Georgia (Pires *et al.*, 2020).

Each phage usually infects a specific type of bacterium. Phages are viruses that infect bacteria and have a restricted host range. People, dogs, and plants may all safely interact with these highly specialized beings (Cooper 2016). Phages regularly come into contact with humans without causing them any harm. Phages are essential to the natural cycle and support the equilibrium of bacteria in the environment (Zbikowska *et al.*, 2020). Following attachment to a specific bacterium, phages either exploit the bacterial protein machinery to multiply inside the host, producing bacterial lysis and the release of newly produced viruses, or they integrate their genetic information into the bacterial chromosome without causing cell death (a process known as the "lysogenic pathway"). The safest approach is to utilize phages with a strict lytic pathway, which causes bacterial death without genome integration into the host DNA, as opposed to phages with a lysogenic pathway, which have the potential to build or spread antibiotic resistance (Sillankorva *et al.*, 2012). Phages can be utilized in a variety of applications in various fields, including aquaculture and agriculture, functioning as antimicrobials or for the detection of bacteria. They can be used in medicine to cure illnesses, infections, and to modify the micro biome through phage therapy (Strathdee *et al.*, 2023). Phage usage as a therapeutic agent has previously been studied in a number of food-related disciplines. It is used to reduce pathogen colonization in living animals during primary production before slaughter, as well as for hygienic purposes to clean instruments and surfaces that come into contact with food (Lavilla *et al.*, 2023). Phages can be used to prevent contamination and the growth of bacterial pathogens through the storage of the finished food items, as well as to regulate the bacterial colonization of food during industrial processing. In the food industry, they provide the following advantages as natural antimicrobials. Self-replication (low doses that proliferate as long as the host is there).

- High specificity of the host
- Effectiveness in doses is decreases
- Safety
- Isolation and propagation (common in nature)
- Prolonging the storage of food and preventing bacterial infection;
- When utilizing purified phages, food organoleptic qualities like structure, taste, color, and aroma are maintained (Vikram *et al.,* 2020).

Phages are found in all forms of nature, have been found in dairy products like cheese and yoghurt, and are frequently found in the stomach of humans. This implies that phages are something that people directly or indirectly come into contact with every day. Furthermore, it is unknown whether giving phages to animals orally might make them sick. Additionally, due to their widespread distribution, phages can offer a valuable and extensive repertory of control agents that can be used to control pathogens in environments with a range of selection pressures, including temperature, salt, pH, and UV radiation (Gangwar *et al.,* 2021). Common environmental viruses called bacteriophages have the ability to target certain kinds of bacteria. Bacteriophages, sometimes referred to as phages, are viruses that target and multiply within bacteria. How they interact with their bacterial hosts requires an awareness of a number of biological features of their function.

2. BIOLOGICAL FACETS OF BACTERIOPHAGE ACTION

The main biological facets of bacteriophage action are as follows:

1. **Attachment and Adsorption:** Receptor Recognition: Bacteriophages possess specific proteins or molecules on their surface that enable them to identify and adhere to receptors on the surface of bacterial cells. These receptors are frequently found in cell walls as proteins or lipopolysaccharides.
2. **Penetration and Injection:** Injection of Genetic Material: After attaching itself, the phage injects its genetic material (DNA or RNA) into the bacterial cell. The bacterial cell wall is usually breached by the phage's tail structure, which permits the genetic material to enter the host cell.
3. **Replication and Transcription:** Expression of Phage Genes: The mechanism of the injected phage cell is taken over by it. Early genes that code for proteins that aid in phage genetic material replication are the first to be transcribed:

   a. **Replication:** The phage genome is duplicated by the phage DNA during this process.
   b. **Transcription:** Phage proteins are made from mRNA, which is translated from the transcribed phage genes.
4. **Assembly of New Phage Particles:** Protein Synthesis: Enzymes and structural proteins needed for the synthesis of fresh phage particles are produced by the transcription and translation of phage genes:
   a. **Assembly:** Complete phage particles are formed from newly replicated viral genetic material and proteins.
5. **Cell Lysis and Release:** Cell Lysis: The phage's enzymes damage the bacterial cell wall, resulting in it lysing (bursting open). The freshly formed phage particles are released during this lysis:
   a. **Release:** Particles of the phage are released into the environment, where they can continue the cycle by infecting fresh bacterial cells.
6. **Lysogenic (Optional):** Integration of Phage DNA into Bacterial Genome: Rather than replicating and lysing quickly, some phages can incorporate their genetic material into the bacterial chromosome to form prophagia:
   a. **Lysogenic Cycle:** The phage DNA that has been integrated replicates with the bacterial DNA as the host cell grows. Under certain conditions, the prophage can exit the bacterial chromosome and enter the lytic cycle.
7. **Evolutionary Implications:** Horizontal Gene Transfer: The exchange of genetic information between bacteria can be facilitated by phages. Within bacterial communities, this horizontal gene transfer may promote genetic diversity and evolution:
   a. **Co-evolution:** Bacteriophages and bacteria frequently engage in an evolutionary arms race where phages evolve to overcome these defenses while bacteria develop new defenses against phage invasion (Nelson *et al.,* 2008).

Their natural and secure bactericidal procedure has piqued interest in minimizing antimicrobial resistance without affecting food safety, sensory quality, or probiotic culture. Since the FDA authorized the use of listeria phages in ready-to-eat beef and poultry products in 2006, these uses have drawn attention. Food safety has improved as a result. The EFSA noted in 2009 that bacteriophages are especially useful when applied to meat and dairy products. However, the agency concluded that any phage or phage cocktail that is meant to be applied to food has to be assessed individually. Later, other commercial goods were developed specifically for the food sector, such as SalmoFresh, Listex P100, Listex Shield, Armament, and Salmonelex (Islam *et al.,* 2021).

3. BACTERIOPHAGES MODE OF ACTION IN FOOD

A bacteriophage depends on its host for its growth and replication. A stage when phage is not in contact with host are called as search stage or extracellular stage. So, it necessary for bacteriophage to survive during this stage and this stage depends on the number of host cell and phage. Adsorption of the bacteriophage with the host cell happens next, which happens when the phage particle collides with the vulnerable host cell. Phage transfers its genome into the host bacterial cell at this point. The metabolic state of the host cell determines the rate of bacteriophage infection, replication, and offspring release (Authority, 2009).

3.1 Mechanism of Phage-Cell Surface Interaction

Availability of fluid for the mobility of bacteriophages are also important factor for proper adsorption of phage with host cell. For the process of adsorption, a hexagonal dome shaped baseplate of bacteriophage helps in the recognition of host bacterium cell. The process start from baseplate and move in a wavelike movement in the whole length of contractile sheath (Harada *et al.*, 2018). In addition to this, Tail fibers that are usually six in numbers originating from dome shaped baseplate determine the specific part of host cell for bindings. The sheath contracts from 925 to 420 Å following the successful attachment of the phage to the host cell, causing the baseplate structure to shift from hexagonal to planar and from dome-shaped to star-shaped (Hameed *et al.,* 2022). The capsid and tube move closer to the host bacterial cell as a result of the sheath contracting, piercing the membrane with the central baseplate spike protein. In this way genetic material of bacteriophage are injected from tube into bacteria cytoplasm. However, ejection of genetic material and contraction of tail sheath are not linked to each other (Leiman *et al.*, 2004). There should be a receptor on the host cell membrane that unlock the tube to release phage genetic material into host cytoplasm (Harada *et al.*, 2018).

3.2 Life Cycle of Bacteriophages

First of all, Phage after identification and attachment with host it releases its genomic material into host bacterial cell. After this phage exhibit a lytic or lysogenic life cycle in host bacterial cell Fig 1. In lytic cycle the genetic material after entering into host cell remains separate and free-floating particles and causes rapid lysis and death of host cell. Actually, this cycle occurs in multi steps 1) Attachment of phage with host 2) Penetration during which phage insert its genome into host bacterial cell 3) Transcription of genetic material 4) Biosynthesis of phage new virion 5) Maturation of phage 6) Lysis where new virions release into environment (Hameed *et al.*, 2022).

Phages that can follow only lytic cycle are called as virulent Bacteriophages. The basic principle in the lytic phages is that that they enter in host cell and causes its lysis and prevent spoilage of food from microorganism to ensure food safety (Farooq *et al.*, 2019). Lysis of host cell can take place in two possible mechanisms explained below in Fig 2.

1. **Internal lysis:** In this process, phage replication causes bacterial cell lysis. Once the phage's genetic material enters the host cell, it replicates several times and modifies the host cell's machinery to create protein needed for the creation of new phages. It uses up all of the host cell's resources to create new phages, which makes the cell weaker until it bursts and releases new phages into the environment. (Authority, 2009).
2. **Lysis from without:** This method allows for lysis to occur without phage replication. In this case large no of phage particles attached with host cell and lysis occurs as result of alteration in electric potential of membrane or due to cell wall degrading enzyme activity (Authority, 2009). Conversely, lysogenic phages, also known as temperate phages, are distinguished by the fact that they either live as plasmids or integrate their DNA into the genome of their host cell. This is how phage-incorporated DNA is reproduced throughout the bacterial genome and transmitted down to the following generation. Until these phage genes return to the lytic cycle and produce fully constructed phages, this integrated viral DNA will be passed down through several generations of phages. Temperate phages are often not thought to be appropriate for phage treatment (Authority, 2009).

Figure 2. Life cycle of bacteriophages
Source: Gencay et al. (2019)

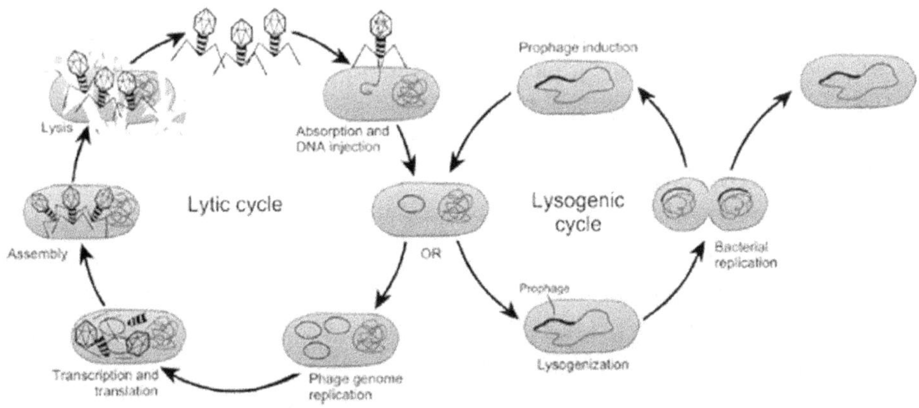

4. FOODBORNE PATHOGENS/PATHOGEN IN FOOD

In the fifth century BC, food was linked to disease for the first time, leading to the observation of the first food infection (Bintsis, 2017). In food contamination and food-borne illnesses, the most prevalent pathogens are viruses, parasites, and bacteria. Of them, *Salmonella, E. Coli, Listeria*, and *Campylobacter jejuni* are the most prevalent and well-known for causing food-borne illnesses (Anvar *et al.*, 2021; Khare *et al.*, 2018; Schirone *et al.*, 2019).

Since most bacteria are mesophilic and thrive best at temperatures between 20 and 25, they may readily live in the human body. *Y. enterocolitica* is one pathogen that can endure colder temperatures, but (Ryan *et al.,* 2011). Food-borne infections can create biofilms, which makes it difficult to treat them since the biofilm inhibits the effectiveness and penetration of disinfectants and antibacterial treatments (Mizan *et al.*, 2020). Food is a favorable material to form biofilm and these pathogens can form biofilm directly on food surface or on equipment used for the processing of food (Bintsis, 2017; Bai *et al.*, 2021).

So there is a need to study the characteristics of food borne pathogens and factors which affect their growth or reproduction.

4.1 Salmonella

Gram-negative Salmonella is a member of the Enterobacteriaceae family of bacteria. Morphologically it is rod shaped bacteria. Salmonella is the of cause of Salmonellosis which is a famous foodborne disease and characterized by abdominal pain, vomiting, fever and headache. This pathogen is transferred by soil, water, feces and feed source. Poultry and other animal meat is most commonly infected sources and transferred to human being. Numerous studies has confirmed that salmonella can be controlled through bacteriophages (Vikram *et al.*, 2021).

Recently a research conducted on salmonella contaminated milk and lettuce showed that bacteriophages reduce the number of pathogenic bacteria (Yan *et al.*, 2020). Research is also being conducted on the detection of salmonella in turkey meat and infant formula (Nguyen *et al.*, 2020).

4.2 Escherichia Coli

All animals naturally have Escherichia coli in their intestines, which can lead to extraintestinal and intestinal diseases in newborns, such as sepsis or meningitis (Alonso *et al.,* 2016). E. Coli has two classes of pathogenic strains. Diarrheal E. Coli is one that can cause diarrhoea or the flu. Enteroinvasive E. Coli (EIEC), Shiga toxin-producing E. Coli (STEC), and enterotoxigenic E. Coli (ETEC) are examples of this kind of bacteria, which is also known as enteric pathogenic. Extraintestinal pathogenic E. coli is the name of the other group, which is responsible for infections outside of the gut (Riley, 2020). This group can spread illness because to their tiny numbers and strong contagiousness, which can be spread by food or water (Bintsis, 2017).

Additionally, helpful for guarding food against E. coli are bacteriophages. A recent study by Vengarai Jagannathan et al. (2021) looked at how to lessen the amount of harmful E. coli on spinach leaves. Using cocktail phage, they discovered that the pathogen on spinach leaves was decreased by 99 percent (Vengarai Jagannathan *et al.,* 2021). Additionally, researchers have discovered and proposed that phages are efficient against E. coli biofilm (Wang *et al.,* 2020).

4.3 Listeria

It is a food-borne, anaerobic, gram-positive bacterium. It has a rod-like morphology as well (Kaptchouang Tchatchouang *et al.,* 2020). It is a zoonotic infection that spreads quickly from humans to vertebrates. It is the cause of listeriosis, a condition that is characterised by meningitis, encephalitis, sepsis, and gastrointestinal disorders. It may also result in miscarriage in expectant mothers (Swaminathan and Gerner-Smidt, 2007).

Recent researches have confirmed that bacteriophages are very effective against Listeria strains that are found in meat and meat products. The best thing is that these phages are not harmful for beneficial microbes of consumers (Matle *et al.*, 2020). Listeria can be effectively controlled by commercially available phages such as LM-103a, LMP-102a, Ply511, and Listex P-100 (Strydom, 2015; Dhama *et al.*, 2015). Bacteriophages are not only effective for food of animal origin research has demonstrated that these are also effective for preventing Listeria contamination on melons and apples (Wojcicki *et al.*, 2019).

4.4 Campylobacter

A significant foodborne pathogen, *Campylobacter jejuni* is a member of the Campylobacteraceae family. It's usually found in the intestines of mammals and birds. It can cause infection in gastrointestinal which is called campylobacteriosis (Bintsis, 2017). Diarrhea is a common symptoms of its infection and its infections occur during drinking or consumption of contaminated food (Harrer *et al.*, 2019).

Bacteriophages are also very effective to control the contamination and food borne disease due to *campylobacter jejuni*. A research conducted using bacteriophages as a tool to prevent its contamination on meat. In this research meat was contaminated with bacteria and then bacteriophages were sprayed on it. The researchers found that bacterial number was decreased and they concluded that bacteriophages can be used against Campylobacter jejuni contaminated meat products (Thung *et al.*, 2020).

4.5 Yersinia

A Gram-negative bacterium called Yersinia is frequently spread by uncooked food and water. It causes fourth common food bone gastrointestinal diseases in human which is known as yersinosis. It has great threat to human and its major symptoms are fever, lymphadenitis, nausea, and abdominal pain (Leon-Velarde *et al.*, 2019). This pathogen has an ability to form biofilm and it is very difficult to clear its contamination (Pyra *et al.*, 2020).

Moreover, Combating these pathogenic bacteria is a big challenge as it has developed resistance against broad lines of antibiotics (Gwak *et al.*, 2018). Knowledge about phages that can lysis of yersinia is less as compare to other pathogenic bacteria. These findings suggested the use of these phages to control the contamination of *Y. enterocolitica* in raw food and kitchen utensils (Jun *et al.*, 2018).

5. CHARACTERIZATION AND ISOLATION OF BACTERIOPHAGES FOR POSSIBLE USE IN FOOD SYSTEMS

Microbiological food contamination is a major problem in the food industry. Contaminated food may contain a variety of microorganisms, such as bacteria that may use food as an energy source and do not cause food deterioration or foodborne illnesses (Nerin *et al.,* 2016). Any changes in a product's sensory properties that make it unfit for ingestion is considered food spoiling. A wide range of metabolic byproducts that result in off-odor, off-flavor, as well as color and texture alterations may cause food loss, having a significant negative impact on the economy and the environment (Rawat 2015).

Bacillus sp. Families that produce spores, such as Bacillus subtilize and Bacillus cereus, are able to withstand high processing temperatures. often found in a variety of spoiled foods, such as milk odour, rice that produces slime, and ropiness in bread (Yang *et al.,* 2017). Even while it could be safe to eat decaying food, certain bacteria can be dangerous and lead to foodborne diseases. Handling, storing, processing, and preparation procedures are frequently linked to bacterial food contamination risk factors. To assist increase food safety, conventional food preservation techniques including pasteurization, high-pressure processing, irradiation, and chemical or biological agents are frequently employed. On the other hand, those therapies often come with toxic-threatening side effects, loss of nutrients, and altered organoleptic features (Dwivedi *et al.,* 2017).

5.1 Isolation of Bacteriophages From Soil and Measurement of Titers

The *Bacillus cereus* phage S1 (BC-S1), and the *Bacillus subtilis* phage S2 (BS-S2) can be isolated from several soil samples collected near an organic waste disposal site. The agar overlay test demonstrated that phages had lysed bacteria by producing visible plaques. The concentration of the isolated bacteriophages can be established by tier determination. Phage BC-S1 had the highest titer as compared to phage BS-S2, phage ETEC-S3 and phage EHEC-S4 (Artawinata *et al.,* 2023).

5.2 Using Bacteriophages on Food Samples

Bacteriophages have ability to reduce the amount of pathogenic bacteria on food samples. *Bacillus cereus* phages can be applied to different food matrices and surfaces like cooked rice, pasteurized milk, meat and lettuce. After overnight incubation at 28 °C it was found that pathogenic bacteria were reduced to safe level and bacteriophages were also dropped (Elsayed *et al.,* 2023). Fig 3 explained the use of bacteriophages to manage Salmonella infection.

5.3 Bacteriophages for Biocontrol of Pathogens in Food

Microbe pesticide tests have a long history of safe use, very easy processing, and strong and targeted antibacterial activity, which allows them to significantly enhance bacterial quality. Table 1 shows pathogen and the reported results of phage actions on food items.

A cocktail of salmonella specific bacteriophages can be used to reduce the number of salmonella strains but a strain of salmonella *Salmonella Paratyphi* B S661 is resistant to these bacteriophages action. There are many alternate options like monoclonal antibodies targeting this mutant variety with novel formulation that can result in significant reduction (Woolston *et al.,* 2013). Combinations of bacteriophages can be used to target specific bacteria, e.g., common bacteria found in specific food processing industries, or bacteria with troublesome mutations that make them resistant to phages. Salmonella may be eliminated from food and surfaces used in food preparation by combining bacteriophages. For example, the salmonella-specific cocktail, when given on its own can reduce the amount of Salmonella in experimentally infected chicken parts. This impact can be enhanced when the bacteriophage is mixed with regular chemical sanitizers. Salmonella populations can be significantly decreased when the bacteriophage cocktail is sprayed on chicken breast steaks or when the fillets are submerged in a vessel holding the virus solution (Sukumaran *et al.,* 2015).

Table 1. Reported results

Pathogens	Reported Results	References
Listeria monocytogenes	Apples and chopped melon are out. blended with nisin. Beef does not exhibit combination phage-nisin activity at 4°C. removal from soft cheese that has a crimson stain that has matured on the surface	(Garcia *et al.*, 2008)
Salmonella spp.	After applying a four-phage cocktail, the number of salmonellas on melon slices decreased by about 3.5 logs at 5 and 10°C and by about 2.5 logs at 20°C; however, there was no bacterial reduction on apple slices.	(Whichard *et al.*, 2003)
Escherichia coli	Reduction in potable water and rectal feeding (cattle)	(Raya *et al.*, 2006)
Shigella sonnei	Several RTE meals, including smoked salmon, lettuce, melon, corned beef, and pre-cooked chicken, were treated with a five-phage cocktail designed specifically to combat Shigella. This cocktail was shown to reduce the recovery of Shigella logs by 1.0–1.4 when compared to the control group.	Soffer *et al.*, (2017)
Campylobacter	Decreased counts of cecal concentration in broilers	(Wagenaar *et al.*, 2005)

Figure 3. Using bacteriophages to manage salmonella infection

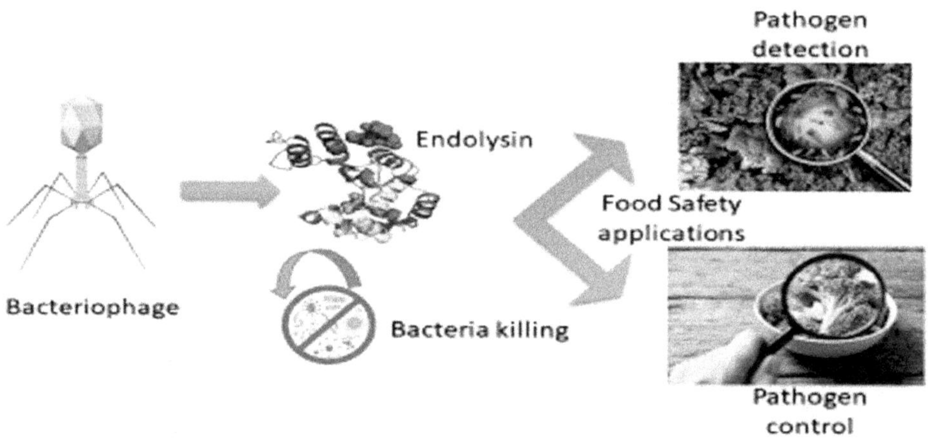

6. BACTERIOPHAGES IN THE FOOD PRODUCTION PROCESS AFTER HARVEST

Concentration of preservatives and nutrients density in food may impact the survival and growth of a variety of bacterial illnesses. Certain phages can help stop the spread of some harmful illnesses. A number of intervention approaches utilizing phages to control infections in postharvest food items have been reported. Table 2 has illustrated the details of postharvest application.

Table 2. Postharvest application

Product	Bacterial Species	Phages	Results	References
Quebec	Salmonella S3, S200, S203, and javianna S200 (105 CFU/mL) Newport S195 Salmonella (105 CFU/mL)	Felix01, HER20, SE13 (2500), F3, and F6	The pathogen very slightly decreased, from 2-3 CFU/cm2, and there was no discernible decrease	(Wong *et al.*, 2020)
Chop Spinach	The CFU/cm2 of *Listeria monocytogenes* is 4.5.	ListShield (2)	A log reduction of 1.51–2.51 CFU/cm2 was seen after 14 days	(Boyacioglu *et al.*, 2016)
Lettuce	*Salmonella typhimurium* (10^7 CFU/mL)	LPST10 (1, 10, 100)	Log reductions of 1.7 and 2.7 CFU/cm2 were seen after five hours	(Huang *et al.*, 2018)
Romaine lettuce	*Salmonella* (10^5 CFU/mL)	Salmo Fresh (1000)	2.43 CFU/g of log decrease was seen. 2.16 CFU/g of log decrease was seen	(Zhang *et al.*, 2019)
Alfalfa seed and sprout	*Escherichia coli* (10^4 CFU/mL)	EcoM-HG2, HG-HG7, and HG-8 (2.5 × 106)	A log reduction of 1.5 CFU/g was seen after 5 days	(Lone *et al.*, 2016)

Phages, or bacteriophages, are viruses that exclusively reproduce and infect bacterial cells. Known to be the most prevalent biological agent on the planet, they exhibit remarkable diversity in terms of size, appearance, and genetic organization. The genetic material of all bacteriophages is protected and delivered to the subsequent host cell by a capsid protein shell made of phage-encoded nucleic acid. The intricate life cycle of bacteriophages includes attachment, penetration, multiplication, and release. Their shape and genetic organization are categorized internationally. Since its initial discovery in 1915, bacteriophages have been the focus of study for more than a century. They have several important characteristics and applications, including:

- **Genetic exchange:** Bacteriophages can transfer genetic material between bacterial cells, leading to genetic diversity and evolution.
- **Population control:** Bacteriophages can change microbial populations by feeding on certain types of bacteria and sparing others.
- **Antimicrobial resistance:** Bacteriophages have been designed to insert genes into certain strains for therapeutic purpose, and they are capable of carrying antimicrobial resistance genes.

Finding bacteriophages in a complicated environmental sample, such water sources contaminated by faeces, can serve as a diagnostic for the existence of the bacterium's host.

- **Research instrument:** Bacteriophages have been important in laboratory research, especially in the investigation of viral replication.

7. IMPORTANCE OF STUDYING BACTERIOPHAGES IN FOOD SCIENCE

Studying bacteriophages is important in food science for several reasons, including:

- **Food safety:** Bacteriophages can be used as biocontrol agents to combat pathogenic bacteria that spoil food, extending the shelf life of products.
- **Specificity:** Bacteriophages have high specificity to target their host determined by bacterial cell wall receptors, which means they can be directed toward pathogens while leaving other bacteria associated with a food untouched.
- **Biomarker detection:** The detection of bacteriophages can be used as a biomarker for the presence of their host in food, which can help identify potential sources of contamination.
- **Alternative to antibiotics:** Bacteriophages can be used as an alternative to antibiotics in food production, which can help reduce the risk of antibiotic resistance.

Overall, bacteriophages have the potential to improve food safety and quality, reduce food waste, and provide an alternative to antibiotics in food production. However, there are also challenges to using bacteriophages in food science, such as the risk of bacterial resistance to phages and the need for further research to optimize their use.

8. PHAGE TECHNOLOGY AND APPLICATIONS

8.1 Understanding Bacteriophages

Applications for bacteriophages in phage technology and contemporary medicine are numerous. They are a possible substitute for antibiotics in the treatment of bacterial infections because of their great specificity to the host bacterium. They can also be applied in biotechnology, agriculture, and food production. Here are some fundamental knowledge points regarding bacteriophages in phage technology and usage:

- Phage intervention Utilizing bacteriophages: **P**hage therapy is a means of treating bacterial illnesses, especially antibiotic-resistant strains. Because phages are very particular to the bacteria they inhabit, they may be targeted at diseases and ignore other bacteria.

- **Biocontrol agents:** To battle pathogenic bacteria that ruin food and prolong product shelf life, bacteriophages can be employed as biocontrol agents. Vaccines and detection: Bacteriophages can be used as vehicles for vaccines, both DNA and protein and for the detection of pathogenic bacterial strains.
- **Display systems:** Bacteriophages can be used as display systems for many proteins and antibodies.
- **Molecular biology and biotechnology:** Bacteriophages have contributed significantly to the field of molecular biology and biotechnology.

8.2 Historical Context and Discovery

The process of discovering bacteriophages and understanding their historical background spanned many decades. The first observations of certain bacteriolytic occurrences were made in the second decade of the twentieth century, and this fact was tied to bacteriophages. The scientific method of gradual evolution led to the discovery of phage rather than a sudden eureka moment. Bacteriophages were separately discovered in 1915 by Frederick William Twort and in 1917 by Felix d'Herelle. D'Herelle's viral idea of the phage was not confirmed until the late 1930s when electron microscopy was used to see bacteriophages.

8.3 Phage Life Cycle and Replication

Phages exhibit two main life cycles, lytic and lysogenic.

The lytic cycle involves viral replication within a host cell. It includes six stages:

- **Attachment**: Phage attaches to the host cell wall via specific receptors.
- **Penetration**: Phage injects genetic material into the host cell.
- **Replication**: Phage genome takes over host cell machinery for replication.
- **Maturation**: Phage parts assemble around replicated genomes.
- **Release**: Host cell lyses, releasing new phage particles.

Table 3. Phage classification and discovery

Sr. No.	Arrangement	Families
1.	Caudovirales	Podoviridae, Myoviridae, Siphoviridae, and Ackermannviridae
2.	Ligamenvirales	Rudiviridae and Lipothrixviridae
3.	Unassigned	The families that make up Ampullaviridae, Bicaudaviridae, Clavaviridae, Corticoviridae, Cystoviridae, Fuselloviridae, Globuloviridae, Guttaviridae, Inoviridae, Leviviridae, Microviridae, Plasmaviridae, Pleolipoviridae, Portogloboviridae, Sphaerolipoviridae, Spiraviridae, Tectiviridae, Tristromaviridae, and Turriviridae

- **Reinfection**: New phage particles attach to other host cells, restarting the cycle.

8.4 Lysogenic Cycle

In this cycle, the phage genome integrates into the host cell's genome, replicating along with it. Stages include attachment, penetration, and integration. The lysogenic cycle can transition to the lytic cycle through phage DNA excision.

Phages and Food Safety: Combatting Bacteria from Farm to Fork

8.5 Phage Therapy in Agriculture

Phages are proposed as alternatives to antibiotics in agriculture, offering specificity, safety, and effectiveness against antibiotic-resistant bacteria. They control bacterial infections in crops and livestock, reducing disease severity. Challenges include sensitivity to environmental factors.

8.6 Controlling Bacterial Pathogens on Farms

Control involves biological agents, hygienic practices, soil and manure management.

8.7 Reducing Contamination During Food Production

Practices like good hygiene, separation of raw meats, and cleaning surfaces help minimize contamination risks.

8.8 Bacteriophages and Their Role in Food Safety

8.8.1 Foodborne Illnesses

Foodborne illnesses result from contaminated foods, causing various symptoms. Prevention involves hygiene and environmental control.

8.8.2 Phages as Natural Antimicrobials

Phages are specific, safe, and effective against bacteria. They combat antibiotic-resistant strains, playing a crucial role in food safety.

8.8.3 Mechanisms of Action Against Foodborne Pathogens

Phages employ lytic infection, growth hindrance, and pathogen reduction, offering specificity against bacterial species and strains.

8.9 Phage Technology in Food Processing

8.9.1 Applications

Phages find use in primary production, biosanitization, biopreservation, decontamination, and detection.

8.9.2 Challenges and Limitations in Food Processing

Challenges include host specificity, environmental factors, regulatory issues, endotoxins, and technical hurdles.

8.10 Future Prospects and Innovations

Ongoing research explores new delivery systems, phage therapy, endolysins, and stability improvement.

8.10.1 Evolutionary Dynamics of Phage-Host Interactions

Phage-host interactions involve complex evolutionary processes, with defense-counterdefense strategies. Bioinformatic tools provide insights into structure-function relationships.

8.10.2 Coevolution

Bacteria and phages The process of reciprocal adaptation and counter-adaptation between the phage and its bacterial host is known as coevolution. Phage resistance and coevolution have played significant roles in the development of bacteriophages into antimicrobials. Bacterial antibiotic resistance can be influenced by phages, and phage evolution can influence bacterial antibiotic resistance.

8.10.3 Implications for Long-Term Antimicrobial Strategies

Here are some implications for long-term antimicrobial strategies in the evolution of bacteriophages as antimicrobials:

- **Phage biobanks:** The concept of phage biobanks is increasing, which can help to address the time-consuming process of developing a novel antibiotic.
- **Combination therapy:** Phage-antibiotic combination therapy has been researched as a potential treatment approach for bacterial infections. On the other hand, other findings suggest that the bacteriophage and antibiotic may potentially interact antagonistically when certain phages and low-dose antibiotic treatments are combined.
- **Phage-sugar eyewear:** Antimicrobial films or coatings containing pullulan or trehalose phage have the potential to stabilize bacteriophages for usage in food over an extended period of time.
- **Phage treatment**: is a popular antibacterial tactic that involves using a particular kind of virus to either suppress or eradicate dangerous bacteria. It offers a chance to create efficient phage therapeutics to combat bacteria resistant to antibiotics.
- **Phage-bacteria coevolution:** In microbiological communities, phage-bacteria coevolution may be a driving force behind ecological and evolutionary processes.

9. APPLICATIONS OF BACTERIOPHAGES AS INDICATORS OF ENVIRONMENTAL SAFETY AND INSTRUMENTS FOR BACTERIAL CONTROL

Bacteriophages, sometimes known as phages, are viruses that reproduce and infect only bacteria. They have been used in environmental research, biotechnology, and medicine for a variety of purposes. In order to combat antibiotic resistance and track environmental health, this article examines the many use of bacteriophages as bacterial control tools and environmental safety indicators.

9.1 Use as a Bacterial Control Tool

9.1.1 Phage Therapy

In phage therapy, bacterial infections in humans, animals, and even plants are treated with bacteriophages. A vision was developed when phages were first found in the early 20th century. Antibiotics may tackle a broad variety of bacteria, but bacteriophages are highly specialised and only infect particular strains or species. This specificity reduces the risk of antibiotic resistance and damage to beneficial microbes, which is a benefit.

Phage treatment has gained more attention recently as concern over antibiotic-resistant bacteria has grown on a worldwide basis. Because superbugs are becoming more and more resistant to various treatments, bacteriophages offer a promising new therapeutic approach.

9.1.2 Food Safety

To reduce bacterial contamination in the food business, bacteriophages are also used. They can be used to combat harmful bacteria that can lead to foodborne illnesses, like Salmonella and E. coli. The growth and spread of dangerous bacteria can be prevented by using phage preparations on food surfaces or during food processing, improving food safety without the use of chemical additives.

9.1.3 Agriculture

Agriculture has looked at using bacteriophages to control bacterial illnesses in cattle and crops. For example, they can be applied to aquaculture to fight bacterial infections, hence lowering the demand for antibiotics in fish farming. Additionally, phages can be used in crop protection techniques to manage bacterially-caused plant diseases, enhancing agricultural output and food security.

9.2 Biomedical Implants and Devices

Bacteriophages are being used to stop bacterial infections caused by biomedical implants and devices including catheters and prosthetic joints. Due to the development of biofilms, which are bacterial colonies enclosed in a protective matrix, infections associated with these devices can be difficult to cure with antibiotics. Phages can more efficiently penetrate biofilms than antibiotics and can be modified to target the precise bacterial strains that cause these diseases.

9.2.1 Veterinary Medicine

Bacteriophages are used in veterinary medicine to treat bacterial illnesses in farm animals and pet animals. This is particularly important in the agricultural industry, as the use of antibiotics in livestock husbandry helps to breed germs that are resistant to them. With phages as an alternative to antibiotics, the selective pressure for antibiotic resistance is lessened while helping to maintain animal health.

9.2.2 Bio Control in Aquaculture

Bacterial infections are frequently encountered in aquaculture, the production of aquatic organisms, which can result in large financial losses. A long-term approach to eradicating these illnesses from aquatic ecosystems is provided by bacteriophages. Phages can enhance the health and production of fish and shellfish farms while reducing the environmental impact of antibiotic use by selectively targeting bacterial pathogens in aquaculture systems.

9.2.3 Bio Preservation of Food

Phages are involved in the process of bio-preservation, which extends the shelf life of food items by using beneficial bacteria. Bacteriophages have the ability to eradicate the need for chemical preservatives by inhibiting the growth of bacteria and illnesses that lead to food deterioration. This strategy not only improves food safety, but it also satisfies customer demands for sustainable and all-natural food preservation techniques.

9.2.4 Phage-Based Bio Pesticides

To tackle bacterial plant diseases, phage-based biopesticides are being developed in agriculture. These biopesticides can help protect crops from infections without leaving damaging residues on the environment or on harvested produce, making them an environmentally benign substitute for chemical pesticides. This invention is very useful for sustainable and organic farming methods.

9.2.5 Combating Biofilm-Related Infections

Healthcare organizations must contend with the problem of bacterial biofilms, which are frequently linked to persistent infections and contaminated medical equipment. Phages have demonstrated potential for destroying these biofilms and boosting the efficiency of antibiotics. Phage-antibiotic synergy, an approach, has the potential to completely alter the way biofilm-associated infections, particularly those brought on by bacteria resistant to antibiotics, are treated.

9.3 Use as an Environmental Safety Indicator

9.3.1 Evaluating the Water Quality

In many different environmental situations, particularly in water bodies like rivers, lakes, and seas, the presence of distinct bacteriophages can act as markers of water quality. One popular substitute for fecal contamination in water systems is coli phages, which infect Escherichia coli (E. coli). Increased levels of coli phages point to the potential presence of pathogens that could be dangerous to humans and aquatic environments, such as enteric bacteria and viruses.

9.3.2 Pollution and Ecological Health Evaluation

Bacteriophages can be used to measure environmental contamination and the health of the ecosystem in natural settings. Phage population changes can shed light on the dynamics of microbial communities, the origins of pollution, and the effects of environmental disturbances. Phage diversity and abundance are used by researchers as biomarkers to assess ecosystem health and pinpoint regions where adverse effects from human activity may be occurring.

9.3.3 Bioremediation

By focusing on bacteria that break down contaminants or take part in biodegradation processes, phages can aid in bioremediation efforts. To improve the effectiveness of bioremediation processes like the cleanup of oil spills or the treatment of contaminated groundwater, researchers are looking into the usage of bacteriophages. Phages can direct microbial communities toward more effective pollutant removal by selectively infecting and killing particular pollutant-degrading bacteria

9.3.4 Fecal Contamination Indicator

In addition to keeping track of water quality, bacteriophages have been extremely helpful in locating the sources of fecal pollution. Coliphages, as previously indicated, are especially helpful in this area. Public health professionals and environmental scientists need to know this information in order to detect potential water contamination hotspots and take the necessary precautions to reduce hazards to both human and environmental health.

9.3.5 Monitoring Genes for Antibiotic Resistance

Antibiotic resistance genes (ARGs) in environmental samples can be monitored using bacterial phages. Because ARGs have the potential to spread to dangerous bacteria and lessen the efficiency of antibiotics, they are a serious concern. Researchers have developed phage-based assays to identify and quantify ARGs in a range of environmental matrices, including soil, water, and sewage. With this knowledge, it becomes simpler to comprehend the prevalence and spread of antibiotic resistance in natural ecosystems and to identify potential hotspots for the dissemination of resistance genes.

9.4 For Environmental Health Bio Indicators

Phages can be used as bio indicators to evaluate the stability and general health of ecosystems. Phage communities can reflect changes in microbial populations and ecological dynamics through changes in diversity and abundance. Environmental stresses like pollution, climate change, or habitat loss may be responsible for these alterations. Researchers can learn more about the ecological effects of these stressors and create plans for conservation and restoration efforts by studying phage populations.

10. BACTERIOPHAGES APPLICATION FOR QUICK IDENTIFICATION OF FOOD-BORNE DISEASES AND BIOCONTROL

Ensuring the quality of food items and avoiding foodborne diseases are critical in today's environment. One of the newest methods in the field of food safety is the use of bacteriophages, viruses that infect and destroy bacteria, to control and identify foodborne diseases. Bacteriophages provide a promising means of enhancing food safety due to their rapid detection and biocontrol of target microbes. This study looks at the application of bacteriophages in biocontrol and the prompt identification of foodborne pathogens to guarantee the security and calibre of our food supply.

10.1 Uses of Biocontrol

10.1.1 Pathogen Targeting and Elimination

Bacteriophages have drawn interest as a highly specialized and all-natural method of eradicating foodborne infections. This focused strategy is especially beneficial for sectors like dairy, meat processing, and fresh produce, where pathogen contamination with Salmonella, Escherichia coli, and Listeria monocytogenes can pose serious health threats.

10.1.2 Getting Rid of Chemical Remains

Chemical residues in food products can be decreased by using bacteriophages as biocontrol agents. Traditional chemical preservatives and disinfectants could leave behind residues or byproducts that are harmful to the quality and safety of food. On the other hand, bacteriophages are organic organisms that do not bring chemical residues into the food supply. They are thus a desirable option for clean-label and organic food items, where there is a great demand for natural, minimally processed ingredients.

10.1.3 Increased Shelf Life

By halting the development of viruses and bacteria that cause spoiling, bacteriophages can help increase the shelf life of a variety of food products. Bacteriophages can slow the deterioration of food quality by specifically targeting the microorganisms that cause food spoiling and inhibiting their proliferation. This can lessen food waste and improve the food industry's ability to sustain its economy.

10.1.4 Non-Chemical Method

A non-chemical method of controlling pathogens is biocontrol using bacteriophages. This is especially important given the rising consumer concern over chemical residues in food goods. The food business may overcome these issues and meet consumer expectations for goods with less processing and clear labels by depending on bacteriophages, which are organic organisms.

10.1.5 Reducing Antibiotic Use

Antibiotics have been employed in various food production environments to reduce bacterial contamination. However, this practice has sparked worries about the threat that antibiotic resistance poses to world health. Bacteriophage biocontrol offers an effective substitute for antibiotics, reducing the dependency on these medications in the food business and preventing the emergence of germs that are resistant to medicines.

10.1.6 Flexibility With Different Food Products

Bacteriophage biocontrol is adaptable to many kinds of food products. Bacteriophages can be adapted to target particular pathogens often associated with each type of product, whether it be dairy products, meats, fresh produce, or even processed foods. They are useful in a variety of food processing and manufacturing situations because of their versatility.

10.1.7 Reducing Economic Losses

Food manufacturers and distributors may suffer large financial losses as a result of food recalls and outbreaks brought on by dangerous diseases. By reducing contamination and the associated requirement for product recalls, bacteriophage biocontrol can assist to mitigate these losses. In addition to preserving public health, this also safeguards the standing and financial success of the food industry.

10.2 Environmental Advantages

Many times, bacteriophage biocontrol is thought to be less harmful to the environment than many traditional chemical disinfection methods. Sustainable food production practices are promoted, and the use of chemicals, which can harm the environment, is reduced. Rapid Detection of Food-Borne Pathogens Using Bacteriophages.

10.2.1 Assays Based on Phages

Bacteriophages can serve as the foundation for the creation of quick and extremely accurate detection assays for foodborne pathogens. Phage-based assays often use genetically modified bacteriophages with reporter genes, which, upon infecting and lysing the target bacterium, create a detectable signal (e.g., fluorescence or luminescence). In contrast to conventional culture-based procedures, which might take days to provide results, these assays offer immediate results, frequently within hours. For outbreaks to be contained and contaminated products from reaching consumers, quick detection is essential.

10.2.2 Enhanced Specificity and Sensitivity

High sensitivity and specificity bacterial phage-based tests are well established. As a result, they are particularly useful for identifying pathogens in food samples at low concentrations, even when they may be present in very minute amounts.

Bacteriophages' specificity guarantees that false-positive results are kept to a minimum, lowering the possibility of needless product recalls or food safety concerns.

10.2.3 Localized Testing

Bacteriophage-based detection techniques have the ability for on-site testing, which is one of its main benefits. This implies that without sending samples to a lab, food manufacturers, distributors, and even consumers can conduct quick and accurate tests. In-person testing can speed up decision-making and enable quick corrective steps, like isolating tainted batches or items.

10.2.4 Multiplex Detection

Multiple foodborne pathogens can be concurrently detected in a single test using bacteriophage-based assays. This capacity is essential since a variety of diseases can contaminate food. In one step, multiplex detection offers a thorough evaluation of food safety, saving time and resources.

10.2.5 Real-Time Monitoring

To ensure food safety, it is crucial to continuously check the settings and supply chains used for food manufacturing. Automated monitoring systems can incorporate bacteriophage-based detection techniques, enabling real-time monitoring of food samples and processing machinery. This technologically-based strategy improves the food industry's responsiveness to possible threats.

10.2.6 Devices for Portable Detection

Technology developments have resulted in the creation of portable bacteriophage-based detection tools. Even in remote or resource-constrained environments, these handheld gadgets are easy to utilize on-site.

10.2.7 New Bacteriophages Research

The repertory of detectable bacteriophages is growing as a result of ongoing work in bacteriophage biology and genetics. Researchers are constantly finding novel bacteriophages with particular host ranges that can be exploited to create ever more specialized and effective detection techniques. Using bacteriophages to quickly identify foodborne pathogens has a number of benefits in terms of sensitivity, speed, multiplexing capacities, and adaptability to various sample types.

10.3 Phage Application Against Common Foodborne Pathogens in Food Systems

During the different processing stages like slaughtering, fermentation, packaging, milking and storage food products, may contaminated by different pathogens (Garcia *et al.*, 2008). By implementation of high standard hygienic procedures, use of different biocides & disinfectants and rational running of process line, food borne pathogens can be eliminated (Sillankorva *et al.*, 2012). Currently adopted methods of pathogen elimination (like steam use, UV light or dry heat) are not more reliable and they can alter the organoleptic characteristics of the food product. There are a few factors (Antimicrobial Agent Selection, Safety, Dosage and Concentration, Regulatory Compliance, Process Integration and Microbial Resistance to consider while using specialized antimicrobial treatments on some products, particularly on vegetables, fruits, and ready-to-eat food items. Microbial resistance in food is a result of overusing sanitizers (Hyla *et al.*, 2022).

Phages stand in contrast to the various antibacterial methods that are routinely employed. All these kinds of viruses can be used to a wide range of the food matrices, Phage procedures also do not alter the product's qualities (Thung *et al.*, 2019). Phages are isolated from a wide range of foods, indicating that this is how they naturally come into touch with humans. Phage-based approaches are a significant substitute to conventional antibacterial practices because to their affordability and customer friendliness (Ramos-Vivas *et al.*, 2021).

Microbiological safety, which directly jeopardizes consumer health, is of very importance in the food making industries. Limitations of current antibacterial techniques are significant. Numerous investigations have demonstrated the effective action of different bacteriophages against various kind of foodborne pathogens, and numerous researchers have noted the effective efficacy of different bacterial viruses that work to kill this specific type of the bacteria (Koch *et al.*, 2017).

Bacteriophage formulations that are appropriate and permitted for usage in the food business have increased since last year. The FDA originally accepted the utilization of a phages cocktail in 2006 to treat meat and poultry products for L. monocytogenes (Moye *et al.*, 2018).

Currently, about 13 bacteriophages that got licensed for the utilization in different food safety applications by various North American or European bodies, the bulk of which are devoted to eradicating Salmonella spp. and E. coli. Other medications work well against L. monocytogenes and Shigella spp (Vikram, Woolston, & Sulakvelidze, 2021). It is important to note that Omnilytix (USA) sells AgriPhage, a phage-based product designed specifically for agricultural uses. You can buy a product that is effective against Pseudomonas syringae pv. tomato and Xanthomonas campestris pv. vesicatoria. It's significant to note that various preparations have been made available for purchase on the Asian market but have not yet been approved by western institutions (Additives *et al.*, 2021).

Phages are isolated from different foods like cheese yogurt or fermented milk, they are ubiquitous in nature and present in human gut naturally (Shkoporov & Hill, 2019). Due to oral administration of the phage no disease occurs in human or animals (Gangwar *et al.*, 2021).

In addition to a phage's biological traits, such as its lytic or lysogenic capacities, several environmental conditions can affect the phage's efficacy and survivability. For example, external environmental factors like as temperature, pH, salinity content, and UV light may all affect phage activity (Jończyk, Kłak, Międzybrodzki, & Górski, 2011). The strength of these parameters dictates the level of inactivation, which varies for various phages (Ackermann *et al.*, 2004).

Bacterial isolates may develop resistance after meeting a phage host. Due to early exposure to environmental selection processes, endemic strains of the target pathogen populations are likely to be phage resistant. *L. monocytogenes* has been reported to produce phenotype-linked subpopulations, also known as the phenotypic switch, in response to unfavorable environmental variables. When creating targeted interventions in the arms of phage race against different foodborne illnesses, it is very crucial to note the proportion of phage-insensitive strains that are present in a population (Metselaar *et al.*, 2013).

Foodborne viruses that are present and persistent on the hide and inside the stomach of farm animals that used for production of food (pre-harvest) are the important cause of food chain cross-contamination. Control methods that are phage-based can be successfully lower target disease populations on and within animals before slaughter (Newell *et al.*, 2011). A target pathogen's prevalence and persistence can differ between individuals, herd members, or even different body sections of the same animal. On some farms, every bird is infected with C. jejuni, which is well known to be common in broiler chickens at the time of slaughter (Newell & Fearnley, 2003). Up to 20% of cattle herds have been found to contain E. coli O157 super-shedders (Matthews et al., 2006). It can be difficult to discern between transitory and super-shedding calves in a herd before phage therapy, necessitating multiple sampling operations over a long period of time. The position inside the digestive tract influences the effectiveness of phage-based therapy because a target pathogen may colonize multiple regions of the cow gut, including the rumen, cecum, colon, and rectum (Williams *et al.*, 2014). Pathogen colonization and the development of biofilms in the rectum can protect targeted cells against phages even in situations where physiological parameters in the rumen, such as pH, could render phages inert (Naylor *et al.*, 2003; Smith *et al.*, 1987).

Numerous research has investigated how well bacteriophages can control Salmonella, Campylobacter, and E. coli O157:H7 in farm animals. In most of the research, the animal received one bacteriophage or a mixture of phages orally. At different phage dosages (106-1012 PFU/mL), a 2-log decrease in the Campylobacter or Salmonella in finisher broiler chicks may help to lessen the number of infections that reach the stage of processing. Despite the studies of (Arthur *et al.*, 2017; Clavijo *et al.*, 2019) that investigations focused on germ-free small-scaled in vivo controlled tests that were carried out in industrial farms and slaughterhouses and targeted C. jejuni, Salmonella, and E. coli O157:H7, respectively. For instance, the quantity of germs present in people is influenced by the amount of viral exposure in the environment. Participants in germ-free small-scale controlled studies, on the other hand, receive the identical dosage at a particular site within the digestive system. Despite the fact that a big challenge dosage may be used to form a specific baseline of the phage-mediated to reduce pathogen, additional factors, such as individual variability in pathogen colonization, immune responses, the availability of nutrients, and then composition of the gut microbiome, could be also be taken into the consideration (Pickard *et al.*, 2017).

11. LIMITATIONS AND CHALLENGES OF BACTERIOPHAGE APPLICATIONS

To effectively use phage-based antibacterial treatments in the food business, it is necessary to produce enormous amounts of high-titer phage suspensions. The procedure for handling phages in laboratories is clearly laid out. However, it is difficult to apply these techniques in industrial settings. Even if many products that are phage-based are currently on the market, one of the biggest obstacles in the area is still how to process phages on a large scale while remaining cost-effective. Additionally, research is necessary to improve the procedure for phage propagation and purification. Future research is required to improve the cost-effective methods, safety, and several other elements of large-scale bacteriophage production. These issues can be handled by collaborative efforts between academic and industrial sectors (Li *et al.*, 2022).

11.1 Risk of the Bacterial Resistance Against Bacteriophages (BPs)

Resistance of bacteria to BPs is theoretically feasible given that the bacteria which already having or can develop a range of specific defensive mechanisms against the viral infections. Prevention of phage replication and release, modification or removal of the receptor, substances secretion that hinder phage attachment to bacterial pathogens, and reinforcement of defenses against phage DNA injection into cells are all potential treatment (Seed, 2015). It has been established that E. coli, S. aureus, Bordetella bronchiseptica, and Vibrio cholerae all have deleted or changed versions of the receptor for changes to membrane proteins. Pseudomonas spp. and Enterobacteriaceae, respectively, secrete glycoconjugates and extracellular polymeric compounds (Drulis-Kawa *et al.*, 2012). One can prevent bacterial resistance to BPs by employing BP mixtures, administering a larger initial BP inoculum, or combining BPs with specific antibiotics. Because BPs can destroy diseases before they spread, a higher inoculum that is associated with a reduced chance of the formation of BP-resistant bacteria. All of these data suggest that it is essential to take into account the potential for each virus to lead to bacterial resistance as well as the dosage necessary to prevent this from happening when choosing a therapeutic BP (Torres-Barceló, 2018).

11.2 Possibility of Bacteriophages (BPs) Adding to the Rise of Antibiotic Resistance

Bacterial genetics contains the DNA of lysogenic viruses. Therefore, they may serve as a means of genetic communication and promote the transmission of genes that confer antibiotic resistance (ARGs). Transduction may theoretically lead to the emergence of whole new bacterial strains or even more severe illnesses (Brabban *et al.*, 2002). Phage involvement in the propagation of ARGs is still not fully understood. These viruses have been found to serve as ARG repositories in several studies that were specifically designed to find out how frequently phages encode ARGs. The DNA of the phage fraction extracted from the intestinal mucus of wild freshwater fish species, human faecal samples, the virome of the very activated sludge and various water samples, as well as the virome of the activated sludge and environmental water samples, have all revealed the presence of genomes such as blaTEM (resistance to -lactams), ermB (resistance to macrolides), qnrS (reduced susceptibility to fluoroquinolones), sulI (resistance to sulphonamides), and tetW (resistance to tetracyclines) (Alemayehu *et al.*, 2012). ARG dissemination can be considerably increased by phage inducers, or chemicals that can cause the excision and spread of temperate BPs or the production of prophage gene products. Among the most crucial antibiotics are fluoroquinolones and a few anticoagulants. When wastewater samples are treated with EDTA or sodium citrate, lysogenic phages are inserted into the lytic cycle. They subsequently produce fresh particles of phage, induce bacterial lyses, expel phage from the cell, and spread phage infection to additional bacteria (Quirós *et al.*, 2014).

In the tissues and secretions of patients who experience recurrent infections brought on by micro-organisms resistant to antibiotics and who have previously undergone multiple courses of treatment with antimicrobial medications, a significant number of phages containing genes linked to antibiotic resistance have been discovered. The study by (Fancello, Gandini, Pelicci, & Mazzarella, 2019) shows that one of the strongest instances in this regard is CF. In their analysis of 1,031 short sequences in the CF virome that may have putatively encoded antibiotic resistance, these researchers discovered 66 efflux pump genes, 15 fluoroquinolone resistance genes, and 9 -lactamase genes. According to studies, the use of BPs may result in the generation and selection of multidrug-resistant organisms with chimeric repertoires. These studies also hypothesize that these observations and others may support this idea (Fancello *et al.*, 2011). Although it is still possible, the likelihood of AGR in phages appears to have been greatly overstated, according to a recent evaluation of the risk of transduction. Some studies yielded false results as a result of the high bacterial DNA concentration in the tested samples. Inadequate methods have also been used to find ARGs in phage genomes (Enault *et al.*, 2017).

11.3 Immune System Reaction-Induced Reduced Activity

It's not shocking that the immune system may detect phages and different metabolites which produce by them because they are not self-antigens. In principle, this might limit the advantages of using bacteriophages. Both animal and humans experiments show an response of immune to the BPs, albeit there are differences that depending on the phage type, method of delivery, and prior exposure level. Soon after birth of the phagocytic cells in the animals swallowed BPs, and all these cells can be eliminated the BPs in less than two hours. It infects most strains of Escherichia coli and relies on these hosts) after 60 minutes of injection, whereas BP titres remained more stable for a considerable amount of the time in experimental animals with severe combined immunodeficiency. Because the phage titres in the B-cell-deficient animals were unaffected, it was thought that certain antibodies were the primary mechanism for phage clearance from blood (Kaźmierczak *et al.,* 2014).

Although none of the participants in the study (Dąbrowska *et al.,* 2014) had undergone phage therapy, their investigation into the antigenic proteins of the E. coli T4 bacteriophage (BP) head surface suggested that specific antibodies could be detected in 80% of the enrolled individuals. While there is a high probability that the immunological response triggered by phages may not significantly affect bacterial mortality resulting from phage treatment, this has not been definitively established. Bacterial lysis precedes the creation of specific antibodies. Moreover, the administration of phages was seldom linked to tissue damage, an elevation in pro-inflammatory cytokines, or an increase in reactive oxygen species (ROS), with some noteworthy exceptions. Dąbrowska et al. (2014) demonstrated that intraperitoneal administration of phage T4 and its head proteins to mice did not specifically impact the production of interleukin (IL)-1, interferon (IFN)-γ, tumor necrosis factor (TNF)-α, monocyte chemoattractant protein (MCP-1), monokine induced by gamma (MIG), granulocyte colony-stimulating factor (GCSF), and granulocyte-macrophage colony-stimulating factor (GM-CSF). Hwang, Kim, Song, and Park (2016) investigated the safety of an E. coli phage cocktail given orally to rats for four weeks, reporting similar findings.

In a study (Carmody *et al.,* 2010), the effectiveness of phage treatment via intranasal inhalation was explored in a mouse model of lung infection caused by B. cenocepacia. Treated mice exhibited significantly reduced levels of Macrophage Inflammatory Protein-2 (MIP-2), TNF-α, and bacterial density compared to untreated controls ($p < 0.05$). Human subject data supporting the safety of BP treatment suggest that, even if an immune response exists, it may not be clinically significant (Principi, Silvestri, & Esposito, 2019).

12. BACTERIOPHAGE BASED FOOD PACKAGING

The categories of traditional or passive packaging materials are active, intelligent, and smart. Since they may actively respond to changes in the internal and external surroundings of the packed meals, active packaging materials stand out more from the list (Kuai *et al.*, 2021). Now, the majority of active packaging materials fall into one of several broad categories, including those based on antioxidants, antimicrobials, gas scavengers/emitters, etc. It is urgent to create antimicrobial substances that have excellent (host) particularity that target just pathogenic microbes without causing damage the population of beneficial bacteria (Cao *et al.*, 2023).

The stability and release of bacteriophages present significant challenges for researchers. Food systems significantly delay the discharge of active the phages form polymer films, coverings, and hydrogels compared to aqueous systems. In edible polymeric films and coatings, phage stability has been observed to vary, and total inhibition or suppression of the bacteriophages has been described. Phage inhibition or inactivation's exact mechanism is uncertain. Many researchers have proposed encasing the phages to address these problems, which would boost their stability and damage resistance (Kasimanickam *et al.*, 2013).

Korehei and Kadla (2013) claim that the electrospinning (suspension) of the bacteriophage T4 significantly reduced viral activity. The T4 phage was merged into the specific fiber core during this procedure, resulting in a core/shell fiber configuration. For several weeks at +4°C, the fiber encapsulated T4 core demonstrated greater bacteriophage survivability.

The best way to encapsulate the bacteriophage K with a stronger shield of acid for the oral delivery was investigated. To increase phage survival under in vitro acidic conditions, calcium carbonate (microparticles) containing phage K were enclosed in alginate microspheres. Phages without encapsulations have been rendered inactive by stomach fluid exposure with a pH of 2.5. With just a 0.17 log drop seen after the exposure of 2 hours to SGF at the pH 2.5, they claimed that the calcium carbonate addition to the specific alginate microspheres prolonged the lifespan of the phage K that was enclosed in the material. In comparison, after one hour of incubation at pH 2.5 in SGF, the survival of the phage K that are alginate-encapsulated dropped to only 2.4 log(Ma et al., 2012). The number of E. coli was significantly reduced by a specific antibacterial film that created by the immobilizing phage T4 on a poly-caprolactone (PCL) film. Manufactured PCL film surpassed the film that included physically adsorbed phage T4 when it came to of microbial inhibitory qualities after being utilized as a packaging material for raw beef infected with E. coli O157:H7. These results indicate that the PCL film created and coupled with phage T4 will likely be used in active packaging of food against *E. coli* (Choi *et al*., 2021).

The creation of the bacteriophage IBB-PF7A, which uses sodium alginate to prevent Pseudomonas fluorescens from causing microbiological meat rotting, was examined by a group of researchers. The bacteriophages, according to them, had been successfully loaded in movies with a lot of vitality. The total number of P. fluorescens microorganisms decreased by 2 log over the initial two days of preservation in the freezer, yet only by 1 log over the next five days, according to their data. To test the film's antibacterial potency, P. fluorescens was artificially injected into chicken breast fillets. Gouvêa et al. 2015 investigated the effectiveness of acetate cellulose film treated with bacteriophage against Salmonella typhimurium and found that the lag phase was increased, showing that bacterial growth was slower in the environment containing the films than in the control (without phage). Bacteriophages were added, but the mechanical and physical characteristics of the films such as thickness, elongation, and puncture resistance—did not significantly change. However, bacteriophages could only be found on the cellulose acetate film for the first 14 days of its existence (Gouvêa *et al*., 2015).

To limit the spread of bacterial diseases in food, scientists created prototypes of bioactive packaging materials based on immobilized bacteriophages. When applied to foods that had been purposefully tainted, phage-based compounds demonstrated potent antibacterial activity. The created bioactive films were able to use free or immobilized lytic bacteriophage cocktails to prevent L. monocytogenes from developing in ready-to-eat beef under a variety of storage circumstances (Wagh, Priyadarshi, & Rhim, 2023). A group of scientists examined the creation of chitosan films with generated phages put in them to reduce E. coli in beef (Wagh *et al.,* 2023).

13. CONCLUDING REMARKS AND FUTURE PERSPECTIVE

13.1 Phages in One Health Approach

At least six out of every ten infectious illnesses that are known to afflict humans are thought to have animal origins. Furthermore, due to the specific pressure exerted on the phytobacteria, a good number of defense mechanisms have emerged, increasing their pathogenicity toward humans, particularly those who are elderly, immunocompromised, or have cancer *(Erken et al.,*2013; Falkinham *et al.*, 2015).

The concept of one health recognizes the connection of human, animal, and environmental health. The One Health initiative proposes merging environmental, veterinary, and human medical research to enhance the quality of life for all living creatures. Agriculture and food safety are working in this comprehensive and multi-sectoral antimicrobial resistance plan (Baum *et al.,* 2017; Hernando-Amado *et al.,* 2019).According to the One Health viewpoint, innovative different therapies and treatments are needed to prevent the spread of antibiotic resistance. Since their discovery, phages have been utilized in the food business, as well as in the microbiomes of humans, animals, and other environments (including wastewater treatment), to control the proliferation of bacteria (Abuladze *et al.*, 2008).

A noteworthy case of the extensive usage of bacteriophages is very wide range of utilization possibilities in the different food manufacturing units, at their they may be utilized at the all steps of the food processing and production, from the slathering to food transportation, and even to enhance the shelf life of final food products. In actual, many phage-based food products that intended for the utilization in food have previously been labeled as generally recognized as safe (GRAS) by the FDA in US. Phages are congruent with the One specific Health approach and control the abuse of antibiotics and the spread of antibiotic resistance to humans because they can be utilized in a number of contexts (such as food, animals, or crops) (Kittler *et al.*, 2017).

13.2 Emerging Approaches

Antibiotics and phages are probably the most natural partners (Chaudhry *et al.*, 2017; Torres-Barceló & Hochberg, 2016). When that used together, phages and the antibiotics are showing synergistic advantages and positive effect against planktonic cells, but the solo therapies were not very effective. Where regular phage usage facilitated biofilm growth, phage and antibiotic use led to biofilm removal (Henriksen *et al.*, 2019).

Biofilm may have undergone structural changes because of one or both medicines, which could account for the improved efficacy. For instance, by eliminating peripheral cells, the phage that can enhance the available resources to the core cells and enhance their metabolic health, to making them more susceptible to specific phages and some type of antibiotics. Additionally, antibiotics may naturally change the biofilm's structure, making it easier for phages to penetrate biofilms (Díaz-Pascual *et al.*, 2019).

Large antibiotics dose can prevent the spread of the phages, and not all phage-antibiotic combinations are effective. This is particularly clear when using antibiotics that prevent the formation of cell proteins. Even though there is no synergistic antibacterial effect detected, there are some situations where utilizing different phages and antibiotics together significantly suppresses or even stops the growth of bacteria that are resistant to both antibiotics and phages (Coulter *et al.,* 2019).

While several research has examined the impact of these kinds of medicines, a logical strategy for assessing how bacteria respond to phage-antibiotic therapy has evolved (Heilpern & Waldor, 2003; Marzari *et al.*, 1997). The isolation of phages that target the outermost layer membrane proteins—which the bacteria use as multidrug efflux pumps, is an illustration of this strategy. The bacteria would have to change their efflux pump in order to become resistant to this phage, which would increase their susceptibility to particular antibiotic classes (Chan *et al.,* 2016).

Enzymes and phages can be administered simultaneously to boost activity. For instance, even when they do not naturally express them, depolymerizes can be introduced to phages to boost their activity against biofilms. DNase enzymes that can be utilized in the concert with the phages to increase phage work and degrade the DNA elements of the biofilm matrix. In other successful cases, probiotics, honey, xylitol, triclosan, chlorhexidine, hydrogen peroxide, and chlorine were used with phages (Woo & Ahn, 2014).

While temperate type phages are also the subject of some engineering experiments and methods for the phage therapy, lytic phages have attracted the majority of engineering attention. The simplest approach is genetically modifying phages to make them exclusively lytic. The genetic module in charge of lysogeny establishment was deleted to achieve this (Dorscht et al., 2009; Kilcher *et al.*, 2018).

By creating virulent mutations of previously temperate phages, the type and variety of the phages which are readily present for therapeutic usage can be rapidly enhanced. The most successful and good therapy of a patient of 15-year-old cystic fibrosis by broad Mycobacterium abscessus infection utilizing a mixture of one naturally occurring lytic phage and two synthetic temperate type of phages is a superb example of the efficacy of this approach (Dedrick *et al.,* 2019).

Additionally, by utilizing their natural capacity for incorporation into the genetic code of the bacteria that are the host, where about phage showed the required chemical, temperate phages may be developed for carrying synthetic gene networks. Bacteriophages have changed to function as the adjuvants to antibiotics by establishing dominant type of antibiotic susceptible genes, CRISPR-Cas systems that restore bacterial resistance to antibiotics, or CRISPR-Cas systems which specifically target the bacterial cells (Park *et al.,* 2017).

All methods of bioengineering may strengthen the phages' antibacterial properties and produce cutting-edge treatments for bacterial diseases. Although the consequences of genetically changing phage genomes need to be carefully investigated, phage methods of engineering are seriously reviewed as therapeutic alternative. More about, synthetic phages are more easy to patent as compared to biological phages, which could result in additional commercial opportunities (Pires *et al.*, 2020).

The preceding details give a glimpse of the wide range of applications for the phages in microbiology and medicine. Phage therapy, phage vaccinations, and phage typing are all methods of using phages for disease diagnosis, prevention, and treatment. Phage-related benefits to humans could take many different forms. It would be simple to treat a range of bacterial diseases that are presently resistant to the latest generation of antibiotics by creating a cock tail of phages. If a phage has the capacity to lyse a cell of bacteria, it can be employed to treat an infection caused by bacteria on its own. We were also able to employ the antibodies to fight the bacteria which were evident on the bacteriophage surface because of how adaptable phages are. A DNA or phage presentation vaccination could also provide a protective antigen. A combination of GM phages would therefore be more helpful in resolving each of these problems. Additionally, phages have demonstrated efficacy in curing infections caused by bacteria in plants, fruits, and food deterioration.

The usage of phages raises a few problems. It covers difficulties with effectiveness, safety, and immunological reactions to injected phages. It's also important to work on phage development and purification methods. It is thought that these organisms (phages), which have a broad distribution across the earth's biosphere, could provide answers to many of the problems that humanity is currently facing because of the quick advances in the domains of molecular science and biotechnology

REFERENCES

Abuladze, T., Li, M., Menetrez, M. Y., Dean, T., Senecal, A., & Sulakvelidze, A. (2008). Bacteriophages reduce experimental contamination of hard surfaces, tomato, spinach, broccoli, and ground beef by Escherichia coli O157: H7. *Applied and Environmental Microbiology*, *74*(20), 6230–6238. doi:10.1128/AEM.01465-08 PMID:18723643

Additives, E. P. o., & Feed, P. o. S. u. i. A., Bampidis, V., Azimonti, G., Bastos, M. d. L., Christensen, H., . . . López Puente, S. (2021). Safety and efficacy of a feed additive consisting on the bacteriophages PCM F/00069, PCM F/00070, PCM F/00071 and PCM F/00097 infecting Salmonella Gallinarum B/00111 (Bafasal®) for all avian species (Proteon Pharmaceuticals SA). *EFSA Journal, 19*(5), e06534. PMID:34025802

Alemayehu, D., Casey, P. G., McAuliffe, O., Guinane, C. M., Martin, J. G., Shanahan, F., Coffey, A., Ross, R. P., & Hill, C. (2012). Bacteriophages ϕMR299-2 and ϕNH-4 can eliminate Pseudomonas aeruginosa in the murine lung and on cystic fibrosis lung airway cells. *mBio, 3*(2), 00029–00012. doi:10.1128/mBio.00029-12 PMID:22396480

Alonso, C. A., González-Barrio, D., Tenorio, C., Ruiz-Fons, F., & Torres, C. (2016). Antimicrobial resistance in faecal Escherichia coli isolates from farmed red deer and wild small mammals. Detection of a multiresistant E. coli producing extended-spectrum beta-lactamase. *Comparative Immunology, Microbiology and Infectious Diseases, 45*, 34–39. doi:10.1016/j.cimid.2016.02.003 PMID:27012919

Anvar, A. A., Ahari, H., & Ataee, M. (2021). Antimicrobial properties of food nanopackaging: A new focus on foodborne pathogens. *Frontiers in Microbiology, 12*, 690706. doi:10.3389/fmicb.2021.690706 PMID:34322104

Artawinata, P. C., Lorraine, S., & Waturangi, D. E. (2023). Isolation and characterization of bacteriophages from soil against food spoilage and foodborne pathogenic bacteria. *Scientific Reports, 13*(1), 9282. doi:10.103841598-023-36591-6 PMID:37286897

Arthur, T. M., Kalchayanand, N., Agga, G. E., Wheeler, T. L., & Koohmaraie, M. (2017). Evaluation of bacteriophage application to cattle in lairage at beef processing plants to reduce Escherichia coli O157: H7 prevalence on hides and carcasses. *Foodborne Pathogens and Disease, 14*(1), 17–22. doi:10.1089/fpd.2016.2189 PMID:27870549

Authority, E. F. S. (2009). The use and mode of action of bacteriophages in food production-Endorsed for public consultation 22 January 2009-Public consultation 30 January–6 March 2009. *EFSA Journal, 7*(5), 1076. doi:10.2903/j.efsa.2009.1076

Azman, N., Khairul, W. M., & Sarbon, N. (2022). A comprehensive review on biocompatible film sensor containing natural extract: Active/intelligent food packaging. *Food Control, 141*, 109189. doi:10.1016/j.foodcont.2022.109189

Bai, X., Nakatsu, C. H., & Bhunia, A. K. (2021). Bacterial biofilms and their implications in pathogenesis and food safety. *Foods, 10*(9), 2117. doi:10.3390/foods10092117 PMID:34574227

Baum, S. E., Machalaba, C., Daszak, P., Salerno, R. H., & Karesh, W. B. (2017). Evaluating one health: Are we demonstrating effectiveness? *One Health*, *3*, 5–10. doi:10.1016/j.onehlt.2016.10.004 PMID:28616496

Bintsis, T. (2017). Foodborne pathogens. *AIMS Microbiology*, *3*(3), 529–563. doi:10.3934/microbiol.2017.3.529 PMID:31294175

Bonaterra, A., Badosa, E., Daranas, N., Francés, J., Roselló, G., & Montesinos, E. (2022). Bacteria as biological control agents of plant diseases. *Microorganisms*, *10*(9), 1759. doi:10.3390/microorganisms10091759 PMID:36144361

Boyacioglu, O., Sulakvelidze, A., Sharma, M., & Goktepe, I. (2016). Effect of a bacteriophage cocktail in combination with modified atmosphere packaging in controlling Listeria monocytogenes on fresh-cut spinach. *Irish Journal of Agricultural and Food Research*, *55*(1), 74–79. doi:10.1515/ijafr-2016-0007

Brabban, A., Hite, E., & Callaway, T. (2005). Evolution of foodborne pathogens via temperate bacteriophage-mediated gene transfer. *Foodborne Pathogens and Disease*, *2*(4), 287–303. doi:10.1089/fpd.2005.2.287 PMID:16366852

Bumunang, E. W., Ateba, C. N., Stanford, K., Niu, Y. D., Wang, Y., & McAllister, T. A. (2020). Activity of bacteriophage and complex tannins against biofilm-forming shiga toxin-producing Escherichia coli from Canada and South Africa. *Antibiotics (Basel, Switzerland)*, *9*(5), 257. doi:10.3390/antibiotics9050257 PMID:32429187

Bumunang, E. W., Zaheer, R., Niu, D., Narvaez-Bravo, C., Alexander, T., McAllister, T. A., & Stanford, K. (2023). Bacteriophages for the Targeted Control of Foodborne Pathogens. *Foods*, *12*(14), 2734. doi:10.3390/foods12142734 PMID:37509826

Campbell, A. (2003). The future of bacteriophage biology. *Nature Reviews. Genetics*, *4*(6), 471–477. doi:10.1038/nrg1089 PMID:12776216

Cao, S., Wang, S., Wang, Q., Lin, G., Niu, B., Guo, R., Yan, H., & Wang, H. (2023). Sodium alginate/chitosan-based intelligent bilayer film with antimicrobial activity for pork preservation and freshness monitoring. *Food Control*, *148*, 109615. doi:10.1016/j.foodcont.2023.109615

Carmody, L. A., Gill, J. J., Summer, E. J., Sajjan, U. S., Gonzalez, C. F., Young, R. F., & LiPuma, J. J. (2010). Efficacy of bacteriophage therapy in a model of Burkholderia cenocepacia pulmonary infection. *The Journal of Infectious Diseases*, *201*(2), 264–271. doi:10.1086/649227 PMID:20001604

Chan, B. K., Sistrom, M., Wertz, J. E., Kortright, K. E., Narayan, D., & Turner, P. E. (2016). Phage selection restores antibiotic sensitivity in MDR Pseudomonas aeruginosa. *Scientific Reports*, *6*(1), 26717. doi:10.1038rep26717 PMID:27225966

Chan, B. K., Turner, P. E., Kim, S., Mojibian, H. R., Elefteriades, J. A., & Narayan, D. (2018). Phage treatment of an aortic graft infected with Pseudomonas aeruginosa. *Evolution, Medicine, and Public Health*, *2018*(1), 60–66. doi:10.1093/emph/eoy005 PMID:29588855

Chanishvili, N. (2012). Phage therapy—History from Twort and d'Herelle through Soviet experience to current approaches. *Advances in Virus Research*, *83*, 3–40. doi:10.1016/B978-0-12-394438-2.00001-3 PMID:22748807

Chaudhry, W. N., Concepcion-Acevedo, J., Park, T., Andleeb, S., Bull, J. J., & Levin, B. R. (2017). Synergy and order effects of antibiotics and phages in killing Pseudomonas aeruginosa biofilms. *PLoS One*, *12*(1), e0168615. doi:10.1371/journal.pone.0168615 PMID:28076361

Choi, I., Chang, Y., Kim, S. Y., & Han, J. (2021). Polycaprolactone film functionalized with bacteriophage T4 promotes antibacterial activity of food packaging toward Escherichia coli. *Food Chemistry*, *346*, 128883. doi:10.1016/j.foodchem.2020.128883 PMID:33434864

Clavijo, V., Baquero, D., Hernandez, S., Farfan, J., Arias, J., Arévalo, A., Donado-Godoy, P., & Vives-Flores, M. (2019). Phage cocktail SalmoFREE® reduces Salmonella on a commercial broiler farm. *Poultry Science*, *98*(10), 5054–5063. doi:10.3382/ps/pez251 PMID:31073590

Clokie, M. R., Millard, A. D., Letarov, A. V., & Heaphy, S. (2011). Phages in nature. *Bacteriophage*, *1*(1), 31–45. Advance online publication. doi:10.4161/bact.1.1.14942 PMID:21687533

Cooper, I. R. (2016). A review of current methods using bacteriophages in live animals, food and animal products intended for human consumption. *Journal of Microbiological Methods*, *130*, 38–47. doi:10.1016/j.mimet.2016.07.027 PMID:27485708

Costa, M. J., Pastrana, L. M., Teixeira, J. A., Sillankorva, S. M., & Cerqueira, M. A. (2023). Bacteriophage delivery systems for food applications: Opportunities and perspectives. *Viruses*, *15*(6), 1271. doi:10.3390/v15061271 PMID:37376571

Coulter, L. B., McLean, R. J., Rohde, R. E., & Aron, G. M. (2014). Effect of bacteriophage infection in combination with tobramycin on the emergence of resistance in Escherichia coli and Pseudomonas aeruginosa biofilms. *Viruses*, *6*(10), 3778–3786. doi:10.3390/v6103778 PMID:25285538

Dąbrowska, K., Miernikiewicz, P., Piotrowicz, A., Hodyra, K., Owczarek, B., Lecion, D., Kaźmierczak, Z., Letarov, A., & Górski, A. (2014). Immunogenicity studies of proteins forming the T4 phage head surface. *Journal of Virology*, *88*(21), 12551–12557. doi:10.1128/JVI.02043-14 PMID:25142581

De Jonge, P. A., Nobrega, F. L., Brouns, S. J., & Dutilh, B. E. (2019). Molecular and evolutionary determinants of bacteriophage host range. *Trends in Microbiology*, *27*(1), 51–63. doi:10.1016/j.tim.2018.08.006 PMID:30181062

Dedrick, R. M., Guerrero-Bustamante, C. A., Garlena, R. A., Russell, D. A., Ford, K., Harris, K., ... Schooley, R. T. (2019). Engineered bacteriophages for treatment of a patient with a disseminated drug-resistant Mycobacterium abscessus. *Nature Medicine*, *25*(5), 730–733. doi:10.103841591-019-0437-z PMID:31068712

Dhama, K., Karthik, K., Tiwari, R., Shabbir, M. Z., Barbuddhe, S., Malik, S. V. S., & Singh, R. K. (2015). Listeriosis in animals, its public health significance (food-borne zoonosis) and advances in diagnosis and control: A comprehensive review. *The Veterinary Quarterly*, *35*(4), 211–235. doi:10.1080/01652176.2015.1063023 PMID:26073265

Díaz-Pascual, F., Hartmann, R., Lempp, M., Vidakovic, L., Song, B., Jeckel, H., ... Link, H. (2019). Breakdown of Vibrio cholerae biofilm architecture induced by antibiotics disrupts community barrier function. *Nature Microbiology*, *4*(12), 2136–2145. doi:10.103841564-019-0579-2 PMID:31659297

Dickey, J., & Perrot, V. (2019). Adjunct phage treatment enhances the effectiveness of low antibiotic concentration against Staphylococcus aureus biofilms in vitro. *PLoS One*, *14*(1), e0209390. doi:10.1371/journal.pone.0209390 PMID:30650088

Domingo-Calap, P., & Delgado-Martínez, J. (2018). Bacteriophages: Protagonists of a post-antibiotic era. [basel]. *Antibiotics (Basel, Switzerland)*, *7*(3), 66. doi:10.3390/antibiotics7030066 PMID:30060506

Dorscht, J., Klumpp, J., Bielmann, R., Schmelcher, M., Born, Y., Zimmer, M., Calendar, R., & Loessner, M. J. (2009). Comparative genome analysis of Listeria bacteriophages reveals extensive mosaicism, programmed translational frameshifting, and a novel prophage insertion site. *Journal of Bacteriology*, *191*(23), 7206–7215. doi:10.1128/JB.01041-09 PMID:19783628

Drulis-Kawa, Z., Majkowska-Skrobek, G., Maciejewska, B., Delattre, A.-S., & Lavigne, R. (2012). Learning from bacteriophages-advantages and limitations of phage and phage-encoded protein applications. *Current Protein & Peptide Science*, *13*(8), 699–722. doi:10.2174/138920312804871193 PMID:23305359

Du Toit, A. (2017). The language of phages. *Nature Reviews. Microbiology*, *15*(3), 135–135. PMID:28190882

Dulger & Özpinar. (2021). Use of bacteriophages to improve food safety. *İstanbul Gelişim Üniversitesi Sağlık Bilimleri Dergisi*, (15), 705-712.

Dwivedi, S., Prajapati, P., Vyas, N., Malviya, S., & Kharia, A. (2017). A review on food preservation: Methods, harmful effects and better alternatives. *Asian Journal of Pharmacy and Pharmacology*, *3*, 193–199.

Elsayed, M. M., Elkenany, R. M., Zakari, A. I., & Badawy, B. M. (2023). Isolation and characterization of bacteriophages for combating multidrug-resistant Listeria monocytogenes from dairy cattle farms in conjugation with silver nanoparticles. *BMC Microbiology*, *23*(1), 146. doi:10.118612866-023-02893-y PMID:37217869

Enault, F., Briet, A., Bouteille, L., Roux, S., Sullivan, M. B., & Petit, M.-A. (2017). Phages rarely encode antibiotic resistance genes: A cautionary tale for virome analyses. *The ISME Journal*, *11*(1), 237–247. doi:10.1038/ismej.2016.90 PMID:27326545

Endersen, L., & Coffey, A. (2020). The use of bacteriophages for food safety. *Current Opinion in Food Science*, *36*, 1–8. doi:10.1016/j.cofs.2020.10.006

Erken, M., Lutz, C., & McDougald, D. (2013). The rise of pathogens: Predation as a factor driving the evolution of human pathogens in the environment. *Microbial Ecology*, *65*(4), 860–868. doi:10.100700248-013-0189-0 PMID:23354181

Fair & Tor. (2014). Antibiotics and bacterial resistance in the 21st century. *Perspectives in Medicinal Chemistry, 6*.

Falkinham, J. O. III, Pruden, A., & Edwards, M. (2015). Opportunistic premise plumbing pathogens: Increasingly important pathogens in drinking water. *Pathogens (Basel, Switzerland)*, *4*(2), 373–386. doi:10.3390/pathogens4020373 PMID:26066311

Fancello, L., Desnues, C., Raoult, D., & Rolain, J. M. (2011). Bacteriophages and diffusion of genes encoding antimicrobial resistance in cystic fibrosis sputum microbiota. *The Journal of Antimicrobial Chemotherapy*, *66*(11), 2448–2454. doi:10.1093/jac/dkr315 PMID:21816767

Fancello, L., Gandini, S., Pelicci, P. G., & Mazzarella, L. (2019). Tumor mutational burden quantification from targeted gene panels: Major advancements and challenges. *Journal for Immunotherapy of Cancer*, *7*(1), 1–13. doi:10.118640425-019-0647-4 PMID:31307554

Farooq, Ali, Shoukat, Paray, & Ali. (2019). *Bacteriophages as biocontrol agents for foodborne diseases*. Academic Press.

Fernández, L., Gutiérrez, D., Rodríguez, A., & García, P. (2018). Application of bacteriophages in the agro-food sector: A long way toward approval. *Frontiers in Cellular and Infection Microbiology*, *8*, 296. doi:10.3389/fcimb.2018.00296 PMID:30186776

Gangwar, M., Rastogi, S., Singh, D., Shukla, A., Dhameja, N., Kumar, D., Kumar, R., & Nath, G. (2021). Study on the effect of oral administration of bacteriophages in charles foster rats with special reference to immunological and adverse effects. *Frontiers in Pharmacology*, *12*, 615445. doi:10.3389/fphar.2021.615445 PMID:33912038

Garcia, P., Martinez, B., Obeso, J., & Rodriguez, A. (2008). Bacteriophages and their application in food safety. *Letters in Applied Microbiology*, *47*(6), 479–485. doi:10.1111/j.1472-765X.2008.02458.x PMID:19120914

Ge, H., Fu, S., Guo, H., Hu, M., Xu, Z., Zhou, X., Chen, X., & Jiao, X. (2022). Application and challenge of bacteriophage in the food protection. *International Journal of Food Microbiology*, *380*, 109872. doi:10.1016/j.ijfoodmicro.2022.109872 PMID:35981493

Gencay, Y. E., & Brøndsted, L. (2019). Bacteriophages for Biological Control of Foodborne Pathogens. *Food Microbiology: Fundamentals And Frontiers*, 755-786.

Gouvêa, D. M., Mendonça, R. C. S., Soto, M. L., & Cruz, R. S. (2015). Acetate cellulose film with bacteriophages for potential antimicrobial use in food packaging. *Lebensmittel-Wissenschaft + Technologie*, *63*(1), 85–91. doi:10.1016/j.lwt.2015.03.014

Gwak, Choi, Lee, Oh, & Park. (2018). *Isolation and characterization of a lytic and highly specific phage against Yersinia enterocolitica as a novel biocontrol agent*. Academic Press.

Hameed, F., Bandral, J. D., Gupta, N., Nayik, G. A., Sood, M., & Rahman, R. (2022). Use of Bacteriophages as a Target Specific Therapy against Food-Borne Pathogens in Food Industry-a Review: Bacteriophage. *Journal of Microbiology, Biotechnology and Food Sciences*, *11*(4), e2949–e2949. doi:10.55251/jmbfs.2949

Haq, I. U., Chaudhry, W. N., Akhtar, M. N., Andleeb, S., & Qadri, I. (2012). Bacteriophages and their implications on future biotechnology: A review. *Virology Journal*, *9*(1), 1–8. doi:10.1186/1743-422X-9-9 PMID:22234269

Harada, L. K., Silva, E. C., Campos, W. F., Del Fiol, F. S., Vila, M., Dąbrowska, K., Krylov, V. N., & Balcão, V. M. (2018). Biotechnological applications of bacteriophages: State of the art. *Microbiological Research*, *212*, 38–58. doi:10.1016/j.micres.2018.04.007 PMID:29853167

Harrer, A., Bücker, R., Boehm, M., Zarzecka, U., Tegtmeyer, N., Sticht, H., Schulzke, J. D., & Backert, S. (2019). Campylobacter jejuni enters gut epithelial cells and impairs intestinal barrier function through cleavage of occludin by serine protease HtrA. *Gut Pathogens*, *11*(1), 1–16. doi:10.118613099-019-0283-z PMID:30805031

Hasan, M., & Ahn, J. (2022). Evolutionary dynamics between phages and bacteria as a possible approach for designing effective phage therapies against antibiotic-resistant bacteria. *Antibiotics (Basel, Switzerland)*, *11*(7), 915. doi:10.3390/antibiotics11070915 PMID:35884169

Heilpern, A. J., & Waldor, M. K. (2003). pIIICTX, a predicted CTXφ minor coat protein, can expand the host range of coliphage fd to include Vibrio cholerae. *Journal of Bacteriology*, *185*(3), 1037–1044. doi:10.1128/JB.185.3.1037-1044.2003 PMID:12533480

Henriksen, K., Rørbo, N., Rybtke, M. L., Martinet, M. G., Tolker-Nielsen, T., Høiby, N., Middelboe, M., & Ciofu, O. (2019). P. aeruginosa flow-cell biofilms are enhanced by repeated phage treatments but can be eradicated by phage–ciprofloxacin combination: —monitoring the phage–P. aeruginosa biofilms interactions. *Pathogens and Disease*, *77*(2), ftz011. doi:10.1093/femspd/ftz011 PMID:30821815

Hernando-Amado, S., Coque, T. M., Baquero, F., & Martínez, J. L. (2019). Defining and combating antibiotic resistance from One Health and Global Health perspectives. *Nature Microbiology*, *4*(9), 1432–1442. doi:10.103841564-019-0503-9 PMID:31439928

Huang, C., Shi, J., Ma, W., Li, Z., Wang, J., Li, J., & Wang, X. J. (2018). Isolation, characterization, and application of a novel specific Salmonella bacteriophage in different food matrices. *Food Research International*, *111*, 631–641. doi:10.1016/j.foodres.2018.05.071 PMID:30007727

Hwang, J.-Y., Kim, J.-E., Song, Y.-J., & Park, J.-H. (2016). Safety of using Escherichia coli bacteriophages as a sanitizing agent based on inflammatory responses in rats. *Food Science and Biotechnology*, *25*(1), 355–360. doi:10.100710068-016-0050-6 PMID:30263278

Hyla, K., Dusza, I., & Skaradzińska, A. (2022). Recent Advances in the Application of Bacteriophages against Common Foodborne Pathogens. *Antibiotics (Basel, Switzerland)*, *11*(11), 1536. doi:10.3390/antibiotics11111536 PMID:36358191

Ishaq, A., Ebner, P. D., Syed, Q. A., & Ur Rahman, H. U. (2020). Employing list-shield bacteriophage as a bio-control intervention for Listeria monocytogenes from raw beef surface and maintain meat quality during refrigeration storage. *Lebensmittel-Wissenschaft + Technologie*, *132*, 109784. doi:10.1016/j.lwt.2020.109784

Islam, M. S., Yang, X., Euler, C. W., Han, X., Liu, J., Hossen, M. I., & Li, J. (2021). Application of a novel phage ZPAH7 for controlling multidrug-resistant Aeromonas hydrophila on lettuce and reducing biofilms. *Food Control*, *122*, 107785. doi:10.1016/j.foodcont.2020.107785

Jończyk, E., Kłak, M., Międzybrodzki, R., & Górski, A. (2011). The influence of external factors on bacteriophages. *Folia Microbiologica*, *56*(3), 191–200. doi:10.100712223-011-0039-8 PMID:21625877

Jun, J. W., Park, S. C., Wicklund, A., & Skurnik, M. (2018). Bacteriophages reduce Yersinia enterocolitica contamination of food and kitchenware. *International Journal of Food Microbiology*, *271*, 33–47. doi:10.1016/j.ijfoodmicro.2018.02.007 PMID:29477807

Kaptchouang Tchatchouang, C.-D., Fri, J., De Santi, M., Brandi, G., Schiavano, G. F., Amagliani, G., & Ateba, C. N. (2020). Listeriosis outbreak in South Africa: A comparative analysis with previously reported cases worldwide. *Microorganisms*, *8*(1), 135. doi:10.3390/microorganisms8010135 PMID:31963542

Kasimanickam, V., Kasimanickam, M., & Kasimanickam, R. (2021). Antibiotics use in food animal production: escalation of antimicrobial resistance: where are we now in combating AMR? *Medical Science*, *9*(1), 14. PMID:33669981

Kasman, L. M., & Porter, L. D. (2022). Bacteriophages. StatPearls Publishing.

Kasman, L. M., & Porter, L. D. (2023). Bacteriophages. StatPearls Publishing.

Kaźmierczak, Z., Piotrowicz, A., Owczarek, B., Hodyra, K., Miernikiewicz, P., Lecion, D., Harhala, M., Górski, A., & Dąbrowska, K. (2014). Molecular imaging of T4 phage in mammalian tissues and cells. *Bacteriophage*, *4*(2), e28364. doi:10.4161/bact.28364 PMID:24653943

Khare, S., Tonk, A., & Rawat, A. (2018). Foodborne diseases outbreak in India: A Review. *International Journal of Food Sciences and Nutrition*, *3*, 9–10.

Kilcher, S., Studer, P., Muessner, C., Klumpp, J., & Loessner, M. J. (2018). Cross-genus rebooting of custom-made, synthetic bacteriophage genomes in L-form bacteria. *Proceedings of the National Academy of Sciences of the United States of America*, *115*(3), 567–572. doi:10.1073/pnas.1714658115 PMID:29298913

Koch, B. J., Hungate, B. A., & Price, L. B. (2017). Food-animal production and the spread of antibiotic resistance: The role of ecology. *Frontiers in Ecology and the Environment*, *15*(6), 309–318. doi:10.1002/fee.1505

Korehei, R., & Kadla, J. (2013). Incorporation of T4 bacteriophage in electrospun fibres. *Journal of Applied Microbiology, 114*(5), 1425–1434. doi:10.1111/jam.12158 PMID:23445225

Kuai, L., Liu, F., Chiou, B.-S., Avena-Bustillos, R. J., McHugh, T. H., & Zhong, F. (2021). Controlled release of antioxidants from active food packaging: A review. *Food Hydrocolloids, 120*, 106992. doi:10.1016/j.foodhyd.2021.106992

Kuek, M., McLean, S. K., & Palombo, E. A. (2022). Application of bacteriophages in food production and their potential as biocontrol agents in the organic farming industry. *Biological Control, 165*, 104817. doi:10.1016/j.biocontrol.2021.104817

Lavelle, K., McDonnell, B., Fitzgerald, G., van Sinderen, D., & Mahony, J. (2023). Bacteriophage-host interactions in Streptococcus thermophilus and their impact on co-evolutionary processes. *FEMS Microbiology Reviews, 47*(4), fuad032. doi:10.1093/femsre/fuad032 PMID:37339909

Lavilla, M., Domingo-Calap, P., Sevilla-Navarro, S., & Lasagabaster, A. (2023). Natural Killers: Opportunities and Challenges for the Use of Bacteriophages in Microbial Food Safety from the One Health Perspective. *Foods, 12*(3), 552. doi:10.3390/foods12030552 PMID:36766081

Leiman, P. G., Chipman, P. R., Kostyuchenko, V. A., Mesyanzhinov, V. V., & Rossmann, M. G. (2004). Three-dimensional rearrangement of proteins in the tail of bacteriophage T4 on infection of its host. *Cell, 118*(4), 419–429. doi:10.1016/j.cell.2004.07.022 PMID:15315755

Leiman & Shneider. (2012). Contractile tail machines of bacteriophages. *Viral Molecular Machines*, 93-114.

Leon-Velarde, C. G., Jun, J. W., & Skurnik, M. (2019). Yersinia phages and food safety. *Viruses, 11*(12), 1105. doi:10.3390/v11121105 PMID:31795231

Leung, V., Groves, L., Szewczyk, A., Hosseinidoust, Z., & Filipe, C. D. (2018). Long-Term Antimicrobial Activity of Phage–Sugar Glasses is Closely Tied to the Processing Conditions. *ACS Omega, 3*(12), 18295–18303. doi:10.1021/acsomega.8b02679

Levin, B. R., Moineau, S., Bushman, M., & Barrangou, R. (2013). The population and evolutionary dynamics of phage and bacteria with CRISPR–mediated immunity. *PLOS Genetics, 9*(3), 1003312. doi:10.1371/journal.pgen.1003312 PMID:23516369

Li, C., Shi, T., Sun, Y., & Zhang, Y. (2022). A novel method to create efficient phage cocktails via use of phage-resistant bacteria. *Applied and Environmental Microbiology, 88*(6), e02323–e21. doi:10.1128/aem.02323-21 PMID:35080902

Li, J., Zhao, F., Zhan, W., Li, Z., Zou, L., & Zhao, Q. (2022). Challenges for the application of bacteriophages as effective antibacterial agents in the food industry. *Journal of the Science of Food and Agriculture, 102*(2), 461–471. doi:10.1002/jsfa.11505 PMID:34487550

Lin, D. M., Koskella, B., & Lin, H. C. (2017). Phage therapy: An alternative to antibiotics in the age of multi-drug resistance. *World Journal of Gastrointestinal Pharmacology and Therapeutics, 8*(3), 162. doi:10.4292/wjgpt.v8.i3.162 PMID:28828194

Ling, H., Lou, X., Luo, Q., He, Z., Sun, M., & Sun, J. (2022). Recent advances in bacteriophage-based therapeutics: Insight into the post-antibiotic era. *Acta Pharmaceutica Sinica. B, 12*(12), 4348–4364. doi:10.1016/j.apsb.2022.05.007 PMID:36561998

Liu, A., Liu, Y., Peng, L., Cai, X., Shen, L., Duan, M., Ning, Y., Liu, S., Li, C., Liu, Y., Chen, H., Wu, W., Wang, X., Hu, B., & Li, C. (2020). Characterization of the narrow-spectrum bacteriophage LSE7621 towards Salmonella enteritidis and its biocontrol potential on lettuce and tofu. *Lebensmittel-Wissenschaft + Technologie, 118*, 108791. doi:10.1016/j.lwt.2019.108791

Liu, C., Hong, Q., Chang, R. Y. K., Kwok, P. C. L., & Chan, H. K. (2022). Phage–Antibiotic therapy as a promising strategy to combat multidrug-resistant infections and to enhance antimicrobial efficiency. *Antibiotics (Basel, Switzerland), 11*(5), 570. doi:10.3390/antibiotics11050570 PMID:35625214

Lone, A., Anany, H., Hakeem, M., Aguis, L., Avdjian, A.-C., Bouget, M., Atashi, A., Brovko, L., Rochefort, D., & Griffiths, M. W. (2016). Development of prototypes of bioactive packaging materials based on immobilized bacteriophages for control of growth of bacterial pathogens in foods. *International Journal of Food Microbiology, 217*, 49–58. doi:10.1016/j.ijfoodmicro.2015.10.011 PMID:26490649

Lu, T. K., & Collins, J. J. (2007). Dispersing biofilms with engineered enzymatic bacteriophage. *Proceedings of the National Academy of Sciences of the United States of America, 104*(27), 11197–11202. doi:10.1073/pnas.0704624104 PMID:17592147

Ma, Y., Pacan, J. C., Wang, Q., Sabour, P. M., Huang, X., & Xu, Y. (2012). Enhanced alginate microspheres as means of oral delivery of bacteriophage for reducing Staphylococcus aureus intestinal carriage. *Food Hydrocolloids, 26*(2), 434–440. doi:10.1016/j.foodhyd.2010.11.017

Marchi, J., Zborowsky, S., Debarbieux, L., & Weitz, J. S. (2023). The dynamic interplay of bacteriophage, bacteria and the mammalian host during phage therapy. *iScience, 26*(2), 106004. doi:10.1016/j.isci.2023.106004 PMID:36818291

Marzari, R., Sblattero, D., Righi, M., & Bradbury, A. (1997). Extending filamentous phage host range by the grafting of a heterologous receptor binding domain. *Gene*, *185*(1), 27–33. doi:10.1016/S0378-1119(96)00623-3 PMID:9034309

Matle, I., Mbatha, K. R., & Madoroba, E. (2020). A review of Listeria monocytogenes from meat and meat products: Epidemiology, virulence factors, antimicrobial resistance and diagnosis. *The Onderstepoort Journal of Veterinary Research*, *87*(1), 1–20. doi:10.4102/ojvr.v87i1.1869 PMID:33054262

Matthews, L., Low, J., Gally, D., Pearce, M., Mellor, D., Heesterbeek, J., ... Reid, S. (2006). Heterogeneous shedding of Escherichia coli O157 in cattle and its implications for control. *Proceedings of the National Academy of Sciences of the United States of America*, *103*(3), 547–552. doi:10.1073/pnas.0503776103 PMID:16407143

Metselaar, K. I., den Besten, H. M., Abee, T., Moezelaar, R., & Zwietering, M. H. (2013). Isolation and quantification of highly acid resistant variants of Listeria monocytogenes. *International Journal of Food Microbiology*, *166*(3), 508–514. doi:10.1016/j.ijfoodmicro.2013.08.011 PMID:24051176

Mizan, M. F. R., Cho, H. R., Ashrafudoulla, M., Cho, J., Hossain, M. I., Lee, D.-U., & Ha, S.-D. (2020). The effect of physico-chemical treatment in reducing Listeria monocytogenes biofilms on lettuce leaf surfaces. *Biofouling*, *36*, 1243–1255. PMID:33401969

Moye, Z. D., Woolston, J., & Sulakvelidze, A. (2018). Bacteriophage applications for food production and processing. *Viruses*, *10*(4), 205. doi:10.3390/v10040205 PMID:29671810

Naylor, S. W., Low, J. C., Besser, T. E., Mahajan, A., Gunn, G. J., Pearce, M. C., McKendrick, I. J., Smith, D. G. E., & Gally, D. L. (2003). Lymphoid follicle-dense mucosa at the terminal rectum is the principal site of colonization of enterohemorrhagic Escherichia coli O157: H7 in the bovine host. *Infection and Immunity*, *71*(3), 1505–1512. doi:10.1128/IAI.71.3.1505-1512.2003 PMID:12595469

Nazir, A., Xu, X., Liu, Y., & Chen, Y. (2023). Phage Endolysins: Advances in the World of Food Safety. *Cells*, *12*(17), 2169. doi:10.3390/cells12172169 PMID:37681901

Nelson, D. L., & Cox, M. M. (2008). *Lehninger Principles of Biochemistry* (5th ed.). W.H. Freeman and Company.

Nerin, C., Aznar, M., & Carrizo, D. (2016). Food contamination during food process. *Trends in Food Science & Technology*, *48*, 63–68. doi:10.1016/j.tifs.2015.12.004

Newell, D., Elvers, K., Dopfer, D., Hansson, I., Jones, P., James, S., ... Connerton, I. (2011). Biosecurity-based interventions and strategies to reduce Campylobacter spp. on poultry farms. *Applied and Environmental Microbiology*, *77*(24), 8605–8614. doi:10.1128/AEM.01090-10 PMID:21984249

Newell, D., & Fearnley, C. (2003). Sources of Campylobacter colonization in broiler chickens. *Applied and Environmental Microbiology*, *69*(8), 4343–4351. doi:10.1128/AEM.69.8.4343-4351.2003 PMID:12902214

Nguyen, M. M., Gil, J., Brown, M., Cesar Tondo, E., Soraya Martins De Aquino, N., Eisenberg, M., & Erickson, S. (2020). Accurate and sensitive detection of Salmonella in foods by engineered bacteriophages. *Scientific Reports*, *10*(1), 17463. doi:10.103841598-020-74587-8 PMID:33060781

O'Shea, Y. A., & Boyd, E. F. (2002). Mobilization of the Vibrio pathogenicity island between Vibrio cholerae isolates mediated by CP-T1 generalized transduction. *FEMS Microbiology Letters*, *214*(2), 153–157. doi:10.1111/j.1574-6968.2002.tb11339.x PMID:12351223

Opperman, C. J., Wojno, J. M., & Brink, A. J. (2022). Treating bacterial infections with bacteriophages in the 21st century. *Southern African Journal of Infectious Diseases*, *37*(1), 346. doi:10.4102ajid.v37i1.346 PMID:35399556

Pandit, M. A., Kumar, J., Gulati, S., Bhandari, N., Mehta, P., Katyal, R., Rawat, C. D., Mishra, V., & Kaur, J. (2022). Major biological control strategies for plant pathogens. *Pathogens (Basel, Switzerland)*, *11*(2), 273. doi:10.3390/pathogens11020273 PMID:35215215

Park, J. Y., Moon, B. Y., Park, J. W., Thornton, J. A., Park, Y. H., & Seo, K. S. (2017). Genetic engineering of a temperate phage-based delivery system for CRISPR/Cas9 antimicrobials against Staphylococcus aureus. *Scientific Reports*, *7*(1), 44929. doi:10.1038rep44929 PMID:28322317

Pickard, J. M., Zeng, M. Y., Caruso, R., & Núñez, G. (2017). Gut microbiota: Role in pathogen colonization, immune responses, and inflammatory disease. *Immunological Reviews*, *279*(1), 70–89. doi:10.1111/imr.12567 PMID:28856738

Pires, D. P., Costa, A. R., Pinto, G., Meneses, L., & Azeredo, J. (2020). Current challenges and future opportunities of phage therapy. *FEMS Microbiology Reviews*, *44*(6), 684–700. doi:10.1093/femsre/fuaa017 PMID:32472938

Połaska, M., & Sokołowska, B. (2019). Bacteriophages—A new hope or a huge problem in the food industry. *AIMS Microbiology*, *5*(4), 324–346. doi:10.3934/microbiol.2019.4.324 PMID:31915746

Principi, N., Silvestri, E., & Esposito, S. (2019). Advantages and limitations of bacteriophages for the treatment of bacterial infections. *Frontiers in Pharmacology*, *10*, 513. doi:10.3389/fphar.2019.00513 PMID:31139086

Pyra, A., Filik, K., Szermer-Olearnik, B., Czarny, A., & Brzozowska, E. (2020). New insights on the feature and function of tail tubular protein b and tail fiber protein of the lytic bacteriophage φYeO3-12 specific for Yersinia enterocolitica serotype O: 3. *Molecules (Basel, Switzerland)*, *25*(19), 4392. doi:10.3390/molecules25194392 PMID:32987777

Quirós, P., Colomer-Lluch, M., Martínez-Castillo, A., Miró, E., Argente, M., Jofre, J., Navarro, F., & Muniesa, M. (2014). Antibiotic resistance genes in the bacteriophage DNA fraction of human fecal samples. *Antimicrobial Agents and Chemotherapy*, *58*(1), 606–609. doi:10.1128/AAC.01684-13 PMID:24165177

Rahimi-Midani, A., Lee, S. W., & Choi, T. J. (2021). Potential solutions using bacteriophages against antimicrobial resistant bacteria. *Antibiotics (Basel, Switzerland)*, *10*(12), 1496. doi:10.3390/antibiotics10121496 PMID:34943708

Ramos-Vivas, J., Elexpuru-Zabaleta, M., Samano, M. L., Barrera, A. P., Forbes-Hernández, T. Y., Giampieri, F., & Battino, M. (2021). Phages and enzybiotics in food biopreservation. *Molecules (Basel, Switzerland)*, *26*(17), 5138. doi:10.3390/molecules26175138 PMID:34500572

Rawat, S. (2015). Food spoilage: Microorganisms and their prevention. *Asian J. Plant Sci. Res.*, *5*, 47–56.

Raya, R. R., Varey, P., Oot, R. A., Dyen, M. R., Callaway, T. R., Edrington, T. S., Kutter, E. M., & Brabban, A. D. (2006). Isolation and characterization of a new T-even bacteriophage, CEV1, and determination of its potential to reduce Escherichia coli O157:H7 levels in sheep. *Applied and Environmental Microbiology*, *72*(9), 6405–6410. doi:10.1128/AEM.03011-05 PMID:16957272

Riley, L. W. (2020). Extraintestinal foodborne pathogens. *Annual Review of Food Science and Technology*, *11*(1), 275–294. doi:10.1146/annurev-food-032519-051618 PMID:32004092

Ryan, E. M., Gorman, S. P., Donnelly, R. F., & Gilmore, B. F. (2011). Recent advances in bacteriophage therapy: How delivery routes, formulation, concentration and timing influence the success of phage therapy. *The Journal of Pharmacy and Pharmacology*, *63*(10), 1253–1264. doi:10.1111/j.2042-7158.2011.01324.x PMID:21899540

Schirone, M., Visciano, P., Tofalo, R., & Suzzi, G. (2019). Foodborne pathogens: Hygiene and safety. Frontiers Media SA.

Seed, K. D. (2015). Battling phages: How bacteria defend against viral attack. *PLoS Pathogens*, *11*(6), e1004847. doi:10.1371/journal.ppat.1004847 PMID:26066799

Sharma, S., Chatterjee, S., Datta, S., Prasad, R., Dubey, D., Prasad, R. K., & Vairale, M. G. (2017). Bacteriophages and its applications: An overview. *Folia Microbiologica*, *62*(1), 17–55. doi:10.100712223-016-0471-x PMID:27718043

Shkoporov, A. N., & Hill, C. (2019). Bacteriophages of the human gut: The “known unknown” of the microbiome. *Cell Host & Microbe*, *25*(2), 195–209. doi:10.1016/j.chom.2019.01.017 PMID:30763534

Sillankorva, S. M., Oliveira, H., & Azeredo, J. (2012). Bacteriophages and their role in food safety. *International Journal of Microbiology*, *2012*, 863945. doi:10.1155/2012/863945 PMID:23316235

Smith, H. W., Huggins, M. B., & Shaw, K. M. (1987). Factors influencing the survival and multiplication of bacteriophages in calves and in their environment. *Microbiology*, *133*(5), 1127–1135. doi:10.1099/00221287-133-5-1127 PMID:3309178

Sohail, H. A., Coffey, A., Debrowska, K., Meyer, I. M., Middelboe, M., Sohail, M., & Clokie, M. R. (2020). Bacteriophages: Emerging applications in medicine, food, and biotechnology. *PHAGE (New Rochelle, N.Y.)*, *1*(2), 75–82. doi:10.1089/phage.2020.29004.has PMID:36147894

Strange, J. E., Leekitcharoenphon, P., Møller, F. D., & Aarestrup, F. M. (2021). Metagenomics analysis of bacteriophages and antimicrobial resistance from global urban sewage. *Scientific Reports*, *11*(1), 1600. doi:10.103841598-021-80990-6 PMID:33452346

Strathdee, S. A., Hatfull, G. F., Mutalik, V. K., & Schooley, R. T. (2023). Phage therapy: From biological mechanisms to future directions. *Cell*, *186*(1), 17–31. doi:10.1016/j.cell.2022.11.017 PMID:36608652

Strydom, A. (2015). *Control of Listeria monocytogenes in an avocado processing facility*. University of the Free State.

Sukumaran, A. T., Nannapaneni, R., Kiess, A., & Sharma, C. S. (2015). Reduction of Salmonella on chicken meat and chicken skin by combined or sequential application of lytic bacteriophage with chemical antimicrobials. *International Journal of Food Microbiology*, *207*, 8–15. doi:10.1016/j.ijfoodmicro.2015.04.025 PMID:25950852

Summers, W. C. (2012). The strange history of phage therapy. *Bacteriophage*, *2*(2), 130–133. doi:10.4161/bact.20757 PMID:23050223

Summers, W. C. (2021). The discovery of bacteriophages and the historical context. Bacteriophages: Biology, Technology. *Therapy*, 387–400.

Svircev, A., Roach, D., & Castle, A. (2018). Framing the future with bacteriophages in agriculture. *Viruses*, *10*(5), 218. doi:10.3390/v10050218 PMID:29693561

Swaminathan, B., & Gerner-Smidt, P. (2007). The epidemiology of human listeriosis. *Microbes and Infection*, *9*(10), 1236–1243. doi:10.1016/j.micinf.2007.05.011 PMID:17720602

Thung, T. Y., Lee, E., Mahyudin, N. A., Anuradha, K., Mazlan, N., Kuan, C. H., Pui, C. F., Ghazali, F. M., Ab Rashid, N.-K. M., & Rollon, W. D. (2019). Evaluation of a lytic bacteriophage for bio-control of Salmonella Typhimurium in different food matrices. *Lebensmittel-Wissenschaft + Technologie*, *105*, 211–214. doi:10.1016/j.lwt.2019.02.033

Thung, T. Y., Lee, E., Mahyudin, N. A., Anuradha, K., Mazlan, N., Kuan, C. H., ... Rollon, W. D. (2019). Evaluation of a lytic bacteriophage for bio-control of Salmonella Typhimurium in different food matrices. *Lebensmittel-Wissenschaft + Technologie*, *105*, 211–214. doi:10.1016/j.lwt.2019.02.033

Thung, T. Y., Lee, E., Mahyudin, N. A., Wan Mohamed Radzi, C. W. J., Mazlan, N., Tan, C. W., & Radu, S. (2020). Partial characterization and in vitro evaluation of a lytic bacteriophage for biocontrol of Campylobacter jejuni in mutton and chicken meat. *Journal of Food Safety*, *40*(2), e12770. doi:10.1111/jfs.12770

Torres-Barceló, C. (2018). Phage therapy faces evolutionary challenges. *Viruses*, *10*(6), 323. doi:10.3390/v10060323 PMID:29895791

Torres-Barceló, C., & Hochberg, M. E. (2016). Evolutionary rationale for phages as complements of antibiotics. *Trends in Microbiology*, *24*(4), 249–256. doi:10.1016/j.tim.2015.12.011 PMID:26786863

Vandamme, E. J., & Mortelmans, K. (2019). A century of bacteriophage research and applications: Impacts on biotechnology, health, ecology and the economy! *Journal of Chemical Technology and Biotechnology*, *94*(2), 323–342. doi:10.1002/jctb.5810

Vengarai Jagannathan, B., Kitchens, S., Priyesh Vijayakumar, P., Price, S., & Morgan, M. (2021). Efficacy of bacteriophage cocktail to control E. coli O157: H7 contamination on baby spinach leaves in the presence or absence of organic load. *Microorganisms*, *9*(3), 544. doi:10.3390/microorganisms9030544 PMID:33800760

Vikram, A., Tokman, J. I., Woolston, J., & Sulakvelidze, A. (2020). Phage Biocontrol Improves Food Safety by Significantly Reducing the Level and Prevalence of Escherichia coli O157:H7 in Various Foods. *Journal of Food Protection*, *83*(4), 668–676. doi:10.4315/0362-028X.JFP-19-433 PMID:32221572

Vikram, A., Woolston, J., & Sulakvelidze, A. (2021). Phage biocontrol applications in food production and processing. *Current Issues in Molecular Biology*, *40*, 267–302. doi:10.21775/cimb.040.267 PMID:32644048

Villalpando-Aguilar, J. L., Matos-Pech, G., López-Rosas, I., Castelán-Sánchez, H. G., & Alatorre-Cobos, F. (2022). Phage therapy for crops: Concepts, experimental and bioinformatics approaches to direct its application. *International Journal of Molecular Sciences*, *24*(1), 325. doi:10.3390/ijms24010325 PMID:36613768

Wagenaar, J. A., Van Bergen, M. A., Mueller, M. A., Wassenaar, T. M., & Carlton, R. M. (2005). Phage therapy reduces Campylobacter jejuni colonization in broilers. *Veterinary Microbiology*, *109*(3–4), 275–283. doi:10.1016/j.vetmic.2005.06.002 PMID:16024187

Wagh, R. V., Priyadarshi, R., & Rhim, J.-W. (2023). Novel bacteriophage-based food packaging: An innovative food safety approach. *Coatings*, *13*(3), 609. doi:10.3390/coatings13030609

Whichard, J., Sriranganathan, M. N., & Pierson, F. W. (2003). Suppression of Salmonella growth by wild-type and large-plaque variants of bacteriophage Felix O1 in liquid culture and on chicken frankfurters. *Journal of Food Protection*, *66*(2), 220–225. doi:10.4315/0362-028X-66.2.220 PMID:12597480

Williams, K., Ward, M., Dhungyel, O., Hall, E., & Van Breda, L. (2014). A longitudinal study of the prevalence and super-shedding of Escherichia coli O157 in dairy heifers. *Veterinary Microbiology*, *173*(1-2), 101–109. doi:10.1016/j.vetmic.2014.07.001 PMID:25064268

Wójcicki, M., Błażejak, S., Gientka, I., & Brzezicka, K. (2019). The concept of using bacteriophages to improve the microbiological quality of minimally processed foods. *Acta Scientiarum Polonorum. Technologia Alimentaria*, *18*, 373–383. PMID:31930789

Wong, C. W., Delaquis, P., Goodridge, L., Lévesque, R. C., Fong, K., & Wang, S. (2020). Inactivation of Salmonella enterica on post-harvest cantaloupe and lettuce by a lytic bacteriophage cocktail. *Current Research in Food Science, 2*, 25–32.

Woo, J., & Ahn, J. (2014). Assessment of synergistic combination potential of probiotic and bacteriophage against antibiotic-resistant Staphylococcus aureus exposed to simulated intestinal conditions. *Archives of Microbiology*, *196*(10), 719–727. doi:10.100700203-014-1013-z PMID:25015717

Yan, T., Liang, L., Yin, P., Zhou, Y., Mahdy Sharoba, A., Lu, Q., Dong, X., Liu, K., Connerton, I. F., & Li, J. (2020). Application of a novel phage LPSEYT for biological control of Salmonella in foods. *Microorganisms*, *8*(3), 400. doi:10.3390/microorganisms8030400 PMID:32178465

Yang, S. C., Lin, C. H., Aljuffali, I. A., & Fang, J. Y. (2017). Current pathogenic *Escherichia coli* foodborne outbreak cases and therapy development. *Archives of Microbiology*, *199*(6), 811–825. doi:10.100700203-017-1393-y PMID:28597303

Żbikowska, K., Michalczuk, M., & Dolka, B. (2020). The use of bacteriophages in the poultry industry. *Animals (Basel)*, *10*(5), 872. doi:10.3390/ani10050872 PMID:32443410

Zhang, X., Niu, Y. D., Nan, Y., Stanford, K., Holley, R., McAllister, T., & Narváez-Bravo, C. (2019). SalmoFresh effectiveness in controlling salmonella on romaine lettuce, mung bean sprouts and seeds. *International Journal of Food Microbiology*, *305*, 108250.

Zia, S., & Alkheraije, K. A. (2023). Recent trends in the use of bacteriophages as replacement of antimicrobials against food-animal pathogens. *Frontiers in Veterinary Science*, *10*, 1162465. doi:10.3389/fvets.2023.1162465 PMID:37303721

Chapter 9
Development of Edible Food Wrappers:
An Eco-Friendly Approach Towards Sustainability

Zunaira Basharat
Jiangsu University, China

Darakhshan Ahmed
Department of Chemistry, UET Lahore, Pakistan

Muhammad Rizwan Tariq
University of the Punjab, Lahore, Pakistan

Shinawar Waseem Ali
University of the Punjab, Lahore, Pakistan

Muhammad Usama Pervaiz
Department of Clinical Sciences and Nutrition, University of Chester, UK

ABSTRACT

Edible food wrappers are safe packaging systems that can be consumed with food. Edible packaging developed using natural renewable resources come under the category of bio-based and biodegradable packaging. The main components of edible wrappers include a biopolymer solubilized in a solvent to form film casting solution and certain additives to improve functional characteristics of packaging. Biopolymers are frequently used in food packaging applications due to their flexibility in film formation and biodegradability. The development of green packaging materials utilizing discarded edible materials will be beneficial to confront the challenges associated with traditional packaging systems, attaining sustainability, and fostering recycling in the food industry. Edible food wrappers present a number of benefits over synthetic materials. These packaging systems can replace and possibly strengthen the outside layers of packed items in order to elude moisture loss, flavors, and bioactive components from the foods as well as between them.

DOI: 10.4018/979-8-3693-0819-6.ch009

INTRODUCTION

The demand for sustainable packaging solutions has increased tremendously due to increased consumer awareness, environmental concerns and natural resources depletion (Mellinas et al., 2016). Conventional packaging is a significant concern for environmental sustainability begins non-biodegradable and non-renewable (Jeevahan & Chandrasekaran, 2019). Extracting biopolymers from these by-products offers marketable benefits such as non-toxicity, biocompatibility, environmental friendliness and barrier to moisture and gasses (Mellinas et al., 2016). Thus, these raw materials can be utilized to develop renewable and sustainable packaging systems (Petkoska et al., 2021).

Edible food wrappers are safe packaging systems that can be consumed with food. They can be diversified into films or coatings without altering their chemical properties. Films create bags, pouches, and wraps, while coatings are ingested directly (Aguirre-Joya et al., 2018). Selecting the right material and fabrication method depends on the food, and the packaging must have sensory compatibility with the enclosed food (Restrepo et al., 2018).

The components used to engineer food wrappers are intended to be edible, natural, biodegradable, and sustainable that offer a legitimate substitute to traditional packaging with high environmental risks (Wei & Yazdanifard, 2013). These wrappers also carry bioactive substances, adding technical and therapeutic advantages. After use, they naturally decompose in digestion or through composting, reducing waste and environmental harm (Patel, 2019).

COMPOSITION OF EDIBLE FOOD WRAPPERS

The main components of edible wrappers include a biopolymer solubilized in a solvent to form film casting solution and certain additives to improve functional characteristics of packaging as shown in Figure 1 (Teixeira-Costa & Andrade, 2012).

Biopolymers

Bio macromolecules including polysaccharides, proteins and lipids are extensively used as a biopolymer is formation of edible wrappers (Teixeira-Costa & Andrade, 2012). Edible packaging developed using natural renewable resources come under the category of bio based and biodegradable packaging. According to Chen et al. (2019), the selection of packaging materials is relying on food characteristics, intended properties and environmental factors to which wrapped product is expected to exposed during distribution and storage.

Figure 1. Composition of edible food wrappers

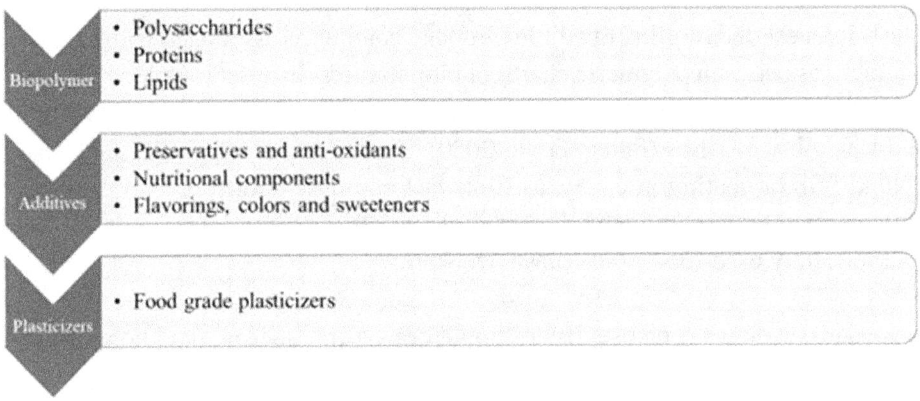

Velickova et al. (2015), classified edible wrappers into three categories on the basis of biopolymers; hydrocolloids, lipids, and composites. Hydrocolloids, are designed using hydrophilic biopolymers, including animal and plant proteins (caseins, gelatin, corn zein, peanut, soybean, and wheat) and polysaccharides (alginate, cellulose, chitosan, carrageenan, pectin, and starch), that assemble into complex framework. Lipids comprise essential oils, oils, resins, and waxes (Shit & Shah, 2014).

Polysaccharide and protein-based films and coating provides excellent gas barrier properties and mechanical strength however, allows moisture transfer (Bharti et al., 2020). Composites or multi-component films are developed to transcend these constraints by blending both lipids and hydrocolloids (Senturk Parreidt et al., 2018).

Additives (Active Components)

The literature suggested that indirect incorporation of bioactive components such as edible food wrappers supplemented with of bioactive compound including antioxidants and antimicrobials could be more efficient than the direct incorporation into the food metrics (Valdés et al., 2017).

Furthermore, the inclusion of bioactive into edible films ensures gradual release and sustained activity for extended duration (Benbettaïeb et al., 2019). The organoleptic properties of food could be improved by developing films and coatings with inclusion of seasoning, sweetener, flavoring and coloring. Additionally, incorporation of nutritional components like vitamins, minerals, prebiotics and probiotics could enhance market value of products (Senturk Parreidt et al., 2018).

Plasticizers

The brittleness and fragility of edible films and coatings is associated with extensive interactions between polymeric chains of biopolymers. In order to reduce polymer–polymer interactions, plasticizers are added to improve toughness, stretchability, and flexibility of films (Suhag et al., 2020). Plasticizers are typically essential to uphold film or coating integrity, as films lacking these additives tend to be rigid with elevated tensile strength (Thakur et al., 2019).

Frequently used food-grade plasticizers are:

- Carbohydrates (glucose, fructose-glucose syrups, sucrose, and honey)
- Polyols (glycerol, sorbitol, glyceryl derivatives, and polyethylene glycols)
- Lipids and their derivatives (phospholipids, FA, oils and waxes)

Classification of Biopolymers

Biopolymers are frequently used in food packaging applications due to their flexibility in film formation and biodegradability. These coatings and films function as an interface between the environment and the food to maintain food quality and expedite shelf life (Petkoska et al., 2021). Biopolymers have been categorized into three distinct groups, based on their origin and processing methods as shown in Figure 2 (Dhall, 2013).

Figure 2. Classification of biopolymers

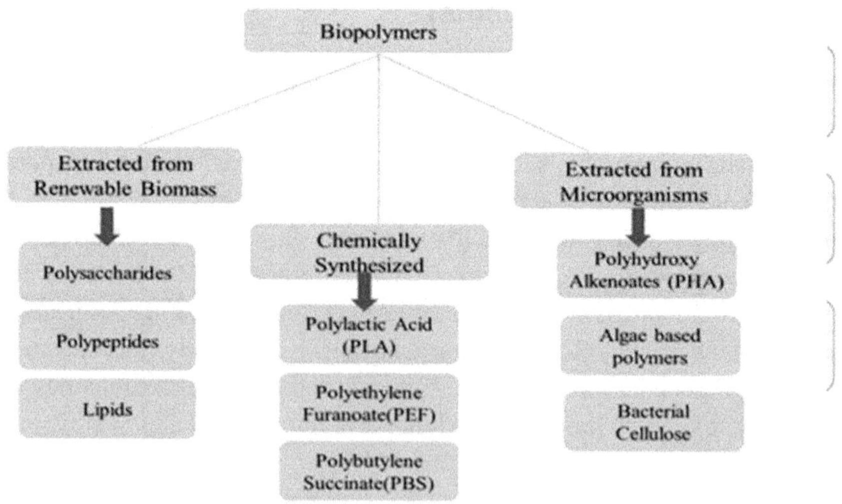

1. Extracted from renewable biomass
2. Chemically synthesized biopolymers
3. Derived from microorganism

Biopolymers are considered as extremely valued food packaging material due to their chemical complexity, mechanical strength, thermal resistance and barrier properties (Li et al., 2019). These materials can be used alone or in combination with other edible materials to obtain anticipated characteristics.

BIOPOLYMERS EXTRACTED FROM RENEWABLE BIOMASS

Polysaccharides

Polysaccharides are intricate bio macromolecules made up of repeated mono- or disaccharide units connected by glycosidic linkages. Alginate, cellulose, chitin and starch are most common framework for active edible food packaging. Polysaccharides derivatives exhibit remarkable oxygen and carbon dioxide barrier properties under low-humidity (Caz´ on et al., 2017). Hence, these films are modified by addition of hydrophobic ingredients to improve the water vapor transference of the final edible wraps. Classification is presented in Figure 3.

- **Cellulose-based food packaging films:** Cellulose derivatives are extensively used to develop biodegradable combinations because of non-toxicity,

Figure 3. Classification of biopolymers extracted from renewable resources

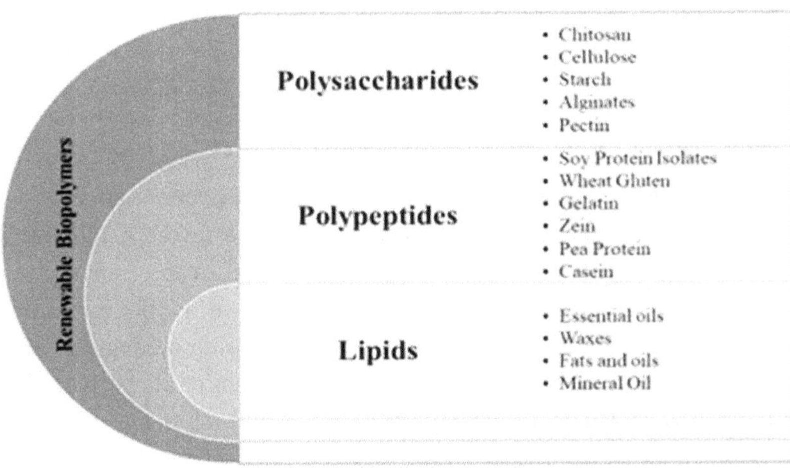

accessibility, excellent mechanical properties, and high chemical stability (Sen Wang et al., 2016). Active films can't be fabricated directly from natural cellulosic materials due to their considerable structural complexity and high crystallinity (Sen Wang et al., 2016). However, hydrophobicity of cellulose-based films is improved through photo-crosslinking hence; enhanced water barrier properties are attained (Shahbazi et al., 2016).

- **Chitosan-based food packaging films:** Chitosan is derived from alkaline N-deacetylation of chitin (Kim et al., 2006) and found extensive industrial implications due to its biodegradability, non-toxicity, and biocompatibility (Dutta et al., 2009). The extrusion process is economical however low degradation temperature of chitosan hinders film formation through extrusion hence solvent casting is considered as efficient approach (Pelissari et al., 2011). Additionally, Chitosan based films always requires a suitable plasticizer to aid smooth fabrication because of their low water solubility (Rangaraj et al., 2021).
- **Starch-based food packaging films:** Starch is considered as the most prevalent polysaccharide derivative, produced from variety of agricultural commodities including beans, cassava, maize, rice, sorghum, and wheat (Pelissari et al., 2019). The conformation of amylopectin and amylose in starch granules has a remarkable influence on film formation for instance; the elevated levels of amylopectin in edible films possess a significant O_2 barrier property. (Caz´ on et al., 2017).

The eminent crystalline nature imparts starch granules with substantial tensile strength and limited elongation, thereby leading to the formation of brittle edible films (Jonhed et al., 2008). In order to reduce the brittleness and improve flexibility various biocompatible synthetic polymers and plasticizers are blended with primary formulation (Yoon, 2014; Mekonnen, et al., 2013). Starch, owing to its strong affinity for natural additives such as antioxidants and nanofillers, offers a versatile foundation for creating a diverse range of edible food wrappers.

- **Alginate-based food packaging films:** Alginate, principally derived from brown seaweeds, is a linear polysaccharide derivative largely exists as sodium salt, composed of monomeric units of 1-4-linked α-D-mannuronate and β-L-guluronate (Pawar & Edgar, 2012). Current investigations are concentrated on exploring alginate's potential for film fabrication, driven by its environment friendly characteristics encompassing biodegradability, non-toxicity, and economic viability (Tavassoli-Kafrani et al., 2016). Generally, sodium alginate is cross-linked with calcium salt through the solution casting to manufacture edible active film with exceptional mechanical properties.

However, alginate-based edible warps are associated with weak water resistance attributed to their hydrophilic nature. Hence, alginate films are modified through amalgamation with polyvalent metal cations (Ca^{2+}) or with polyamines (chitosan) to yield crosslinked edible film with excellent moisture resistance (Sharma et al., 2012).

Polypeptides

Polypeptides are biopolymers characterized by intricate structures featuring diverse bonds that confer distinct functional attributes, and their interaction with different biopolymer derivatives can yield high-quality composite films with exceptional properties. Furthermore, the film-forming ability of the polypeptide derivatives is function of their charge, molecular weight, structure, and solubility (Calva-Estrada et al., 2019) whereas, mechanical properties of the edible films are associated with arrangement of amino acids in the protein chain and extent of intra-protein interactions (Koshy et al., 2015). Polypeptide films exhibit excellent O_2 and CO_2 barrier properties, which restrict flavour loss and functional component migration hence, ameliorate food shelf-life (Murrieta-Martínez et al., 2018). Furthermore, these films are biodegradable and serve as biofertilizers because of high nitrogen content (Chen et al., 2019). Plant-based proteins are of significant interest for both food manufacturers and consumers seeking natural food sources and substitutes to accommodate vegetarian, vegan, and dietary restrictions related to food allergies.

- **Gelatin based food packaging films:** Gelatin is a water-soluble biopolymer, compatible with different antioxidants and antimicrobial agents, which possess excellent film- forming ability. Gelatin exhibits low O_2 permeability, and it is primarily used for developing active food packaging films and coatings (Ramos et al., 2016). Gelatin is prepared through partial hydrolysis of insoluble collagen protein, mainly isolated from bones, skin, and connective tissues of the biowaste produced during animal thrashing (Nur Hanani et al., 2014). The mechanical and barrier properties of gelatin films are function of their amino acid sequence, structure and molecular weight (G´ omez-Guill´en et al., 2011).

Gelatin films have high moisture sensitivity interlinked to its hygroscopic behavior (Ramos et al., 2016), hence, cross linking is essential to manipulate intermolecular forces to form an edible film with a perfect interpenetrated network structure (Lin et al., 2019). Gelatin can be integrated with other biopolymers, such as chitosan (Hosseini et al., 2013), rice bran protein (Shin et al., 2011), starch (Al-Hassan &

Norziah, 2012), zein protein (Xia et al. 2019), etc., to develop hybrid edible food wrappers with enhanced mechanical and barrier properties.

- **Wheat gluten based food packaging films:** Wheat gluten (WG) can be used as raw material to develop edible packaging systems attributed to its divergent viscoelastic and film-forming properties. Usually, WG-based edible films are manufactured employing the solution casting technique in an aqueous ethanol media (Umaraw & Verma, 2017). Uniformity of edible films can be regulated by adjusting casting solution pH (Angellier-Coussy et al., 2011), controlled disulfide bond disruption and re-oxidation during hot solution treatment and subsequent evaporation lead to a well-structured WG films (Umaraw & Verma, 2017).
- **Soy protein-based food packaging films:** Soy protein isolates (SPIs) are extracted from soy flour, are appropriate precursor for formulating edible film wraps because of abundant availability, biodegradability, economic viability and non-toxicity (Carpin´e, et al., 2015). Generally, SPI edible films are manufactured using the solution casting method under acidic and basic conditions.

Lipids

Addition of lipids in basic formulation is interlinked to provision of flexibility, hydrophobicity, and cohesion in the film. Lipid based food coverings act as excellent moisture barriers and, to some extent oxygen barrier ultimately, they maintain long-term food quality (Milani & Nemati, 2022). Lipids based polymers includes fats and oils, essential oils and plant extracts (eugenol, mint, cinnamon, and oregano), waxes (beeswax, jojoba, and paraffin), and resins (chicle and olibanum).

- **Fats and Oils:** Rodrigues et al. (2016), developed palm oil-based film with superior moisture permeation resistance, transparency, and elasticity. On the other hand, Hassani et al. (2012), used rice bran oil coating for extended shelf life of kiwi fruit that significantly preserved the taste, color, and firmness of fruits.
- **Essential Oils:** Essential oils are reservoirs of various hydrophobic and volatile materials such as terpenoids, terpenes, and other aromatic compounds owing exceptional anti-microbial properties (Han, 2014). Different citrus essential oils have been explored for their anti-microbial activity when incorporated with biopolymer films; results showed enhanced anti-microbial activity however, water permeability value and tensile strength were not affected (Liu et al., 2020; Randazzo et al., 2016).

- **Waxes:** Waxes are higher molecular weight compounds originated from plants or animals; serve as protective covering attributed to high hydrophobicity (Sikorski & Kolakowska, 2011). Previously, Saucedo-Pompa et al. (2007) developed an edible coating using Aloe Vera gel and candelilla wax and applied on fresh-cut fruit to determine its efficacy as a substitute of fruit preservation, improved firmness, and better appearance was observed in comparison to uncoated fruits. Similarly, another film was crafted using carnauba wax incorporated with cassava starch, stearic acid, and glycerol. The developed edible wrap showed excellent mechanical, thermal and barrier properties (Chiumarelli & Hubinger, 2014).
- **Resins:** Resins are the substances formed and released by plants as a defensive mechanism in response to an infection, however, some insects can also produce them, such as Laccifer lacca produces shellac resin (Baldwin et al., 2012). Chauhan et al. (2015) and Chitravathi et al. (2014), reported the application of shellac based edible coatings on green chilies and tomatoes, the coatings enhanced shelf life by 12 days and prohibited senescence attributed to excellent moisture barriers properties. Furthermore, films were transparent and glossy with quick drying ability and significant emulsion stability.

CHEMICALLY SYNTHESIZED BIOPOLYMERS

Chemically synthesized biopolymers are class of biopolymers that are formed as a result of polymerization and cross-linking interactions of bio monomers. These bio monomers are generally extracted from renewable resource or industrial bio-waste as a consequence of microbial fermentation.

Polylatic Acid (PLA)

PLA_s are considered as an ideal substitute of aliphatic plastics due to their lower toxicity, high transparency, biosorbability, and easy processability (Tawakkal et al., 2014). PLAs can be synthesized either by carbohydrate fermentation or lactic acid (LA) polymerization (Mehta et al., 2005). PLA films are fabricated through melt extrusion, followed by thermoforming method. These materials are brittle, exhibit weak O_2 and moisture barrier properties, and low melt strength at high temperatures hence; these materials are modified through blending of compatible polymers and integration of nanofillers and plasticizers (Tawakkal et al., 2014).

Polyethylene Furanoate (PEF)

PEF is a 100% bio-based polyester formed by polymerizing 2,5-furandicarboxylic acid (FDCA) and monoethylene glycol whereas FDCA is obtained through sugar-based bio-fermentation (Mendieta et al., 2022). PEF is comparable to PET in terms of permeability and thermal resistance. Furthermore, PEF provides a water vapor barrier that is twice as high as PET (Rosenboom et al., 2018), and ESFA declared safe under the migration levels 5 mg/kg food. Based on these grounds PEF can be suggested as an eco-friendly substitute of PET with additional focus directed towards further modifications in properties to enhance gas and moisture barriers and regulate biodegradability (Reichert et al., 2020).

Polybutylene Succinate (PBS)

PBS belongs to group of succinate-derived polymers, obtained from anhydration reaction between a diacid or acid anhydride and diols (Groh et al., 2019). Previously, PBS were derived from petroleum and natural gas, however, now PBS can also be manufactured using renewable materials like sugarcane, cassava, and corn through bacterial fermentation (Otoni et al., 2017).

The processability of PBS highly depends on its molecular weight, Mw below 100,000 suitable for extrusion and injection molding, while higher Mw allows film blowing and casting (Kennouche et al., 2016). High crystallization rate and good thermal stability are crucial factors for smooth processing. However, nucleating agents can be employed to improve transparency and crystallinity of films (Mallegni et al., 2018). Siracusa et al. (2015), evaluated PBS and PBSA films in terms of permeability behavior and the changes after food contact; results confirmed the potential of these materials for food packaging systems as no significant variations or damage was observed (Rutz et al., 2016).

BIOPOLYMERS DERIVED FROM MICROORGANISM

Polyhydroxy Alkenoates (PHA)

PHA_s are biodegradable polyesters of different hydroxy alkenoate derivatives extracted from the microbial fermentation (Fabra et al., 2013). PHB films are extremely brittle due to high glass transition rate and melting temperatures (Erceg et al., 2010) hence, PHB is copolymerized with hydroxy valerate to produce poly hydroxy butyl valerate (PHBV) which is comparatively flexible but exhibits low barrier properties. PHB and PHBV can be amalgamated with other biopolymers to augment their barrier properties however, these polymers exhibit less structural compatibility with protein and polysaccharides (Fabra et al., 2014). The interfacial polymerization notably augmented surface adhesion between starch and PHBV, leading to composite films displaying elevated mechanical and barrier characteristics (Xu et al., 2017; Fabra et al., 2013).

Algae Based Biopolymers

Over the past few years, algae have raised as a highly sustainable substitute for bioplastic production, owing to their rapid growth rate, resilience to challenging conditions, and ability to thrive on non-arable land. A wide range of natural polymers can be extracted from algae including alginates, agar, carrageenan, algal proteins, fucoidan, funoran, and ulvan. These isolated biopolymers are merged with additives and solvents to fabricate packaging films which showcase outstanding mechanical and thermal properties with high biodegradability (Yap, et al., 2023).

Food Industry By-Product: Sustainable Source of Edible Food Packaging Systems

By-products of food processing industries contain considerable amounts of bioactive components and biopolymers that can be retrieved to replace utilization of virgin components. Implication of such approaches not only fosters circular economy and food chain sustainability but also recycle bio-waste contributing to reduced environmental hazards (Schieber, 2017).

Marine Processing By-Products

- **Residual material from fish and shellfish:** Byproducts from shellfish processing, could exceed 80% and consist primarily of heads and shells (Suresh et al., 2018). On the other hand, canning of tuna result in 70% waste entailing backbone, belly flaps, dark meat, head, and skin (Sasidharan & Venugopal, 2020). Stevens et al. (2018) these byproducts can be valorized to into high-value products for human use like nutraceuticals, pharmaceuticals, oil supplements, and protein powders. These valorized products can be used for direct human consumption or as a raw material of other industries.
- **Residual material from Algae:** Macroalgae are an excellent reservoir of macro-biomolecules including polysaccharides, proteins, lipids, and minerals together with pigments and polyphenols. The most predominant polysaccharides isolated from seaweeds are alginates, agar, and carrageenan with some minor storage polysaccharides including fucoidans, xylans, and

laminarin (Patel, 2019; Ganesan et al., 2018). The fabrication of edible films and coatings involved blending microalgae with additional biopolymers, utilizing intact microalgae cells or extracting compounds like protein concentrates from *Chlorella vulgaris* or *Spirulina platensis* (Morales-Jiménez et al., 2020; Stejskal et al., 2020).

Agricultural Processing By-Products

- **Residual material from fruits and vegetables:** Vegetables and fruit processing industry contribute to a significant amount (20%–40%) of bio-waste in the form of stones/pits, pomace, peels, and pulps. However, these byproducts are excellent source of bioactive compounds and biopolymers (Bas-Bellver et al., 2020). Wastes from fruits and vegetables have the potential to develop edible films and coatings with superior properties. Ben Othman et al. (2020), reported that by-products such seeds and peels of various entails >15% more phenolics as compared to pulp. Furthermore, these residual compounds (dietary fibers, polyphenols, pigments, and essential oils), can enhance the performance of food packaging systems by imparting certain nutritional and functional benefits interlinked to their biological (Dilucia et al., 2020).

Animal Processing By-Products

- **Residual material from meat:** Gelatin found extensive application in food industry as an additive, commonly commercial gelatins are sourced from cattle and pigs, alternative sources encompass fish and even chicken feet (Santana et al., 2020). Gelatins are extracted from collagens, which are prevalent in animal skin, bones, and hooves (Saha & Bhattacharya, 2010). Gelatin serves as a significant biopolymer in edible packaging, exhibiting favorable barrier traits however, due to its hygroscopic nature, synergistic blends with other biopolymers are preferred to enhance packaging functionality and extend the shelf life of food items (Ramos et al., 2016).
- **Residual material from dairy:** Dairy industry generates huge volumes of proteinaceous waste (caseins and whey proteins). Milk proteins have excellent barrier and filmogenic properties, which render them appropriate biomaterial for the synthesis of edible packaging (Hamed et al., 2020).

FABRICATION OF EDIBLE WRAPPERS

Film Formation Methods

Casting Method

Solution casting is most commonly used technique for film formation utilizing a range of animal and plant-derived materials as the primary reservoir of biopolymers (Otoni et al., 2018). The casting process encompasses three stages (Suhag et al., 2020):

1. Preparation of film forming solution by dissolving biopolymers in a suitable solvent (acetic acid, ethanol, lactic acid, or water) with the incorporation of bioactive components and plasticizers.
2. Molding of film forming solution (acrylic, silicon, teflon, or glass).
3. Drying at defined conditions implicating hot air oven, microwave, or tray or vacuum driers.

Subsequently, the fabricated dried film with moisture of 5-8% is peeled from the surface (Shahidi & Hossain, 2020). Prior to peeling, degassing is performed using vacuum degassing, centrifugation, and ultrasonication to obtain a uniform and defect free film with optimum functionalities (Suhag et al., 2020). Film casting can be influenced by a range of factors such as combination of temperature, time, equipment, and atmospheric conditions used for film formation (Yadav et al., 2020). Furthermore, solution casting offers the benefits of simplicity, operational convenience, cost-effectiveness, and eco-friendliness (Wang et al., 2021)

Extrusion

Extrusion is based on the thermoplastic behavior of polymers when they are plasticized and heated over their glass transition temperature (Verbeek & van den Berg, 2010). The process includes three phases:

1. Feeding: Raw materials are degassed by applying compressions and filled up in feed zone.
2. Kneading: Materials are subjected to elevated temperature and pressure to attain particular physical attributes.
3. Heating: Final heating with maximum temperature, pressure and shear rate.

Physicochemical characteristics of film formed are function of heating zone whereas screw configuration, feeding rate, screw speed and moisture percentage can be optimized as per intended use of films (Rowat et al., 2021).

Coatings Formation Methods

Edible coatings are directly applied to surface of food in form of thin layers through brushing, dipping, and spraying (Atieno et al., 2019; Dhumal & Sarkar 2018; Dhall, 2013).

- Brushing: Coating solution is applied on food material using brush
- Dipping: Food items are dipped in coating solution followed by draining and drying; frequently used approach.
- Spraying: Low viscosity liquids (droplets) are sprayed on product surface through air spray atomization or electrostatic spraying.

Applications

The emergence of food packaging technologies is driven by consumer behavior, market trends, and the need to mitigate the environmental impact of food production while improving its safety and quality (Carlucci et al., 2015). The development of green packaging materials utilizing discarded edible materials will be beneficial to confront the challenges associated with traditional packaging systems, attaining sustainability, and fostering recycling in the food industry (Han et al., 2018). Edible food wrappers are suitable packaging solution for various food commodities including meat and its products, dairy products, fruits and vegetables, nuts, grains and bakery products (Sharma et al., 2020; Jeevahan & Chandrasekaran, 2019; Han et al., 2018). The selection of appropriate edible polymer is function of type of packaged food, storage time and conditions (Sharma et al., 2020; Jeevahan & Chandrasekaran, 2019; Aguirre-Joya et al., 2018). Tables 1-4 provides an overview of the recent studies on development and application of edible food wrappers prepared from renewable resources.

Advantages

Edible food wrappers crafted from renewable resources as well as through valorization of food by-products offer multitude of advantages. Firstly, these can be used as sustainable alternate to traditional plastic packaging which is a constant threat to environment. On contrary, edible wrappers are biodegradable and economical approach. Furthermore, these wrappers can be designed to improve food quality,

Table 1. Application of fabricated food wrappers

Packaged Food	Fabricated Edible Wrapper	Reported Effects	Reference
Fruits			
Banana	Rice starch coating with sucrose esters	• Shelf-life extension by 12 days • Improved post-harvest quality • Delayed ethylene bio-synthesis	Thakur et al. (2019)
Papaya 12 days storage	Carnuba wax nanoemuslion with *Cymbopogon martinii* extract	• Delayed ripening • Reduced weight loss • Preserved firmness and colour loss • Prevented microbial incidence	Oliveira Filho et al. (2022)
Plum fruits 19 ^{0}C for 8 days	Pectin coating	• Augmented anti-oxidant activity • Enhanced nutritional value associated with high amounts of vitamin C and polyphenols	Panahirad et al. (2020)
Mango 8±1 ^{0}C for 16 days	Alginate-pectin coating incorporated with ascorbic acid	• Extended shelf life • Improved quality aspects such as colour retention	Silva et al. (2018)
Lime fruit	Pectin coating	• Gradual increase in wilting and green colour loss, accelerated by higher temperatures	Maftoonazad and Ramaswamy (2019)
Strawberry (Fresh Cuts)	Gellan coatings infused with geraniol	• Reduced microbial counts • Improved microbial stability • Better firmness loss	Tomandoni et al. (2018)
Sweet cherries 4 ^{0}C for 25 days & 20 ^{0}C for 15 days	Chitosan based coatings	• Reduced respiration rate • Improved microbial stability • Shelf-life extension	Tokath and (2020)
Strawberry	Edible films fabricated of banana starch-chitosan and Aloe vera gel	• Shelf-life extension up to 15 days with improved firmness and colour retention • Reduced fungal decay • Restricted water evaporation; 5% reduction in weight loss	Pinzon et al. (2020)
Vegetables			
Carrots 5 ± 1oC for 21 days	CMC-candeilla wax coating infused with ascorbic acid	• Improved water transfer barrier • Retained freshness • Augmented oxidative stability associated with limited polyphenol oxidase activity	Kowalczyk et al. (2020)
Tomatoes 23oC for 15 days	Carnuba wax nanoemulsions	• Showed excellent instrumental gloss as compared to control • Reduced water loss	Miranda et al. (2022)
Tomatoes	Chitosan-*Opuntia ficus-indica* mucilage film	• In-vitro and in-situ properties against *Rhizopus stolonifer* • Miantained quality parameters (firmness, colour, texture) • Mitigation of phytopathogenic fungi	Olicón-Hernándeza et al. (2019)

continued on following page

Table 1. Continued

Packaged Food	Fabricated Edible Wrapper	Reported Effects	Reference
Cherry tomatoes	Thymol nano emulsion incorporated in quinoa protein/chitosan based edible film Chitosan based edible film with encapsulated Artemisia Annua oil liposome.	• Preservation effect against *Botrytis cinerea* • Inactivation of *E. coli* O157:H7	Robledo et al. (2018) Cui et al. (2017)
Okra	Nanoemulsified basil oil incorporated in alginate based edible films	• Anti-fungal effect (*Aspergillus flavus* and *Penicillium chrysogenum*)	Gundewadi et al. (2018)
Cucumber 25oC and 10oC	Edible coating fabricated from corn starch and mint extract	• Improved quality indices • Prolonged shelf life	Raghav and Saini (2018)

Table 2. Application of edible food wrappers on meat and meat products

Packaged Food	Edible Wrapper	Reported Effect	Reference
Meat and meat products			
Chicken Fillets 4oC	CMC composite with cellulose nanofiber (CNF) and inulin	• Improved storage stability for supplementary 6 days storage	Zabihollahi et al. (2020)
Beef	Chitosan-starch film impregnated with thymus essential oil and pomegranate peel extract	• Enhanced Storage stability	Mehdizadeh et al. (2020)
Lamb meat 7 oC for 2 days	Carrageenan films with olive leaves extracts	• Reduced water permeability • Excellent anti-oxidant effect • Higher microbial stability	Martiny et al. (2020)
Lamb meat	Postbiotics enriched *Malva sylvestris* mucilage films	• Supressed formation of primary and secondary oxidation products • Preserved moisture value, hardness and pH • Regulated sensory properties • Delayed growth of microorganisms • Augmented shelf life (> 10 days)	Abbasi et al. (2023)
Fresh Pork 4 ± 1°C for 16 days	Nanoemulsion coatings with chitosan and *Schizonepeta tenuifolia* essential oil	• Extended shelf life • Retarded microbial spoilage and lipid oxidation	Zhang et al. (2021)
Pork Slices	Chitosan-gelatin/encapsulated tarragon essential oil	• Regulated release of bioactive components associated with increased shelf life	Zhang et al. (2020)
Rainbow trout (*Oncorhynchus mykiss*) 4 ± 1oC for 10 days	Gelatin films enriched with garlic peel extract	• Improved physiochemical parameters (sensory and anti-oxidant properties) • Delayed microbial proliferation • Improved shelf life	Uçak (2019)

continued on following page

Table 2. Continued

Packaged Food	Edible Wrapper	Reported Effect	Reference
Fresh Pork	Chitosan-gelatin based coating supplemented with grape seed extract	• Repressed meat oxidation • Reduced microbial spoilage • Enhanced anti-oxidant activity	Xiong et al. (2020)
Bluefin tuna (*Thunnus thynnus*) fillets	Soy protein films incorporated with MMT clay and clove essential oil	• Storage life enhanced up to 7 days at 2oC	Echeverria et al. (2018)
Deep water pink shrimps (*Parapenaeus longirostris*) Vacuum Packed, 4oC for 15 days	Citrus peel essential oil infused chitosan films	• Augmented anti-microbial and anti-oxidant properties • Delayed melanosis • Sensory stability throughout storage	Alparslan & Baygar (2017)
Sliced Hams	Edible sodium alginate films incorporated with probiotics bacteria Sodium alginate edible films enriched with oregano essential oil	• Efficient delivery of probiotics • Mitigated Listeria growth • Product aroma was improved • Negative impact on colour	Pandhi et al. (2019); Pavli et al. (2019)
Harbin red Sausages	Chitosan based edible coatings	• Delayed microbial proliferation • Regulated storage stability such as pH and water retention	Dong et al. (2020)

safety and nutrition (Galus & Kadzinska, 2019; Otoni et al., 2017). Edible packaging can also provide different colors and flavors as well as customizing surface features (such as hydrophobicity and hydrophilicity) that lead to enhancing the organoleptic properties of packaged foods. Furthermore, these can serve as a conduit for functional elements with possible additional benefits to your health or well-being. (Jeevahan & Chandrasekaran, 2019; Pooja Saklani et al., 2019). Typically, the hydrocolloid, protein, and lipid layers of composite films are supported by a polysaccharide or lipid material spread in a protein or polysaccharide matrix. (Yousuf & Qadri, 2020; Pooja Saklani et al., 2019; Chakravartula et al., 2019; Dhumal & Sarkar, 2018).

Safety Regulations

It is vital to assess the components that are used, the methods that are followed during manufacture, and any potential allergic reactions in order to determine whether or not they meet the requirements for food safety. Despite the emphasis on sustainability, environmental restrictions might need to be considered. For instance, environmental protection rules may apply when specific materials are harvested or grown for edible wrappers. In addition to standards governing food safety, there are rules governing how items, including edible wrappers, should be packaged, labelled, and promoted.

Table 3. Application of edible food wrappers on nuts and bakery products

Packaged Food	Edible Wrapper	Reported Effect	Reference
Nuts			
Peanuts 25oC for 90 days	Thymol and carvacol infused corn-zein coatings	• Mitigated aflatoxin formation • Anti-microbial impact • Magnificent moisture and oxygen barrier	Afshari et al. (2018)
Cashew nuts	Mango leaf extract infused chitosan films	• 56% Improved oxidation resistance for 28 days storage	K et al. (2019)
Fresh Walnuts Kernels	Chitosan film with green tea extract	• Improved anti-oxidant and anti-fungal effects	Sabaghi et al. (2015)
Baked Products			
Bakery items	Edible coatings enriched with essential oils (cinnamon, thyme, lemon grass and oregano)	• Retarded growth of pathogenic microbes • Prolonged shelf life with amplified product safety	Gavahian et al. (2020)
Mini bread loafs	Polysaccharide based coatings with *Malva sylvestris* extracts	• Imparted fungicidal and yeasticidal effect • Excellent moisture retention • Reduced microbial loads for 3 day storage	Alexieva et al. (2022)
Muffins	Edible coatings fabricated from triticale	• Delayed staling	Pandhi et al. (2019)
Fruit bars	Edible coating made of Na-alginate, CMC and whey protein isolate	• Restricted moisture loss • Sustained texture • Improved oxidative stability	Eyiz et al. (2020)
Sour dough bread	Edible coatings reinforced with encapsulated essential oils	• Extended shelf life from 3 days to 10 days • Delayed staling • Suppressed microbial spoilage • Reduced colour loss • High sensory acceptability	Bautista-Espinoza et al. (2023)

Table 4. Application of edible food wrappers on dairy products

Packaged Food	Edible Wrapper	Reported Effect	Reference
Dairy Products			
Cheddar cheese	Gum-based edible coating (Guar/Tragacanth)	• Improved nutrition value; increased content of fat, tryptophan and tyrosine • Amended acidity and pH values	Pourmolaie et al. (2018)
Cheese	Na-alginate based edible coatings incorporated with probiotics	• Reduced total coliform counts • Edible coating delivers probiotic bacteria to product	Olivo et al. (2020)
Nabulsi Cheese	Chitosan film incorporated with bitter vetch protein	• Regulated pH in fresh cheese • Delayed microbial growth • Improved texture profile (hardness, gumminess, chewiness)	Sabbah et al. (2019)
Queso Fresco Cheese	Chitosan coating with hydrogen peroxide, Lauric arginate, sodium caprylate, and acidified calcium sulphate with lactic acid	• Augmented control of *Listeria monocytogenes* • Active ingredients contributed to listeristatic and listericidal effect up to 35 days storage	Brown et al. (2018)

Future Perspective

With the advancement of material science and food technology research, it is conceivable to foresee the development of even more environmentally friendly and flexible edible wrapper materials. The future development of edible wrappers is likely to be heavily influenced by regulatory organizations. Cooperation between industry stakeholders and government will be necessary to meet food safety and environmental criteria.

CONCLUSION

Edible food wrappers are those that can be used to wrap food. When made into coatings or films, they keep their chemical characteristics. While coatings are eaten, films are employed to make bags, pouches, and wraps. The development of edible food wrappers is a crucial step in doing things correctly. These wraps offer a practical way to preserve food while also addressing the issues of plastic pollution and food waste. By fusing the strength of natural ingredients, cutting-edge technologies, and biodegradable materials, they create an inventive and ecological packaging solution. Edible wrappers might also be advantageous for reasons other than environmental sustainability.

REFERENCES

Abbasi, A., Sabahi, S., Bazzaz, S., Tajani, A. G., Lahouty, M., Aslani, R., & Hosseini, H. (2023). An edible coating utilizing Malva sylvestris seed polysaccharide mucilage and postbiotic from Saccharomyces cerevisiae var. boulardii for the preservation of lamb meat. *International Journal of Biological Macromolecules*, *246*, 125660. doi:10.1016/j.ijbiomac.2023.125660 PMID:37399877

Afshari, A., Piroozehpoor, M., Jamshidi, A., Azizzadeh, M., & Sarafan Sadeghi, A. (2018). Effect of novel zein coating combined with different antioxidants (thymol and carvacrol) on the aflatoxin production of peanut. Journal of Nutrition. *Fasting and Health*, *6*(1), 23–29.

Aguirre-Joya, J. A., De Leon-Zapata, M. A., & Alvarez-Perez, O. B. (2018). Basic and applied concepts of edible packaging for foods. In Food packaging and preservation (pp. 1–61). Academic Press.

Ahmadi, P., Jahanban-Esfahlan, A., Ahmadi, A., Tabibiazar, M., & Mohammadifar, M. (2020). Development of ethyl cellulose-based formulations: A perspective on the novel technical methods. *Food Reviews International*, 1–48.

Al-Hassan, A. A., & Norziah, M. H. (2012). Starch–gelatin edible films: Water vapor permeability and mechanical properties as affected by plasticizers. *Food Hydrocolloids*, *26*(1), 108–117. doi:10.1016/j.foodhyd.2011.04.015

Alexieva, I., Baeva, M., Popova, A., Fidan, H., Goranova, Z., & Milkova-Tomova, I. (2022). Development and Application of Edible Coatings with Malva sylvestris L. Extract to Extend Shelf-Life of Small Loaf. *Foods*, *11*(23), 3831. doi:10.3390/foods11233831 PMID:36496640

Alparslan, Y., & Baygar, T. (2017). Effect of chitosan film coating combined with orange peel essential oil on the shelf life of deepwater pink shrimp. *Food and Bioprocess Technology*, *10*(5), 842–853. doi:10.100711947-017-1862-y

Angellier-Coussy, H., Guillard, V., Guillaume, C., & Gontard, N. (2011). 23-Wheat gluten (WG)-based materials for food packaging. In J.-M. Lagaron (Ed.), Multifunctional and nanoreinforced polymers for food packaging (pp. 649–668). Woodhead Publishing.

Atieno, L., Owino, W., Ateka, E. M., & Ambuko, J. (2019). Influence of coating application methods on the postharvest quality of cassava. *International Journal of Food Sciences*, *2019*, 1–16. doi:10.1155/2019/2148914 PMID:30863776

Baldwin, E. A., Hagenmaier, R., & Bai, J. (2012). *Edible Coatings and Films to Improve Food Quality* (2nd ed.). CRC Press.

Bas-Bellver, C., Barrera, C., Betoret, N., & Seguí, L. (2020). Turning agri-food cooperative vegetable residues into functional powdered ingredients for the food industry. *Sustainability (Basel)*, *12*(4), 1284. doi:10.3390u12041284

Bautista-Espinoza, P. I., Di Pierro, P., Amaya-Llano, S. L., García-Almendarez, B. E., Mares-Mares, E., Escamilla-García, M., Granados-Arvizu, J. A., Velázquez de la Cruz, G., & Regalado-González, C. (2023). Impact of edible coatings reinforced with mesoporous silica nanoparticles added with cinnamon and lemongrass essential oils applied on sourdough bread. *Food Bioscience*, *55*, 102992. doi:10.1016/j.fbio.2023.102992

Ben-Othman, S., Jõudu, I., & Bhat, R. (2020). Bioactives from agrifood wastes: Present insights and future challenges. *Molecules (Basel, Switzerland)*, *25*(3), 510. doi:10.3390/molecules25030510 PMID:31991658

Benbettaïeb, N., Debeaufort, F., & Karbowiak, T. (2019). Bioactive edible films for food applications: Mechanisms of antimicrobial and antioxidant activity. *Critical Reviews in Food Science and Nutrition*, *59*(21), 3431–3455. doi:10.1080/10408398.2018.1494132 PMID:29999405

Bharti, S. K., Pathak, V., Alam, T., Arya, A., Basak, G., & Awasthi, M. G. (2020). Materiality of edible film packaging in muscle foods: A worthwhile conception. *Journal of Packaging Technology and Research*, *4*(1), 117–132. doi:10.100741783-020-00087-9

Brown, S. R., Kozak, S. M., & D'Amico, D. J. (2018). Applications of edible coatings formulated with antimicrobials inhibit Listeria monocytogenes growth on Queso Fresco. *Frontiers in Sustainable Food Systems*, *2*, 1. doi:10.3389/fsufs.2018.00001

Calva-Estrada, S. J., Jim'enez-Fernandez, ´. M., & Lugo-Cervantes, E. (2019). Protein-based films: Advances in the development of biomaterials applicable to food packaging. *Food Engineering Reviews*, *11*(2), 78–92. doi:10.100712393-019-09189-w

Carlucci, D., Nocella, G., De Devitiis, B., Viscecchia, R., Bimbo, F., & Nardone, G. (2015). Consumer purchasing behaviour towards fish and seafood products. Patterns and insights from a sample of international studies. *Appetite*, *84*, 212–227. doi:10.1016/j.appet.2014.10.008 PMID:25453592

Carpin'e, D., Dagostin, J. L. A., Bertan, L. C., & Mafra, M. R. (2015). Development and characterization of soy protein isolate emulsion-based edible films with added coconut oil for olive oil packaging: Barrier, mechanical, and thermal properties. *Food and Bioprocess Technology*, *8*(8), 1811–1823. doi:10.100711947-015-1538-4

Cazón, P., Velazquez, G., Ramírez, J. A., & Vázquez, M. (2017). Polysaccharide-Based Films and Coatings for Food Packaging: A Review. *Food Hydrocolloids*, *68*, 136–148. doi:10.1016/j.foodhyd.2016.09.009

Chakravartula, S. S. N., Cevoli, C., & Balestra, F. (2019). Evaluation of drying of edible coating on bread using NIR spectroscopy. *Journal of Food Engineering*, *240*, 29–37. doi:10.1016/j.jfoodeng.2018.07.009

Chauhan, O. P., Nanjappa, C., Ashok, N., Ravi, N., Roopa, N., & Raju, P. S. (2015). Shellac and Aloe vera gel-based surface coating for shelf life extension of tomatoes. *Journal of Food Science and Technology*, *52*(2), 1200–1205. doi:10.100713197-013-1035-6 PMID:25694740

Chen, H., Wang, J., Cheng, Y., Wang, C., Liu, H., Bian, H., Pan, Y., Sun, J., & Han, W. (2019). Application of protein-based films and coatings for food packaging: A Review. *Polymers, 11*(12), 2039. doi:10.3390/polym11122039 PMID:31835317

Chitravathi, K., Chauhan, O.P., & Raju, P.S. (2014). *Postharvest biology and technology postharvest shelf-life extension of green chillies (Capsicum annuum L.) using shellac-based edible surface coatings*. Academic Press.

Chiumarelli, M., & Hubinger, M. D. (2014). Evaluation of edible films and coatings formulated with cassava starch, glycerol, carnauba wax and stearic acid. *Food Hydrocolloids, 38*, 20–27. doi:10.1016/j.foodhyd.2013.11.013

Cui, H., Yuan, L., Li, W., & Lin, L. (2017). Edible film incorporated with chitosan and Artemisia annua oil nanoliposomes for inactivation of Escherichia coli O157: H7 on cherry tomato. *International Journal of Food Science & Technology, 52*(3), 687–698. doi:10.1111/ijfs.13322

Dhall, R. K. (2013). Advances in edible coatings for fresh fruits and vegetables: A review. *Critical Reviews in Food Science and Nutrition, 53*(5), 435–450. doi:10.1080/10408398.2010.541568 PMID:23391012

Dhumal, C. V., & Sarkar, P. (2018). Composite edible films and coatings from food-grade biopolymers. *Journal of Food Science and Technology, 55*(11), 4369–4383. doi:10.100713197-018-3402-9 PMID:30333633

Dilucia, F., Lacivita, V., Conte, A., & Nobile, M. A. D. (2020). Sustainable use of fruit and vegetable by-products to enhance food packaging performance. *Foods, 9*(7), 857. doi:10.3390/foods9070857 PMID:32630106

Dong, C., Wang, B., Li, F., Zhong, Q., Xia, X., & Kong, B. (2020). Effects of edible chitosan coating on Harbin red sausage storage stability at room temperature. *Meat Science, 159*, 1–9. doi:10.1016/j.meatsci.2019.107919 PMID:31472934

Dutta, P. K., Tripathi, S., Mehrotra, G. K., & Dutta, J. (2009). Perspectives for chitosan based antimicrobial films in food applications. *Food Chemistry, 114*(4), 1173–1182. doi:10.1016/j.foodchem.2008.11.047

Echeverría, I., Lopez-Caballero, ´. M. E., Gomez-Guill, ´., ´en, M. C., Mauri, A. N., & Montero, M. P. (2018). Active nanocomposite films based on soy proteins montmorillonite- clove essential oil for the preservation of refrigerated bluefin tuna (Thunnus thynnus) fillets. *International Journal of Food Microbiology, 266*, 142–149. doi:10.1016/j.ijfoodmicro.2017.10.003 PMID:29216554

Erceg, M., Kovačić, T., & Perinović, S. (2010). Isothermal degradation of poly(3-hydroxybutyrate)/organically modified montmorillonite nanocomposites. *Polymer Composites*, *31*(2), 272–278. doi:10.1002/pc.20798

Eyiz, V., Tontul, ˙. I., & Türker, S. (2020). The effect of edible coatings on physical and chemical characteristics of fruit bars. *Journal of Food Measurement and Characterization*, *2020*(14), 1775–1783. doi:10.100711694-020-00425-0

Fabra, M. J., Lopez-Rubio, A., & Lagaron, J. M. (2013). High barrier polyhydroxyalcanoate food packaging film by means of nanostructured electrospun interlayers of zein. *Food Hydrocolloids*, *32*(1), 106–114. doi:10.1016/j.foodhyd.2012.12.007

Fabra, M. J., Pardo, P., Martínez-Sanz, M., Lopez-Rubio, A., & Lagarón, J. M. (2016). Combining polyhydroxyalkanoates with nanokeratin to develop novel biopackaging structures. *Journal of Applied Polymer Science*, *133*(2), app.42695. doi:10.1002/app.42695

Flaris, V., & Singh, G. (2009). Recent developments in biopolymers. *Journal of Vinyl and Additive Technology*, *15*(1), 1–11. doi:10.1002/vnl.20171

Galus, S., & Kadzinska, J. (2019). Gas barrier and Wetting properties of whey protein isolates based emulsion films. *Polymer Engineering and Science*, *59*(1), 375–383. doi:10.1002/pen.24992

Ganesan, A. R., Shanmugam, M., & Bhat, R. (2018). Producing novel edible films from semi refined carrageenan (SRC) and ulvan polysaccharides for potential food applications. *International Journal of Biological Macromolecules*, *112*, 1164–1170. doi:10.1016/j.ijbiomac.2018.02.089 PMID:29454950

Gavahian, M., Chu, Y.-H., Lorenzo, J. M., Mousavi Khaneghah, A., & Barba, F. J. (2020). Essential oils as natural preservatives for bakery products: Understanding the mechanisms of action, recent findings, and applications. *Critical Reviews in Food Science and Nutrition*, *60*(2), 310–321. doi:10.1080/10408398.2018.1525601 PMID:30431327

Gomez-Guill, ´., ´en, M. C., Turnay, J., Fernandez-Diaz, M. D., Ulmo, N., Lizarbe, M. A., & Montero, P. (2002). Structural and physical properties of gelatin extracted from different marine species: A comparative study. *Food Hydrocolloids*, *16*(1), 25–34. doi:10.1016/S0268-005X(01)00035-2

Groh, K. J., Backhaus, T., Carney-Almroth, B., Geueke, B., Inostroza, P. A., Lennquist, A., Leslie, H. A., Maffini, M., Slunge, D., Trasande, L., Warhurst, A. M., & Muncke, J. (2019). Overview of known plastic packaging-associated chemicals and their hazards. *The Science of the Total Environment*, *651*, 3253–3268. doi:10.1016/j.scitotenv.2018.10.015 PMID:30463173

Gundewadi, G., Rudra, S. G., Sarkar, D. J., & Singh, D. (2018). Nanoemulsion based alginate organic coating for shelf life extension of okra. *Food Packaging and Shelf Life*, *18*, 1–12. doi:10.1016/j.fpsl.2018.08.002

Hamed, I., Jakobsen, A. N., & Lerfall, J. (2022). Sustainable edible packaging systems based on active compounds from food processing byproducts: A review. *Comprehensive Reviews in Food Science and Food Safety*, *21*(1), 198–226. doi:10.1111/1541-4337.12870 PMID:34907649

Han, J. H. (2014). *Innovations in Food Packaging* (2nd ed.). Elsevier Ltd.

Han, J. W., Ruiz-Garcia, L., Qian, J. P., & Yang, X. T. (2018). Foodpackaging: A comprehensive review and future trends. *Comprehensive Reviews in Food Science and Food Safety*, *17*(4), 860–877. doi:10.1111/1541-4337.12343 PMID:33350114

Hanani, N. Z. A., Roos, Y. H., & Kerry, J. P. (2014). Use and application of gelatin as potential biodegradable packaging materials for food products. *International Journal of Biological Macromolecules*, *71*, 94–102. doi:10.1016/j.ijbiomac.2014.04.027 PMID:24769086

Hosseini, S. F., Rezaei, M., Zandi, M., & Ghavi, F. F. (2013). Preparation and functional properties of fish gelatin–chitosan blend edible films. *Food Chemistry*, *136*(3), 1490–1495. doi:10.1016/j.foodchem.2012.09.081 PMID:23194553

Jeevahan, J., & Chandrasekaran, M. (2019). Nanoedible films for food packaging: A review. *Journal of Materials Science*, *54*(19), 12290–12318. doi:10.100710853-019-03742-y

Jonhed, A., Andersson, C., & J¨arnstrom, ¨. L. (2008). Effects of film forming and hydrophobic properties of starches on surface sized packaging paper. *Packaging Technology & Science*, *21*(3), 123–135. doi:10.1002/pts.783

K, R., G, B., Banat, F., Show, P. L., & Cocoletzi, H. H. (2019). Mango leaf extract incorporated chitosan antioxidant film for active food packaging. *International Journal of Biological Macromolecules*, *126*, 1234–1243.

Kennouche, S., Le Moigne, N., Kaci, M., Quantin, J. C., Caro-Bretelle, A. S., Delaite, C., & Lopez-Cuesta, J. M. (2016). Morphological characterization and thermal properties of compatibilized poly (3-hydroxybutyrate-co-3-hydroxyvalerate)(PHBV)/poly (butylene succinate)(PBS)/halloysite ternary nanocomposites. *European Polymer Journal, 75*, 142–162. doi:10.1016/j.eurpolymj.2015.12.009

Kim, K. M., Son, J. H., Kim, S.-K., Weller, C. L., & Hanna, M. A. (2006). Properties of chitosan films as a function of pH and solvent type. *Journal of Food Science, 71*(3), E119–E124. doi:10.1111/j.1365-2621.2006.tb15624.x

Koh, J. J., Zhang, X., & He, C. (2018). Fully biodegradable poly(lactic acid)/starch blends: A review of toughening strategies. *International Journal of Biological Macromolecules, 109*, 99–113. doi:10.1016/j.ijbiomac.2017.12.048 PMID:29248552

Koshy, R. R., Mary, S. K., Thomas, S., & Pothan, L. A. (2015). Environment friendly green composites based on soy protein isolate – a review. *Food Hydrocolloids, 50*, 174–192. doi:10.1016/j.foodhyd.2015.04.023

Kowalczyk, D., Skrzypek, T., & Łupina, K. (2020). Effect of carboxymethyl cellulose/candelilla wax edible coating incorporated with ascorbic acid on the physicochemical and sensory qualities of prepackaged minimally processed carrots (Daucus carota L.) during cold storage. *Journal of Food Processing and Preservation, 44*(9), e14713. doi:10.1111/jfpp.14713

Li, F., Yu, H.-Y., Wang, Y.-Y., Zhou, Y., Zhang, H., Yao, J.-M., Abdalkarim, S. Y. H., & Tam, K. C. (2019). Natural biodegradable poly(3-hydroxybutyrate-co-3-hydroxyvalerate) nanocomposites with multifunctional cellulose nanocrystals/graphene oxide hybrids for high-performance food packaging. *Journal of Agricultural and Food Chemistry, 67*(39), 10954–10967. doi:10.1021/acs.jafc.9b03110 PMID:31365242

Liang, J., Yan, H., Zhang, J., Dai, W., Gao, X., Zhou, Y., Wan, X., & Puligundla, P. (2017). Preparation and Characterization of Antioxidant Edible Chitosan Films Incorporated with Epigallocatechin Gallate Nanocapsules. *Carbohydrate Polymers, 171*, 300–306. doi:10.1016/j.carbpol.2017.04.081 PMID:28578967

Lin, J., Pan, D., Sun, Y., Ou, C., Wang, Y., & Cao, J. (2019). The modification of gelatin films: Based on various cross-linking mechanism of glutaraldehyde at acidic and alkaline conditions. *Food Sciences and Nutrition, 7*(12), 4140–4146. PMID:31890193

Liu, Q., Zhang, M., Bhandari, B., Xu, J., & Yang, C. (2020). Effects of nanoemulsion-based active coatingswith composite mixture of star anise essential oil, polylysine, and nisin on the quality and shelf life of ready-to-eat Yao meat products. *Food Control, 107*, 106771. doi:10.1016/j.foodcont.2019.106771

Ma, T., Zhao, H., Wang, J., & Sun, B. (2019). Effect of processing conditions on the morphology and oxidative stability of lipid microcapsules during complex coacervation. *Food Hydrocolloids*, *87*, 637–643. doi:10.1016/j.foodhyd.2018.08.053

Maftoonazad, N., & Ramaswamy, H. S. (2019). Application and evaluation of a pectin based edible coating process for quality change kinetics and shelf-life extension of lime fruit (Citrus aurantifolium). *Coatings*, *9*(5), 1–14. doi:10.3390/coatings9050285

Mallegni, N., Phuong, T. V., Coltelli, M. B., Cinelli, P., & Lazzeri, A. (2018). Poly (lactic acid)(PLA) based tear resistant and biodegradable flexible films by blown film extrusion. *Materials (Basel)*, *11*(1), 148. doi:10.3390/ma11010148 PMID:29342099

Martiny, T. R., Pacheco, B. S., Pereira, C. M., Mansilla, A., Astorga-España, M. S., Dotto, G. L., Moraes, C. C., & Rosa, G. S. (2020). A novel biodegradable film based on κ-carrageenan activated with olive leaves extract. *Food Science & Nutrition, 8*, 3147–3156.

Mehdizadeh, T., Tajik, H., Langroodi, A. M., Molaei, R., & Mahmoudian, A. (2020). Chitosan-starch film containing pomegranate peel extract and Thymus kotschyanus essential oil can prolong the shelf life of beef. *Meat Science, 163*, 108073.

Mehta, R., Kumar, V., Bhunia, H., & Upadhyay, S. N. (2005). Synthesis of poly(lactic acid): A review. Journal of Macromolecular Science. *Part C*, *45*(4), 325–349.

Mekonnen, T., Mussone, P., Khalil, H., & Bressler, D. (2013). Progress in bio-based plastics and plasticizing modifications. *Journal of Materials Chemistry. A, Materials for Energy and Sustainability*, *1*(43), 13379–13398. doi:10.1039/c3ta12555f

Mellinas, C., Vald'es, A., Ramos, M., Burgos, N., Garrigos, M. D. C., & Jim'enez, A. (2016). Active edible films: Current state and future trends. *Journal of Applied Polymer Science*, *133*(2), 1–15. doi:10.1002/app.42631

Mendieta, C. M., González, G., Vallejos, M. E., & Area, M. C. (2022). Bio-polyethylene furanoate (Bio-PEF) from lignocellulosic biomass adapted to the circular bioeconomy. *BioResources*, *17*(4), 7313–7337. doi:10.15376/biores.17.4.Mendieta

Milani, J. M., & Nemati, A. (2022). Lipid-based edible films and coatings: A review of recent advances and applications. *Journal of Packaging Technology and Research*, *6*(1), 11–22. doi:10.100741783-021-00130-3

Miranda, M., Marilene De Mori, M. R., Spricigo, P. C., Pilon, L., Mitsuyuki, M. C., Correa, D. S., & Ferreira, M. D. (2022). Carnauba wax nanoemulsion applied as an edible coating on fresh tomato for postharvest quality evaluation. *Heliyon*, *8*(7), e09803. doi:10.1016/j.heliyon.2022.e09803 PMID:35800251

Morales-Jiménez, M., Gouveia, L., Yáñez-Fernández, J., Castro-Muñoz, R., & Barragán-Huerta, B. (2020). Production, preparation and characterization of microalgae-based biopolymer as a potential bioactive film. *Coatings*, *10*(2), 120. doi:10.3390/coatings10020120

Murrieta-Martínez, C. L., Soto-Valdez, H., Pacheco-Aguilar, R., Torres-Arreola, W., Rodríguez-Felix, F., Marquez, ´., & Ríos, E. (2018). Edible protein films: Sources and behavior. *Packaging Technology & Science*, *31*(3), 113–122. doi:10.1002/pts.2360

Olicón-Hernández, D. R., Acosta-Sánchez, Á., Monterrubio-López, R., & Guerra-Sánchez, G. (2019). Chitosan and Opuntia ficus indica mucilage as the base of a polymeric edible film for the protection of tomatoes against Rhizopus stolonifer. *TIP Revista Especializada en Ciencias Químico-Biológicas, 22*(1), 1–9.

Oliveira Filho, J. G. D., Silva, G. D. C., Oldoni, F. C. A., Miranda, M., Florencio, C., Oliveira, R. M. D. D., Gomes, M. P., & Ferreira, M. D. (2022). Edible coating based on carnauba wax nanoemulsion and Cymbopogon martinii essential oil on papaya postharvest preservation. *Coatings*, *12*(11), 1700. doi:10.3390/coatings12111700

Olivo, P. M., da Silva Scapim, M. R., Maia, L. F., Miazaki, J., Rodrigues, B. M., Madrona, G. S., ... dos Santos Pozza, M. S. (2020). Probiotic coating for ripened cheeses with lactobacillus acidophilus and lactobacillus helveticus inclusion. *Journal of Agricultural Studies*, *8*(3), 152–170. doi:10.5296/jas.v8i3.16052

Otoni, C. G., Avena-Bustillos, R. J., Azeredo, H. M., Lorevice, M. V., Moura, M. R., Mattoso, L. H., & McHugh, T. H. (2017). Recent advances on edible films based on fruits and vegetables—A review. *Comprehensive Reviews in Food Science and Food Safety*, *16*(5), 1151–1169. doi:10.1111/1541-4337.12281 PMID:33371612

Otoni, C. G., Lodi, B. D., Lorevice, M. V., Leitão, R. C., Ferreira, M. D., de Moura, M. R., & Mattoso, L. H. (2018). Optimized and scaled-up production of cellulose-reinforced biodegradable composite films made up of carrot processing waste. *Industrial Crops and Products*, *121*, 66–72. doi:10.1016/j.indcrop.2018.05.003

Pandhi, S., Kumar, A., & Alam, T. (2019). Probiotic edible films and coatings: Concerns, applications and future prospects. *Journal of Packaging Technology and Research*, *3*(3), 261–268. doi:10.100741783-019-00069-6

Patel, P. (2019). *Edible packaging*. Academic Press.

Pavli, F., Argyri, A. A., Skandamis, P., Nychas, G.-J., Tassou, C., & Chorianopoulos, N. (2019). Antimicrobial activity of oregano essential oil incorporated in sodium alginate edible films: Control of Listeria monocytogenes and spoilage in ham slices treated with high pressure processing. *Materials (Basel)*, *12*(22), 1–26. doi:10.3390/ma12223726 PMID:31718078

Pawar, S. N., & Edgar, K. J. (2012). Alginate derivatization: A review of chemistry, properties and applications. *Biomaterials*, *33*(11), 3279–3305. doi:10.1016/j.biomaterials.2012.01.007 PMID:22281421

Pelissari, F. M., Ferreira, D. C., Louzada, L. B., dos Santos, F., Corr^ea, A. C., Moreira, F. K. V., & Mattoso, L. H. (2019). Starch-based edible films and coatings: An eco-friendly alternative for food packaging. In M. T. P. Silva Clerici & M. Schmiele (Eds.), Starches for food application (pp. 359–420). Academic Press.

Petkoska, A. T., Daniloski, D., D'Cunha, N. M., Naumovski, N., & Broach, A. T. (2021). Edible packaging: Sustainable solutions and novel trends in food packaging. *Food Research International*, *140*, 109981. doi:10.1016/j.foodres.2020.109981 PMID:33648216

Pinzon, M. I., Sanchez, L. T., Garcia, O. R., Gutierrez, R., Luna, J. C., & Villa, C. C. (2020). Increasing shelf life of strawberries (Fragaria ssp) by using a banana starch-chitosan-Aloe vera gel composite edible coating. *International Journal of Food Science & Technology*, *55*(1), 92–98. doi:10.1111/ijfs.14254

Pooja Saklani, P., nath, S., Kishor Das, S., & Singh, S. M. (2019). A review of edible packaging for foods. *International Journal of Current Microbiology and Applied Sciences*, *8*(7), 2885–2895. doi:10.20546/ijcmas.2019.807.359

Pourmolaie, H., Khosrowshahi Asl, A., Ahmadi, M., Zomorodi, S., & Naghizadeh Raeisi, S. (2018). The effect of Guar and Tragacanth gums as edible coatings in Cheddar cheese during ripening. *Journal of Food Safety*, *38*(6), 1–9. doi:10.1111/jfs.12529

Raghav, P. K., & Saini, M. (2018). Development of mint (Mentha viridis L.) herbal edible coating for shelf life enhancement of cucumber (Cucumis sativus). *International Journal of Green and Herbal Chemistry*, *7*(2), 379–391.

Ramos, M., Vald'es, A., Beltr'an, A., & Garrigos, ´. C. M. (2016). Gelatin-based films and coatings for food packaging applications. *Coatings*, *6*(4), 41. doi:10.3390/coatings6040041

Ramos, M., Valdes, A., Beltran, A., & Garrigós, M. C. (2016). Gelatinbased films and coatings for food packaging applications. *Coatings*, *6*(4), 41. doi:10.3390/coatings6040041

Rangaraj, V. M., Rambabu, K., Banat, F., & Mittal, V. (2021). Effect of date fruit waste extract as an antioxidant additive on the properties of active gelatin films. *Food Chemistry*, *355*, 129631. doi:10.1016/j.foodchem.2021.129631 PMID:33799252

Reichert, C. L., Bugnicourt, E., Coltelli, M. B., Cinelli, P., Lazzeri, A., Canesi, I., Braca, F., Martínez, B. M., Alonso, R., Agostinis, L., Verstichel, S., Six, L., Mets, S. D., Gómez, E. C., Ißbrücker, C., Geerinck, R., Nettleton, D. F., Campos, I., Sauter, E., ... Schmid, M. (2020). Bio-based packaging: Materials, modifications, industrial applications and sustainability. *Polymers*, *12*(7), 1558. doi:10.3390/polym12071558 PMID:32674366

Restrepo, A. E., Rojas, J. D., García, O. R., Sánchez, L. T., Pinzón, M. I., & Villa, C. C. (2018). Mechanical, barrier, and color properties of banana starch edible films incorporated with nanoemulsions of lemongrass (Cymbopogon citratus) and rosemary (Rosmarinus officinalis) essential oils. *Food Science & Technology International*, *24*(8), 705–712. doi:10.1177/1082013218792133 PMID:30060685

Robledo, N., Vera, P., Lopez, L., Yazdani-Pedram, M., Tapia, C., & Abugoch, L. (2018). Thymol nanoemulsions incorporated in quinoa protein/chitosan edible films; antifungal effect in cherry tomatoes. *Food Chemistry*, *246*, 211–219. doi:10.1016/j.foodchem.2017.11.032 PMID:29291841

Rodrigues, D. C., Cunha, A. P., de Brito, E. S., Gallao, M., De Azeredo, H. M. C., & Rodrigues, D. C. (2016). Mesquite seed gum and palm fruit oil emulsion edible films: Influence of oil content and sonication. *Food Hydrocolloids*, *56*, 227–235. doi:10.1016/j.foodhyd.2015.12.018

Rojas-Graü, M. A., Soliva-Fortuny, R., & Martín-Belloso, O. (2009). Edible coatings to incorporate active ingredients to fresh-cut fruits: A review. *Trends in Food Science & Technology*, *20*(10), 438–447. doi:10.1016/j.tifs.2009.05.002

Rosenboom, J. G., Hohl, D. K., Fleckenstein, P., Storti, G., & Morbidelli, M. (2018). Bottle-grade polyethylene furanoate from ring-opening polymerisation of cyclic oligomers. *Nature Communications*, *9*(1), 2701. doi:10.103841467-018-05147-y PMID:30042388

Rowat, S. J., Legge, R. L., & Moresoli, C. (2021). Plant protein in material extrusion 3D printing: Formation, plasticization, prospects, and challenges. *Journal of Food Engineering*, *308*, 110623. doi:10.1016/j.jfoodeng.2021.110623

Rutz, J. K., Borges, C. D., Zambiazi, R. C., da Rosa, C. G., & da Silva, M. M. (2016). Elaboration of microparticles of carotenoids from natural and synthetic sources for applications in food. *Food Chemistry*, *202*, 324–333. doi:10.1016/j.foodchem.2016.01.140 PMID:26920301

Sabaghi, M., Maghsoudlou, Y., Khomeiri, M., & Ziaiifar, A. M. (2015). Active edible coating from chitosan incorporating green tea extract as an antioxidant and antifungal on fresh walnut kernel. *Postharvest Biology and Technology*, *110*, 224–228. doi:10.1016/j.postharvbio.2015.08.025

Sabbah, M., Di Pierro, P., Dell'Olmo, E., Arciello, A., & Porta, R. (2019). Improved shelf-life of Nabulsi cheese wrapped with hydrocolloid films. *Food Hydrocolloids*, *96*, 29–35. doi:10.1016/j.foodhyd.2019.05.010

Saha, D., & Bhattacharya, S. (2010). Hydrocolloids as thickening and gelling agents in food: A critical review. *Journal of Food Science and Technology*, *47*(6), 587–597. doi:10.100713197-010-0162-6 PMID:23572691

Sangroniz, A., Zhu, J.-B., Tang, X., Etxeberria, A., Chen, E. Y. X., & Sardon, H. (2019). Packaging materials with desired mechanical and barrier properties and full chemical recyclability. *Nature Communications*, *10*(1), 3559. doi:10.103841467-019-11525-x PMID:31395871

Santana, J. C., Gardim, R. B., Almeida, P. F., Borini, G. B., Quispe, A. P., Llanos, S. A., Heredia, J. A., Zamuner, S., Gamarra, F. M. C., Farias, T. M. B., Ho, L. L., & Berssaneti, F. T. (2020). Valorization of chicken feet by-product of the poultry industry: High qualities of gelatin and biofilm from extraction of collagen. *Polymers*, *12*(3), 529. doi:10.3390/polym12030529 PMID:32121646

Sasidharan, A., & Venugopal, V. (2020). Proteins and co-products from seafood processing discards: Their recovery, functional properties and applications. *Waste and Biomass Valorization*, *11*(11), 5647–5663. doi:10.100712649-019-00812-9

Saucedo-Pompa, S., Rojas-Molina, R., Aguilera-Carbó, A. F., Sáenz-Galindo, A., & de La Garza, H. (2009). Edible film based on candelilla wax to improve the shelf life and quality of avocado. *Food Research International*, *42*(4), 511–515. doi:10.1016/j.foodres.2009.02.017

Schieber, A. (2017). Side streams of plant food processing as a source of valuable compounds: Selected examples. *Annual Review of Food Science and Technology*, *8*(1), 97–112. doi:10.1146/annurev-food-030216-030135 PMID:28068488

Senturk Parreidt, T., Müller, K., & Schmid, M. (2018). Alginate-based edible films and coatings for food packaging applications. *Foods*, *7*(10), 170. doi:10.3390/foods7100170 PMID:30336642

Shahbazi, M., Ahmadi, S. J., Seif, A., & Rajabzadeh, G. (2016). Carboxymethyl cellulose film modification through surface photo-crosslinking and chemical crosslinking for food packaging applications. *Food Hydrocolloids*, *61*, 378–389. doi:10.1016/j.foodhyd.2016.04.021

Shahidi, F., & Hossain, A. (2020). Preservation of aquatic food using edible films and coatings containing essential oils: A review. *Critical Reviews in Food Science and Nutrition*, 1–40. PMID:32847401

Sharma, S., Sanpui, P., Chattopadhyay, A., & Ghosh, S. S. (2012). Fabrication of antibacterial silver nanoparticle—Sodium alginate–chitosan composite films. *RSC Advances*, *2*(13), 5837–5843. doi:10.1039/c2ra00006g

Shin, Y. J., Jang, S.-A., & Song, K. B. (2011). Preparation and mechanical properties of rice bran protein composite films containing gelatin or red algae. *Food Science and Biotechnology*, *20*(3), 703–707. doi:10.100710068-011-0099-1

Shit, S. C., & Shah, P. M. (2014). Edible polymers: Challenges and opportunities. *JournalofPolymers*, *2014*, 1–13.

Sikorski, Z. Z. E., & Kolakowska, A. (2011). *Chemical, Biological, and Functional Aspects of Food Lipids* (2nd ed.). CRC Press.

Silva, F. A., Finkler, L., & Finkler, C. L. L. (2018). Effect of edible coatings based on alginate/pectin on quality preservation of minimally processed 'Espada' mangoes. *Journal of Food Science and Technology*, *55*(12), 5055–5506. doi:10.100713197-018-3444-z PMID:30483001

Siracusa, V., & Lotti, N. (2018). Biobased Plastics for Food Packaging. In *Reference Module in Food Science* (pp. 1–19). Elsevier. doi:10.1016/B978-0-08-100596-5.22413-X

Siracusa, V., Lotti, N., Munari, A., & Dalla Rosa, M. (2015). Poly (butylene succinate) and poly (butylene succinate-co-adipate) for food packaging applications: Gas barrier properties after stressed treatments. *Polymer Degradation & Stability*, *119*, 35–45. doi:10.1016/j.polymdegradstab.2015.04.026

Stejskal, N., Miranda, J. M., Martucci, J. F., Ruseckaite, R. A., Barros-Velázquez, J., & Aubourg, S. P. (2020). Quality enhancement of refrigerated hake muscle by active packaging with a protein concentrate fromSpirulina platensis. *Food and Bioprocess Technology, 13*(7), 1110–1118. doi:10.100711947-020-02468-z

Stevens, J. R., Newton, R. W., Tlusty, M., & Little, D. C. (2018). The rise of aquaculture by-products: Increasing food production, value, and sustainability through strategic utilisation. *Marine Policy, 90*, 115–124. doi:10.1016/j.marpol.2017.12.027

Suhag, R., Kumar, N., Petkoska, A. T., & Upadhyay, A. (2020). Film formation and deposition methods of edible coating on food products: A review. *Food Research International, 136*, 109582. doi:10.1016/j.foodres.2020.109582 PMID:32846613

Suresh, P. V., Kudre, T. G., & Johny, L. C. (2018). Sustainable valorization of seafood processing by-product/discard. In R. R. Singhania, R. A. Agarwal, R. P. Kumar, & R. K. Sukumaran (Eds.), *Waste to wealth* (pp. 111–139). Springer. doi:10.1007/978-981-10-7431-8_7

Tavassoli-Kafrani, E., Shekarchizadeh, H., & Masoudpour-Behabadi, M. (2016). Development of edible films and coatings from alginates and carrageenans. *Carbohydrate Polymers, 137*, 360–374. doi:10.1016/j.carbpol.2015.10.074 PMID:26686140

Tawakkal, I. S. M. A., Cran, M. J., Miltz, J., & Bigger, S. W. (2014). A review of poly(lactic acid)-based materials for antimicrobial packaging. *Journal of Food Science, 79*(8), R1477–R1490. doi:10.1111/1750-3841.12534 PMID:25039867

Teixeira-Costa, B. E., & Andrade, C. T. (2021). Natural polymers used in edible food packaging—History, function and application trends as a sustainable alternative to synthetic plastic. *Polysaccharides, 3*(1), 32–58. doi:10.3390/polysaccharides3010002

Thakur, R., Pristijono, P., Bowyer, M., Singh, S. P., Scarlett, C. J., Stathopoulos, C. E., & Vuong, Q. V. (2019). A starch edible surface coating delays banana fruit ripening. *Lebensmittel-Wissenschaft + Technologie, 100*, 341–347. doi:10.1016/j.lwt.2018.10.055

Thakur, R., Pristijono, P., Scarlett, C. J., Bowyer, M., Singh, S. P., & Vuong, Q. V. (2019). Starch-based films: Major factors affecting their properties. *International Journal of Biological Macromolecules, 132*, 1079–1089. doi:10.1016/j.ijbiomac.2019.03.190 PMID:30926503

Tokatlı, K., & Demirdöven, A. (2020). Effects of chitosan edible filmcoatings on the physicochemical and microbiological qualities of sweet cherry (Prunus avium L.). *Scientia Horticulturae, 259*, 108656. doi:10.1016/j.scienta.2019.108656

Tomadoni, B., Moreira, M. D. R., Pereda, M., & Ponce, A. G. (2018). Gellan-based coatings incorporated with natural antimicrobials in fresh-cut strawberries: Microbiological and sensory evaluation through refrigerated storage. *Lebensmittel-Wissenschaft + Technologie*, *97*, 384–389. doi:10.1016/j.lwt.2018.07.029

Uçak, I. (2019). Physicochemical and antimicrobial effects of gelatin based edible films incorporated with garlic peel extract on the rainbow trout fillets. *Progress in Nutrition*, *21*(1), 232–240.

Umaraw, P., & Verma, A. K. (2017). Comprehensive review on application of edible film on meat and meat products: An eco-friendly approach. *Critical Reviews in Food Science and Nutrition*, *57*(6), 1270–1279. doi:10.1080/10408398.2014.986563 PMID:26090546

Valdés, A., Ramos, M., Beltrán, A., Jiménez, A., & Garrigós, M. C. (2017). State of the art of antimicrobial edible coatings for food packaging applications. *Coatings*, *7*(4), 56. doi:10.3390/coatings7040056

Velickova, E., Winkelhausen, E., Kuzmanova, S., Moldao-Martins, M., & Alves, V. D. (2015). Characterization of multilayered and composite edible films from chitosan and beeswax. *Food Science & Technology International*, *21*(2), 83–93. doi:10.1177/1082013213511807 PMID:24285830

Verbeek, C. J., & van den Berg, L. E. (2010). Extrusion processing and properties of protein-based thermoplastics. *Macromolecular Materials and Engineering*, *295*(1), 10–21. doi:10.1002/mame.200900167

Vilela, C., Kurek, M., Hayouka, Z., Röcker, B., Yildirim, S., Antunes, M. D. C., Nilsen-Nygaard, J., Pettersen, M. K., & Freire, C. S. (2018). A concise guide to active agents for active food packaging. *Trends in Food Science & Technology*, *80*, 212–222. doi:10.1016/j.tifs.2018.08.006

Wang, H., Ding, F., Ma, L., & Zhang, Y. (2021). Edible films from chitosan-gelatin: Physical properties and food packaging application. *Food Bioscience*, *40*, 100871. doi:10.1016/j.fbio.2020.100871

Wang, S., Lu, A., & Zhang, L. (2016). Recent advances in regenerated cellulose materials. *Progress in Polymer Science*, *53*, 169–206. doi:10.1016/j.progpolymsci.2015.07.003

Wei, L. T., & Yazdanifard, R. (2013). Edible food packaging as an eco-friendly technology using green marketing strategy. *Global Journal of Commerce & Management Perspective*, *2*(6), 8–11.

Xia, C., Wang, W., Wang, L., Liu, H., & Xiao, J. (2019). Multilayer zein/gelatin films with tunable water barrier property and prolonged antioxidant activity. *Food Packaging and Shelf Life*, *19*, 76–85. doi:10.1016/j.fpsl.2018.12.004

Xiong, Y., Chen, M., Warner, R. D., & Fang, Z. (2020a). Incorporating nisin and grape seed extract in chitosan-gelatine edible coating and its effect on cold storage of fresh pork. *Food Control*, *110*, 1–8. doi:10.1016/j.foodcont.2019.107018

Xu, P., Zeng, Q., Cao, Y., Ma, P., Dong, W., & Chen, M. (2017). Interfacial modification on polyhydroxyalkanoates/starch blend by grafting in-situ. *Carbohydrate Polymers*, *174*, 716–772. doi:10.1016/j.carbpol.2017.06.048 PMID:28821123

Yadav, S., Mehrotra, G. K., Bhartiya, P., Singh, A., & Dutta, P. K. (2020). Preparation, physicochemical and biological evaluation of quercetin-based chitosan-gelatin film for food packaging. *Carbohydrate Polymers*, *227*, 115348. doi:10.1016/j.carbpol.2019.115348 PMID:31590881

Yap, X. Y., Gew, L. T., Khalid, M., & Yow, Y. Y. (2023). Algae-Based Bioplastic for Packaging: A Decade of Development and Challenges (2010–2020). *Journal of Polymers and the Environment*, *31*(3), 833–851. doi:10.100710924-022-02620-0

Yoon, S.-D. (2014). Cross-linked potato starch-based blend films using ascorbic acid as a plasticizer. *Journal of Agricultural and Food Chemistry*, *62*(8), 1755–1764. doi:10.1021/jf4024855 PMID:23909738

Zabihollahi, N., Alizadeh, A., Almasi, H., Hanifian, S., & Hamishekar, H. (2020). Development and characterization of carboxymethyl cellulose based probiotic nanocomposite film containing cellulose nanofiber and inulin for chicken fillet shelf-life extension. *International Journal of Biological Macromolecules*, *160*, 409–417. doi:10.1016/j.ijbiomac.2020.05.066 PMID:32416305

Zhang, H., Li, X., Kang, H., & Peng, X. (2021). Antimicrobial and antioxidant effects of edible nanoemulsion coating based on chitosan and Schizonepeta tenuifolia essential oil in fresh pork. *Journal of Food Processing and Preservation*, *45*(11), e15909. doi:10.1111/jfpp.15909

Zhang, H., Liang, Y., Li, X., & Kang, H. (2020). Effect of chitosan-gelatin coating containing nano-encapsulated tarragon essential oil on the preservation of pork slices. *Meat Science*, *166*, 108137. doi:10.1016/j.meatsci.2020.108137 PMID:32272381

Chapter 10
Extraction of Protein and Polyphenols From Agro-Industrial Waste Through Eco-Innovative Technologies

Arisha Ahmad
NUR International University, Pakistan

Ali Imran
The University of Melbourne, Australia

Rutaba Nadeem
NUR International University, Pakistan

Fakhar Islam
https://orcid.org/0000-0001-6935-5924
Government College University, Faisalabad, Pakistan

Saima Naz
NUR International University, Pakistan

Hafiz Suleria
The University of Melbourne, Australia

ABSTRACT

The recovery of valuable sources from waste material is an excellent and encouraging approach towards the ecological significances and food security. Agro-industrial surplus produced by agricultural and food processing activities signifies a rich source of beneficial compounds like minerals, vitamins, and other essential components that are helpful in different industries for different purposes. The demands of plant-

DOI: 10.4018/979-8-3693-0819-6.ch010

based protein extraction are very high. Along with proteins, various bioactive components, like carotenoids, lycopene, polyphenols, sugars, and polysaccharide, have also been extracted with the help of eco-friendly technologies. The purpose of this chapter is to study the latest researches related to extraction of protein and polyphenols or also to highlight the significances of green-technologies. These eco-friendly technologies play their vital role in producing nutrient rich products from waste residues to overcome the food shortage and to maintain food wastage.

INTRODUCTION

The world's population is increasing day by day and this explosion creates a demand for an increase, in the need for various food and energy resources, causing a significant surge which has led to advancements, in agriculture and related industries (Awasthi et al., 2020). This increasing demand prompted the rapid industrialization and modernization of agriculture and food-related sectors, resulting in enhanced production diversity and the overall improvement of both product quality and quantity (Arun et al., 2020). During various unit operations from different sources like industrial and domestic kitchens, cafeterias restaurants, and different food processing plants the organic waste (biodegradable material that comes from animal or plant sources) was generated at a heavy scale (Melikoglu et al., 2013).

According to FAO, 1.3 billion tons' approx. vegetables, dairy products, fruits, bakery, and other food products are lost during the food supply chain (FAO, 2012). From 2005-2025, in Asian countries, the amount of food wastage rose from 278 to 416 million tons approximately every year. Over the next 25 years, due to demographic and economic growth in Asian countries food wastage is supposed to be increased (FAO, 2014).

The major source of municipal solid waste is yard and kitchen waste. The major portion of kitchen waste comprises peels, seeds, leaves, and extra portions of food products like peels of fruits and vegetables, egg shells, etc. which are not consumed by humans. Instead of Landfilling and throwing away the waste, it may be converted into beneficial products by using Green technologies and could be further utilized for the welfare of human beings and to overcome food wastage and shortage (Hansen et al., 2006).

The peels are considered a valuable source of various bioactive components such as cutin, Protein, and polyphenols (Lu et al., 2019). The extraction of these compounds from peel extracts presents an opportunity for both economic gain and environmental enhancement (Benítez et al., 2018).

By using organic solvents like acetone, ethanol, or methanol, for the extraction of phenolic compounds from agro-food wastes various extraction techniques have been used. Along with proteins, various bioactive components, like carotenoids, lycopene, polyphenols, sugars, and polysaccharides have also been extracted by using advanced extraction techniques, such as pulsed electric field (Andreou et al. 2020) and high-pressure homogenization (Juri´c et al., 2019).

With low environmental impact, ecological and green techniques are a growing need, leads to extract protein and rich polyphenols. From waste materials, the extraction of protein and functional polyphenolic compounds is done through the most innovative techniques (Panzella et al., 2020). The most commonly eco-friendly, innovative extraction technique is the use of Green technologies that consist of Ultrasound waves, Microwaves and Supercritical Fluid along with Green solvents like deep eutectic solvent (Fermoso et al., 2018).

VALORIZATION OF AGRO-INDUSTRIAL WASTE

A relatively new conception called 'valorization' has been promoting in the field of industrial wastage management that is responsible for the promotion of sustainable environment (Federici et al., 2009). The main aim of the valorization of food waste is to produce the appropriate environment friendly product from wastes (Tiwari et al., 2021). This approach encourages the advancement of a "zero-waste" bio economy, by producing economically viable products utilizing the agro-industrial waste as a primary raw material (Mikula et al., 2022).

The intention behind creating the bio refineries, is offering an environmentally sustainable approach to generate valuable bio-based products such as food for humans and feed for animals chemicals, and materials, as well as biofuels like bioethanol, bio butanol and biodiesel. The purpose of bio refineries is to focus on utilizing biomass, particularly organic waste, as a renewable energy source and also to produce biogas (Rojas et al., 2022). Reducing food waste through its conversion into valuable compounds has become a significant achievement in the field of bio refineries, present substantial benefits to the environment, society and economy (Capanoglu et al., 2021). A sustainable amount of wastes is generated through food and agriculture sector which is further used as a raw material to produce valuable product.

In developing environment the main purpose of researches are to overcome the wastage of food and they focus on the wastes called 'agro-industrial waste' that includes leaves, peels, pulps, seeds, press cakes and the pits (Freitas et al., 2021). A huge range of secondary plant metabolites an important part of bioactive compounds that have an antioxidant activity present in plant tissues called phenolic components are present in it.

The most broadly consumed edible oil is known as olive oil, whose 98% worldwide production is done through Greece, Spain, Italy and Portugal. Two different wastes that are generated during the industrial extraction of olive oil are OMW (olive mill waste water) and its solid residue (Henkel et al., 2012). The olive mill waste residue that is generated during the extraction of Olive oil is further valorized and used as a raw material. It is mainly water approx. 83-94% and its remaining components are carbohydrates, phenolic compounds, lipids, minerals, organic compounds, pectin and tannins (Dermeche et al., 2013). The olives variation, agricultural methods climatic situations and most important the oil extraction techniques through which extraction was done are the major factors that effects the quality and quantity of content of olive mill. This can involve either a discontinuous pressing method or a continuous extraction process, such as the two-phase or three-phase processes. The three-phase extraction process is the most prevalent and involves more water usage for oil extraction compared to the two-phase process, leading to the production of larger quantities of OMW wastewater (Gonçalves et al., 2009).

The residue comes from the processing of animal products and the agricultural production of plants can undergo enzymatic treatment to yield valuable products (Marín et al., 2019). i.e., coffee peel has been hydrolyzed by cellulose that results in the formation of fiber which is suitable for human health and utilize in the animal feed. Coffee peel is a reservoir of lignocellulosic material and rich in bioactive compounds, that acts as a substrate for various enzymes like pectinase, cellulose and xylanase. With the process of enzymatic hydrolysis, the residue can be converted into sugar, alcohol, chlorogenic acid and oligosaccharides (Dong et al., 2020).

METHODS OF EXTRACTION OF PROTEIN FROM AGRO-INDUSTRIAL WASTE

Protein and its building blocks called amino acids, have various function that comprises the cell composition and tissue structure and considered as an essential nutrients for human body and serves as a constituents of enzymes. Approx.7 million tons of protein till 2025 is necessary for human consumption and it is not possible to fulfill the demand by only livestock industry. To fulfill the gap, plant based protein and many other alternative sources are mandatory. Several process were done to extract protein from plant based sources, like high pressure and temperature, toxic reactants and extreme pH. Green extraction technologies are emerged as a great alternative to traditional methods for recovering high-value compounds from agricultural and industrial wastes because they result in high-quality and pure final extracts (Segatto et al., 2022).

From vegetable matrix the main step of protein extraction is done through cell lysis that is followed by the isolation and concentration of desired product. The extraction of protein consists of chemical, physical and biochemical process to break the cell wall, and it allows to releasing of target protein. This procedure calls for an extraction medium that encourages cell lysis and the mass move of the protein out of the cell. This process is frequently facilitated by a technology that aids in the disruption of the cells, speeds up mass transfer, and increases the solubility of the protein in the extraction medium. When using procedures that require protein precipitation, the extraction medium should be left alone so that undesirable substances can be selectively solubilized without dissolving the desired proteins. These procedures are frequently followed by a subsequent step that needs a solvent that can dissolve proteins (Pojić et al., 2018).

Supercritical fluids extraction is regarded as beneficial technology for increasing the benefits of agro-industrial wastes. Since it facilitates the extraction of numerous chemicals with antioxidant activity that have potential uses in a variety of industries (Lizcano et al., 2019).

In terms of cost UAE is one of the reliable technique among the non-conventional extraction methods. Therefore, in order to be practical on an industrial scale, it has undergone significant laboratory investigations. In order to facilitate the release of target components, accelerate diffusion, improve mass transfer, and allow for greater solvent penetration into the ultrasound-radiated matrix, cell walls are disrupted using ultrasonic waves (Chakraborty et al., 2020).

To utilize discarded tomato, watermelon, and apple peels, a cascade extraction method based on ultrasound-assisted extraction (UAE) was performed in this study. The first extraction stage produced the protein and antioxidant components. The largest levels of total phenolics and proteins were found in the watermelon peels (WP), however the ABTS, FRAP, and DPPH assays revealed that apple peels had the highest levels of antioxidant activity. Later on cutin was obtained by extracting the leftover residue from the initial extraction (Mellinas et al., 2022).

In comparison to ultrasound-assisted extraction, microwave-assisted extraction is regarded as a relatively new technology. But numerous labs have conducted studies on it, signifying the huge potential of this unconventional energy source for extraction applications. Microwaves are electromagnetic radiation that ranges in frequency from 300 MHz (radio radiation) to 300 GHz, according to the definition (Chen et al., 2017).

METHODS OF EXTRACTION AND CHARACTERIZATION OF POLYPHENOLS FROM AGRO-INDUSTRIAL WASTE

The processing methods of food and agro-industries generate a large amount of side-stream and waste materials, which are a great sources of functional bioactive molecules like phenolic compounds. Recovering these materials can be advantageous and very beneficial for human health as well as for food sustainability for many industrial uses like flavoring agents and therapeutic applications like anti-inflammatory, anti-bacterial and antimicrobial activities.

The recovery of bioactive chemicals is significantly impacted by the treatments and techniques used for extraction. In addition to the traditional extraction procedures, various innovative methods have been developed and improved to enable the extraction of bioactive components more effectively and efficiently (Abbasi et al., 2022). In various studies, agro-industrial waste was valorized broadly to obtain bioactive components through conventional extraction techniques, and they are considered as main active biomolecules.

The first and essential step in recovering natural bioactive components from byproducts is extraction. The extraction effectiveness in different biomasses can be simulated by different extraction components. Initially, the temperature, duration, and solvents concentration during the extraction are significant factors. Since phenolic compounds are highly sensitive to pH, an ideal pH should be taken into consideration to ensure the intact extraction of phenolic compounds (Amendola et al., 2010).

Hydro-soluble compounds, like phenolic acids and flavonoids, are often extracted using water, alcoholic solvents, acetone, ethyl acetate, or a combination of these in various ratios. The method of extraction and the choice of solvent has a major impact on how the phenolic components separate afterward. The potential of the green extraction techniques in terms of phenolic yield extraction and state has been observed by many researchers (Ameer et al., 2017). The use of a moderately acidified solvent increases the stability of phenolic components in extracts (Fontana et al., 2013).

The phenolic compounds are subjected to the characterization procedures once they have been extracted. High-performance liquid chromatography (HPLC), which can typically be coupled with electrochemical detection techniques (Khoddami et al., 2013) like UV, mass spectrometry, fluorescence, and nuclear magnetic resonance, is one of the most common techniques used to separate and characterize various classes of phenolic in by-products (Ignat et al., 2011).

Nuclear magnetic resonance is essential for characterizing the structure of flavonoids in complex matrices, despite its drawbacks, which include a high cost and the possibility that it may not be appropriate for all applications (li et al., 2022). But today, the most popular technique for identifying and characterizing flavonoids is mass spectrometry (MS). Recently, MS has also been used with ultra-high pressure liquid chromatography (UHPLC) to detect and identify precise phenolic structures (de Villiers et al., 2016).

Nowadays, modern chromatography-mass spectrometry (LC-MS) procedures are the best analytical method for evaluating polyphenols in vegetal samples and are the most useful tool for the study of anthocyanin structure. The characterization of the sugar moiety and anthocyanin aglycone is made conceivable by the MS/MS method, which is a very potent tool. Procyanidins, tannins, prodelphinidins, and proanthocyanidins, are examples of complex molecular structures that can be characterized by LC-MS. These compounds provide experimental evidence of support for previously postulated structures.

Both SPME (solid-phase microextraction) and LC-MS have been used to measure the levels of resveratrol in wine, a significant polyphenol well-known for its health benefits. The former method produced the best findings in the context of a sensitivity (Flamini, 2003).

FUTURE PERSPECTIVES FOR THE VALORIZATION OF WASTES THROUGH GREEN TECHNOLOGY

For an ecological and environmentally pleasant future, the valorization of agro industrial waste through green technology is very necessary. Incorporating agricultural and industrial waste into a circular economy model green technologies will a vital role. Waste material can be considered as a useful resource for the production of chemicals, energy production, and bio-based products, rather than being a disposable trash. Additionally, it will make it possible to create platforms for converting waste from the agro-industrial sector into useful goods with little harm to the environment. It can be used as a feedstock for the creation of bioenergy, including the manufacturing of biofuels like biogas, bioethanol, and biodiesel.

The future will likely see advancements in bioprocessing techniques and the development of high-yielding bioenergy crops to maximize energy generation from waste. The concept of agro-industrial biorefineries will expand. These facilities will not only produce bioenergy but also a wide range of value-added products such as biomaterials, chemicals, bio-based and bioplastics from waste creeks.

In the future, the main focus of research is on recovering valuable resources such as nutrients and minerals from agro-industrial residue. These can be used as supplements, drugs, and fertilizers that contribute to the sustainability of agriculture.

With the increasing demand for eco-friendly packaging materials or to reduce plastic pollution, green technology will lead to the production of biodegradable packaging materials from agro-industrial wastes. International organizations

Table 1. Application of green technology on agro-industrial waste to obtain proteins, bioactive compounds, and other beneficial components

Agro-Industrial Waste	Extraction Method	Extracted Compounds	References
Jamun seed	MAE	Polysaccharides	(Abdullah & Ponmurugan, 2019)
Citrus plant (Orange and lemon peel)	MAE	Essential oil	(Auta et al., 2018)
Distillate liquor residues of strawberry	SFE	Phenolic compounds	(Alexandre et al., 2020)
Mango peel	SFE	Carotenoids	(Ballesteros et al., 2019)
Pineapple peel, crown & core	MAE, UAE, Submerged and solid-state fermentation	Polyphenols, catechins, ferulic acid gallic acid, epicatechins	(Polania et al., 2023)
Soybean waste (pressing oil extraction)	SFE	Phenolic compounds and flavonoids	(Alvarez et al., 2019)
Custard apple peel	UAE	Pectin	(Shivamathi et al., 2019)
Pistachio hulls	SWE	Phenolic compounds	(Ersan et al., 2018)
Tomato processing residue	UAE	lycopene	(Rahimi & Mikani, 2019)
Cocoa bean shell	MAE	Proteins, polysaccharides and polyphenols	(Mellinas et al., 2020)
Onion wastes	SFE	Oleoresin	(Devani et al., 2019)
Sour orange peel	UAE	Pectin	(Hosseini et al., 2019)
Apple pomace	UAE, MEA Sox let extraction	Nutraceuticals pectin.	(Putra et al 2023)
Mango peel and seed	Pulsed electric fields, high-voltage electrical discharges and ultra-sonication	polyphenols and carotenoids	(García et al., 2023)
Satsuma mandarin peel	SWE	Flavonoids	(ko et al., 2016)
Carrot juice processing waste	MAE	Carotenoids	(Elik et al., 2020)
Grape skin		Phenolic acids, flavonoids, gallate, p-coumaric acid, Gallic acid	(Liu et al., 2018)
Grape marc, Orange peel, Strawberry, Citrus pulp, Camelina cake	FRAP ABTS	Phenolic compounds	(Castrica et al., 2019)
Mango peel	UAE & maceration techniques	Polyphenols	(Safdar et al., 2017)

and Governments may probably implement stricter rules and regulations on the management of waste materials coming from industries, households, agriculture, etc., and also promote the use of green technologies. Incentives and subsidies may also be provided to motivate the implementation of sustainable waste valorization practices.

CONCLUSION

In conclusion from the agro-industrial waste the extraction of proteins and polyphenols using eco-innovative technologies are a vital step toward the production of nutrient rich products and sustainable waste management to overcome food shortage and wastage. By extracting valuable components from waste materials the efficiency of food production may also increases. The residue comes from the processing can undergo various eco-innovative treatments to yield valuable products for the welfare of society. For researchers, policymakers and industry participants looking to maximize the economic and environmental benefits of agro-industrial waste in a circular and sustainable economy, this comprehensive review is a helpful resource.

ACKNOWLEDGMENT

The author acknowledges the financial support provided by the Punjab Higher Education Department, Punjab, Pakistan (PHEC) in terms of a postdoc international fellowship for the University of Melbourne, Australia to the corresponding author.

REFERENCES

Abbasi-Parizad, P., Scarafoni, A., Pilu, R., Scaglia, B., De Nisi, P., & Adani, F. (2022). The recovery from agro-industrial wastes provides different profiles of anti-inflammatory polyphenols for tailored applications. *Frontiers in Sustainable Food Systems*, *6*, 996562. doi:10.3389/fsufs.2022.996562

Abdullah Al-Dhabi, N., & Ponmurugan, K. (2019). Microwave assisted extraction and characterization of polysaccharide from waste jamun fruit seeds. *International Journal of Biological Macromolecules*, *152*, 1157–1163. doi:10.1016/j.ijbiomac.2019.10.204 PMID:31751731

Alexandre, A.M.R.C., Serraa, A.T., Matiasa, A.A., Duarte, C.M.M., & Bronze, M.R. (2020). Supercritical fluid extraction of Arbutus unedo distillate residues – impact of process conditions on antiproliferative response of extracts. *J. CO2 Util.*, *37*, 29–38.

Alvarez, M. V., Cabred, S., Ramirez, C. L., & Fanovich, M. A. (2019). Valorization of an agroindustrial soybean residue by supercritical fluid extraction of phytochemical compounds. *The Journal of Supercritical Fluids*, *143*, 90–96. doi:10.1016/j.supflu.2018.07.012

Ameer, K., Shahbaz, H. M., & Kwon, J. H. (2017). Green extraction methods for polyphenols from plant matrices and their byproducts: A review. *Comprehensive Reviews in Food Science and Food Safety*, *16*(2), 295–315. doi:10.1111/1541-4337.12253 PMID:33371540

Amendola, D., De Faveri, D. M., & Spigno, G. (2010). Grape marc phenolics: Extraction kinetics, quality and stability of extracts. *Journal of Food Engineering*, *97*(3), 384–392. doi:10.1016/j.jfoodeng.2009.10.033

Andreou, V., Dimopoulos, G., Dermesonlouoglou, E., & Taoukis, P. (2020). Application of pulsed electric fields to improve product yield and waste valorization in industrial tomato processing. *Journal of Food Engineering*, *270*, 109778. doi:10.1016/j.jfoodeng.2019.109778

Arun, K. B., Madhavan, A., Sindhu, R., Binod, P., Pandey, A., Reshmy, R., & Sirohi, R. (2020). Remodeling agro-industrial and food wastes into value-added bioactives and biopolymers. *Industrial Crops and Products*, *154*, 112621. doi:10.1016/j.indcrop.2020.112621

Auta, M., Musa, U., Tsado, D. G., Faruq, A. A., Isah, A. G., Raji, S., & Nwanisobi, C. (2018). Optimization of citrus peels D-limonene extraction using solvent-free microwave green technology. *Chemical Engineering Communications*, *205*(6), 789–796. doi:10.1080/00986445.2017.1419206

Awasthi, S. K., Sarsaiya, S., Awasthi, M. K., Liu, T., Zhao, J., Kumar, S., & Zhang, Z. (2020). Changes in global trends in food waste composting: Research challenges and opportunities. *Bioresource Technology*, *299*, 122555. doi:10.1016/j.biortech.2019.122555 PMID:31866141

Ballesteros-Vivas, D., Alvarez-Rivera, G., & Medina, S. J. M. (2019). An integrated approach for the valorization of mango seed kernel: Efficient extraction solvent selection, phytochemical profiling and antiproliferative activity assessment. *Food Research International*, *126*, 108616. doi:10.1016/j.foodres.2019.108616 PMID:31732074

Benítez, J. J., Castillo, P. M., Del Río, J. C., León-Camacho, M., Domínguez, E., Heredia, A., & Heredia-Guerrero, J. A. (2018). Valorization of tomato processing by-products: Fatty acid extraction and production of bio-based materials. *Materials (Basel)*, *11*(11), 2211. doi:10.3390/ma11112211 PMID:30405081

Capanoglu, E., Nemli, E., & Tomas-Barberan, F. (2021). Novel approaches in the valorization of agricultural wastes and their applications. *Journal of Agricultural and Food Chemistry, 2022*, 6787–6804. PMID:35195402

Castrica, M., Rebucci, R., Giromini, C., Tretola, M., Cattaneo, D., & Baldi, A. (2019). Total phenolic content and antioxidant capacity of agri-food waste and by-products. *Italian Journal of Animal Science, 18*(1), 336–341. doi:10.1080/1828051X.2018.1529544

Chakraborty, P. K., Adhikari, J., & Saha, P. (2018). Facile fabrication of electrospun regenerated cellulose nanofiber scaffold for potential bone-tissue engineering application. *International Journal of Biological Macromolecules, 122*, 644–652. doi:10.1016/j.ijbiomac.2018.10.216 PMID:30391587

Chen, C., Zhang, B., Huang, Q., Fu, X., & Liu, R. (2017). Microwave-assisted extraction of polysaccharides from Moringa oleifera Lam. leaves: Characterization and hypoglycemic activity. *Industrial Crops and Products, 100*, 1–11. doi:10.1016/j.indcrop.2017.01.042

de Villiers, A., Venter, P., & Pasch, H. (2016). Recent advances and trends in the liquid-chromatography–mass spectrometry analysis of flavonoids. *Journal of Chromatography. A, 1430*, 16–78. doi:10.1016/j.chroma.2015.11.077 PMID:26718188

Dermeche, S., Nadour, M., Larroche, C., Moulti-Mati, F., & Michaud, P. (2013). Olive mill wastes: Biochemical characterizations and valorization strategies. *Process Biochemistry (Barking, London, England), 48*(10), 1532–1552. doi:10.1016/j.procbio.2013.07.010

Devani, B. M., Jani, B. L., Balani, P. C., & Akbari, S. H. (2019). Optimization of supercritical CO2 extraction process for oleoresin from rotten onion waste. *Food and Bioproducts Processing, 119*, 287–295. doi:10.1016/j.fbp.2019.11.014

Dong, W., Wang, D., Hu, R., Long, Y., & Lv, L. (2020). Chemical composition, structural and functional properties of soluble dietary fiber obtained from coffee peel using different extraction methods. *Food Research International, 136*, 109497. doi:10.1016/j.foodres.2020.109497 PMID:32846578

Elik, A., Yanik, D. K., & Göğüş, F. (2020). Microwave-assisted extraction of carotenoids from carrot juice processing waste using flaxseed oil as a solvent. *Lebensmittel-Wissenschaft + Technologie, 123*, 109100. doi:10.1016/j.lwt.2020.109100

Ersan, S., Üstündag, ˇ. O. G., Carle, R., & Schweiggert, R. M. (2018). Subcritical water extraction of phenolic and antioxidant constituents from pistachio (Pistacia vera L.) hulls. *Food Chemistry, 253*, 46–54. doi:10.1016/j.foodchem.2018.01.116 PMID:29502842

FAO. (2012). *Towards the Future We Want: End hunger and make the transition to sustainable agricultural and food systems*. Food and Agriculture Organization of the United Nations Rome.

FAO. (2014). *Food wastage footprint full-cost accounting*. Food Wastage Footprint.

Federici, F., Fava, F., Kalogerakis, N., & Mantzavinos, D. (2009). Valorisation of agro-industrial by-products, effluents and waste: concept, opportunities and the case of olive mill wastewaters. *Journal of Chemical Technology & Biotechnology: International Research in Process, Environmental & Clean Technology, 84*(6), 895–900.

Fermoso, F. G., Serrano, A., Alonso-Fariñas, B., Fernández-Bolaños, J., Borja, R., & Rodríguez-Gutiérrez, G. (2018). Valuable compound extraction, anaerobic digestion, and composting: A leading biorefinery approach for agricultural wastes. *Journal of Agricultural and Food Chemistry, 66*(32), 8451–8468. doi:10.1021/acs.jafc.8b02667 PMID:30010339

Flamini, R. (2003). Mass spectrometry in grape and wine chemistry. Part I: Polyphenols. *Mass Spectrometry Reviews, 22*(4), 218–250. doi:10.1002/mas.10052 PMID:12884388

Fontana, A. R., Antoniolli, A., & Bottini, R. (2013). Grape pomace as a sustainable source of bioactive compounds: Extraction, characterization, and biotechnological applications of phenolics. *Journal of Agricultural and Food Chemistry, 61*(38), 8987–9003. doi:10.1021/jf402586f PMID:24004377

Freitas, L. C., Barbosa, J. R., da Costa, A. L. C., Bezerra, F. W. F., Pinto, R. H. H., & de Carvalho, R. N. Jr. (2021). From waste to sustainable industry: How can agro-industrial wastes help in the development of new products? *Resources, Conservation and Recycling, 169*, 105466. doi:10.1016/j.resconrec.2021.105466

García-Mahecha, M., Soto-Valdez, H., Carvajal-Millan, E., Madera-Santana, T. J., Lomelí-Ramírez, M. G., & Colín-Chávez, C. (2023). Bioactive compounds in extracts from the agro-industrial waste of mango. *Molecules (Basel, Switzerland), 28*(1), 458. doi:10.3390/molecules28010458 PMID:36615647

Gonçalves, C., Lopes, M., Ferreira, J. P., & Belo, I. (2009). Biological treatment of olive mill wastewater by non-conventional yeasts. *Bioresource Technology, 100*(15), 3759–3763. doi:10.1016/j.biortech.2009.01.004 PMID:19231162

Hansen, T. L., Bhander, G. S., Christensen, T. H., Bruun, S., & Jensen, L. S. (2006). Life cycle modelling of environmental impacts of application of processed organic municipal solid waste on agricultural land (EASEWASTE). *Waste Management & Research, 24*(2), 153–166. doi:10.1177/0734242X06063053 PMID:16634230

Henkel, M., Müller, M. M., Kügler, J. H., Lovaglio, R. B., Contiero, J., Syldatk, C., & Hausmann, R. (2012). Rhamnolipids as biosurfactants from renewable resources: Concepts for next-generation rhamnolipid production. *Process Biochemistry (Barking, London, England), 47*(8), 1207–1219. doi:10.1016/j.procbio.2012.04.018

Hosseini, S. S., Khodaiyan, F., Kazemi, M., & Najari, Z. (2019). Optimization and characterization of pectin extracted from sour orange peel by ultrasound assisted method. *International Journal of Biological Macromolecules, 125*, 621–629. doi:10.1016/j.ijbiomac.2018.12.096 PMID:30543886

Ignat, I., Volf, I., & Popa, V. I. (2011). A critical review of methods for characterisation of polyphenolic compounds in fruits and vegetables. *Food Chemistry, 126*(4), 1821–1835. doi:10.1016/j.foodchem.2010.12.026 PMID:25213963

Juri'c, S., Ferrari, G., Velikov, K. P., & Donsì, F. (2019). High-pressure homogenization treatment to recover bioactive compounds from tomato peels. *Journal of Food Engineering, 262*, 170–180. doi:10.1016/j.jfoodeng.2019.06.011

Khoddami, A., Wilkes, M. A., & Roberts, T. H. (2013). Techniques for analysis of plant phenolic compounds. *Molecules (Basel, Switzerland), 18*(2), 2328–2375. doi:10.3390/molecules18022328 PMID:23429347

Ko, M.-J., Kwon, H.-L., & Chung, M.-S. (2016). Pilot-scale subcritical water extraction of flavonoids from satsuma mandarin (Citrus unshiu Markovich) peel. *Innovative Food Science & Emerging Technologies, 38*, 175–181. doi:10.1016/j.ifset.2016.10.008

Li, J., Pettinato, M., Campardelli, R., De Marco, I., & Perego, P. (2022). High-Pressure Technologies for the Recovery of Bioactive Molecules from Agro-Industrial Waste. *Applied Sciences (Basel, Switzerland), 12*(7), 3642. doi:10.3390/app12073642

Liu, Q., Tang, G. Y., Zhao, C. N., Feng, X. L., Xu, X. Y., Cao, S. Y., Meng, X., Li, S., Gan, R.-Y., & Li, H. B. (2018). Comparison of antioxidant activities of different grape varieties. *Molecules (Basel, Switzerland), 23*(10), 2432. doi:10.3390/molecules23102432 PMID:30249027

Lizcano, S. C., Davila, ´. J. A., & Hernandez, ´. V. (2019). Fruit agroindustrial wastes for preparing beverages for medicinal purposes by supercritical fluid extraction technology: Andes berry (Rubus glaucus benth) case. *Prod. Manage. Bev.*, *1*, 151–177. doi:10.1016/B978-0-12-815260-7.00005-5

Lu, Z., Wang, J., Gao, R., Ye, F., & Zhao, G. (2019). Sustainable valorisation of tomato pomace: A comprehensive review. *Trends in Food Science & Technology*, *86*, 172–187. doi:10.1016/j.tifs.2019.02.020

Marín, M., Artola, A., & Sánchez, A. (2019). Optimization of down-stream for cellulases produced under solid-state fermentation of coffee husk. *Waste and Biomass Valorization*, *10*(10), 2761–2772. doi:10.100712649-018-0327-5

Melikoglu, M., Lin, C. S. K., & Webb, C. (2013). Analysing global food waste problem: Pinpointing the facts and estimating the energy content. *Central European Journal of Engineering*, *3*, 157–164.

Mellinas, A. C., Jim'enez, A., & Garrigos, ´. M. C. (2020). Optimization of microwave-assisted extraction of cocoa bean shell waste and evaluation of its antioxidant, physicochemical and functional properties. *Lebensmittel-Wissenschaft + Technologie*, *127*, 109361. doi:10.1016/j.lwt.2020.109361

Mellinas, C., Solaberrieta, I., Pelegrín, C. J., Jiménez, A., & Garrigós, M. C. (2022). Valorization of Agro-Industrial Wastes by Ultrasound-Assisted Extraction as a Source of Proteins, Antioxidants and Cutin: A Cascade Approach. *Antioxidants*, *11*(9), 1739. doi:10.3390/antiox11091739 PMID:36139813

Mikula, K., Gersz, A., Witek-Krowiak, A., Skrzypczak, D., Izydorczyk, G., & Chojnacka, K. (2022). Agrochemicals in view of circular economy. In *Smart Agrochemicals for Sustainable Agriculture* (pp. 57–80). Academic Press. doi:10.1016/B978-0-12-817036-6.00004-2

Panzella, L., Moccia, F., Nasti, R., Marzorati, S., Verotta, L., & Napolitano, A. (2020). Bioactive phenolic compounds from agri-food wastes: An update on green and sustainable extraction methodologies. *Frontiers in Nutrition*, *7*, 60. doi:10.3389/fnut.2020.00060 PMID:32457916

Pojić, M., Mišan, A., & Tiwari, B. (2018). Eco-innovative technologies for extraction of proteins for human consumption from renewable protein sources of plant origin. *Trends in Food Science & Technology*, *75*, 93–104. doi:10.1016/j.tifs.2018.03.010

Polanía, A. M., Londoño, L., Ramírez, C., Bolivar, G., & Aguilar, C. N. (2023). Valorization of pineapple waste as novel source of nutraceuticals and biofunctional compounds. *Biomass Conversion and Biorefinery*, *13*(5), 3593–3618. doi:10.100713399-022-02811-8

Putra, N. R., Rizkiyah, D. N., Abdul Aziz, A. H., Che Yunus, M. A., Veza, I., Harny, I., & Tirta, A. (2023). Waste to Wealth of Apple Pomace Valorization by Past and Current Extraction Processes: A Review. *Sustainability (Basel)*, *15*(1), 830. doi:10.3390u15010830

Rahimi, S., & Mikani, M. (2019). Lycopene green ultrasound-assisted extraction using edible oil accompany with response surface methodology (RSM) optimization performance: Application in tomato processing wastes. *Microchemical Journal*, *146*, 1033–1042. doi:10.1016/j.microc.2019.02.039

Rojas, L. F., Zapata, P., & Ruiz-Tirado, L. (2022). Agro-industrial waste enzymes: Perspectives in circular economy. *Current Opinion in Green and Sustainable Chemistry*, *34*, 100585. doi:10.1016/j.cogsc.2021.100585

Safdar, M. N., Kausar, T., & Nadeem, M. (2017). Comparison of ultrasound and maceration techniques for the extraction of polyphenols from the mango peel. *Journal of Food Processing and Preservation*, *41*(4), e13028. doi:10.1111/jfpp.13028

Segatto, M. L., Stahl, A. M., Zanotti, K., & Zuin, V. G. (2022). Green and sustainable extraction of proteins from agro-industrial waste: An overview and a closer look to Latin America. *Current Opinion in Green and Sustainable Chemistry*, *37*, 100661. doi:10.1016/j.cogsc.2022.100661

Shivamathi, C. S., Moorthy, I. G., Kumar, R. V., Soosai, M. R., Maran, J. P., Kumar, R. S., & Varalakshmi, P. (2019). Optimization of ultrasound assisted extraction of pectin from custard apple peel: Potential and new source. *Carbohydrate Polymers*, *225*, 115240. doi:10.1016/j.carbpol.2019.115240 PMID:31521287

Tiwari, A., & Khawas, R. (2021). *Food Waste and Agro By-Products: A Step towards Food Sustainability*. IntechOpen. doi:10.5772/intechopen.96177

Chapter 11
Nanoclay Packaging Materials as Biodegradable

Zain Mushtaq
Department of Soil Science, Faculty of Agricultural Sciences, University of the Punjab, Pakistan

Maleeha Islam
Department of Soil Science, Faculty of Agricultural Sciences, University of the Punjab, Pakistan

Salwa Saeed
Department of Soil Science, Faculty of Agricultural Sciences, University of the Punjab, Pakistan

Korkmaz Bellitürk
Tekirdag Namık Kemal University, Turkey

Predrag Ilić
https://orcid.org/0000-0002-0316-5615
PSRI Institute for Protection and Ecology of the Republic of Srpska, Bosnia and Herzegovina

Amir Parnian
Independent Researcher, Iran

ABSTRACT

Nanoclays are nanomaterials consisting of layered silicate clay minerals and have various applications in food packaging. They can improve the mechanical, thermal, and barrier properties of both petroleum-based and bio-based polymers, as well as provide other functions such as antimicrobial activity, colorimetric indication, biodegradability enhancement, additive partitioning, and control and release of active ingredients. This chapter reviews the types, properties, and manufacturing methods of nanoclay-based polymer nanocomposites for food packaging. In addition, the latest achievements and challenges related to the use of nanoclays as food packaging materials are discussed, and an overview of the current state of knowledge and future perspectives for the development of nanoclay technology in the food industry is provided.

DOI: 10.4018/979-8-3693-0819-6.ch011

INTRODUCTION

Packaging for food is crucial because it protects the product until it reaches the customer. Packaging not only makes the product easier to handle, store, and transport, but it also protects the product from environmental hazards like bacteria and chemicals that might shorten its shelf life. Packaging must be able to preserve the food's color, flavor, weight, and texture because these aspects have come to get increased attention in recent years (Perera et al., 2023).

Plastic has long been the go-to for packaging; for example, yoghurt and sour cream containers are typically made from polypropylene, while polyamide, and ethylene vinyl alcohol are also frequently used. The food packaging industry extensively utilizes plastic-based packaging materials due to their remarkable mechanical and barrier qualities (Kausar et al., 2023). However, recent developments show a shift away from these materials and toward more environmentally friendly options like bioplastics and biopolymers. The rising environmental awareness of both customers and manufacturers has contributed to this shift, as more people are becoming aware of the harmful effects of plastic packaging. Additionally, nations including Kenya, Morocco, Oman, Moldova, and Ireland have passed laws prohibiting the use of single-use plastics. In addition, green procurement, educational campaigns, and economic mechanisms have all been employed to reduce plastic consumption (Nourmohammadi et al., 2023).

Thus, research into bioplastics and biopolymers, two types of packaging material that have very little environmental impact, is now underway in the food packaging business. The term "bionanocomposites" is currently employed to designate composite materials that incorporate at least one constituent derived from biological sources and particles with dimensions ranging from 1 to 100 nm (Majeed et al., 2023). These materials provide a burgeoning alternative due to their enhanced sustainability compared to plastic, while yet maintaining their useful properties. Given their ability to enhance the mechanical, thermal, and gas barrier properties of biopolymer materials while retaining their biodegradability, nanoparticles provide a potential path towards the development of a plastic packaging substitute that is acceptable to the environment. The mechanical and physical characteristics of materials used in food packaging are greatly improved by the nanoscale dispersion capabilities of nanocomposites (Khalaj et al., 2016). Researchers are interested in nanoparticles because of their antibacterial properties; examples of these are copper oxide, magnesium oxide, and zinc oxide. The European Union (EU) has granted permission for their utilization within specified limits. In accordance with EU laws, maximum permissible migration levels have been established for several nanoparticles, as outlined below: According to Nouri et al. (2018), Cu content in food or food simulant is 5 mg/kg, and Zn concentration is also 5 mg/kg. Furthermore, Mg concentrations in food or food simulants have been found to be 0.6 mg/kg (Majeed et al., 2023).

Due to their useful mechanical qualities and barrier capabilities, nanoclays have recently attracted attention for usage in food packaging. Nanoclay is more significant than other nanomaterials used to develop materials science and technology because it is plentiful, inexpensive, biocompatible, and has a varied shape and different chemical compositions. Nanoclay exhibits a wide range of significant applications beyond its utilization in the food packaging sector (Guo et al., 2018). Energy storage and conversion, electrochemical energy, water purification, biomedical research, tissue engineering, the creation of COVID-19 antibodies, cancer therapy, environmental cleanup, nanorobotics, photocatalysis, nanocatalysis, automotive technologies, and additive manufacturing are all examples of nanomaterial applications. Nanoclays can be distinguished by their distinctive platelet morphology, flaky consistency, relatively low density, and high aspect ratio, characterized by their nanoscale thickness. According to Scarfato et al. (2017), the utilization of these materials in polymeric packaging can serve the dual purpose of enhancing structural integrity and minimizing gas permeability.

A polymer nanocomposite can be produced by combining an organic polymer matrix with an inorganic nanoclay. The composite material under consideration exhibits superior properties compared to conventional micro composites, while also possessing a reduced mass in comparison to typical nanoclay materials. According to Yahiaoui et al. (2015), nanoclay has gained significant popularity as a nanomaterial for food packaging since its initial debut as the pioneering polymer of nanocomposites. The global nanoclay market exhibited a valuation of $2.1 billion in 2022, with projections indicating a substantial increase to $6.4 billion by 2032. Nanoclay is a prominent component found in the products of Cabot Corp., RTO Company, UBE Industries, Kowa Co., Sun Chem., Nanocor Inc., Statnano, and Nanoshell LLC. Nanocomposites derived from nanoclay have the potential to be utilized in the fabrication of various structures, extending beyond the conventional rigid forms observed in carbonated beverage containers (Padil et al., 2022). The utilization of Durethan® KU2-2601, a commercially available nanoclay-engineered polyamide film developed by Bayer, was employed in the packaging of beverages. The utilization of flexible film nanocomposites has been observed in the packaging of both perishable and nonperishable food items. According to Pirsa et al. (2020), a variety of products incorporate nanoclay nanocomposites in their packaging. These nanocomposites, namely nanoMax®-PP-nH, nanoMax®-PP, nanoMax®-LDPE, and nanoMax®-HDPE, are among the products developed by the company. The goods exhibited characteristics such as mechanical resistance, impact resistance, fireproofing, and stiffness. According to Yahiaoui et al. (2015), nanoclays have demonstrated efficacy as antibacterial agents, as well as the ability to enhance biodegradability and function as a colorimetric indicator template. These various applications are now being explored for potential use within the food packaging industry.

Nanoclays find application in the packaging of a diverse range of products owing to their diverse characteristics. The classification of natural clays can be accomplished by utilizing the layer type, which allows for the categorization into two main groups: (1) 1:1 (consisting of a tetrahedral plate and an octahedral plate as the reference plate, examples include kaolinite and halloysite) and (2) 2:1 (composed of two tetrahedral plates and an octahedral plate as the reference plate, examples include montromorillonite, saponite, and BT). The nanoclay structure consists of core cations, predominantly silicon (IV) (Si^{4+}), which are coordinated by four oxide anions (O_2) positioned at the corners of each tetrahedron. Every tetrahedron is composed of three O_2 groups located at its base and one O_2 group positioned at its apex (Chan et al., 2011). The three O_2-base groups that are commonly shared can interact with one other, resulting in the polymerization of tetrahedrons into a tetrahedral sheet of varying dimensions. The apical oxygen (O_2) refers to the specific location where it meets an octahedral sheet. The aluminum cation, denoted as Al^{3+}, is commonly observed as the central metal cation (Mn^+) within each octahedron. It forms coordination bonds with six O_2 groups located at the vertices. When two adjacent octahedra establish edge-sharing interactions, including two O_2 or OH groups, a planar octahedral sheet is generated (Rhim et al., 2009). In certain cases, lower valence ions possessing a comparable atomic radius have the potential to replace Si^{4+} at the core of the tetrahedral sheet, as well as Al^{3+} at the core of the octahedral sheet. The cations magnesium (Mg^{2+}), lithium (Li^+), and ferrous (Fe^{2+}) have the potential to replace either Si^{4+} or Al^{3+} cations. MMT, Laponite®, and halloysite are widely recognized and extensively researched nanoclays. Notably, MMT and organophilic MMT (OMMT) are particularly preferred owing to their cost-effectiveness. MMT (Montmorillonite) and OMMT (Organ modified Montmorillonite) are frequently employed in the realm of food packaging because to their notable attributes, including their substantial surface area, considerable aspect ratio ranging from 50 to 1000, and remarkable compatibility with a diverse array of organic thermoplastics (Awad et al., 2009). The clay is composed of tactoids, which possess a thickness of merely 1 nm and aggregate to create its structure. The disentanglement of these tactoids is challenging due to the presence of van der Waals contacts. The swelling effect of MMT enhances both clay dispersion and polymer chain intercalation. The integration of delaminating silicate layers from MMT with a polymer matrix enables the development of a nanocomposite material that exhibits enhanced tensile properties (Jawaid et al., 2016; Subramaniyan and Sun, 2006).

NANOCLAY IN FOOD PACKAGING MATERIAL

The demand for packaged food has expanded over the last few decades, encouraging food scientists to develop methods to package food with little loss of nutrients and sophisticated packaging materials (Ahari et al., 2021).

Nanoclays are nanoparticles or nanomaterials which are composed of layered silicate clay minerals typically with a particles size of less than 100nm and are universally the most applicable nanomaterials (Uddin et al., 2008; Chan et al., 2016). These are cost effective nanomaterials which are often used as reinforcement material and improve thermal, mechanical and barrier properties of the packaging material (Bumbudsanpharoke et al., 2019; Jawaid et al., 2016). Nanoclays possess low specific gravity, a flaky soft structure, and a characteristic platelet form (Ganguly et al., 2011). There are four major classes of clay materials which are most used in food packaging materials which are chlorite, MMT/smectite, illite, and kaolinite (Ahari et al., 2021). Among these, MMT has gained the most attention as nanomaterial in food packaging because of its high surface area and its capability with various thermoplastics (Golebiewski et al., 2008; Farhoodi et al., 2016).

The miscibility of MMT is limited to hydrophilic polymers, specifically poly vinyl alcohol and polylactic acid. However, it is important to note that a significant proportion of food packaging materials are sourced from polymers generated from petroleum, such as polyethylene and polypropylene (Javadi et al., 2010). These polymers include inherent hydrophobic characteristics. To optimize their compatibility with an organophilic host matrix, it is necessary to chemically modify these substances utilizing organic cations to achieve a hydrophilic surface (Majeed et al., 2013). The combination of an inorganic clay with an organic polymer matrix results in the formation of a hybrid material known as a polymer nanocomposite. According to Pereira de Abreu et al. (2007), when the nanoclay content is reduced, nanocomposites exhibit enhanced characteristics and reduced weight compared to standard micro composites. For example, solution intercalation, in situ intercalative polymerization, in situ direct synthesis, and melt intercalation are four different processes that may be applied in the production of nanocomposites. The utilization of melt intercalation has gained significant acceptance in several fields owing to its straightforward implementation, practicality, cost-efficiency, and positive environmental impact (Parvinzadeh et al., 2010; Albdiry et al., 2013). The procedure of this methodology entails the amalgamation of nanoclay particles with a polymer matrix at a temperature surpassing the polymer's softening point. The application of shear force induces a reorientation of nanoclay particles inside the polymer matrix, hence affecting their dispersion characteristics. For reinforcing, the organization of nanoclay platelets inside the polymer matrix is critical. Alternatively, when the nanoclay platelets are adequately isolated and uniformly distributed throughout the matrix in a random manner, the resulting composite is referred to as an exfoliated nanocomposite (Chiou et al., 2006; Trifol et al., 2016).

Nanocomposites can be produced on a large scale in the industrial setting by the utilization of nanoclay pellet form, namely nanoclay masterbatches, which are commercially accessible through a range of trademarked goods (Zhu and Xanthos, 2004). Nanocomposites are utilized in the packaging of beverages, where they are fabricated as containers or hard bottles. Conversely, for the packaging of fresh or dried food, nanocomposites are manufactured in the form of flexible films (Trifol et al., 2016).

The utilization of melt intercalation has gained significant acceptance in several fields owing to its straightforward implementation, practicality, cost-efficiency, and positive environmental impact (Parvinzadeh et al., 2010; Albdiry et al., 2013). Cloisite® additives are clay-derived substances that are employed to enhance a range of physical and barrier characteristics in halogen-free flame-retardant thermoplastics. According to Bensadoun et al. (2011), Cloisite® 10A is characterized by the presence of a benzene ring and demonstrates a relatively lower degree of hydrophobicity. On the other hand, Cloisite® 30B has a hydroxyl group and exhibits a lower level of hydrophobicity. In contrast, Cloisite® 15A is distinguished by its larger basal spacing between layers and displays a higher degree of hydrophobicity (Albdiry et al., 2013).

CHARACTERISTICS OF BIO-BASED PACKAGING WITH NANOCLAYS

Mechanical Properties of Nanoclays-Containing Bio-Based Packaging Materials

During the various phases of handling, storing, and shipping, the mechanical characteristics of packaging materials have a significant influence in enhancing food quality and extending shelf life. Bio-based food packaging materials can be designed such that their mechanical attributes match the requirements of the food product they are intended to hold as well as their intended function. To enhance the mechanical qualities of the packaging material, a nanoparticle serving as a reinforcing agent is included into a biopolymer matrix. Young's modulus (YM), percentage elongation at break (EB), and tensile strength (TS) are used to assess the mechanical properties of nanocomposite films. Films depend on the interfacial interaction between the polymer matrix and nano-sized fillers; hence the polymer matrix must include evenly scattered clay layers. According to Albdiry et al. (2013), nanoclay's mechanical properties can be enhanced by increasing its aspect ratio.

Table 1. Application of nanoclay in food packaging

Nanoclay	Packaging-Material	Properties of Packaging-Material	Benefits Compared With Conventional Methods
BT	Poly-lactic acid	Active	Properties, including mechanical, barrier, and active ones
BT (32A)	Carnauba-wax	Coating for retaining flavor and nutrition	Eco-friendly, long-lasting, and decomposable
MMT	Rice flour (gelatin)	Antimicrobial packaging has been shown to increase food stability.	Antimicrobial, mechanical, barrier, and biodegradable characteristics
MMT Cloisite C30B	Chitosan (Rosehip seed oil)	Packaging film with antimicrobial properties	Antimicrobial, mechanical, barrier, and biodegradable characteristics
MMT-Na^+	Polyvinyl alcohol	Waterproofing of Packages to Prevent Moisture Absorption	Insulating qualities
Nanoclay–hydrophilic bentonite	Polyvinyl alcohol (red cabbage extract)	Flexible, pH-sensitive, electronic packaging film	Eco-friendly, biodegradable, has mechanical and barrier qualities
Organically modified MMT	k-carrageenan–cellulose nanocrystals	Packaging film that works actively	Eco-friendly, biodegradable, has mechanical and barrier qualities
Nanoclay Cloisite® Na+	Pectin	Smart active film detects butter oxidation	Mechanical properties, barrier properties, and intelligent properties; degradability; environmental friendliness

BT; bentonite
MMT; montmorillonite

Yeasmin et al. (2020) investigated the use of a solution casting approach to create a nanohybrid film including cellulose nanofibrils, MMT, and pullulan. The films were fabricated using different concentrations of Montmorillonite (MMT), ranging from 0% to 5%. The integration of MMT led to enhancements in the thermal stability, water barrier characteristics, and tensile strength of the film, while concurrently decreasing its susceptibility to moisture. The film containing 5 wt% MMT exhibited a maximum tensile strength (TS) of 45.9 MPa, indicating a presumed enhancement of 10 MPa in comparison to the film devoid of MMT content (0 wt% MMT). An empirical observation revealed a decline in the capacity of moisture absorption, as evidenced by a decrease of roughly 3% between the film devoid of MMT (0% MMT) and the film containing 5% MMT. Notably, these films demonstrated a substantial degree of biodegradability, undergoing breakdown within a timeframe of 18 days (Majeed et al., 2013).

Tensile strength (TS) of food packaging materials is increased when nanoclay is added to biopolymers as a reinforcing agent. It is possible to attribute the observed phenomena to the efficient dispersion of nanoclay particles inside the biopolymer matrix as well as the alignment of these particles with the tensioned polymer chains. In addition, the use of clay and polymers offers further strength, as seen in Figure 1. Additionally, the addition of nanoclay to bio-based food packaging materials lowers the films' EB value (Perera et al., 2023). As a result, the nanomaterial's integration causes the nanocomposite's overall elasticity to diminish. The phenomenon under investigation can be ascribed to the stiffening and reinforcing properties of nanoclay. These properties give rise to the development of a polymeric phase network and a reduction in the mobility of polymer chains. This phenomenon has been demonstrated in the study conducted by Radfar et al. (2020) and is illustrated in Table 2.

Since the nanoclay volume fraction was above the critical amount, there was also a decrease in the material's Young's modulus (YM). The consequence was a transition from fully exfoliated to partially exfoliated-intercalated structures. Consequently, increasing the mechanical capabilities of various packaging materials by using nanoclays has great promise. The concentration of nanoclays at which they are most beneficial has not been identified but is thought to vary according to the type of nanoclay used. According to Davachi and Shekarabi (2018), elongation at break, tensile strength, and Young's modulus all improved when the nanoclay concentration was raised from 0.2 to 2.0 percent. However, all mechanical metrics except Young's modulus declined with the addition of 3 weight percent nanoclay. Tensile strength (TS) of the packing material decreased because of the development of nanoparticle agglomerates inside the film matrix due to an increase in nanoclay

Table 2. The impact of nanoclay on the mechanical characteristics of bio-based packaging materials

Nanoclay	Characteristics	Nanoclay (%)	Elongation at Break (%)	Young's Module	Tensile Strength
MMT	Starch blend (Sodium caseinate)	0% (W/W)	30.48±0.54	N/A	7.74±0.88
		10% (W/W)	70.82±1.24	N/A	7.64±1.37
Nanoclay	Flour	0% (W/W)	85.64±0.98	N/A	3.68±0.08
		5% (W/W)	95.64±0.55	N/A	3.92±0.98
Halloysite	Xylan	0% (W/W)	32.92±1.24	52.24±0.78	8.89±1.48
		1% (W/W)	36.45±1.05	45.47±0.59	12.34±1.57
		2% (W/W)	28.45±1.34	45.67±1.57	8.95±0.68
		3% (W/W)	21.95±0.98	44.64±0.99	7.54±1.57
Cloisite 30B	Gelatin	0% (W/W)	37.54±0.74	N/A	5.54±0.78
		5% (W/W)	26.58±0.37	N/A	9.38±0.98

Source: Perera et al. (2023)

Figure 1. Nanoclay and polymer interaction

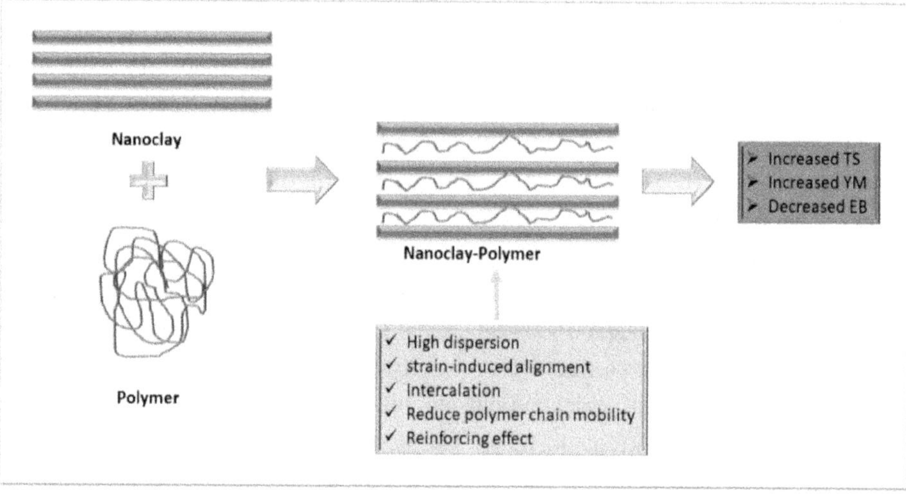

(Figures reference: Perera et al., 2023)

TS: Tensile-strength; **EB**: Elongation-break; **YM**: Young's-modulus

concentration. Table 2 illustrates how the addition of nanoclays enhances the mechanical characteristics of packing matrices, as further research has proven.

Polymers for Food Packaging Nanoclay Composites

Petroleum-derived polymers have gained widespread usage in the food sector due to their advantageous characteristics, including cost-effectiveness, chemical inertness, processability, and mechanical strength. Various polymers are utilized in the packaging of diverse food and beverage products. For example, polyethylene terephthalate (PET) and polystyrene (PS) are extensively utilized materials to produce rigid and transparent containers. On the other hand, high density polyethylene (HDPE) is frequently employed for the manufacturing of milk bottles and bags. Translucent containers and pliable pouches are commonly manufactured using linear low-density polyethylene, polypropylene polymers and low density polyethylene. Furthermore, it is usual practice to employ polyvinyl chloride for the purpose of overwrapping (Duncan, 2011).

The disposal issue associated with conventional plastics has emerged as a significant environmental concern. The demand for its replacement is substantial due to the non-biodegradable nature of these items and the detrimental impact they have on the ecology when disposed of. Bio-based or biodegradable polymers can be produced by the synthesis of renewable carbon resources, such as plants and animals, or through the complete polymerization of materials utilizing chemical or biological techniques that utilize renewable resources (Kim et al., 2013). These polymers possess the ability to undergo biodegradation upon disposal (Jim et al., 2011). The primary sources of renewable resources to produce biodegradable polymers include starch, cellulose, sugar, protein, lignin, chitosan, and vegetable oil (Iwata, 2015).

Polyethylene Terephthalate (PET)

Polyethylene terephthalate (PET) is a type of polyester that exhibits a semicrystalline structure. It is commonly seen in the form of transparent films, sheets, as well as bottles and jars used for food packaging purposes. According to Sudesh and Iwata (2008), this alternative packaging exhibits enhanced potential for displacing conventional packaging due to its reduced weight and decreased energy requirements throughout the manufacturing process. Polyethylene terephthalate (PET) containers are widely employed in the packaging industry for various beverages such as water, juices, beer, tea, soft drinks, and edible oil. This preference is primarily due to PET's notable attributes, including its chemical resistance, optical clarity, permeability, and impact strength. Injection blow moulding (IBM), injection stretch blow moulding (ISBM), and thermoforming are the manufacturing processes used to produce PET (Reddy et al., 2013; Bumbudsanpharoke, 2019).

Nanoclays induction in the manufacturing process serves as a means of reinforcement, aiming to boost resistance to stress cracking during production. Additionally, nanoclays are utilized to improve barrier qualities against various elements such as moisture, oxygen, carbon dioxide, UV radiation, and volatiles (Frounchi and Dourbash, 2009).

Ethylene-Vinyl Alcohol (EVOH)

EVOH are semicrystalline copolymers which are outstanding gas barriers, highly resistant to chemicals and possess high transparency. These are used in food packaging as they act as an ultimate barrier for oxygen and organic compounds (Rastogi et al., 2004). Nevertheless, a significant limitation of EVOH is its susceptibility to moisture, which ultimately compromises its gas barrier performance due to the hydroxyl groups derived from vinyl alcohol. EVOH is limited in its applicability to multilayer constructions due to the requirement of sandwiching the EVOH layer between extremely hydrophobic materials. This configuration leads to several drawbacks, including high production costs, reduced transparency, and challenges in recycling (Echegoyen et al., 2016). To address this limitation, the inclusion of nanoclay within ethylene vinyl alcohol (EVOH) can be employed. The technique results in a reduction in the molecular diffusion of gases and moisture, as well as an enhancement in the thermal barrier characteristics of EVOH. This improvement can be attributed to the tortuosity caused by the effective dispersion of nanoclay particles (Perera et al., 2023).

Polyolefin

Cheap, hydrophobic, non-polar polymers called polyolefins are frequently used in bottles, shrink wrap, pouches, bags, and flexible film for labelling. Although they are resistant to moisture, they have poor grease and gas barrier qualities (Bumbudsanpharoke and Ko, 2015). Polyolefins can benefit from the incorporation of a small quantity of nanoclay, such as OMMT, which effectively functions as an impermeable lamellar to increase mechanical and thermal characteristics as well as barrier qualities (Kim and Cha, 2014).

Polylactide (PLA)

The substance in question is a thermoplastic polyester with a linear aliphatic structure. Lactic acid, which is produced by fermentation of starch and sugar feedstock, is polymerized to create it. This thermoplastic exhibit considerable promise and possesses the ability to undergo processing in standard plastic machinery. Due to its moderate oxygen and moisture barrier qualities and limited heat stability (Arora et al., 2011), this material is primarily suitable for applications such as packaging fresh foods, disposable food containers, and short-term storage bottles. Several investigations have indicated that the use of nanoclay as a possible reinforcing agent might considerably improve the barrier qualities of polylactic acid (PLA), making it more appropriate for usage in the food and beverage packaging industry (Mokwena and Tang, 2012).

Starch

Starch is becoming a widely accepted bio-based material in food packaging because it is cost effective, has high transparency, is flavorless, tasteless, and environmentally compatible (Cabedo et al., 2004). It possesses better mechanical barrier properties than any other polysaccharides (Raj et al., 2005). However, as compared to plastic, starch has high water permeability because of the hydrophilic nature of polysaccharides and low mechanical barrier properties. The starch and nanoclays can be combined to get around this problem. Numerous investigations have demonstrated that adding unmodified hydrophilic nanoclay to starch significantly improves its mechanical and barrier qualities (Hong and Rhim, 2012).

Advancements in Nanoclay

Besides barrier modification and reinforcement functions in both bio-based or petroleum-based polymers, nanoclay possesses many more functions which extend its applications in food and beverage packaging. The advancements in nanoclay as food packaging material such as antimicrobial agents, colorimetric indicator system, biodegradability stimulation, additive partitioning and control and release substance are discussed below (Chan et al., 2011).

Antimicrobial Agents

Antimicrobial function of nanoclay has been recognized by several studies in various industries including food packaging sector. The antibacterial characteristic of nanoclay is highly effective due to its elevated surface-to-volume ratio and enhanced surface reactivity of nanoscale antimicrobial agents, resulting in considerable inactivation of microorganisms (Malucelli et al., 2007; Dadbin et al., 2008). Studies have revealed that unmodified clay and strong hydrophilic modified clay show better antimicrobial performance than other nanocomposites. This aspect of antimicrobial function was originated from the hydrophilicity of nanoclay as the antimicrobial activity increases with the increase of hydrophilicity (Arunvisut et al., 2007). Another study has shown that only a small amount of OMMT can inhibit the growth of a pathogenic yeast called Candida albicans.

Colorimetric Indicator System

Nanoclay has promising application as a colorimetric indicator for intelligent food and beverage packaging. This was developed when blueberry extract was added into the interlayer or nanoclay platelets. There was a good response to the change in pH because of the presence of MMT and OMMT powder, in both acidic and alkaline mediums. The stronger color intensity was observed because the larger basal spacing allowed the penetration of blueberry extract (Scarfato et al., 2016). This shows that a potential colorimetric indicator system for fresh foods can be obtained by adding MMT and OMMT powder in the interlayer of polymer film. Biodegradable colorimetric indicator systems for milk spoilage have also been developed (Zehetmeyer et al., 2012).

Additive Partitioning

The torturous pattern of polymer matrix in nanocomposite not only has its applications in gas and moisture barrier properties but also it has many other benefits. This lamella structure reduces the migration and diffusion of chemicals from packaging in nonfat food stimulation while having no negative impact on the migration of additives contained in the packaging to the food. Research has shown that when samples are treated with 10% ethanol and 3% acetic acid w/v%, there is a considerable decrease in the diffusion of additives into the food in the presence of nanoclays; however, this effect is less pronounced for isooctane (Khalaj et al., 2016).

The Management and Dissemination of Active Ingredients in a Controlled Manner

In addition to employing nanoclay as a reinforcing agent, a control and release mechanism was devised, wherein OMMT was utilized as an active antibacterial agent. The melt intercalation technique was employed to disseminate organically modified montmorillonite (OMMT) within a low-density polyethylene (LDPE) film. The study examined the antibacterial efficacy of the substance against Escherichia coli and Listeria monocytogenes over a period of 500 days. The antibacterial activity of LDPE films diminishes within one month of manufacturing, but LDPE OMMT films maintain their antimicrobial activity for a duration of up to one year. Halloysite nanoclay functions to control and release active compounds, making it a promising material for use in active packaging applications. The structure in issue has a tubular topology, with an inner diameter ranging from 15 to 100 nm and a length ranging from 500 to 1000 nm, according to Bumbudsanpharoke et al. (2017).

Biodegradability Enhancement

According to Siracusa et al. (2008), individuals typically consume food three times daily, but food packaging accounts for over two-thirds of the overall volume of packaging trash. Biodegradability is highly preferred because of less energy requirement and less degradation time (Molinaro et al., 2013). Thermoplastics are inert to environment which causes serious disposal and no degradability issues. The biodegradability of petroleum-based as well as bio-based polymers can be significantly accelerated by the incorporation of OMMT's (Rhim et al., 2019). The presence of nanoclay in polymer nanocomposites in polyolefin enhance its UV absorption capability by improving the oxidation/reduction degradation (Ayana et al., 2014). Moreover, biodegradability in bio-based and biodegradable polymers (e.g. PLA, PBAT) is stimulated by the presence of microorganisms and moisture. The degradation rate is also governed by hydrophilicity of nanocomposites. The greater the hydrophilicity, the higher will be the degradation rate (Rocha et al., 2016). Finally, it can be concluded that nanoclays can enhance the biodegradability of both petroleum-based and bio-based polymers by increasing their UV absorption, hydrophilicity, and susceptibility to microorganisms. Nanoclay is a promising additive for developing eco-friendly packaging materials (Javadi et al., 2010).

Risk of Nanoclay Packing

Although humans have always been exposed to clay to some degree as a biological species, this exposure has undoubtedly risen over the past 20 years due to the widespread usage of clay minerals in technology and medicine. Most of the exposure occurs through inhalation, with the remaining portions occurring via intragastric and subcutaneous routes. Exposure to occupational dust mainly happens during mineral extraction, in farming, and in aluminum manufacturing facilities. There is some discrepancy in the data on oral exposure to MMT, kaolinite, and other clays. The spontaneous exposure to clays resulting from their introduction into polymer materials necessitates the evaluation of the migration of both nano-structured and microparticle clays from packing materials into food products. Moreover, during a product's life cycle, NCs can enter the environment in significant amounts. For instance, when packaging trash and packing materials are burned at combustion plants, NCs can enter the environment. In affluent countries, food additives and pharmaceuticals are important sources of oral exposure to clays. The antacids and antidiarrhea drugs bentonite, montmorrilonite, kaolinite, and palygorskite should be included here. In areas where endemic hazards for mycotoxicosis and aflatoxicosis are relatively high, certain clays are administered to people to prevent these illnesses (Gmoshinski et al., 2020).

CONCLUSION

Promising nanomaterials called nanoclays have the potential to improve food packaging materials' functionality and performance. They can create nanocomposites with various architectures and morphologies to enhance the mechanical, thermal, and barrier qualities of petroleum- and bio-based polymers. They can also provide other benefits such as antimicrobial activity, colorimetric indication, biodegradability enhancement, additive partitioning, and control and release of active ingredients. Various organic cations can be employed to modify nanoclays, hence enhancing their compatibility with organophilic host matrices. Nanoclays have the potential to serve as a packaging material for a diverse array of food and beverage items, including water, juices, beer, tea, soft drinks, edible oil, fresh foods, dried goods, and milk. Nanoclays exhibit considerable potential as intelligent materials for food and beverage packaging, offering the ability to assess and communicate the quality and safety of food products. Nanoclays exhibit characteristics that render them economically viable, ecologically sustainable, and readily accessible nanomaterials, hence presenting considerable promise for application within the food sector. Nevertheless, there exist some obstacles and constraints that necessitate attention, including the optimization of nanoclay content, the regulation of nanoclay migration, the assessment of nanoclay toxicity, and the standardization of nanoclay characterization.

REFERENCES

Ahari, H., Anvar, A. A., Ataee, M., & Naeimabadi, M. (2021). Employing Nanosilver, Nanocopper, and Nanoclays in Food Packaging Production: A Systematic Review. *Coatings, 11*(5), 509. doi:10.3390/coatings11050509

Albdiry, M. T., Yousif, B. F., Ku, H., & Lau, K. T. (2013). A critical review on the manufacturing processes in relation to the properties of nanoclay/polymer composites. *Journal of Composite Materials, 47*(9), 1093–1115. doi:10.1177/0021998312445592

Arora, A., Choudhary, V., & Sharma, D. K. (2011). Effect of clay content and clay/surfactant on the mechanical, thermal and barrier properties of polystyrene/organoclay nanocomposites. *Journal of Polymer Research, 18*(4), 843–857. doi:10.100710965-010-9481-6

Arunvisut, S., Phummanee, S., & Somwangthanaroj, A. (2007). Effect of clay on mechanical and gas barrier properties of blown film LDPE/clay nanocomposites. *Journal of Applied Polymer Science, 106*(4), 2210–2217. doi:10.1002/app.26839

Awad, W. H., Beyer, G., Benderly, D., Ijdo, W. L., Songtipya, P., del Mar Jimenez-Gasco, M., Manias, E., & Wilkie, C. A. (2009). Material properties of nanoclay PVC composites. *Polymer, 50*(8), 1857–1867. doi:10.1016/j.polymer.2009.02.007

Ayana, B., Suin, S., & Khatua, B. B. (2014). Highly exfoliated eco-friendly thermoplastic starch (TPS)/poly (lactic acid)(PLA)/clay nanocomposites using unmodified nanoclay. *Carbohydrate Polymers, 110*, 430–439. doi:10.1016/j.carbpol.2014.04.024 PMID:24906776

Bensadoun, Kchit, Billotte, Trochu, & Ruiz. (2011). A comparative study of dispersion techniques for nanocomposite made with nanoclays and an unsaturated polyester resin. *Journal of Nanomaterials*.

Bumbudsanpharoke, N., & Ko, S. (2015). Nano-food packaging: An overview of market, migration research, and safety regulations. *Journal of Food Science, 80*(5), R910–R923. doi:10.1111/1750-3841.12861 PMID:25881665

Bumbudsanpharoke, N., & Ko, S. (2019). Nanoclays in Food and Beverage Packaging. *Journal of Nanomaterials, 2019*, 8927167. doi:10.1155/2019/8927167

Bumbudsanpharoke, N., Lee, W., Choi, J. C., Park, S. J., Kim, M., & Ko, S. (2017). Influence of montmorillonite nanoclay content on the optical, thermal, mechanical, and barrier properties of low-density polyethylene. *Clays and Clay Minerals, 65*(6), 387–397. doi:10.1346/CCMN.2017.064071

Chan, M. L., Lau, K. T., Wong, T. T., & Cardona, F. (2011). Interfacial bonding characteristic of nanoclay/polymer composites. *Applied Surface Science, 258*(2), 860–864. doi:10.1016/j.apsusc.2011.09.016

Chan, M. L., Lau, K. T., Wong, T. T., Ho, M. P., & Hui, D. (2011). Mechanism of reinforcement in a nanoclay/polymer composite. *Composites. Part B, Engineering, 42*(6), 1708–1712. doi:10.1016/j.compositesb.2011.03.011

Chiou, B. S., Yee, E., Wood, D., Shey, J., Glenn, G., & Orts, W. (2006). Effects of processing conditions on nanoclay dispersion in starch-clay nanocomposites. *Cereal Chemistry, 83*(3), 300–305. doi:10.1094/CC-83-0300

Dadbin, S., Noferesti, M., & Frounchi, M. (2008). Oxygen barrier LDPE/LLDPE/organoclay nano-composite films for food packaging. *Macromolecular Symposia, 274*(1), 22–27. doi:10.1002/masy.200851404

Duncan, T. V. (2011). Applications of nanotechnology in food packaging and food safety: Barrier materials, antimicrobials and sensors. *Journal of Colloid and Interface Science, 363*(1), 1–24. doi:10.1016/j.jcis.2011.07.017 PMID:21824625

Echegoyen, Y., Rodriguez, S., & Nerin, C. (2016). Nanoclay migration from food packaging materials. *Food Additives & Contaminants. Part A, Chemistry, Analysis, Control, Exposure & Risk Assessment, 33*(3), 530–539. doi:10.1080/19440049.2015.1136844 PMID:26751017

Farhoodi, M. (2016). Nanocomposite materials for food packaging applications: Characterization and safety evaluation. *Food Engineering Reviews, 8*(1), 35–51. doi:10.100712393-015-9114-2

Frounchi, M., & Dourbash, A. (2009). Oxygen barrier properties of poly (ethylene terephthalate) nanocomposite films. *Macromolecular Materials and Engineering, 294*(1), 68–74. doi:10.1002/mame.200800238

Ganguly, S., Dana, K., Mukhopadhyay, T. K., Parya, T. K., & Ghatak, S. (2011). Organophilic nano clay: A comprehensive review. *Transactions of the Indian Ceramic Society, 70*(4), 189–206. doi:10.1080/0371750X.2011.10600169

Gmoshinski, I. V., Bagryantseva, O. V., Arnautov, O. V., & Khotimchenko, S. A. (2020). Nanoclays in food products: Benefits and possible risks (literature review). *Health Risk Analysis*, (1), 142–164. doi:10.21668/health.risk/2020.1.16.eng

Golebiewski, J., Rozanski, A., Dzwonkowski, J., & Galeski, A. (2008). Low density polyethylene–montmorillonite nanocomposites for film blowing. *European Polymer Journal*, *44*(2), 270–286. doi:10.1016/j.eurpolymj.2007.11.002

Guo, F., Aryana, S., Han, Y., & Jiao, Y. (2018). A review of the synthesis and applications of polymer–nanoclay composites. *Applied Sciences (Basel, Switzerland)*, *8*(9), 1696. doi:10.3390/app8091696

Hong, S. I., & Rhim, J. W. (2012). Preparation and properties of melt-intercalated linear low density polyethylene/clay nanocomposite films prepared by blow extrusion. *Lebensmittel-Wissenschaft + Technologie*, *48*(1), 43–51. doi:10.1016/j.lwt.2012.03.009

Iwata, T. (2015). Biodegradable and bio-based polymers: Future prospects of eco-friendly plastics. *Angewandte Chemie International Edition in English*, *54*(11), 3210–3215. doi:10.1002/anie.201410770 PMID:25583677

Javadi, A., Srithep, Y., Lee, J., Pilla, S., Clemons, C., Gong, S., & Turng, L. S. (2010). Processing and characterization of solid and microcellular PHBV/PBAT blend and its RWF/nanoclay composites. *Composites. Part A, Applied Science and Manufacturing*, *41*(8), 982–990. doi:10.1016/j.compositesa.2010.04.002

Jawaid, M., Qaiss, A. & Bouhfid, R. (2016). *Nanoclay reinforced polymer composites*. Academic Press.

Jawaid, M., Qaiss, A., & Bouhfid, R. (2016). *Nanoclay Reinforced Polymer Composites*. Springer. doi:10.1007/978-981-10-1953-1

Joo, M., Lewandowski, N., Auras, R., Harte, J., & Almenar, E. (2011). Comparative shelf life study of blackberry fruit in bio-based and petroleum-based containers under retail storage conditions. *Food Chemistry*, *126*(4), 1734–1740. doi:10.1016/j.foodchem.2010.12.071 PMID:25213952

Kausar, A., Ahmad, I., Aldaghri, O., Ibnaouf, K. H., & Eisa, M. H. (2023). Nanoclay-Reinforced Nanocomposite Nanofibers—Fundamentals and State-of-the-Art Developments. *Minerals (Basel)*, *13*(6), 817. doi:10.3390/min13060817

Khalaj, M. J., Ahmadi, H., Lesankhosh, R., & Khalaj, G. (2016). Study of physical and mechanical properties of polypropylene nanocomposites for food packaging application: Nano-clay modified with iron nanoparticles. *Trends in Food Science & Technology*, *51*, 41–48. doi:10.1016/j.tifs.2016.03.007

Kim, S. G., Lofgren, E. A., & Jabarin, S. A. (2013). Dispersion of nanoclays with poly (ethylene terephthalate) by melt blending and solid state polymerization. *Journal of Applied Polymer Science*, *127*(3), 2201–2212. doi:10.1002/app.37796

Kim, S. W., & Cha, S. H. (2014). Thermal, mechanical, and gas barrier properties of ethylene–vinyl alcohol copolymer-based nanocomposites for food packaging films: Effects of nanoclay loading. *Journal of Applied Polymer Science*, *131*(11), 40281–40288. doi:10.1002/app.40289

Lluís Cabedo, E. (2004). Development of EVOH-kaolinite nanocomposites. *Polymer*, *45*(15), 5233–5238. doi:10.1016/j.polymer.2004.05.018

Majeed, K., Jawaid, M., Hassan, A., Abu Bakar, A., Abdul Khalil, H. P. S., Salema, A. A., & Inuwa, I. (2013). Potential materials for food packaging from nanoclay/natural fibres filled hybrid composites. *Materials & Design*, *46*, 391–410. doi:10.1016/j.matdes.2012.10.044

Majeed, K., Jawaid, M., Hassan, A. A. B. A. A., Bakar, A. A., Khalil, H. A., Salema, A. A., & Inuwa, I. (2013). Potential materials for food packaging from nanoclay/natural fibres filled hybrid composites. *Materials & Design (1980-2015)*, *46*, 391-410.

Malucelli, G., Ronchetti, S., Lak, N., Priola, A., Dintcheva, N. T., & la Mantia, F. P. (2007). Intercalation effects in LDPE/o-montmorillonites nanocomposites. *European Polymer Journal*, *43*(2), 328–335. doi:10.1016/j.eurpolymj.2006.11.024

Mokwena, K. K., & Tang, J. (2012). Ethylene vinyl alcohol: A review of barrier properties for packaging shelf stable foods. *Critical Reviews in Food Science and Nutrition*, *52*(7), 640–650. doi:10.1080/10408398.2010.504903 PMID:22530715

Molinaro, S., Cruz Romero, M., Boaro, M., Sensidoni, A., Lagazio, C., Morris, M., & Kerry, J. (2013). Effect of nanoclay-type and PLA optical purity on the characteristics of PLA-based nanocomposite films. *Journal of Food Engineering*, *117*(1), 113–123. doi:10.1016/j.jfoodeng.2013.01.021

Nouri, A., Yaraki, M. T., Ghorbanpour, M., & Wang, S. (2018). Biodegradable κ-carrageenan/nanoclay nanocomposite films containing Rosmarinus officinalis L. extract for improved strength and antibacterial performance. *International Journal of Biological Macromolecules*, *115*, 227–235. doi:10.1016/j.ijbiomac.2018.04.051 PMID:29660461

Nourmohammadi, A., Azar, H. H., & Aminzare, M. (2023). The effects of whey protein/nanoclay biocomposite containing Thymus fedtschenkoi Ronneger essential oil and resveratrol on the shelf life of liqvan cheese during refrigerated storage. *Lebensmittel-Wissenschaft + Technologie*, *187*, 115175. doi:10.1016/j.lwt.2023.115175

Padil, V. V., Kumar, K. A., Murugesan, S., Torres-Mendieta, R., Wacławek, S., Cheong, J. Y., Černík, M., & Varma, R. S. (2022). Sustainable and safer nanoclay composites for multifaceted applications. *Green Chemistry*, *24*(8), 3081–3114. doi:10.1039/D1GC03949K

Parvinzadeh, M., Moradian, S., Rashidi, A., & Yazdanshenas, M. E. (2010). Effect of the addition of modified nanoclays on the surface properties of the resultant polyethylene terephthalate/clay nanocomposites. *Polymer-Plastics Technology and Engineering*, *49*(9), 874–884. doi:10.1080/03602551003664628

Pereira de Abreu, D. A., Paseiro Losada, P., Angulo, I., & Cruz, J. M. (2007). Development of new polyolefin films with nanoclays for application in food packaging. *European Polymer Journal*, *43*(6), 2229–2243. doi:10.1016/j.eurpolymj.2007.01.021

Perera, K. Y., Hopkins, M., Jaiswal, A. K., & Jaiswal, S. (2023). Nanoclays-containing bio-based packaging materials: Properties, applications, safety, and regulatory issues. *Journal of Nanostructure in Chemistry*, 1–23. doi:10.100740097-023-00525-5 PMID:36747507

Pirsa, S., Mohtarami, F., & Kalantari, S. (2020). Preparation of biodegradable composite starch/tragacanth gum/nanoclay film and study of its physicochemical and mechanical properties. *Chemical Review and Letters*, *3*(3), 98–103.

Raj, B., Jagadish, R. S., Srinivas, P., & Siddaramaiah. (2005). Crosslinking of low-density polyethylene by diisocyanates for superior barrier properties. *Journal of Applied Polymer Science*, *96*(no. 4), 1193–1199. doi:10.1002/app.21355

Rastogi, R., Vellinga, W. P., Rastogi, S., Schick, C., & Meijer, H. E. H. (2004). The three-phase structure and mechanical properties of poly (ethylene terephthalate). *Journal of Polymer Science. Part B, Polymer Physics*, *42*(11), 2092–2106. doi:10.1002/polb.20096

Reddy, M. M., Vivekanandhan, S., Misra, M., Bhatia, S. K., & Mohanty, A. K. (2013). Biobased plastics and bionanocomposites: Current status and future opportunities. *Progress in Polymer Science*, *38*(10-11), 1653–1689. doi:10.1016/j.progpolymsci.2013.05.006

Rhim, J.-W., Hong, S.-I., & Ha, C.-S. (2009). Tensile, water vapor barrier and antimicrobial properties of PLA/nanoclay composite films. *Lebensmittel-Wissenschaft + Technologie*, *42*(2), 612–617. doi:10.1016/j.lwt.2008.02.015

Rocha, J. M., Lorite, G. S., & Miilumäki, N. (2016). Novel polylactic acid (PLA)-based active packaging with incorporation of nanoparticles and its performance throughout shelf-life of fresh-cut fruit. In *XIX National Congress of Biochemistry*. Sociedade Portuguesa de Bioquímica, University of Minho.

Scarfato, P., Di Maio, L., Milana, M. R., Giamberardini, S., Denaro, M., & Incarnato, L. (2017). Performance properties, lactic acid specific migration and swelling by simulant of biodegradable poly (lactic acid)/nanoclay multilayer films for food packaging. *Food Additives & Contaminants. Part A, Chemistry, Analysis, Control, Exposure & Risk Assessment*, *34*(10), 1730–1742. doi:10.1080/19440049.2017.1321786 PMID:28434378

Scarfato, P., Incarnato, L., Di Maio, L., Dittrich, B., & Schartel, B. (2016). Influence of a novel organo-silylated clay on the morphology, thermal and burning behavior of low density polyethylene composites. *Composites. Part B, Engineering*, *98*(1), 444–452. doi:10.1016/j.compositesb.2016.05.053

Siracusa, V., Rocculi, P., Romani, S., & Rosa, M. D. (2008). Biodegradable polymers for food packaging: A review. *Trends in Food Science & Technology*, *19*(12), 634–643. doi:10.1016/j.tifs.2008.07.003

Subramaniyan, A. K., & Sun, C. T. (2006). Enhancing compressive strength of unidirectional polymeric composites using nanoclay. *Composites. Part A, Applied Science and Manufacturing*, *37*(12), 2257–2268. doi:10.1016/j.compositesa.2005.12.027

Sudesh, K., & Iwata, T. (2008). Sustainability of biobased and biodegradable plastics. *Clean (Weinheim)*, *36*(5-6), 433–442. doi:10.1002/clen.200700183

Trifol, J., Plackett, D., Sillard, C., Szabo, P., Bras, J., & Daugaard, A. E. (2016). Hybrid poly (lactic acid)/nanocellulose/nanoclay composites with synergistically enhanced barrier properties and improved thermomechanical resistance. *Polymer International*, *65*(8), 988–995. doi:10.1002/pi.5154

Uddin, F. (2008). Clays, Nanoclays, and Montmorillonite Minerals. *Metallurgical and Materials Transactions. A, Physical Metallurgy and Materials Science*, *39*(12), 2804–2814. doi:10.100711661-008-9603-5

Yahiaoui, F., Benhacine, F., Ferfera-Harrar, H., Habi, A., Hadj-Hamou, A. S., & Grohens, Y. (2015). Development of antimicrobial PCL/nanoclay nanocomposite films with enhanced mechanical and water vapor barrier properties for packaging applications. *Polymer Bulletin*, *72*(2), 235–254. doi:10.100700289-014-1269-0

Zehetmeyer, G., Soares, R. M. D., Brandelli, A., Mauler, R. S., & Oliveira, R. V. B. (2012). Evaluation of polypropylene/montmorillonite nanocomposites as food packaging material. *Polymer Bulletin*, *68*(8), 2199–2217. doi:10.100700289-012-0722-1

Zhu, L., & Xanthos, M. (2004). Effects of process conditions and mixing protocols on structure of extruded polypropylene nanocomposites. *Journal of Applied Polymer Science*, *93*(4), 1891–1899. doi:10.1002/app.20658

Chapter 12
Recent Advances in the Utilisation of Artificial Intelligence in the Food Industry

Moses Ojukwu
https://orcid.org/0000-0003-2244-9135
Department of Food Science and Technology, Federal University of Technology, Owerri, Nigeria

Chigozie Emmanuel Ofoedu
Department of Food Science and Technology, Federal University of Technology, Owerri, Nigeria

Chijioke M. Osuji
Department of Food Science and Technology, Federal University of Technology, Owerri, Nigeria

Ogbonnaya Okoro Aja
Department of Agricultural Extension, Federal University of Technology, Owerri, Nigeria

ABSTRACT

Global population growth raises food security concerns. AI, like computer vision and ML, revolutionizes agriculture. It monitors crop health, predicts yields, and automates tasks. Mobile apps diagnose crop diseases swiftly, boosting productivity. AI-driven machine vision improves food processing by flaw identification and quality control. AI sorting enhances item classification, increasing productivity. In beer production, AI ensures fermentation quality. It optimizes distribution, predicts demand, and reduces waste. AI detects contaminants for food safety. Chatbots offer personalized recommendations. Data privacy and ethics matter, requiring explainable AI (XAI) and legal frameworks must be addressed despite these advancements. AI promises a secure, efficient, and sustainable food ecosystem.

DOI: 10.4018/979-8-3693-0819-6.ch012

1. INTRODUCTION

Artificial Intelligence (AI) has revolutionized the food industry in a variety of ways. AI technologies like data analytics, machine learning, and computer vision are being utilized to boost and enhance relationships with stores, sales reps, and consumers, improve accuracy in food processing, and enhance food safety standards. AI is also being used to sort food (Goyache *et al.*, 2001), manage the supply chain(Addanki *et al.*, 2022), and ensure employees follow personal hygiene standards(Stoitsis *et al.*, 2023), identifying proteins for healthy foods and creating food recipes (Ramirez-Asis *et al.*, 2022).

The food industry will benefit from AI's capacity to work and adapt like human beings to carry out a number of duties swiftly and in an instantaneous fashion, including sanitation, guaranteeing cleanliness, making food and drink, identifying possible hazards during manufacturing, and distinguishing food based on various variables. (Mavani *et al.*, 2022). Given the continually increasing global population, leveraging technology within the food production ecosystem is imperative to ensure food security for all. Artificial Intelligence (AI) has emerged as a vital tool in addressing poverty by enhancing agricultural practices and boosting food production (Kutyauripo *et al.*, 2023).

The scientific field of artificial intelligence (AI) aims to create intelligent systems that mimic certain aspects of human behaviour using methods that are inspired by natural processes. It is a new tool in the food industry for processing different kinds of foods using robots or machines. (Addanki *et al.*, 2022)..

Artificial intelligence has ushered in significant advancements in agriculture, spanning crop cultivation, harvesting, and marketing (Ivanov & Webster, 2023). As a sophisticated technology, AI offers a viable alternative to human intelligence for problem-solving and decision-making. When integrated with the Internet of Things (IoT), AI becomes a powerful tool that can be harnessed across the entire food system, encompassing agriculture and managing food waste (Palacios *et al.*, 2019).Numerous facets of the food industry, including production, processing, distribution, and even consumer experience, have the potential to be revolutionised by AI applications. This chapter examines recent developments in AI usage in the food industry and their effects on various sectors.

2. ARTIFICIAL INTELLIGENCE IN THE FOOD INDUSTRY

AI is vital in the food industry, playing a key role in production, processing, quality control, and various aspects like manufacturing, packaging, and cleaning. The illustration in Figure 1 demonstrates how AI is beneficial in the food industry.

Figure 1. The work of AI in the food industry

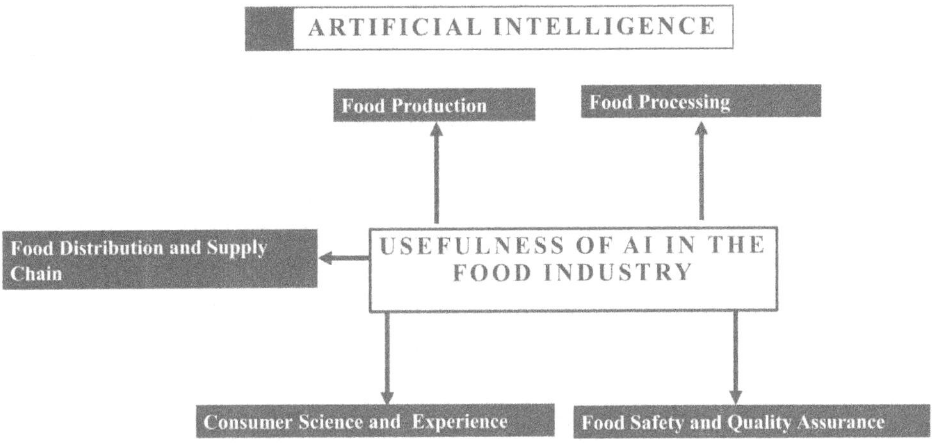

2.1 AI in Food Production

AI is enhancing food production in various industries through its applications in food processing, manufacturing, and supply chain management. Some examples of how AI is used to improve food production include:

- Automation: The arms of robots and machines are set up to carry out exact and routine duties as efficiently as possible, reducing the need for manual labour and improving productivity (Kakani *et al.*, 2020).
- Data-driven optimization: AI algorithms are capable of analyzing vast quantities of data in order to enhance operations, cut waste, and improve consistency in products., leading to better decision-making and improved efficiency (Ramirez-Asis *et al.*, 2022).
- Predictive maintenance: AI enables predictive maintenance, helping businesses maintain their equipment and facilities more effectively, reducing downtime and maintenance costs.

By improving agricultural techniques, AI has made an immense effect on food production. Farmers can monitor the health of their crops, forecast yield, and opt for the optimal moment to harvest using methods such as computer vision and machine learning. Drones and robots with AI capabilities have also been used to automate labour-intensive tasks like planting, weeding, and harvesting, which has increased productivity and decreased costs.

In Pakistan, the agricultural sector has faced a significant decline in productivity due to the difficulties in promptly and precisely identifying crop diseases among farmers. A cutting-edge mobile application designed for Android mobile devices was created to address this issue. This intelligent mobile app is intended to diagnose wheat diseases, a prevalent problem in the country. The application leverages a sophisticated fuzzy inference system to provide accurate disease diagnoses. Its effectiveness was rigorously validated by subjecting it to testing with 100 real-world crop challenges, resulting in an impressive accuracy rate of 99% (Kutyauripo *et al.*, 2023; Toseef & Khan, 2018).

Also, Artificial intelligence has found application in the early identification of crop diseases in banana trees. This involves a fusion of Recurrent Neural Networks (RNN) and Convolutional Neural Networks (CNN), forming a robust hybrid known as Gated Recurrent Convolutional Networks (GRCN)(Nandhini *et al.*, 2022). GRCN represent a type of neural network architecture that combines both recurrent and convolutional layers. typically refers to the use of gating mechanisms, like those in long short-term memory (LSTM) networks or gated recurrent units (GRUs). These gating mechanisms allow the network to control information flow, through the sequences, enabling it to capture long-term dependencies more effectively. The combination of both recurrent and convolutional layers in GRCNs aims to leverage the strengths of both architectures for a more versatile and robust model. This hybrid architecture can be used in tasks, like video analysis, natural language processing, and other tasks involving sequential or spatial information (Wang & Hu, 2021). The system employs a camera positioned within the banana tree plantation to capture images of the banana trees. These images are then analysed for colour changes, much like the human eye. To further investigate the causes of these colour changes, the system utilises CNN and Gated Recurrent Unit (GRU). The CNN is responsible for classifying the images, while the GRU processes and interprets them. The outcome of this approach is an accurate classification of diseases affecting plantain trees and, notably, the ability to detect these diseases at their early stages.

Furthermore, as an AI process, Machine Vision (MV) is currently used to manage weeds in pepper and artificial plant cultivation precisely. This technology achieves precise weed control by spraying only on the targeted weeds while avoiding non-target plants using an intelligent sprayer integrated with artificial intelligence and machine vision. This method ensures the efficient use of agrochemicals, minimizes crop damage, and reduces agricultural chemical waste. (Partel *et al.*, 2019). The weed recognition system is linked to the spraying system, which uses accurate sprinkling techniques, and both are integrated into a system that generates weed maps. The results of this novel system show a significant reduction in agrochemical residues. (Monteiro *et al.*, 2021)

AI has also been used in postharvest management and harvesting. Some examples are summarised in Table 1.

2.2 AI in Food Processing

Food processing includes complex activities such as packaging, sorting, and ensuring quality. AI has enhanced these processes by applying advanced techniques like deep learning and computer vision. (Makkar *et al.*, 2018). Machine vision systems powered by AI can recognize defects, categorize products, and ensure the quality and safety of their contents. Sorting systems driven by AI can boost efficiency and reduce waste by precisely categorizing fruits, vegetables, and other food items according to their size, shape, and quality.(Palacios *et al.*, 2019). The use of AI in the production of specific foods are highlighted below.

2.2.1 AI in Milk Production and Transportation

Milk gets delivered to milk plants by milk automobiles from various milk-producing centers. The cooling process is carried out in the milk facility, and the vehicle is directed to the collection place before the milk spoils. An efficient framework, such as a network of wireless sensors with driven artificial intelligence ideas, must be developed to address this problem (Arshad *et al.*, 2022). The milking of cows by a manual workforce consumes much time. Consequently, As a result, Automated Milking Machines (AMS) or milking robots are utilized to cut down on this time. The invention became available in the United States in the year 2000, but it originated in Europe in the year 1992. (Sandey *et al.*, 2017b).

Also, with the application of AI, Programmed Cleaning-In-Place (CIP) carrys-out washing arrangements in the food industry. These CIPs contain two programs:The

Table 1. Usefulness of artificial intelligence in postharvest management

Crop	Description	Reference
Corn	A decision-tree-based supervised learning method was used to assess the extent of damage to stored maize grain caused by *Prostephanus truncatus.*	(Nyabako *et al.*, 2020)
Water-melon	To analyze acoustic resonance testing data, a deep convolutional neural network was used, allowing the machine to track and assess the level of watermelon ripeness all through postharvest storage..	(Albert-Weiß *et al.*, 2020)
Banana	Ripening was investigated using a laser scan with a light detection and ranging (LiDAR) sensor placed 0.9 meters from the fruit itself..	(Saha & Zude-Sasse, 2022)
Apple	Convolutional Neural Networks, specifically the Deeplab and U-net models, were employed to identify and quantify apple spoilage during postharvest storage.	(Kutyauripo *et al.*, 2023)

Source: (Kutyauripo *et al.*, 2023).

rotary washing program encompasses machinery for heating parts and pasteurisers, while turning sanitizing CIP applications consisting of tanks for collecting pure milk. (Memisi *et al.*, 2015). The uncommon gripper permits the cheddar squares to be picked up and placed on a vehicle for additional preparation (Sandey *et al.*, 2017a). Furthermore, each milk product sector's milk quality is assessed using an Artificial Neural Network (ANN). The milk's temperature is continually tracked and ought to be in the lower temperature range (2o-4° C); if not the milk will spoil. It can be challenging to assess the milk's quality quickly, but the technology keeps track of raw milk and prevents deterioration with the help of a neural network. (O'connell *et al.*, 2016). Lactose is the main sugar in milk, but it also contains protein and lipids. Artificial Neural Networks have additionally been employed to eliminate lactose via a process known as which employs mathematical models generated by ANN and carried out on set bed columns. (Balieiro *et al.*, 2016). Because the adsorbent is in constant contact with the solution, these techniques increase adsorption rates. These two models were created by scientists using MATLAB, as described by Leite *et al.* (2019).

2.2.2 AI in Cheese Production

AI has also been used to assess the quality of cheese. Inside quality characteristics include feel, nutrient content, pesticide residue, spoiled material, and so on. The outside quality characteristics include alterations in size, form, and colour when storage, shelf-life estimation, browning, and melting characteristics. Food evaluations will be performed by trained or inexperienced testers, as well as by equipment. There are three types of Computer Vision Systems (CVS) for external inspection and verifying of products: conventional CVS, hyperspectral CVS, and multispectral CVS. (Lakshmi *et al.*, 2017). The CVS operations include taking pictures, analyzing them, and digitisation. (Lukinac *et al.*, 2018). Pictures are created by looking at objects using a special method called image analysis. This involves using light that we can't see with our eyes, and the camera's lenses capture this light, turning it into electrical signals with the help of an image sensor. These signals are then turned into numbers in a process called digitization. Computer vision systems (CVS) can use these digital images to measure things about the objects, helping to make sure the products are good quality through automated checks (Addanki *et al.*, 2022; Lukinac *et al.*, 2018).

2.2.3 AI in Beverages and Beer Production

AI is being used in making beer too. Beer, enjoyed by a lot of people around the world, is made through a process where yeast is added to the wort. To get the beer just right, we need to keep an eye on factors like oxygen, temperature, and pitching rate during fermentation. Traditional methods make it hard to monitor these factors all the time, but with the utilization of AI and its tools, technology can keep track and control the beer-making process more effectively. (Vassileva & Mileva, 2010). AI techniques such as computer vision, fuzzy logic, and neural networks can help improve the way we make beer. By using fuzzy rules based on current data from the beer-making process, we can estimate the beer's quality by identifying specific elements like higher alcohols, vicinal diketones (VDK), and fatty acids. Neural networks are also used to monitor how the beer is fermenting. (Addanki *et al.*, 2022; Vassileva & Mileva, 2010)

Visual aspects of drinks, like the color of beer, how clear it is, the amount and lasting of foam, and how it looks, along with the taste factors like the feel in your mouth, the smell, and the bitterness, can be checked using AI. Normally, doing this task requires a lot of people and time, but with the computer vision system (CVS) method, it can be done instantaneously. CVS uses a camera, light, and a computer to take a close look at the exterior of the beer through pictures. It performs tasks like taking photos, processing them, and understanding what's in the images. There are three basic procedures in this assignment; first, they take pictures in the fields where grapes are grown a week before picking them to check how closely packed the grape clusters are. This helps to identify any bad grapes, and various methods or algorithms like K-means, Gaussian, and cross-validation are used for this process (Palacios *et al.*, 2019).

A kind of computer program called Deep Convolutional Neural Network (DCNN) is useful in figuring out the nutrition in soft drinks to help with weight control. This method can guess what's in the drink by looking at the ratio of the size of the bottle to the cap. It works for fizzy drinks and other fruit-based drinks. To find out the nutrition, it uses a process called image processing, which is a part of CNN. This process helps take away the background from the picture and gives us the information we need. (Hafiz *et al.*, 2022).

An electronic nose, or e-nose, is a technology that mimics the sense of smell in mammals. It uses gas sensors to analyze odours and can be applied in industries like wine and coffee to assess product quality. In the food and beverage sector, electronic noses are used to monitor and maintain the quality of various products. Similarly, an electronic tongue, or e-tongue, evaluates the quality of drinks such as milk, coffee, tea, wine, and beer. A pulse voltametric e-tongue, for instance, can identify the type of tea based on its age. This information is typically obtained through a simpler analysis method compared to UV-VIS spectrophotometer-based equipment. (Roy *et al.*, 2019).

2.2.4 AI in Fruits and Vegetables Processing

Artificial intelligence is helping sort and cut fruits and vegetables more efficiently. Due to problems like bad farming methods, lack of care, and losses during harvesting, along with expensive labor, the production of these essential crops is decreasing. To address this, automation is being employed to improve efficiency and minimize losses. In this automated system, computer vision, a part of AI, uses software to identify and sort fruits and vegetables based on their quality and detect any damage they might have (Gill *et al.*, 2014; Makkar *et al.*, 2018).

2.2.5 AI in the Food Service Sector

Just in factories but also in services. Waiter robots, made up of a base and an automatic dumb waiter, have three stages and are taking over tasks done by human waiters. Smaller-sized particles better analyse the robot's position (Cheong *et al.*, 2016; Ivanov & Webster, 2023). In a different scenario where the job is to guide the customer to a specific table, the human waiter selects the "lead" option from the main menu and types in the table number. This number stays visible on the robot as it moves. Initially, waiters clean the table. When the robot arrives at the assigned table, the customer puts down plates, containers, and tableware. Facial recognition is handy in restaurants, identifying people by their face pictures (Berezina *et al.*, 2019).

2.3 AI in Food Distribution and Supply Chain

AI is changing how food gets from producers to consumers, making it faster, greener, and better at meeting what people want. In the food delivery and supply chain, AI helps with challenges like keeping track of stock, guessing how much people will want, and planning the best ways to move things around. By looking at huge amounts of information, like past sales and things like weather and holidays, AI can solve these problems. Smart AI programs can figure out what customers will likely want, manage how much stuff to keep in storage, and create efficient routes for delivery, saving money and reducing waste. AI helps businesses learn useful things from what happened before, so they can improve how they work and make smart choices (Trabelsi *et al.*, 2023).

AI technologies, such as predictive analytics and demand forecasting models, help businesses optimize their inventory levels, reducing excess stock, preventing shortages, and minimizing waste (Marvin *et al.*, 2022) AI does a lots of things like making sure we have the right crops, getting food from the farm to the store, delivering it to people, and keeping machines in good shape. This makes things better for customers, helps us manage how we get food, makes things work smoother, and saves money. AI also watches over the whole process of getting food, making sure it follows the rules to keep it safe. It even helps fix big machines before they break a lot, making them cost less to run and do more. AI also makes deliveries and logistics smarter, so we waste less and get meals from restaurants faster. In simple words, AI makes food processes work better, making things faster, better, and safer. (Qian *et al.*, 2022).

We can bring artificial intelligence into the food supply chain, but there are challenges we can tackle. We can do this by helping farmers understand money and technology better and by sharing good ideas among everyone involved. The growth of artificial intelligence can be divided into three stages: looking at data, learning from people, and acting like a person. We can use AI all through the process of getting food, working with things like robots, drones, and smart machines. AI helps make sure food gets where it needs to go the best way, considering things like traffic and weather, which is good for the environment (Marvin *et al.*, 2022).

AI can make transportation routes more efficient and eco-friendly by considering factors like road conditions and weather (Ramirez-Asis *et al.*, 2022). AI can help maintain transparency and accountability throughout the food supply chain, ensuring that products are sourced and produced responsibly.

AI is also being used to address challenges and ethnic considerations in the food supply chain. For example, AI can help food businesses in developing countries adapt to uncertain global economic conditions, ensuring food security and sustainability. However, there are still research gaps and barriers to AI application in the food and value supply chain, such as data privacy and security concerns, and the potential overreliance on AI technology (Trabelsi *et al.*, 2023).

2.4 AI in Food Safety and Quality Assurance

In the food industry, making sure that the food we eat is safe and of good quality is really important. Artificial Intelligence (AI) tools, like computer programs that mimic the human brain, can be super helpful in quickly checking if our food is good and safe to eat. These tools can analyze a ton of data to find patterns and possible risks. For example, they can even read news or social media to detect if there's a problem with certain foods making people sick. AI tools such as artificial neural networks (ANN), deep learning (DL), fuzzy logic (FL), support vector machines (SVM), and random forest (RF) can quickly evaluate the quality of food items. They are capable of analyzing vast amounts of data to identify patterns and potential risks efficiently (O'Shea *et al.*, 2023).

AI techniques can be used in various ways for ensuring food safety and quality. A key benefit of employing AI in the food industry is its capability to mimic intelligent behavior and extract valuable knowledge from existing data. (Stoitsis *et al.*, 2023). This aids in evaluating the quality of food items and identifying the characteristics that impact their quality. AI algorithms can also automate and oversee the monitoring of food quality, offering reliable and consistent assessments. Additionally, AI techniques can help standardize objective tests for evaluating sensory quality and assist researchers in obtaining practical insights from their work. Nevertheless, the specific uses of AI in ensuring food safety and quality may vary based on industry conditions and market needs.

In the food industry, Artificial Intelligence (AI) techniques are utilized to judge the overall quality of food products and pinpoint the specific features crucial for categorizing them according to market and consumer requirements. Machine Learning (ML), a key aspect of AI, involves organizing knowledge from unorganized sources to create computer programs, like expert systems, capable of performing valuable tasks (Goyache *et al.*, 2001).

ML algorithms assist researchers and technicians in pulling out useful human knowledge from available sources. They create clear rules for sorting samples, even when dealing with the complexity of human behaviour. ML also figures out how much each characteristic of the food being assessed matters in the final decision, helping experts make better judgments. In the food industry, ML technologies have been used to design and enhance systems that judge animal production and to estimate how people perceive the taste of food more accurately than traditional statistical methods. (Shi *et al.*, 2023)

2.5 AI in Consumer Experience

AI has changed how consumers engage with the food industry. Artificial intelligence (AI)-powered chatbots and virtual assistants can offer tailored recommendations and respond to customer questions, improving the shopping experience. AI algorithms can examine consumer preferences, dietary restrictions, and purchase history to recommend customised products. This increases customer satisfaction and loyalty.

AI is transforming the consumer experience in various industries, including the food industry. AI is being used to personalize customer experiences, improve customer service, and enhance the overall shopping experience. Some of the ways AI is enhancing the consumer experience in the food industry include:

- Personalization: AI can create personalized nutrition plans and food recipes based on customer preferences and dietary restrictions (Chaturvedi & Verma, 2022)

- Recommendations: AI algorithms can analyze customer data to provide personalized recommendations for food and drink products, improving customer satisfaction (Edelman & Abraham, 2022)
- Chatbots: AI-powered chatbots can provide customer service and support, answering customer questions and resolving issues in real-time (Ameen *et al.*, 2021a).
- Virtual assistants: AI-powered virtual assistants like Amazon's Alexa and Google Assistant can help customers order food, make reservations, and access information about food and drink products (Ameen *et al.*, 2021a)
- Smart shelves: AI-powered smart shelves can detect when products are running low and automatically reorder them, improving inventory management and reducing the likelihood of out-of-stock items (Ameen *et al.*, 2021a)
- Mobile apps: AI-powered mobile apps can provide personalized recommendations, offer discounts and promotions, and allow customers to order food and drink products for delivery or pickup (Ameen *et al.*, 2021b)

While AI has numerous advantages in enhancing the consumer experience, there are also some challenges and concerns. One of the main challenges is the need to balance personalization with privacy concerns, as AI relies on vast amounts of customer data. Additionally, there is a risk of overreliance on AI technology, which can lead to a lack of human expertise and decision-making, potentially leading to errors and customer dissatisfaction.

It is important for businesses to carefully consider the advantages and disadvantages of AI in enhancing the consumer experience and to implement AI technology in a responsible and ethical manner. In conclusion, AI is transforming the consumer experience in the food industry by providing personalized recommendations, chatbots, virtual assistants, smart shelves, and mobile apps. While there are challenges and concerns associated with AI, businesses can leverage AI technology to improve customer satisfaction and loyalty.

3. ADVANTAGES AND DISADVANTAGES OF THE USE OF AI IN THE FOOD INDUSTRY

AI has numerous advantages in the food industry. However, there are also some disadvantages. It is important to weigh these advantages and disadvantages carefully when considering the implementation of AI technology in the food industry. Below are the advantages and disadvantages (Table 2).

Table 2. Advantages and disadvantages of the use of AI in the food industry

S/N	Advantages	Disadvantages
1.	**Improved efficiency:** AI can perform tasks quickly and accurately, reducing the need for manual labor and improving productivity (Ramirez-Asis *et al.*, 2022)	**Cost:** Implementing AI technology can be expensive, and smaller businesses may not have the resources to invest in it(Ramirez-Asis *et al.*, 2022)
2.	**Reduced waste:** AI can optimize food delivery routes and schedules, manage inventory more efficiently, and sort food according to various parameters, reducing waste and improving sustainability (Tegmark, 2016).	**Lack of human touch**: AI cannot replace the human touch in food production, and some consumers may prefer food that is prepared by humans. (Vassileva & Mileva, 2010)
3.	**Enhanced food safety**: AI can monitor and enforce personal hygiene standards, detect potential risks during food production, and ensure compliance with food safety regulations (Mavani *et al.*, 2022).	**Data privacy and security**: AI relies on vast amounts of data, and there are concerns about data privacy and security in the food industry (Qian *et al.*, 2022).
4.	**Improved quality control**: AI algorithms and tools can be trained and customized to check for various parameters in food processing, ensuring that quality standards are maintained (Lakshmi *et al.*, 2017)	**Reliance on technology**: Overreliance on AI technology can lead to a lack of human expertise and decision-making, which can be problematic in certain situations (O'Shea *et al.*, 2023).
5	**Personalization**: AI can create personalized nutrition plans and food recipes based on customer preferences and dietary restrictions	**Risk of errors**: AI algorithms are not infallible and can make mistakes, leading to potential risks in food production and safety.

4. GUIDELINES FOR CHOOSING AN AI MODEL

Choosing the right algorithm is crucial when creating an AI model. It helps users achieve accurate and efficient results while saving costs. The first step is defining the goal of using AI in research or implementation. Common objectives in the food industry include prediction, classification, quality control, detecting adulterants, and estimation. Once the objective is clear, the next decision is whether to use sensors like E-tongue, E-nose, to collect data from samples (Mavani *et al.*, 2022).

Integration with these sensors is often done to gather parameters and characteristics of samples for AI algorithms. Users then compare and choose a suitable algorithm for their study. Commonly used AI algorithms include ANN and ML. Users should consider the complexity of their research when selecting an algorithm. Once the algorithm is chosen, data is integrated with it, and testing and validation are conducted using metrics like R^2 and MSE to analyze the model's performance. If the validation is successful, the AI model is created. If not, users should go back, reevaluate their choice of algorithm, and make adjustments (Kondakci & Zhou, 2017).

5. TRENDS ON THE USE OF AI IN THE FOOD INDUSTRY

In recent years, we've seen a significant increase in the use of AI in the food industry, and this trend is expected to continue for the next decade. This rise is closely tied to the introduction of Industrial Revolution 4.0 (IR 4.0), which involves integrating intelligent systems into traditional industries, often referred to as smart factories (Morrar *et al.*, 2017). AI, a key component of IR 4.0, focuses on creating machines that can perform tasks intelligently, akin to humans. This has had a notable impact on the food industry, particularly in addressing product recalls and quality concerns. With the implementation of AI integrated into sensors, errors during the manufacturing process can be quickly identified and rectified, improving overall efficiency.

IR 4.0 also plays a crucial role in addressing consumer concerns about food quality. In the digital age, people frequently seek information about food online. AI helps enhance food quality and streamline the production process, meeting the rising demand for safe and high-quality food. The year 2020 witnessed a peak in the application of AI in the food industry, driven by increased research activities (Bai *et al.*, 2020). This trend is expected to persist in the coming years due to growing food demand and heightened safety concerns.

A comparison between AI integration with and without sensors for real-time monitoring in the food industry indicates a higher percentage of integration with external sensors. These sensors collect data from samples, which is then used in AI algorithms for tasks like classification, prediction, and quality control. However, it's worth noting that in 2017, there was a higher percentage of AI research conducted without external sensors, reflecting the evolving nature of technology and research preferences (Blagoveshchenskiy *et al.*, 2020).

Looking ahead, as the AI landscape advances, we can confidently predict a continued rise in its use in the food industry. This is because AI offers advantages such as time, cost, and energy savings, along with improved accuracy in predicting factors affecting the food industry (Blasi, 2018). Notably, the recent Covid-19 pandemic has further accelerated this trend, as companies seek AI solutions to cut costs and enhance performance during challenging times. Small and medium-sized enterprises (SMEs) are turning to AI to address financial challenges and ensure smooth operations in the face of increased demand and strict operating procedures.

6. CHALLENGES AND FUTURE DIRECTIONS

Several difficulties persist despite the significant advancements in the adoption of AI within the food industry. It is necessary to address privacy issues, data security, and the ethical implications of implementing AI. Additionally, it may be required to make significant investments and upskill the workforce to integrate AI technologies into current infrastructures and procedures. Future directions include creating AI-powered models for predicting food safety, increasing food production procedure automation, and applying AI to environmentally friendly and sustainable practices. Also, the principle of explainable AI (XAI) has become critical to secure stakeholder's trust in AI-based systems in the food industry. Another future direction will be legal and policy framework formulations for adjudication in matters enabled by AI in the food industry.

The anticipation of the expiration date of packaged food through sensor technology is a notable challenge. This allows people to detect food spoilage, preventing the risk of food-related illnesses (Misra et al., 2020). However, achieving standardization in these methods is difficult. Each research article typically focuses exclusively on a specific flaw or fraud detection, a singular food component, or a particular instrumental parameter. Additionally, different validation methods and unique learning algorithms are employed, limiting the broader applicability of these technologies beyond scientific research investigations. Future studies should concentrate on standardizing effective techniques to facilitate widespread implementation.

Limited efforts have been made to integrate various instrumental methods. Merging multiple sensor types is a complex task, involving the collection of data from diverse sensors and a thorough examination of the advantages and disadvantages of each sensor in specific applications. Sophisticated AI algorithms have not extensively explored this research area. The accurate measurement of damage and abnormalities remains a persistent challenge in the agriculture sector, emphasizing the need for future emphasis on the utilization of learning algorithms.

7. CONCLUSION

With the global population on the rise and a growing shift from rural to urban living, the demand for food and food-related products is increasing daily. Recent advancements in artificial intelligence (AI) technology have brought about a transformative shift in the food industry, presenting significant potential for enhancing productivity, efficiency, and customer satisfaction. AI is poised to play a crucial role in optimizing various factors to create a nutritious and hygienic diet that meets consumer preferences. The application of AI technologies has the potential to enhance production, processing, distribution, and safety measures in the food industry. However, realizing the full potential of AI in this sector requires addressing associated challenges and ethical considerations.

REFERENCES

Addanki, M., Patra, P., & Kandra, P. (2022). Recent advances and applications of artificial intelligence and related technologies in the food industry. *Applied Food Research*, *2*(2), 100126. doi:10.1016/j.afres.2022.100126

Albert-Weiß, D., Hajdini, E., Heinrich, M., & Osman, A. (2020). *Cnn for ripeness classfication of watermelon fruits based on acoustic testing*. Paper presented at the Proceedings of the Virtual 3rd International Symposium on Structural Health Monitoring and Nondestructive Testing (SHM-NDT 2020), Quebeq, QU, Canada.

Ameen, N., Tarhini, A., Reppel, A., & Anand, A. (2021a). Customer experiences in the age of artificial intelligence. *Computers in Human Behavior*, *114*, 106548. doi:10.1016/j.chb.2020.106548 PMID:32905175

Arshad, J., Rehman, A. U., Othman, M. T. B., Ahmad, M., Tariq, H. B., Khalid, M. A., Moosa, M. A. R., Shafiq, M., & Hamam, H. (2022). Deployment of wireless sensor network and iot platform to implement an intelligent animal monitoring system. *Sustainability (Basel)*, *14*(10), 6249. doi:10.3390u14106249

Bai, C., Dallasega, P., Orzes, G., & Sarkis, J. (2020). Industry 4.0 technologies assessment: A sustainability perspective. *International Journal of Production Economics*, *229*, 107776. doi:10.1016/j.ijpe.2020.107776

Balieiro, A., Santos, R., Pereira, M., Figueiredo, R., Freitas, L., De Alsina, O., & ... . (2016). Adsorption process of molecularly imprinted silica for extraction of lactose from milk. *Brazilian Journal of Chemical Engineering*, *33*(2), 361–372. doi:10.1590/0104-6632.20160332s20140089

Berezina, K., Ciftci, O., & Cobanoglu, C. (2019). *Robots, artificial intelligence, and service automation in travel, tourism and hospitality*. Emerald Publishing Limited.

Blagoveshchenskiy, I., Blagoveshchenskiy, V., Besfamilnaya, E., & Sumerin, V. (2020). *Development of databases of intelligent expert systems for automatic control of product quality indicators*. Paper presented at the Journal of physics: conference series. 10.1088/1742-6596/1705/1/012019

Blasi, A. (2018). Scheduling food industry system using fuzzy logic. *Journal of Theoretical and Applied Information Technology*, *96*(19), 6463–6473.

Chaturvedi, R., & Verma, S. (2022). *Artificial Intelligence-Driven Customer Experience: Overcoming The Challenges*. California Management Review Insights.

Cheong, A., Lau, M., Foo, E., Hedley, J., & Bo, J. W. (2016). Development of a robotic waiter system. *IFAC-PapersOnLine*, *49*(21), 681–686. doi:10.1016/j.ifacol.2016.10.679

Edelman, D. C., & Abraham, M. (2022). Customer Experience in the Age of AI. *Harvard Business Review*, *100*(3-4), 116–125.

Gill, J., Sandhu, P. S., & Singh, T. (2014). *A review of automatic fruit classification using soft computing techniques*. Paper presented at the Int. Conf. Comput. Syst. Electron. Eng.

Goyache, F., Bahamonde, A., Alonso, J., López, S., Del Coz, J., Quevedo, J., Ranilla, J., Luaces, O., Alvarez, I., Royo, L. J., & Diez, J. (2001). The usefulness of artificial intelligence techniques to assess subjective quality of products in the food industry. *Trends in Food Science & Technology*, *12*(10), 370–381. doi:10.1016/S0924-2244(02)00010-9

Hafiz, R., Haque, M. R., Rakshit, A., & Uddin, M. S. (2022). Image-based soft drink type classification and dietary assessment system using deep convolutional neural network with transfer learning. *Journal of King Saud University. Computer and Information Sciences*, *34*(5), 1775–1784. doi:10.1016/j.jksuci.2020.08.015

Ivanov, S., & Webster, C. (2023). *Restaurants and robots: public preferences for robot food and beverage services*. Academic Press.

Kakani, V., Nguyen, V. H., Kumar, B. P., Kim, H., & Pasupuleti, V. R. (2020). A critical review on computer vision and artificial intelligence in food industry. *Journal of Agriculture and Food Research*, *2*, 100033. doi:10.1016/j.jafr.2020.100033

Kondakci, T., & Zhou, W. (2017). Recent applications of advanced control techniques in food industry. *Food and Bioprocess Technology*, *10*(3), 522–542. doi:10.100711947-016-1831-x

Kutyauripo, I., Rushambwa, M., & Chiwazi, L. (2023). Artificial intelligence applications in the agrifood sectors. *Journal of Agriculture and Food Research*, *11*, 100502. doi:10.1016/j.jafr.2023.100502

Lakshmi, S., Pandey, A., Ravi, N., Chauhan, O., Gopalan, N., & Sharma, R. (2017). Non-destructive quality monitoring of fresh fruits and vegetables. *Defence Life Science Journal*, *2*(2), 103–110. doi:10.14429/dlsj.2.11379

Leite, M. S., Santos, M. A., Costa, E. M., Balieiro, A., Lima, Á. S., Sanchez, O. L., & Soares, C. (2019). Modeling of milk lactose removal by column adsorption using artificial neural networks: Mlp and Rbf. *Chemical Industry & Chemical Engineering Quarterly*, *25*(4), 369–382. doi:10.2298/CICEQ180606015L

Lukinac, J., Jukić, M., Mastanjević, K., & Lučan, M. (2018). Application of computer vision and image analysis method in cheese-quality evaluation: A review. *Ukrainian Food Journal*, *7*(7, Issue 2), 192–214. doi:10.24263/2304-974X-2018-7-2-4

Makkar, T., & Verma, S., Yogesh, & Dubey, A. K. (2018). *Analysis and detection of fruit defect using neural network.* Paper presented at the Data Science and Analytics: 4th International Conference on Recent Developments in Science, Engineering and Technology, REDSET 2017, Gurgaon, India. 10.1007/978-981-10-8527-7_46

Marvin, H. J., Bouzembrak, Y., Van der Fels-Klerx, H., Kempenaar, C., Veerkamp, R., & Chauhan, A. (2022). *Digitalisation and Artificial Intelligence for sustainable food systems.* Academic Press.

Mavani, N. R., Ali, J. M., Othman, S., Hussain, M., Hashim, H., & Rahman, N. A. (2022). Application of artificial intelligence in food industry—A guideline. *Food Engineering Reviews*, *14*(1), 134–175. doi:10.100712393-021-09290-z

Memisi, N., Moracanin, S. V., Milijasevic, M., Babic, J., & Djukic, D. (2015). CIP cleaning processes in the dairy industry. *Procedia Food Science*, *5*, 184–186. doi:10.1016/j.profoo.2015.09.052

Monteiro, A. L., Freitas Souza, M., Lins, H. A., Teófilo, T. M. S., Barros Júnior, A. P., Silva, D. V., & Mendonça, V. (2021). A new alternative to determine weed control in agricultural systems based on artificial neural networks (ANNs). *Field Crops Research*, *263*, 108075. doi:10.1016/j.fcr.2021.108075

Morrar, R., Arman, H., & Mousa, S. (2017). The fourth industrial revolution (Industry 4.0): A social innovation perspective. *Technology Innovation Management Review*, *7*(11), 12–20. doi:10.22215/timreview/1117

Nandhini, M., Kala, K. U., Thangadarshini, M., & Madhusudhana Verma, S. (2022). Deep Learning model of sequential image classifier for crop disease detection in plantain tree cultivation. *Computers and Electronics in Agriculture*, *197*, 106915. doi:10.1016/j.compag.2022.106915

Nyabako, T., Mvumi, B. M., Stathers, T., Mlambo, S., & Mubayiwa, M. (2020). Predicting Prostephanus truncatus (Horn) (Coleoptera: Bostrichidae) populations and associated grain damage in smallholder farmers' maize stores: A machine learning approach. *Journal of Stored Products Research*, *87*, 101592. doi:10.1016/j.jspr.2020.101592

O'connell, A., Ruegg, P., Jordan, K., O'brien, B., & Gleeson, D. (2016). The effect of storage temperature and duration on the microbial quality of bulk tank milk. *Journal of Dairy Science*, *99*(5), 3367–3374. doi:10.3168/jds.2015-10495 PMID:26947309

O'Shea, N., Greene, D., & Fenelon, M. A. (2023). *Artificial Intelligence in Food Safety Reference Module in Food Science*. Elsevier.

Palacios, F., Diago, M. P., & Tardaguila, J. (2019). A non-invasive method based on computer vision for grapevine cluster compactness assessment using a mobile sensing platform under field conditions. *Sensors (Basel), 19*(17), 3799. doi:10.339019173799 PMID:31480754

Partel, V., Charan Kakarla, S., & Ampatzidis, Y. (2019). Development and evaluation of a low-cost and smart technology for precision weed management utilizing artificial intelligence. *Computers and Electronics in Agriculture, 157*, 339–350. doi:10.1016/j.compag.2018.12.048

Qian, J., Dai, B., Wang, B., Zha, Y., & Song, Q. (2022). Traceability in food processing: Problems, methods, and performance evaluations—A review. *Critical Reviews in Food Science and Nutrition, 62*(3), 679–692. doi:10.1080/10408398.2020.1825925 PMID:33016094

Ramirez-Asis, E., Vilchez-Carcamo, J., Thakar, C. M., Phasinam, K., Kassanuk, T., & Naved, M. (2022). A review on role of artificial intelligence in food processing and manufacturing industry. *Materials Today: Proceedings, 51*, 2462–2465. doi:10.1016/j.matpr.2021.11.616

Roy, R. B., Tudu, B., Bandyopadhyay, R., & Bhattacharyya, N. (2019). Application of electronic nose and tongue for beverage quality evaluation. Engineering tools in the beverage industry, 229-254.

Saha, K. K., & Zude-Sasse, M. (2022). Estimation of chlorophyll content in banana during shelf life using LiDAR laser scanner. *Postharvest Biology and Technology, 192*, 112011. doi:10.1016/j.postharvbio.2022.112011

Sandey, K., Qureshi, M., Meshram, B., Agrawal, A., & Uprit, S. (2017). Robotics—An Emerging Technology in Dairy Industry. *Int. J. Eng. Trends Technol, 43*, 58–62. doi:10.14445/22315381/IJETT-V43P210

Shi, C., Zhao, Z., Jia, Z., Hou, M., Yang, X., Ying, X., & Ji, Z. (2023). Artificial neural network-based shelf life prediction approach in the food storage process: A review. *Critical Reviews in Food Science and Nutrition*, 1–16. doi:10.1080/10408398.2023.2245899 PMID:37688408

Stoitsis, G., Papakonstantinou, M., Karvounis, M., & Manouselis, N. (2023). The role of Big Data and Artificial Intelligence in food risk assessment and prediction. In M. E. Knowles, L. E. Anelich, A. R. Boobis, & B. Popping (Eds.), *Present Knowledge in Food Safety* (pp. 1032–1044). Academic Press. doi:10.1016/B978-0-12-819470-6.00041-X

Tegmark, M. (2016). Benefits and risks of artificial intelligence. *Future of Life*, 29-31.

Toseef, M., & Khan, M. J. (2018). An intelligent mobile application for diagnosis of crop diseases in Pakistan using fuzzy inference system. *Computers and Electronics in Agriculture*, *153*, 1–11. doi:10.1016/j.compag.2018.07.034

Trabelsi, M., Casprini, E., Fiorini, N., & Zanni, L. (2023). Unleashing the value of artificial intelligence in the agri-food sector: Where are we? *British Food Journal*, *125*(13), 482–515. doi:10.1108/BFJ-11-2022-1014

Vassileva, S., & Mileva, S. (2010). AI-based software tools for beer brewing monitoring and control. *Biotechnology, Biotechnological Equipment*, *24*(3), 1936–1939. doi:10.2478/V10133-010-0060-0

Wang, J., & Hu, X. (2021). Convolutional neural networks with gated recurrent connections. *IEEE Transactions on Pattern Analysis and Machine Intelligence*, *44*(7), 3421–3435. doi:10.1109/TPAMI.2021.3054614 PMID:33497326

Chapter 13
Effects of Gluten on Health:
Pseudocereals as Gluten Substitutes

Ipek Bayrakci
Alanya Alaaddin Keykubat University, Turkey

Tugba Aktar
Alanya Alaaddin Keykubat University, Turkey

ABSTRACT

This chapter discusses the role of carbohydrates in nutrition, with particular emphasis on grains and their importance as a source of protein, fiber, and micronutrients. It also explores the negative effects that gluten, a storage protein found in grains, can have on some individuals, leading to disorders such as celiac disease, wheat ataxia, wheat allergy, and dermatitis herpetiformis. The primary treatment for such conditions is to remove gluten from the diet, but this can lead to a reduction in fiber intake and an increase in fat intake. The chapter then looks at the important role gluten plays in dough rheology, particularly in baking, and considers the challenges faced by those who must follow a gluten-free diet. Finally, it examines the potential of pseudocereals such as quinoa, amaranth, and buckwheat as substitutes for gluten-containing grains with their potential usage in the food industry and daily nutrition routines.

INTRODUCTION

Carbohydrates are the major source of energy for mammalians and a food constituent that has a high proportion in a regular diet. Grains are the basic source of carbohydrates, and some of the significant ones are wheat, rye, oats, barley, corn, and rice. In

DOI: 10.4018/979-8-3693-0819-6.ch013

addition to their carbohydrate content, grains are considered an important source of protein and fiber and contain some micronutrients such as B vitamins, zinc, vitamin E, and magnesium. Thus, grains are important in avoiding or suspending most cardiovascular diseases and intestinal problems due to their nutrient contents.

However, in some individuals, grain consumption can cause various disorders, which are mainly attributable to a storage protein called gluten. Gluten is characterized as a mixture of various proteins, primarily gliadin and glutenin; it is found in oats (avenin), barley (hordein), wheat (gliadin), and rye (secalin) and is generally referred to as gluten. Gluten, as a storage protein, is associated with many diseases such as celiac disease, wheat ataxia, wheat allergy, and dermatitis herpetiformis. The primary treatment routine in gluten-related diseases is the entire removal of gluten from the daily diet; however, it is predicted that a gluten-free diet will increase fat intake and decrease fiber intake. Furthermore, gluten plays an important role in dough rheology, especially in bakery products. Gliadin in the content of gluten provides viscosity and extensibility to the dough, while gluten provides adhesion and elasticity to the dough (Hayıt & Gül, 2017). Additionally, gluten traps carbon dioxide during the fermentation of fermented bakery products such as bread, causing the dough to rise and creating a porous structure after baking. Hence, individuals who must follow a gluten-free diet are forced to consume not only nutritionally inadequate but also less acceptable products in terms of their texture and taste. In this context, dairy products, legumes such as chickpeas, lentils, and peas, and hydrocolloid compounds such as nutritional fibers, agar, gums, alginates, and pseudocereals such as sorghum, quinoa, amaranth, and buckwheat are added to gluten-free products to enrich them and provide the technological, textural, and sensory properties of gluten protein in these products (Arendt & Dal Bello, 2011; Pellegrini & Agostoni, 2015).

Recently, pseudocereals have become more and more attractive due to their contents of essential amino acids, dietary fiber, minerals, and vitamins compared to grains. Pseudocereals, which are grain-like crops consisting of buckwheat, amaranth, and quinoa, contain high amounts of starch and can be used for grain purposes, despite being non-grains (Alvarez-Jubete et al., 2010). In this chapter, the effects of gluten on health will be examined, and how the use of pseudocereals, which have high potential as gluten substitutes, affects products in terms of texture, flavor, and nutrition will be discussed.

GLUTEN

Gluten is a storage protein found in grains, including wheat, barley, rye, and controversially, oat. Wheat has the highest gluten content, typically composed of 60-70% starch, 8-15% storage protein, 10-15% moisture, 4% ash, and 1-2% fat. Of the total 8-15% storage protein, 10-15% is comprised of albumin and globulin, while 85-90% is gluten (Biesiekierski, 2017).

Gluten primarily consists of glutenin and gliadin, as well as some minor levels of other proteins. The proteins are referred to as polymerized glutenins and monomeric prolamins, with prolamins differing based on the grain, including zeins in corn, gliadins in wheat, secalins in rye, avenins in oats, and hordeins in barley (Ciclitira et al., 2005). These proteins are not water soluble (Türksoy & Özkaya, 2006). The ratios of prolamins and glutenins in grains such as wheat, rye, oats, and barley are equal, and when combined with water, they form a structure known as the gluten complex. Corn, rice, and einkorn wheat are referred to as gluten-free grains due to their lack of gluten, attributed to the different ratios of prolamins and glutenins (Metin, 2016). Pseudocereals, including buckwheat, quinoa, and amaranth, consist of grain-like foods that do not naturally contain gluten and are referred to as pseudocereal (Alvarez-Jubete et al., 2010).

Gluten is an important component in cereal technology, contributing to the appearance and internal structure of dough. The gliadin fraction provides viscosity and extendibility to the dough, while the glutenin fraction is responsible for adhesion and elasticity (Hayit and Gül, 2017). In addition to contributing to the dough's adhesive and elastic properties, gluten also traps carbon dioxide during fermentation, helping the dough to rise and form a porous structure, thus determining the dough's rheological properties. When gluten is removed, the resulting dough has a watery consistency. As a result, products lacking gluten often have a crumbly texture, lower volume, and reduced color and quality after baking, leading to issues such as fast staling (İşleroğlu et al., 2009). Additionally, gluten absorbs excess water, preventing the starch granules from becoming exposed and increasing the bread's shelf life (Padalino et al., 2016). Therefore, quality issues, such as texture, volume, and shelf life, arise in gluten-free products, which seek solutions industrially and scientifically.

GLUTEN-RELATED DISEASES

Gluten-related diseases are defined as various health problems triggered by the consumption of gluten and have a global prevalence of approximately 5% (Elli et al., 2015). The most effective advice for these diseases is to eliminate gluten from the daily diet either completely or partially. Gluten-related diseases are associated with many diseases such as depression, migraines, autism, gastrointestinal, and skin diseases besides celiac disease (Biesiekierski, Muir, et al., 2013; Biesiekierski, Peters, et al., 2013; Marios Hadjivassiliou et al., 2015). These syndromes and diseases need gluten-free dietary patterns for the patient which would also provide benefits for the treatment or prevention of diseases such as diabetes, rheumatoid arthritis, inflammatory bowel disease, ulcerative colitis, obesity, weight gain, insulin resistance, and fatigue (Biesiekierski, Peters, et al., 2013; Menta et al., 2019; Soares et al., 2013).

Diseases associated with gluten are classified into three main categories as shown in Figure 1: autoimmune, allergic, and non-autoimmune or non-allergic. Autoimmune diseases are referred to as celiac disease, gluten ataxia, and dermatitis herpetiformis, whereas, allergic diseases are indicated as wheat allergy. Non-autoimmune or non-allergic diseases include non-celiac gluten sensitivity, neurological diseases, autism, migraines, schizophrenia, depression, and obesity (Sapone et al., 2012).

As can be seen from Figure 1, autoimmune diseases are one of the gluten related disorders. Celiac disease, a condition characterized by malabsorption in the small intestine, is associated with damage to the intestinal lining caused by gluten, a protein found in wheat, barley, and rye (Türksoy & Özkaya, 2006). Celiac disease is a multisystemic autoimmune disorder triggered by gluten found in these grains (Mulder et al., 1993). The proteins in wheat, barley, and rye that cause celiac disease, known as gluten fractions, are resistant to deamination by the intestinal enzyme transglutaminase (Bethune & Khosla, 2008; Green & Cellier, 2007). This results in the inhibition of intestinal absorption and malabsorption. As a result, celiac disease is a chronic inflammatory condition that manifests as malabsorption (Serin & Akbulut, 2017). The pathogenesis of celiac disease is complex and not fully understood (Göral et al., 2007). Environmental factors such as breastfeeding,

Figure 1. Terminology and classification of gluten-related health problems
Source: Sapone et al. (2012)

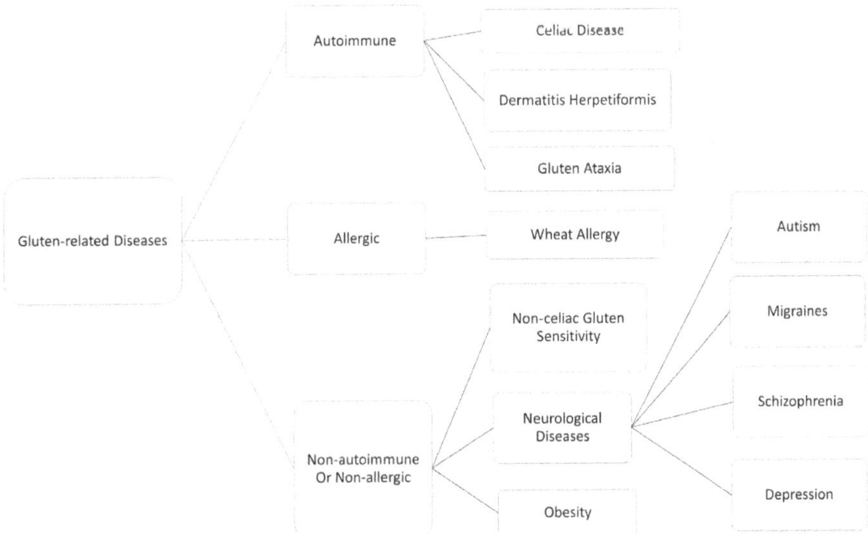

the introduction of gluten-containing foods, and daily gluten intake contribute to its development (Malekzadeh et al., 2005). According to estimates, the prevalence of celiac disease based on serological and histopathological tests ranges from 0.7% to 1.4% worldwide (Singh et al., 2018). The symptoms of celiac disease can be atypical and may not manifest until years later. Some celiac disease patients do not show any symptoms at all (Türksoy & Özkaya, 2006). Celiac disease patients may experience weight loss or failure to gain weight due to malabsorption (Nehra et al., 2014). Other symptoms include dental caries or enamel hypoplasia in children due to nutritional deficiencies of minerals such as iron or magnesium or vitamins such as niacin or thiamin or B12 or riboflavin or folate (Saturni et al., 2010). Dermatitis herpetiformis, a skin condition characterized by blisters and itching, is associated with gluten sensitivity and has been recognized as a subtype of celiac disease. Individuals with dermatitis herpetiformis are less likely to exhibit gastrointestinal symptoms commonly observed in celiac disease (Marios Hadjivassiliou et al., 2003). The specific symptoms of dermatitis herpetiformis include blisters on the elbows, knees, and buttocks (Schober, 2009). he lesions may appear as vesicles or bullae and are often accompanied by skin inflammation. The prevalence of dermatitis herpetiformis is estimated to be between 100,000 and 0.8 cases per 100,00 individuals in northern Europe and the United States (Hervonen et al., 2016). The elimination of gluten from the diet has been shown to improve symptoms in individuals with dermatitis herpetiformis. Gluten ataxia, which is also known as gluten related neurological disorder, is neurological condition associated with gluten sensitivity that is commonly observed in individuals with gluten sensitivity. It is also classified as an immunological disorder triggered by consuming gluten in genetically predisposed individuals (Pellecchia et al., 1999). Gluten ataxia is a type of ataxia that affects approximately 30-40% of all ataxia cases (M Hadjivassiliou et al., 1998). It is commonly considered as a subtype of celiac disease. Individuals with gluten ataxia should avoid consuming foods containing gluten as part of their daily diet (Marios Hadjivassiliou et al., 2003).

Second gluten related disease group is the allergic reactions to the gluten itself. This case is specified with its effects on the respiratory, digestive, and nervous systems where the symptoms may include hives, swelling, itching, difficulty breathing, and anaphylaxis. Wheat allergy is classified as a classic food allergy that affects the soluble system, mucous membranes, or intestinal system. In some individuals, it may cause shock, angioedema, or severe sensitivity reactions following the consumption of gluten. It may also lead to skin conditions such as dermatitis herpetiformis or eczema. Wheat allergy can result in respiratory conditions such as rhinitis, asthma, or bronchitis. The prevalence of wheat allergy in adults is estimated between 1% and 3% which is reported to be one of the most common food allergies in developed countries (Vierk et al., 2007; Zuidmeer et al., 2008). Individuals who avoid wheat in their diet may experience improvement in symptoms when reintroduced to gluten (Anonymous, 2010). Unlike celiac disease, wheat allergy may resolve over time. In a study conducted on children, the resolution rate for wheat allergy was found to be 29% at age four and 65% at age twelve (Keet et al., 2009).

Last group of gluten-related diseases are non-autoimmune or non-allergic diseases. One of these diseases is non-celiac gluten sensitivity, also known as gluten intolerance, is a condition characterized by symptoms such as headache, fatigue, bloating, and diarrhea in response to consuming gluten-containing foods in individuals who do not have celiac disease (Leonard et al., 2017). Noteworthy, he prevalence is higher than that of celiac disease (Boettcher & Crowe, 2013). The pathogenesis of non-celiac gluten sensitivity is not fully understood and the case is not associated with villous atrophy or other histopathological changes (Catassi et al., 2015). The response to a gluten-free diet in individuals with non-celiac gluten sensitivity suggests that the condition may be related to gluten intake rather than gluten sensitivity. However, the duration required for a gluten-free diet to provide symptom relief is not clearly defined (Catassi et al., 2015). Neurological diseases are the second category of the non-autoimmune or non-allergic diseases. Autism, is characterized by social and communication deficits as well as repetitive behaviors, is a developmental disorder that can be identified during early childhood (Fombonne, 2003). Some studies have suggested that gluten and casein-containing foods may contribute to the development of autism (Bandini et al., 2010; Harrington et al., 2006). Moreover, a gluten-free diet may be beneficial for individuals with autism in terms of reducing abnormal behaviors (Volta et al., 2015). Migraine is defined as unilateral headache with or without associated symptoms such as nausea, vomiting, and visual disturbances. Although not the primary cause of migraines, foods such as chocolate, coffee, oranges, tomatoes, eggs, fermented dairy products, alcohol, and fatty foods can trigger migraines (Cady et al., 2012). Studies have shown that individuals with celiac disease are more likely to experience migraines (Gabrielli et al., 2003; van Hemert et al., 2014). Another non-autoimmune or non-allergic disease is schizophrenia, which is characterized by cognitive deficits in areas such as processing speed, memory, learning, and attention, is a significant neurological disorder. Celiac disease, has been linked to an increased risk of developing schizophrenia (Niland & Cash, 2018). A study has shown that reducing gluten intake in the diet of individuals with schizophrenia may lead to a decrease in symptoms (Elder et al., 2006). Last disorder of non-autoimmune or non-allergic disease is depression. It is characterized by changes in mood, loss of interest, fatigue, decreased activity, sleep disturbances, and concentration difficulties, is a common and debilitating disorder. Depression has been associated with celiac disease for several decades. Recent studies have suggested that a healthy diet may also have a beneficial effect on depression (Opie et al., 2015). All these discussed diseases either direct or indirectly are associated with a gluten-free diet which will be investigated in the next topic.

GLUTEN-FREE DIET

In the literature, the gluten-free diet is the easiest solution and is a proven dietary treatment for diseases such as gluten ataxia, celiac disease, wheat allergy, and dermatitis herpetiformis. In addition to these diseases, it is also stated that a gluten-free diet can reduce symptoms of some neurological diseases such as migraines, depression, autism, and schizophrenia. Individuals can benefit from the antidiabetic, anti-obesity, and anti-inflammatory properties of a gluten-free diet (Soares et al., 2013). However, the effects of a gluten-free diet on health in non-disease conditions are not yet fully clear (Newberry et al., 2017).

A gluten-free diet is defined as a diet that does not contain any consumption of gluten protein. In a gluten-free diet, as seen in Table 1, restricted foods include kamut, malt, wheat and its derivatives, rye, spelt, triticale, barley, and products made from these foods. Instead of gluten-containing grains, individuals can consume rice, corn, and soy (Gallagher et al., 2003). They can also opt for naturally gluten-free foods such as fruits, vegetables, legumes, meats, eggs, dairy products, fish, unroasted nuts, butter, vegetable oils, and specially-made gluten-free products. Additionally, pseudo cereals such as buckwheat, amaranth, quinoa, and other products can be used as gluten substitutes (Melini & Melini, 2019; Serin & Akbulut, 2017; Yildirim, 2020). On the other hand, oat consumption is controversial for individuals with celiac disease because there is a risk of cross-contamination in gluten-free product production facilities and it is emphasized that gluten can be transferred through the air. Therefore, only for oat there is a guide limit for the consumption which is recommended to be consumed between 20-25 g daily for children, and 50-75 g daily for adults (Zimmer, 2011).

Table 1. Food recommendations for a gluten-free diet

Foods That Should Not Be Consumed	Foods Allowed to Be Consumed	Suspicious Foods to Be Consumed
All types of wheat (semolina, couscous, bran, marjoram, etc.), all kinds of foods containing wheat (flour, bulgur, pasta, etc.), wheat starch, barley, rye, triticale, oats, although controversial.	Rice, potatoes, millet, corn, amaranth, buckwheat, quinoa, legumes (lentils, chickpeas, beans, etc.), sorghum, meat, chicken, fish, fruits and vegetables, soybeans, sugar, although controversial, oats	Food starches, icing sugar, some fat substitutes, some medications, salad dressings, white vinegar, malt extracts (syrup, extract, sweetener, etc.)

Sources: Cady et al. (2012); Serin and Akbulut (2017)

Since gluten is a protein, a gluten-free diet is often considered to be insufficient in terms of protein. Gluten-free diets are high in calories and high in the amount of saturated fat, while the fiber content is lower compared to standard diet models. It is stated that individuals with celiac disease consume high amounts of energy, sugar, and fat (Saturni et al., 2010). Studies illustrate that adult celiac patients' calorie intake from carbohydrates was low, while the calorie intake from fat was higher. In addition, it is stated that the gluten-free diet model is not varied, balanced, or adequate (Kupper, 2005). Moreover, a gluten-free diet may lead to deficiencies in micronutrients such as iron, folate, and B vitamins (Bulka et al., 2017; Patel & Lacy, 2018). In this context, it is necessary to make improvements in the gluten-free diet to address nutrient deficiencies. Pseudocereals (buckwheat, amaranth, quinoa), animal proteins, dark leafy greens, and nuts can make an impact on the micronutrients specifically for iron. On the other hand, for magnesium deficiency, dark leafy greens, legumes, and nuts can be added to the gluten-free diet. In addition, calcium-rich foods such as dairy products, nuts, molasses, and dark leafy greens can prevent calcium deficiency in individuals on a gluten-free diet (Özkaya & Özkaya, 2018; See & Murray, 2006). Lastly, for B12 vitamin deficiency, it is recommended to consume foods such as meat, liver, chicken, fish, seafood, eggs, and milk in the daily diets of individuals on a gluten-free diet (Kreutz et al., 2020).

Another important issue for gluten-free nutrition is the cross-contamination. It is described as a situation that occurs through the interaction of a gluten-containing food with a gluten-free food through a living or non-living environment and equipment, which leads to the gluten-free food containing gluten. Particularly due to the low density of flour, it is highlighted as a matter that must be taken into great consideration in the food industry for cross-contamination. In this context, it is important that inspections to be carried out in food industries are conducted with great care, and the same care is taken in preparing food at home for gluten-free diet followers.

Despite the nutrition perspective, gluten-free products have some disadvantageous properties in terms of technology and sensory. Generally, gluten-free products are reported to have weak functional, lower nutritional, and sensory-wise weak acceptable properties. The absence of gluten in products such as pasta, cakes, bread, etc., and the raw materials used in the production of these products result in products that become stale quickly, have a bad texture, crumble easily, and have low volume and density (Sivaramakrishnan et al., 2004). This situation poses a challenge not only for the consumer but also for the experts in nutrition, food, and technology in industrial applications (Gallagher et al., 2003). The legislation specifies the glute-free products in two categories. The first category is gluten-free (especially for no gluten consumers) which contains a maximum of 20 mg/kg of gluten. The second category is low gluten which can contain up to 100 mg/kg of gluten (Codex Alimentarius Commission, 2009).

Gluten removal in industrial applications leads to changes in the micro and macro components. Gluten-free products are usually fortified for their negative sensory and nutritional values. In practice, the component analysis showed that they have higher sugar and saturated fat content for texture modifications (Wu et al., 2015). Moreover, gluten-free products have lower protein and fiber contents and higher fat and sodium contents compared to their gluten-containing counterparts such as cakes, biscuits, and snack bars. Studies have shown that the fat content of gluten-free products is twice as high. On the other hand, gluten-free products have a higher glycemic index and lower amounts of fiber, niacin, iron, riboflavin, thiamine, and folate compared to their gluten-containing equivalents. Therefore, methods such as adding calcium, iron, folate, vitamin D, and B vitamins to gluten-free products as supplements are applied to improve the nutritional value of gluten-free products (Penagini et al., 2013; Saturni et al., 2010).

Nutritionally the usage of pseudocereals (amaranth, quinoa, and buckwheat) should be motivated which are contented in nutritional aspects due to protein and essential amino acid content, dietary fiber, vitamins, and minerals (Hayıt & Gül, 2017; Pellegrini & Agostoni, 2015).

PSEUDOCEREALS

Pseudocereals are a double-seeded type of plant, which are referred to as edible seeds. They resemble grains in their physical appearance and high starch content, hence they are referred to as pseudocereals, or grain-like foods. They naturally do not contain gluten and are considered a healthy alternative to grains. Pseudocereals can grow in all kinds of climates, from temperate to tropical climates, and adapt to any kind of weather conditions (Joshi et al., 2019). Buckwheat (*Fagopyrum sp.*), quinoa (*Chenopodium quinoa Willd*), and amaranth (*Amaranthus sp.*) are some examples of pseudocereals (Martínez-Villaluenga et al., 2020). Pseudocereals are not only rich in nutrients but also high in dietary fiber, vitamins, minerals, essential amino acids, and unsaturated fatty acids. Compared to rice, corn, and starch-based low-nutrient and low-fiber gluten-free products used as gluten substitutes, pseudocereals can enhance the quality and nutrient content, and increase the dietary fiber content of the products they are added to. Therefore, they are frequently used in developing and improving gluten-free products (Baykut, 2021; Theethira & Dennis, 2015).

Wheat contains 10% to 12% protein, and therefore individuals who cannot consume gluten may have a deficiency in their protein intake. Thus, the substitution product that replaces gluten in daily nutrition should have a high protein content. For this reason, pseudocereals are suggested as a good alternative due to their high protein content (Green & Cellier, 2007). Additionally, pseudocereals contain bioactive compounds such as minerals, essential fatty acids, vitamins, phytosterols, and polyphenols (as seen in Table 2).

In addition to enriching the content of gluten-free products, pseudocereals also reduce the risk of diseases such as diabetes, cancer, high blood pressure, and cardiovascular diseases (Dębski et al., 2013). Pseudocereals are currently used in the food industry due to their suitability for food processing systems and easy production (Hayıt and Gül, 2017). In recent years, many studies have been conducted on pseudocereals to expand the range of gluten-free products (Alvarez-Jubete et al., 2010).

BUCKWHEAT

Buckwheat, a seedless grain from the Fagopyrum genus of the *Polygonaceae* family, is known as a dicotyledonous plant. It is an ancient plant originally from Central Asia and has been cultivated for many years. In recent years, with the emergence of pseudocereals, buckwheat has been used in biscuits, beer, bread, pasta, and spirit production. According to the literature, buckwheat flour improves the baking properties of gluten-free products (Naqash et al., 2017). The nutrient components of

Table 2. Energy and nutrient content of buckwheat, amaranth, quinoa, and wheat (per 100g)

	Buckwheat	Quinoa	Amaranth	Wheat
Protein (g/100g)	13,25	14,12	13,56	11,20
Fat (g/100g)	3,4	6,07	7,02	1,95
Carbohydrate (g)	71,5	64,2	65,25	74,5
Ash (g)	2,1	2,38	2,88	1,53
Dietary fiber (g)	10	7	6,7	13,10
Calcium (mg)	18	47	159	33
Magnesium (mg)	231	197	248	117
Iron (mg)	2,2	4,57	7,61	3,71
Zinc (mg)	2,4	3,1	2,87	2,96
Vitamin B2 (mg)	0,43	0,32	0,2	0,19
Vitamin B1 (mg)	0,1	0,36	0,12	0,3

Source: USDA (2011)

buckwheat, as seen in Table 2, contain an average of 1.4-4.7% fat, 10-12.5% protein, 55-75% carbohydrates, and 7-10.7% fiber. About 54% of the total carbohydrates are in the form of starch. Buckwheat is known to have a higher water-holding capacity compared to corn and wheat starch due to its grain size and starch fraction (De Arcangelis et al., 2020; Sandri et al., 2017). 100 grams of buckwheat contains approximately 7.14 grams of insoluble and 9.01 grams of soluble fiber which is higher than the other sources (Boukid et al., 2018). Soluble fiber shows a prebiotic effect for some strains of *Bifidobacteria* and lactic acid bacteria, which reduces total blood cholesterol, lowers glycemic index, and decreases the risk of diabetes and obesity. Meanwhile, the Insoluble fiber, depending on its effect on viscosity and water-holding capacity, improves bowel movements (Yaver & Bilgiçli, 2020). In terms of protein content, corn contains 8.3 g/100 g protein and rice contains 6.9 g/100 g protein, while buckwheat contains 13.9 g/100 g protein, which is expected to improve the nutrient content (De Arcangelis et al., 2020). The main storage protein in buckwheat is known as globulins, which are also known as prolamins in grains. Buckwheat contains a trace amount of prolamins; therefore, these individuals can consume buckwheat comfortably (Ekici et al., 2019). Buckwheat has advantageous protein content due to the essential amino acid content, and its protein content is considered an important source in terms of biological value that is equivalent to about 93% of egg protein (Bilgiçli, 2020). On the other hand, lipids are concentrated in the embryo of buckwheat, and it has a rich content of polyunsaturated fatty acids. Their stearic, oleic, behenic, linoleic, arachidonic, and palmitic acid contents constitute 93% of the total fatty acid content (Aydar et al., 2019). Compared to other pseudocereals and gluten-free grains, buckwheat contains about 2-3 g/100 g less fat. It has high amounts of minerals such as copper, potassium, magnesium, phosphorus, and zinc compared to other grains. Additionally, it has a rich content of antioxidants, which have beneficial effects such as anti-inflammatory, antitumor, antidiabetic, antioxidant, and hepatoprotective. It contains therapeutic components such as phytosterols, phenylpropanoid glycosides, quercetin, vitexin, proanthocyanidins, flavonoids, D-chiroinositol, myo-inositol, fagopyrins, and orientin. Myo-inositol, fagopyritols, and D-chiro-inositol, important components in insulin sensitivity found in buckwheat's content, play a role in regulating blood glucose levels and reducing blood pressure (Sandri et al., 2017). Thus, using buckwheat as a gluten substitute has a positive impact on health.

Based on technological investigations of buckwheat as gluten substituting carbohydrate source usage, textural and sensory studies were carried out. One of them found that protein cross-linking, known as transglutaminase, was more active in an alkaline added environment of buckwheat flour (Guo et al., 2017). Another recent study investigated the gluten-free cookie formulations with other sources including buckwheat and emphasized that transglutaminase had a significant effect on the texture of gluten-free biscuits (Altındağ et al., 2015). Bread is also one of the mostly studied gluten-free product and a study explored the addition of mineral to full buckwheat flour. It highlighted that the density was decreased yet the sensory scores were increased compared to wheat sourced bread (Sayed et al., 2016). There are various studies showing the increasing nutrient content of the products with buckwheat partition; specifically with cakes containing other gluten-free sources and buckwheat (Gambuś et al., 2009), gluten-free pasta formulation (Bilgicli, 2008; De Arcangelis et al., 2020), gluten-free bread (Costantini et al., 2014).

AMARANTH

The amaranth plant (*Ammanthus spp.*), is native to South America and has become popular in recent years due to its functional and nutrient properties. However, the production of amaranth plants is not yet widespread; it is known to be a plant that is resistant to climate conditions such as drought and heat. As seen in Table 2, the amaranth plant contains approximately 5-11% fat, 13-21% protein, and 48-69% starch (Kilincceker & Büyük, 2019). Starch is indicated as the main component of amaranth; the amount of starch contained in amaranth varies depending on its species. It is found at a rate of 62% in *A. hypochondriacus* species, while it is found at a rate of 48% in *A. cruentus* species (Baykut, 2021). It is known that the starch contained in amaranth provides functional properties in food and exhibits a very high water-absorbing capacity thanks to its very small granular structure (Caselato-Sousa & Amaya-Farfán, 2012). The dietary fiber content of amaranth also varies depending on its species, ranging from 7.3% to 20.6%. The content of insoluble dietary fiber is 75%, while 25% consists of soluble dietary fiber (Srichuwong et al., 2017). Furthermore, it contains 25% of insoluble β-glucan which is higher than in pseudocereals and other grains (Venskutonis & Kraujalis, 2013). Amaranth contains an average of 13-14% protein, which is higher than other grains. Amaranth has a rich essential amino acid content and has a nutritional value like that of egg protein. It contains amino acids such as phenylalanine, isoleucine, proline, serine, methionine, alanine, leucine, arginine, tryptophan, and valine. It is also important for infant nutrition as it contains essential amino acids for babies such as arginine and histidine. The protein content in amaranth is distributed as 2-3% prolamin, 40% albumin, 25-30% glutenin, and 20% globulin. The globulins in amaranth have emulsifying properties. In other grain types such as corn and rice, the protein in the seed is 80% found in the endosperm, while in amaranth, it is found in the embryo (Baykut, 2021). Amaranth contains an average of 7% fat, which is 2-3 times higher than the fat content of other grains. The fatty acid profile of its content is rich in unsaturated fatty acids, and it is stated that the ratio of saturated fatty acids to unsaturated fatty acids is 1:4 (Maurya & Arya, 2018). It is noted that amaranth contains an average of twice as much mineral as other grain products. It is known to have high amounts of iron, phosphorus, iron, calcium, potassium, zinc, sodium, and magnesium, and the phosphorus in it is known to be in the form of phytic acid. The percentage of phytic acid in amaranth varies between 0.3% and 0.6%, and phytic acid is known to reduce cholesterol levels in the body. Additionally, the amaranth plant is known as a rich source of vitamins. It contains higher levels of vitamin C and B2 than grains (Baykut, 2021).

Recent studies on the health effects of amaranth have focused on its reducing effect on the risk of developing chronic diseases, thanks to its phenolic compounds and antioxidants. The outer layer of amaranth seeds contains a high amount of phytochemicals (Tang & Tsao, 2017). The amount of phytic acid and tannins in amaranth plant content is like that of grains, and it is known that its chymotrypsin and trypsin activities are lower compared to grains. In the studies conducted on amaranth so far, it has been reported that the antibacterial components in its content are at levels that will not cause any health problems (Baykut, 2021). Due to its nutritional and functional superiority in the production of gluten-free products, amaranth is frequently used in the production of gluten-free products such as bread, pasta, crackers, and biscuits. Therefore, in recent years, the use of amaranth flour in the formulation of gluten-free foods has been increasing (Ballabio et al., 2011).

Based on technological investigations of amaranth containing products studies are carried out specifically on; textural properties of gluten-free pasta formulation (Bastos et al., 2016; Cabrera-Chávez et al., 2012; Makdoud & Rosentrater, 2017; Schoenlechner et al., 2010), amaranth added bread formulation (Alencar et al., 2015). In terms of the sensory properties specific studies are on; bakery products of amaranth (Gebreil et al., 2020), gluten free bread (Gambus et al., 2010), and gluten-free energy bars (Bhavani et al., 2018). For the nutrient content and quality, amaranth usage was tested on the products of; bakery products (Gebreil et al., 2020), bread (Gambus et al., 2010), cake mixtures (Gambuś et al., 2009), and cookies (de la Barca et al., 2010).

QUINOA

Quinoa is a pseudocereal from the *Chenopodiaceae* family. Nowadays, quinoa is frequently preferred in gluten-free products due to its nutritional and functional properties. The composition of quinoa varies according to the variety. As seen in Table 2, it consists of approximately 4-8% fat, 54-64% carbohydrates, and 10-18% protein (Keskin & Evlice, 2015). The main carbohydrate in quinoa is starch, which makes up between 52% and 69% of its structure (Filho et al., 2017;

Vilcacundo & Hernández-Ledesma, 2017). Quinoa exhibits a high water-holding capacity and can be used as a thickener in the food industry due to this property (Dağ & Özkan, 2019). Quinoa is an excellent source of soluble and insoluble fiber, with a total fiber content of between 7% and 21.6%. Soluble fiber makes up 22% of the total content, while 78% of it is insoluble fiber. The soluble fiber content is higher than that of corn and wheat. Quinoa has a low glycemic index, which is attributed to its fiber content (Bastidas et al., 2016; Lutz & Bascuñán-Godoy, 2017). Quinoa is known for its rich content of essential amino acids and is referred to as having a high-quality protein content similar to that of casein protein. On average, quinoa contains 15% protein, while rice contains 8.8%, corn contains 10.5%, barley contains 11%, sorghum contains 12.4%, wheat contains 14.8%, and rye contains 11.6% protein (Filho et al., 2017). Quinoa is a gluten-free food due to its low prolamine content and its distant phylogenetic relationship to gluten-containing grains (Lutz & Bascuñán-Godoy, 2017). Quinoa is considered an alternative food to oil seeds due to its lipid fraction content. The n-6/n-3 ratio in quinoa falls within an ideal range associated with cardiovascular diseases, obesity, and diabetes, among other ailments. On average, quinoa contains 7% fat, which is higher than the 4.7% found in corn, but lower than the 19% found in soybeans. Quinoa and soybeans have similar unsaturated fatty acid profiles when compared with linoleic acid, α-linolenic acid, and oleic acid (Tang & Tsao, 2017). Quinoa is a rich source of vitamins, namely vitamins E, B, and C, with α-tocopherol content giving it an anti-oxidative property that protects against lipid oxidation. Quinoa also contains minerals such as magnesium, potassium, phosphorus, and calcium in bioavailable forms. Additionally, quinoa has a higher iron content compared to other traditional grains (Filho et al., 2017; Lutz & Bascuñán-Godoy, 2017; Vilcacundo & Hernández-Ledesma, 2017). Quinoa also contains bioactive components such as saponins, peptides, isoflavones, phytosterols, lectins, squalenes, and carotenoids (İlkay & Mutlu, 2020). Quinoa flour or seeds are used in the production of many gluten-free foods, including bread, biscuits, pasta, and beer. The use of quinoa in gluten-free products enriches their technological and nutritional properties (Caperuto et al., 2001). However, the high amount of saponins in the outer layer of quinoa seeds may limit their usage, which can be considered a threat (De Frutos et al., 2018; Solíz-Guerrero et al., 2002).

Technological observations of product based studies specifically focuses on textural properties of; bread formulations (Brito et al., 2015), sourdough bread (Axel et al., 2015). Meanwhile on sensory properties of; gluten-free beer (Cela et al., 2022), pasta (Caperuto et al., 2001), bread (Kahlon et al., 2019), and cookies (Kaur & Kaur, 2017) were tested to observe its effects.

CONCLUSION

Finally, science and technology still research the development of gluten-free products is ongoing as they aim to prevent or treat disease. However, it should be emphasized that current gluten-free products in the market contain low levels of protein and fiber, and high levels of carbohydrates, fat, and sugar. Another important fact is that gluten-free products may not be a healthy choice for individuals who do not have any gluten sensitivity.

As a commercial product for gluten-free packed foods, it is essential to improve the sensory, textural, and nutritional properties of gluten-free products due to their low acceptance and liking score with the sensory findings. In this context, instead of adding grains such as corn and rice to gluten-free products, which have lower nutritional values, pseudocereals with higher nutritional value should be considered. Therefore, the recipe alternatives, raw material selection, or partition should be investigated in terms of pseudocereals for gluten-free products in the market. In conclusion, to minimize the textural, nutritional, and sensory losses in gluten-free products, there is an urgent necessity to seek alternative raw materials and possible products in terms of economic and sustainable studies on the use of pseudocereals.

REFERENCES

Alencar, N. M. M., Steel, C. J., Alvim, I. D., de Morais, E. C., & Bolini, H. M. A. (2015). Addition of quinoa and amaranth flour in gluten-free breads: Temporal profile and instrumental analysis. *Lebensmittel-Wissenschaft + Technologie*, *62*(2), 1011–1018. doi:10.1016/j.lwt.2015.02.029

Altındağ, G., Certel, M., Erem, F., & İlknur Konak, Ü. (2015). Quality characteristics of gluten-free cookies made of buckwheat, corn, and rice flour with/without transglutaminase. *Food Science & Technology International*, *21*(3), 213–220. doi:10.1177/1082013214525428 PMID:24613830

Alvarez-Jubete, L., Arendt, E. K., & Gallagher, E. (2010). Nutritive value of pseudocereals and their increasing use as functional gluten-free ingredients. *Trends in Food Science & Technology*, *21*(2), 106–113. doi:10.1016/j.tifs.2009.10.014

Anonymous. (2010). Guidelines for the diagnosis and management of food allergy in the United States: Report of the NIAID-sponsored expert panel. *The Journal of Allergy and Clinical Immunology*, *126*(6), S1–S58. doi:10.1016/j.jaci.2010.10.007 PMID:21134576

Arendt, E., & Dal Bello, F. (2011). *Gluten-free cereal products and beverages*. Elsevier.

Axel, C., Röcker, B., Brosnan, B., Zannini, E., Furey, A., Coffey, A., & Arendt, E. K. (2015). Application of Lactobacillus amylovorus DSM19280 in gluten-free sourdough bread to improve the microbial shelf life. *Food Microbiology*, *47*, 36–44. doi:10.1016/j.fm.2014.10.005 PMID:25583336

Aydar, A. Y., Akgün, A., & Dengiz, İ. T. (2019). Glutensiz diyetde kullanilan alternatif hammaddeler. *Geleceğin Dünyasinda*, 20.

Ballabio, C., Uberti, F., Di Lorenzo, C., Brandolini, A., Penas, E., & Restani, P. (2011). Biochemical and immunochemical characterization of different varieties of amaranth (Amaranthus L. ssp.) as a safe ingredient for gluten-free products. *Journal of Agricultural and Food Chemistry*, *59*(24), 12969–12974. doi:10.1021/jf2041824 PMID:22073907

Bandini, L. G., Anderson, S. E., Curtin, C., Cermak, S., Evans, E. W., Scampini, R., Maslin, M., & Must, A. (2010). Food selectivity in children with autism spectrum disorders and typically developing children. *The Journal of Pediatrics*, *157*(2), 259–264. doi:10.1016/j.jpeds.2010.02.013 PMID:20362301

Bastidas, E., Roura, R., Rizzolo, D., Massanés, T., & Gomis, R. (2016). Quinoa (Chenopodium quinoa Willd), from nutritional value to potential health benefits: An integrative review. *Journal of Nutrition & Food Sciences*, *6*(3).

Bastos, G. M., Júnior, M. S. S., Caliari, M., de Araujo Pereira, A. L., de Morais, C. C., & Campos, M. R. H. (2016). Physical and sensory quality of gluten-free spaghetti processed from amaranth flour and potato pulp. *Lebensmittel-Wissenschaft + Technologie*, *65*, 128–136. doi:10.1016/j.lwt.2015.07.067

Baykut, E. (2021). Bazı tahıl benzeri ürünlerin besin içeriği ve gıda endüstrisinde kullanımı. *Avrupa Bilim ve Teknoloji Dergisi*, *23*, 89–98.

Bethune, M. T., & Khosla, C. (2008). Parallels between pathogens and gluten peptides in celiac sprue. *PLoS Pathogens*, *4*(2), e34. doi:10.1371/journal.ppat.0040034 PMID:18425213

Bhavani, B., Reddy, D. K., & Waghray, K. (2018). Energy bars made with popped with amranth. *International Journal of Food and Fermentation Technology*, *8*(2), 223–228. doi:10.30954/2277-9396.02.2018.13

Biesiekierski, J. R. (2017). What is gluten? *Journal of Gastroenterology and Hepatology*, *32*(S1), 78–81. doi:10.1111/jgh.13703 PMID:28244676

Biesiekierski, J. R., Muir, J. G., & Gibson, P. R. (2013). Is gluten a cause of gastrointestinal symptoms in people without celiac disease? *Current Allergy and Asthma Reports, 13*(6), 631–638. doi:10.100711882-013-0386-4 PMID:24026574

Biesiekierski, J. R., Peters, S. L., Newnham, E. D., Rosella, O., Muir, J. G., & Gibson, P. R. (2013). No effects of gluten in patients with self-reported non-celiac gluten sensitivity after dietary reduction of fermentable, poorly absorbed, short-chain carbohydrates. *Gastroenterology, 145*(2), 320–328. doi:10.1053/j.gastro.2013.04.051 PMID:23648697

Bilgicli, N. (2008). *Utilization of buckwheat flour in gluten-free egg noodle production.* Academic Press.

Bilgiçli, N. (2020). Tahıl benzeri ürünler: Bileşimi, beslenme-sağlık üzerine etkileri ve tahıl ürünlerinde kullanımı. *Food and Health, 6*(1), 41–56.

Boettcher, E., & Crowe, S. E. (2013). Celiac Disease: Dispelling Misconceptions and Myths. In Celiac Disease (pp. 255–273). Springer.

Boukid, F., Folloni, S., Sforza, S., Vittadini, E., & Prandi, B. (2018). Current trends in ancient grains-based foodstuffs: Insights into nutritional aspects and technological applications. *Comprehensive Reviews in Food Science and Food Safety, 17*(1), 123–136. doi:10.1111/1541-4337.12315 PMID:33350067

Brito, I. L., de Souza, E. L., Felex, S. S. S., Madruga, M. S., Yamashita, F., & Magnani, M. (2015). Nutritional and sensory characteristics of gluten-free quinoa (Chenopodium quinoa Willd)-based cookies development using an experimental mixture design. *Journal of Food Science and Technology, 52*(9), 5866–5873. doi:10.100713197-014-1659-1 PMID:26345002

Bulka, C. M., Davis, M. A., Karagas, M. R., Ahsan, H., & Argos, M. (2017). The unintended consequences of a gluten-free diet. *Epidemiology (Cambridge, Mass.), 28*(3), e24–e25. doi:10.1097/EDE.0000000000000640 PMID:28166100

Cabrera-Chávez, F., de la Barca, A. M. C., Islas-Rubio, A. R., Marti, A., Marengo, M., Pagani, M. A., Bonomi, F., & Iametti, S. (2012). Molecular rearrangements in extrusion processes for the production of amaranth-enriched, gluten-free rice pasta. *Lebensmittel-Wissenschaft + Technologie, 47*(2), 421–426. doi:10.1016/j.lwt.2012.01.040

Cady, R. K., Farmer, K., Dexter, J. K., & Hall, J. (2012). The bowel and migraine: Update on celiac disease and irritable bowel syndrome. *Current Pain and Headache Reports, 16*(3), 278–286. doi:10.100711916-012-0258-y PMID:22447132

Caperuto, L. C., Amaya-Farfan, J., & Camargo, C. R. O. (2001). Performance of quinoa (Chenopodium quinoa Willd) flour in the manufacture of gluten-free spaghetti. *Journal of the Science of Food and Agriculture*, *81*(1), 95–101. doi:10.1002/1097-0010(20010101)81:1<95::AID-JSFA786>3.0.CO;2-T

Caselato-Sousa, V. M., & Amaya-Farfán, J. (2012). State of knowledge on amaranth grain: A comprehensive review. *Journal of Food Science*, *77*(4), R93–R104. doi:10.1111/j.1750-3841.2012.02645.x PMID:22515252

Catassi, C., Elli, L., Bonaz, B., Bouma, G., Carroccio, A., Castillejo, G., Cellier, C., Cristofori, F., De Magistris, L., Dolinsek, J., Dieterich, W., Francavilla, R., Hadjivassiliou, M., Holtmeier, W., Körner, U., Leffler, D., Lundin, K., Mazzarella, G., Mulder, C., ... Fasano, A. (2015). Diagnosis of non-celiac gluten sensitivity (NCGS): The Salerno experts' criteria. *Nutrients*, *7*(6), 4966–4977. doi:10.3390/nu7064966 PMID:26096570

Cela, N., Galgano, F., Perretti, G., Di Cairano, M., Tolve, R., & Condelli, N. (2022). Assessment of brewing attitude of unmalted cereals and pseudocereals for gluten free beer production. *Food Chemistry*, *384*, 132621. doi:10.1016/j.foodchem.2022.132621 PMID:35257999

Ciclitira, P. J., Ellis, H. J., & Lundin, K. E. A. (2005). Gluten-free diet—What is toxic? *Best Practice & Research. Clinical Gastroenterology*, *19*(3), 359–371. doi:10.1016/j.bpg.2005.01.003 PMID:15925842

Codex Alimentarius Commission. (2009). *Commission regulation (EC) No 41/2009 of 20 January 2009 concerning the composition and labelling of foodstuffs suitable for people intolerant to gluten.* Author.

Costantini, L., Lukšič, L., Molinari, R., Kreft, I., Bonafaccia, G., Manzi, L., & Merendino, N. (2014). Development of gluten-free bread using tartary buckwheat and chia flour rich in flavonoids and omega-3 fatty acids as ingredients. *Food Chemistry*, *165*, 232–240. doi:10.1016/j.foodchem.2014.05.095 PMID:25038671

Dağ, Ş. R. O., & Özkan, A. M. G. (2019). Kinoa (Chenopodium quinoa Willd.) üzerine bir derleme. *Journal of Faculty of Pharmacy of Ankara University*, *43*(3), 309–333.

De Arcangelis, E., Cuomo, F., Trivisonno, M. C., Marconi, E., & Messia, M. C. (2020). Gelatinization and pasta making conditions for buckwheat gluten-free pasta. *Journal of Cereal Science*, *95*, 103073. doi:10.1016/j.jcs.2020.103073

De Frutos, M. F., Fotschki, B., Musoles, R. F., & Llopis, J. M. L. (2018). *Gluten-free cereals and pseudocereals: nutrition and health. In Bioactive Molecules in Food*. Springer.

de la Barca, A. M. C., Rojas-Martínez, M. E., Islas-Rubio, A. R., & Cabrera-Chávez, F. (2010). Gluten-free breads and cookies of raw and popped amaranth flours with attractive technological and nutritional qualities. *Plant Foods for Human Nutrition (Dordrecht, Netherlands)*, *65*(3), 241–246. doi:10.100711130-010-0187-z PMID:20734143

Dębski, B., Gralak, M. A., Bertrandt, J., & Kłos, A. (2013). Minerals and polyphenols content of quinoa (Chenopodium quinoa Willd.) plant. *Probl. Hig. Epidemiol*, *94*(2), 300–304.

Ekici, L., İnanır, C., & Albayrak, S. (2019). Karabuğdayın fitokimyası, farmakolojisi ve biyofonksiyonel özellikleri. *Avrupa Bilim ve Teknoloji Dergisi*, *16*, 713–722.

Elder, J. H., Shankar, M., Shuster, J., Theriaque, D., Burns, S., & Sherrill, L. (2006). The gluten-free, casein-free diet in autism: Results of a preliminary double blind clinical trial. *Journal of Autism and Developmental Disorders*, *36*(3), 413–420. doi:10.100710803-006-0079-0 PMID:16555138

Elli, L., Branchi, F., Tomba, C., Villalta, D., Norsa, L., Ferretti, F., Roncoroni, L., & Bardella, M. T. (2015). Diagnosis of gluten related disorders: Celiac disease, wheat allergy and non-celiac gluten sensitivity. *World Journal of Gastroenterology*, *21*(23), 7110–7119. doi:10.3748/wjg.v21.i23.7110 PMID:26109797

Filho, A. M. M., Pirozi, M. R., Borges, J. T. D. S., Pinheiro Sant'Ana, H. M., Chaves, J. B. P., & Coimbra, J. S. D. R. (2017). Quinoa: Nutritional, functional, and antinutritional aspects. *Critical Reviews in Food Science and Nutrition*, *57*(8), 1618–1630. doi:10.1080/10408398.2014.1001811 PMID:26114306

Fombonne, E. (2003). Epidemiological surveys of autism and other pervasive developmental disorders: An update. *Journal of Autism and Developmental Disorders*, *33*(4), 365–382. doi:10.1023/A:1025054610557 PMID:12959416

Gabrielli, M., Cremonini, F., Fiore, G., Addolorato, G., Padalino, C., Candelli, M., De Leo, M. E., Santarelli, L., Giacovazzo, M., & Gasbarrini, A. (2003). Association between migraine and celiac disease: Results from a preliminary case-control and therapeutic study. *The American Journal of Gastroenterology*, *98*(3), 625–629. doi:10.1111/j.1572-0241.2003.07300.x PMID:12650798

Gallagher, E., Gormley, T. R., & Arendt, E. K. (2003). Crust and crumb characteristics of gluten free breads. *Journal of Food Engineering*, *56*(2–3), 153–161. doi:10.1016/S0260-8774(02)00244-3

Gambuś, H., Gambuś, F., Pastuszka, D., Wrona, P., Ziobro, R., Sabat, R., Mickowska, B., Nowotna, A., & Sikora, M. (2009). Quality of gluten-free supplemented cakes and biscuits. *International Journal of Food Sciences and Nutrition, 60*(sup4), 31–50.

Gambus, H., Pastuszka, D., Gambus, F., Wrona, P., Ziobro, R., Mickowska, B., & Sikora, M. (2010). Nutritional and dietary value of gluten-free rolls enriched in amaranth flour. *Journal of Pre-Clinical and Clinical Research*, *4*(2).

Gebreil, S. Y., Ali, M. I. K., & Mousa, E. A. M. (2020). Utilization of amaranth flour in preparation of high nutritional value bakery products. *Food and Nutrition Sciences*, *10*(05), 336–354. doi:10.4236/fns.2020.105025

Göral, V., Yildirim, N., Kaplan, A., Dede, Ş., & Çelik, M. (2007). Gluten enteropatisi sıklığı. *Akademik Gastroenteroloji Dergisi*, *6*(3), 144–148.

Green, P. H. R., & Cellier, C. (2007). Celiac disease. *The New England Journal of Medicine*, *357*(17), 1731–1743. doi:10.1056/NEJMra071600 PMID:17960014

Guo, X.-N., Wei, X.-M., & Zhu, K.-X. (2017). The impact of protein cross-linking induced by alkali on the quality of buckwheat noodles. *Food Chemistry*, *221*, 1178–1185. doi:10.1016/j.foodchem.2016.11.041 PMID:27979076

Hadjivassiliou, M. (2003). Gluten ataxia in perspective: Epidemiology, genetic susceptibility and clinical characteristics. *Brain*, *126*(3), 685–691. doi:10.1093/brain/awg050 PMID:12566288

Hadjivassiliou, M., Grünewald, R. A., Chattopadhyay, A. K., Davies-Jones, G. A. B., Gibson, A., Jarratt, J. A., Kandler, R. H., Lobo, A., Powell, T., & Smith, C. M. L. (1998). Clinical, radiological, neurophysiological, and neuropathological characteristics of gluten ataxia. *Lancet*, *352*(9140), 1582–1585. doi:10.1016/S0140-6736(98)05342-2 PMID:9843103

Hadjivassiliou, M., Sanders, D. D., & Aeschlimann, D. P. (2015). Gluten-related disorders: Gluten ataxia. *Digestive Diseases (Basel, Switzerland)*, *33*(2), 264–268. doi:10.1159/000369509 PMID:25925933

Harrington, J. W., Patrick, P. A., Edwards, K. S., & Brand, D. A. (2006). Parental beliefs about autism: Implications for the treating physician. *Autism*, *10*(5), 452–462. doi:10.1177/1362361306066609 PMID:16940312

Hayıt, F., & Gül, H. (2017). *Tam karabuğday unu ve transglutaminaz ilavesinin kısmi pişirilerek dondurulmuş ekşi mayalı ekmeklerin fiziksel ve tekstürel özellikleri üzerine etkisi*. Academic Press.

Hervonen, K., Salmi, T. T., Ilus, T., Paasikivi, K., Vornanen, M., Laurila, K., Lindfors, K., Viiri, K., Saavalainen, P., Collin, P., Kaukinen, K., & Reunala, T. (2016). Dermatitis herpetiformis refractory to gluten-free dietary treatment. *Acta Dermato-Venereologica*, *96*(1), 82–86. doi:10.2340/00015555-2184 PMID:26084552

İlkay, H. Ö., & Mutlu, A. A. (2020). Fonksiyonel bir besin: Kinoa. *Beslenme ve Diyet Dergisi*, *48*(3), 92–101.

İşleroğlu, H., Dirim, S. N., & Ertekin, F. K. (2009). Gluten içermeyen, hububat esaslı alternatif ürün formülasyonları ve üretim teknolojileri. *Gida*, *34*(1), 29–36.

Joshi, D. C., Chaudhari, G. V., Sood, S., Kant, L., Pattanayak, A., Zhang, K., Fan, Y., Janovská, D., Meglič, V., & Zhou, M. (2019). Revisiting the versatile buckwheat: Reinvigorating genetic gains through integrated breeding and genomics approach. *Planta*, *250*(3), 783–801. doi:10.100700425-018-03080-4 PMID:30623242

Kahlon, T. S., Avena-Bustillos, R. J., Brichta, J. L., & Kahlon, A. K. (2019). High-protein nutritious flatbreads and an option for gluten-sensitive individuals. *Foods*, *8*(11), 591. doi:10.3390/foods8110591 PMID:31752409

Kaur, S., & Kaur, N. (2017). Development and sensory evaluation of gluten free bakery products using quinoa (Chenopodium Quinoa) flour. *Journal of Applied and Natural Science*, *9*(4), 2449–2455. doi:10.31018/jans.v9i4.1552

Keet, C. A., Matsui, E. C., Dhillon, G., Lenehan, P., Paterakis, M., & Wood, R. A. (2009). The natural history of wheat allergy. *Annals of Allergy, Asthma & Immunology*, *102*(5), 410–415. doi:10.1016/S1081-1206(10)60513-3 PMID:19492663

Keskin, Ş., & Evlice, A. K. (2015). Fırın ürünlerinde kinoa kullanımı. *Tarla Bitkileri Merkez Araştırma Enstitüsü Dergisi*, *24*(2), 150–156.

Kilincceker, O., & Büyük, G. (2019). Amarant Amaranthus spp'ın Bazı Özellikleri ve Et Ürünlerinde Kullanımı. *Adyutayam Dergisi*, *7*(2), 36–42.

Kreutz, J. M., Adriaanse, M. P. M., van der Ploeg, E. M. C., & Vreugdenhil, A. C. E. (2020). Narrative review: Nutrient deficiencies in adults and children with treated and untreated celiac disease. *Nutrients*, *12*(2), 500. doi:10.3390/nu12020500 PMID:32075276

Kupper, C. (2005). Dietary guidelines and implementation for celiac disease. *Gastroenterology, 128*(4), S121–S127. doi:10.1053/j.gastro.2005.02.024 PMID:15825119

Leonard, M. M., Sapone, A., Catassi, C., & Fasano, A. (2017). Celiac disease and nonceliac gluten sensitivity: A review. *Journal of the American Medical Association, 318*(7), 647–656. doi:10.1001/jama.2017.9730 PMID:28810029

Lutz, M., & Bascuñán-Godoy, L. (2017). The revival of quinoa: a crop for health. *Superfood and Functional Food—an Overview of Their Processing and Utilization.* InTech.

Makdoud, S., & Rosentrater, K. A. (2017). Development and Testing of Gluten-Free Pasta Based on Rice, Quinoa and Amaranth Flours. Iowa State University. *Agricultural and Biosystems Engineering, Food Science and Human Nutrition. Journal of Food Research, 6*(4), 887–1927. doi:10.5539/jfr.v6n4p91

Malekzadeh, R., Sachdev, A., & Ali, A. F. (2005). Coeliac disease in developing countries: Middle East, India and North Africa. *Best Practice & Research. Clinical Gastroenterology, 19*(3), 351–358. doi:10.1016/j.bpg.2005.01.004 PMID:15925841

Martínez-Villaluenga, C., Peñas, E., & Hernández-Ledesma, B. (2020). Pseudocereal grains: Nutritional value, health benefits and current applications for the development of gluten-free foods. *Food and Chemical Toxicology, 137*, 111178. doi:10.1016/j.fct.2020.111178 PMID:32035214

Maurya, N. K., & Arya, P. (2018). Amaranthus grain nutritional benefits: A review. *Journal of Pharmacognosy and Phytochemistry, 7*(2), 2258–2262.

Melini, V., & Melini, F. (2019). Gluten-free diet: Gaps and needs for a healthier diet. *Nutrients, 11*(1), 170. doi:10.3390/nu11010170 PMID:30650530

Menta, P. L. R., Andrade, M. E. R., Leocádio, P. C. L., Fraga, J. R., Dias, M. T. S., Cara, D. C., Cardoso, V. N., Borges, L. F., Capettini, L. S. A., Aguilar, E. C., & Alvarez-Leite, J. I. (2019). Wheat gluten intake increases the severity of experimental colitis and bacterial translocation by weakening of the proteins of the junctional complex. *British Journal of Nutrition, 121*(4), 361–373. doi:10.1017/S0007114518003422 PMID:30554574

Metin, S. (2016). Çölyak hastalığında nutrisyon. *Güncel Gastroenteroloji, 20*(3), 259–262.

Mulder, C. J. J., Van Bergeijk, J. D., Jansen, T. L. T. A., & Uil, J. J. (1993). Coeliac Disease. *Scandinavian Journal of Gastroenterology, 28*(sup200), 42–47.

Naqash, F., Gani, A., Gani, A., & Masoodi, F. A. (2017). Gluten-free baking: Combating the challenges-A review. *Trends in Food Science & Technology, 66*, 98–107. doi:10.1016/j.tifs.2017.06.004

Nehra, V., Marietta, E. V, & Murray, J. A. (2014). Celiac disease and its therapy: current approaches and new advances. *Wheat and Rice in Disease Prevention and Health*, 143–155.

Newberry, C., McKnight, L., Sarav, M., & Pickett-Blakely, O. (2017). Going gluten free: The history and nutritional implications of today's most popular diet. *Current Gastroenterology Reports*, *19*(11), 1–8. doi:10.100711894-017-0597-2 PMID:28948465

Niland, B., & Cash, B. D. (2018). Health benefits and adverse effects of a gluten-free diet in non–celiac disease patients. *Gastroenterology & Hepatology*, *14*(2), 82. PMID:29606920

Opie, R. S., O'Neil, A., Itsiopoulos, C., & Jacka, F. N. (2015). The impact of whole-of-diet interventions on depression and anxiety: A systematic review of randomised controlled trials. *Public Health Nutrition*, *18*(11), 2074–2093. doi:10.1017/S1368980014002614 PMID:25465596

Özkaya, V., & Özkaya, Ş. Ö. (2018). Çölyak Hastalığına Diyetetik Yaklaşım. *Selcuk University Medical Journal, 34*(4).

Padalino, L., Conte, A., & Del Nobile, M. A. (2016). Overview on the general approaches to improve gluten-free pasta and bread. *Foods*, *5*(4), 87. doi:10.3390/foods5040087 PMID:28231182

Patel, N. K., & Lacy, B. E. (2018). Another reason to avoid the gluten-free fad? *Clinical Gastroenterology and Hepatology*, *16*(2), 184–185. doi:10.1016/j.cgh.2017.10.002 PMID:28987505

Pellecchia, M. T., Scala, R., Perretti, A., De Michele, G., Santoro, L., Filla, A., Ciacci, C., & Barone, P. (1999). Cerebellar ataxia associated with subclinical celiac disease responding to gluten-free diet. *Neurology*, *53*(7), 1606. doi:10.1212/WNL.53.7.1606-a PMID:10534283

Pellegrini, N., & Agostoni, C. (2015). Nutritional aspects of gluten-free products. *Journal of the Science of Food and Agriculture*, *95*(12), 2380–2385. doi:10.1002/jsfa.7101 PMID:25615408

Penagini, F., Dilillo, D., Meneghin, F., Mameli, C., Fabiano, V., & Zuccotti, G. V. (2013). Gluten-free diet in children: An approach to a nutritionally adequate and balanced diet. *Nutrients*, *5*(11), 4553–4565. doi:10.3390/nu5114553 PMID:24253052

Sandri, L. T. B., Santos, F. G., Fratelli, C., & Capriles, V. D. (2017). Development of gluten-free bread formulations containing whole chia flour with acceptable sensory properties. *Food Science & Nutrition*, *5*(5), 1021–1028. doi:10.1002/fsn3.495 PMID:28948020

Sapone, A., Bai, J. C., Ciacci, C., Dolinsek, J., Green, P. H. R., Hadjivassiliou, M., Kaukinen, K., Rostami, K., Sanders, D. S., Schumann, M., Ullrich, R., Villalta, D., Volta, U., Catassi, C., & Fasano, A. (2012). Spectrum of gluten-related disorders: Consensus on new nomenclature and classification. *BMC Medicine*, *10*(1), 1–12. doi:10.1186/1741-7015-10-13 PMID:22313950

Saturni, L., Ferretti, G., & Bacchetti, T. (2010). The gluten-free diet: Safety and nutritional quality. *Nutrients*, *2*(1), 16–34. doi:10.3390/nu2010016 PMID:22253989

Sayed, H. S., Sakr, A. M., & Hassan, N. M. M. (2016). Effect of pseudo cereal flours on technological, chemical and sensory properties of pan bread. *World J. Dairy Food Sci*, *11*(1), 10–17.

Schober, T. J. (2009). Manufacture of gluten-free specialty breads and confectionery products. *Gluten-free. Food Science and Technology (Campinas)*, 130–180.

Schoenlechner, R., Drausinger, J., Ottenschlaeger, V., Jurackova, K., & Berghofer, E. (2010). Functional properties of gluten-free pasta produced from amaranth, quinoa and buckwheat. *Plant Foods for Human Nutrition (Dordrecht, Netherlands)*, *65*(4), 339–349. doi:10.1007/s11130-010-0194-0 PMID:20972627

See, J., & Murray, J. A. (2006). Gluten-free diet: The medical and nutrition management of celiac disease. *Nutrition in Clinical Practice*, *21*(1), 1–15. doi:10.1177/011542650602100101 PMID:16439765

Serin, Y., & Akbulut, G. (2017). Çölyak hastalığı ve glutensiz diyet tedavisine güncel yaklaşım. *Journal of the Turkish Society of Intensive Care*, *2*(3).

Singh, P., Arora, A., Strand, T. A., Leffler, D. A., Catassi, C., Green, P. H., Kelly, C. P., Ahuja, V., & Makharia, G. K. (2018). Global prevalence of celiac disease: Systematic review and meta-analysis. *Clinical Gastroenterology and Hepatology*, *16*(6), 823–836. doi:10.1016/j.cgh.2017.06.037 PMID:29551598

Sivaramakrishnan, H. P., Senge, B., & Chattopadhyay, P. K. (2004). Rheological properties of rice dough for making rice bread. *Journal of Food Engineering*, *62*(1), 37–45. doi:10.1016/S0260-8774(03)00169-9

Soares, F. L. P., de Oliveira Matoso, R., Teixeira, L. G., Menezes, Z., Pereira, S. S., Alves, A. C., Batista, N. V., de Faria, A. M. C., Cara, D. C., Ferreira, A. V. M., & Alvarez-Leite, J. I. (2013). Gluten-free diet reduces adiposity, inflammation and insulin resistance associated with the induction of PPAR-alpha and PPAR-gamma expression. *The Journal of Nutritional Biochemistry*, *24*(6), 1105–1111. doi:10.1016/j.jnutbio.2012.08.009 PMID:23253599

Solíz-Guerrero, J. B., de Rodriguez, D. J., Rodríguez-García, R., Angulo-Sánchez, J. L., & Méndez-Padilla, G. (2002). Quinoa saponins: concentration and composition analysis. International Journal of Trends in New Crops and New Uses.

Srichuwong, S., Curti, D., Austin, S., King, R., Lamothe, L., & Gloria-Hernandez, H. (2017). Physicochemical properties and starch digestibility of whole grain sorghums, millet, quinoa and amaranth flours, as affected by starch and non-starch constituents. *Food Chemistry*, *233*, 1–10. doi:10.1016/j.foodchem.2017.04.019 PMID:28530552

Tang, Y., & Tsao, R. (2017). Phytochemicals in quinoa and amaranth grains and their antioxidant, anti-inflammatory, and potential health beneficial effects: A review. *Molecular Nutrition & Food Research*, *61*(7), 1600767. doi:10.1002/mnfr.201600767 PMID:28239982

Theethira, T. G., & Dennis, M. (2015). Celiac disease and the gluten-free diet: Consequences and recommendations for improvement. *Digestive Diseases (Basel, Switzerland)*, *33*(2), 175–182. doi:10.1159/000369504 PMID:25925920

Türksoy, S., & Özkaya, B. (2006). Gluten ve çölyak hastalığı. Türkiye, 9, 24–26.

USDA. (2011). *National Nutrient Database for Standard Reference. Food Composition Database United States Department of Agriculture*. USDA.

van Hemert, S., Breedveld, A. C., Rovers, J. M. P., Vermeiden, J. P. W., Witteman, B. J. M., Smits, M. G., & de Roos, N. M. (2014). Migraine associated with gastrointestinal disorders: Review of the literature and clinical implications. *Frontiers in Neurology*, *5*, 241. doi:10.3389/fneur.2014.00241 PMID:25484876

Venskutonis, P. R., & Kraujalis, P. (2013). Nutritional components of amaranth seeds and vegetables: A review on composition, properties, and uses. *Comprehensive Reviews in Food Science and Food Safety*, *12*(4), 381–412. doi:10.1111/1541-4337.12021 PMID:33412681

Vierk, K. A., Koehler, K. M., Fein, S. B., & Street, D. A. (2007). Prevalence of self-reported food allergy in American adults and use of food labels. *The Journal of Allergy and Clinical Immunology*, *119*(6), 1504–1510. doi:10.1016/j.jaci.2007.03.011 PMID:17451802

Vilcacundo, R., & Hernández-Ledesma, B. (2017). Nutritional and biological value of quinoa (Chenopodium quinoa Willd.). *Current Opinion in Food Science*, *14*, 1–6. doi:10.1016/j.cofs.2016.11.007

Volta, U., Caio, G., De Giorgio, R., Henriksen, C., Skodje, G., & Lundin, K. E. (2015). Non-celiac gluten sensitivity: A work-in-progress entity in the spectrum of wheat-related disorders. *Best Practice & Research. Clinical Gastroenterology*, *29*(3), 477–491. doi:10.1016/j.bpg.2015.04.006 PMID:26060112

Wu, J. H. Y., Neal, B., Trevena, H., Crino, M., Stuart-Smith, W., Faulkner-Hogg, K., Louie, J. C. Y., & Dunford, E. (2015). Are gluten-free foods healthier than non-gluten-free foods? An evaluation of supermarket products in Australia. *British Journal of Nutrition*, *114*(3), 448–454. doi:10.1017/S0007114515002056 PMID:26119206

Yaver, E., & Bilgiçli, N. (2020). Pseudocereals: Composition, effect on nutrition-health and usage in cereal products. *Food and Health*, *6*(1), 41–56. doi:10.3153/FH20006

Yildirim, E. (2020). Çölyak hastalığı ve glutensiz besleme. *Genel Sağlık Bilimleri Dergisi*, *2*(3), 175–187.

Zimmer, K.-P. (2011). Nutrition and celiac disease. *Current Problems in Pediatric and Adolescent Health Care*, *41*(9), 244–247. doi:10.1016/j.cppeds.2011.04.004 PMID:21939908

Zuidmeer, L., Goldhahn, K., Rona, R. J., Gislason, D., Madsen, C., Summers, C., Sodergren, E., Dahlstrom, J., Lindner, T., Sigurdardottir, S. T., McBride, D., & Keil, T. (2008). The prevalence of plant food allergies: A systematic review. *The Journal of Allergy and Clinical Immunology*, *121*(5), 1210–1218. doi:10.1016/j.jaci.2008.02.019 PMID:18378288

Compilation of References

Abbasi, A., Sabahi, S., Bazzaz, S., Tajani, A. G., Lahouty, M., Aslani, R., & Hosseini, H. (2023). An edible coating utilizing Malva sylvestris seed polysaccharide mucilage and postbiotic from Saccharomyces cerevisiae var. boulardii for the preservation of lamb meat. *International Journal of Biological Macromolecules*, *246*, 125660. doi:10.1016/j.ijbiomac.2023.125660 PMID:37399877

Abbasi-Parizad, P., Scarafoni, A., Pilu, R., Scaglia, B., De Nisi, P., & Adani, F. (2022). The recovery from agro-industrial wastes provides different profiles of anti-inflammatory polyphenols for tailored applications. *Frontiers in Sustainable Food Systems*, *6*, 996562. doi:10.3389/fsufs.2022.996562

Abdullah Al-Dhabi, N., & Ponmurugan, K. (2019). Microwave assisted extraction and characterization of polysaccharide from waste jamun fruit seeds. *International Journal of Biological Macromolecules*, *152*, 1157–1163. doi:10.1016/j.ijbiomac.2019.10.204 PMID:31751731

Abuladze, T., Li, M., Menetrez, M. Y., Dean, T., Senecal, A., & Sulakvelidze, A. (2008). Bacteriophages reduce experimental contamination of hard surfaces, tomato, spinach, broccoli, and ground beef by Escherichia coli O157: H7. *Applied and Environmental Microbiology*, *74*(20), 6230–6238. doi:10.1128/AEM.01465-08 PMID:18723643

Acosta, E. (2009). Bioavailability of nanoparticles in nutrient and nutraceutical delivery. *Current Opinion in Colloid & Interface Science*, *14*(1), 3–15. doi:10.1016/j.cocis.2008.01.002

Addanki, M., Patra, P., & Kandra, P. (2022). Recent advances and applications of artificial intelligence and related technologies in the food industry. *Applied Food Research*, *2*(2), 100126. doi:10.1016/j.afres.2022.100126

Additives, E. P. o., & Feed, P. o. S. u. i. A., Bampidis, V., Azimonti, G., Bastos, M. d. L., Christensen, H., . . . López Puente, S. (2021). Safety and efficacy of a feed additive consisting on the bacteriophages PCM F/00069, PCM F/00070, PCM F/00071 and PCM F/00097 infecting Salmonella Gallinarum B/00111 (Bafasal®) for all avian species (Proteon Pharmaceuticals SA). *EFSA Journal*, *19*(5), e06534. PMID:34025802

Adeyeye, S. A. O., & Fayemi, O. E. (2019). Nanotechnology and food processing: Between innovations and consumer safety. *Journal of Culinary Science & Technology*, *17*(5), 435–452. doi:10.1080/15428052.2018.1476276

Afshari, A., Piroozehpoor, M., Jamshidi, A., Azizzadeh, M., & Sarafan Sadeghi, A. (2018). Effect of novel zein coating combined with different antioxidants (thymol and carvacrol) on the aflatoxin production of peanut. Journal of Nutrition. *Fasting and Health*, *6*(1), 23–29.

Aguirre-Joya, J. A., De Leon-Zapata, M. A., & Alvarez-Perez, O. B. (2018). Basic and applied concepts of edible packaging for foods. In Food packaging and preservation (pp. 1–61). Academic Press.

Ahari, H., Anvar, A. A., Ataee, M., & Naeimabadi, M. (2021). Employing Nanosilver, Nanocopper, and Nanoclays in Food Packaging Production: A Systematic Review. *Coatings*, *11*(5), 509. doi:10.3390/coatings11050509

Ahmadi, P., Jahanban-Esfahlan, A., Ahmadi, A., Tabibiazar, M., & Mohammadifar, M. (2020). Development of ethyl cellulose-based formulations: A perspective on the novel technical methods. *Food Reviews International*, 1–48.

Ahmad, T., Aadil, R. M., Ahmed, H., Ur Rahman, U., Soares, B. C., Souza, S. L., ... Esmerino, E. A. (2019). Treatment and utilisation of dairy industrial waste: A review. *Trends in Food Science & Technology*, *88*, 361–372. doi:10.1016/j.tifs.2019.04.003

Aki, K., & Richards, P. G. (1980). Quantitative seismology. W.H. Freeman and Co.

Alarcon-Rojo, A. D., Carrillo-Lopez, L. M., Reyes-Villagrana, R., Huerta-Jiménez, M., & Garcia-Galicia, I. A. (2019). Ultrasound and meat quality: A Review. *Ultrasonics Sonochemistry*, *55*, 369–382. doi:10.1016/j.ultsonch.2018.09.016 PMID:31027999

Albdiry, M. T., Yousif, B. F., Ku, H., & Lau, K. T. (2013). A critical review on the manufacturing processes in relation to the properties of nanoclay/polymer composites. *Journal of Composite Materials*, *47*(9), 1093–1115. doi:10.1177/0021998312445592

Albert-Weiß, D., Hajdini, E., Heinrich, M., & Osman, A. (2020). *Cnn for ripeness classfication of watermelon fruits based on acoustic testing*. Paper presented at the Proceedings of the Virtual 3rd International Symposium on Structural Health Monitoring and Nondestructive Testing (SHM-NDT 2020), Quebeq, QU, Canada.

Alemayehu, D., Casey, P. G., McAuliffe, O., Guinane, C. M., Martin, J. G., Shanahan, F., Coffey, A., Ross, R. P., & Hill, C. (2012). Bacteriophages ϕMR299-2 and ϕNH-4 can eliminate Pseudomonas aeruginosa in the murine lung and on cystic fibrosis lung airway cells. *mBio*, *3*(2), 00029–00012. doi:10.1128/mBio.00029-12 PMID:22396480

Alencar, N. M. M., Steel, C. J., Alvim, I. D., de Morais, E. C., & Bolini, H. M. A. (2015). Addition of quinoa and amaranth flour in gluten-free breads: Temporal profile and instrumental analysis. *Lebensmittel-Wissenschaft + Technologie*, *62*(2), 1011–1018. doi:10.1016/j.lwt.2015.02.029

Alexandre, A.M.R.C., Serraa, A.T., Matiasa, A.A., Duarte, C.M.M., & Bronze, M.R. (2020). Supercritical fluid extraction of Arbutus unedo distillate residues – impact of process conditions on antiproliferative response of extracts. *J. CO2 Util.*, *37*, 29–38.

Alexieva, I., Baeva, M., Popova, A., Fidan, H., Goranova, Z., & Milkova-Tomova, I. (2022). Development and Application of Edible Coatings with Malva sylvestris L. Extract to Extend Shelf-Life of Small Loaf. *Foods, 11*(23), 3831. doi:10.3390/foods11233831 PMID:36496640

Al-Hassan, A. A., & Norziah, M. H. (2012). Starch–gelatin edible films: Water vapor permeability and mechanical properties as affected by plasticizers. *Food Hydrocolloids, 26*(1), 108–117. doi:10.1016/j.foodhyd.2011.04.015

Ali, M. H., Batai, S., & Sarbassov, D. (2019). 3D printing: A critical review of current development and future prospects. *Rapid Prototyping Journal, 25*(6), 1108–1126. doi:10.1108/RPJ-11-2018-0293

Alonso, C. A., González-Barrio, D., Tenorio, C., Ruiz-Fons, F., & Torres, C. (2016). Antimicrobial resistance in faecal Escherichia coli isolates from farmed red deer and wild small mammals. Detection of a multiresistant E. coli producing extended-spectrum beta-lactamase. *Comparative Immunology, Microbiology and Infectious Diseases, 45*, 34–39. doi:10.1016/j.cimid.2016.02.003 PMID:27012919

Alonso-Riaño, P., Sanz Diez, M. T., Blanco, B., Beltrán, S., Trigueros, E., & Benito-Román, O. (2020). Water ultrasound-assisted extraction of polyphenol compounds from Brewer's spent grain: Kinetic study, extract characterization, and concentration. *Antioxidants, 9*(3), 265. doi:10.3390/antiox9030265 PMID:32210202

Alparslan, Y., & Baygar, T. (2017). Effect of chitosan film coating combined with orange peel essential oil on the shelf life of deepwater pink shrimp. *Food and Bioprocess Technology, 10*(5), 842–853. doi:10.100711947-017-1862-y

Alpas, H., Kalchayanand, N., Bozoglu, F., Sikes, A., Dunne, C. P., & Ray, B. (1999). Variation in resistance to hydrostatic pressure among strains of food-borne pathogens. *Applied and Environmental Microbiology, 65*(9), 4248–4251. doi:10.1128/AEM.65.9.4248-4251.1999 PMID:10473446

Alraih, S., Shayea, I., Behjati, M., Nordin, R., Abdullah, N. F., Abu-Samah, A., & Nandi, D. (2022). Revolution or evolution? Technical requirements and considerations towards 6G mobile communications. *Sensors (Basel), 22*(3), 762. doi:10.339022030762 PMID:35161509

Altındağ, G., Certel, M., Erem, F., & İlknur Konak, Ü. (2015). Quality characteristics of gluten-free cookies made of buckwheat, corn, and rice flour with/without transglutaminase. *Food Science & Technology International, 21*(3), 213–220. doi:10.1177/1082013214525428 PMID:24613830

Alvarez-Jubete, L., Arendt, E. K., & Gallagher, E. (2010). Nutritive value of pseudocereals and their increasing use as functional gluten-free ingredients. *Trends in Food Science & Technology, 21*(2), 106–113. doi:10.1016/j.tifs.2009.10.014

Alvarez, M. V., Cabred, S., Ramirez, C. L., & Fanovich, M. A. (2019). Valorization of an agroindustrial soybean residue by supercritical fluid extraction of phytochemical compounds. *The Journal of Supercritical Fluids, 143*, 90–96. doi:10.1016/j.supflu.2018.07.012

Alverez, R. (1990). Structure of the basin of Mexico City and its relation to destruction in the earthquake of 1985. *Geotechnical and Environmental Geophysics*, *3*, 263–279. doi:10.1190/1.9781560802785.ch21

Ameen, N., Tarhini, A., Reppel, A., & Anand, A. (2021a). Customer experiences in the age of artificial intelligence. *Computers in Human Behavior*, *114*, 106548. doi:10.1016/j.chb.2020.106548 PMID:32905175

Ameer, K., Shahbaz, H. M., & Kwon, J. H. (2017). Green extraction methods for polyphenols from plant matrices and their byproducts: A review. *Comprehensive Reviews in Food Science and Food Safety*, *16*(2), 295–315. doi:10.1111/1541-4337.12253 PMID:33371540

Amendola, D., De Faveri, D. M., & Spigno, G. (2010). Grape marc phenolics: Extraction kinetics, quality and stability of extracts. *Journal of Food Engineering*, *97*(3), 384–392. doi:10.1016/j.jfoodeng.2009.10.033

Ameta, S. K., Rai, A. K., Hiran, D., Ameta, R., & Ameta, S. C. (2020). Use of nanomaterials in food science. *Biogenic nano-particles and their use in agro-ecosystems*, 457-488.

Anandharamakrishnan, C. (2013). Nanoencapsulation of food bioactive compounds. *Techniques for Nanoencapsulation of Food Ingredients*, 1–6. doi:10.1007/978-1-4614-9387-7_1

Andreou, V., Dimopoulos, G., Dermesonlouoglou, E., & Taoukis, P. (2020). Application of pulsed electric fields to improve product yield and waste valorization in industrial tomato processing. *Journal of Food Engineering*, *270*, 109778. doi:10.1016/j.jfoodeng.2019.109778

Andrus, R. D., & Stokoe, K. H. II. (2000). Liquefaction resistance of soils from shear-wave velocity. *Journal of Geotechnical and Geoenvironmental Engineering*, *126*(11), 1015–1025. doi:10.1061/(ASCE)1090-0241(2000)126:11(1015)

Angellier-Coussy, H., Guillard, V., Guillaume, C., & Gontard, N. (2011). 23-Wheat gluten (WG)-based materials for food packaging. In J.-M. Lagaron (Ed.), Multifunctional and nanoreinforced polymers for food packaging (pp. 649–668). Woodhead Publishing.

Anonymous. (2010). Guidelines for the diagnosis and management of food allergy in the United States: Report of the NIAID-sponsored expert panel. *The Journal of Allergy and Clinical Immunology*, *126*(6), S1–S58. doi:10.1016/j.jaci.2010.10.007 PMID:21134576

Ansari, A. A., Alhoshan, M., Alsalhi, M. S., & Aldwayyan, A. S. (2010). Prospects of nanotechnology in clinical immunodiagnostics. *Sensors (Basel)*, *10*(7), 6535–6581. doi:10.3390100706535 PMID:22163566

Ansari, M. A. (2023). Nanotechnology in food and plant science: Challenges and future prospects. *Plants*, *12*(13), 2565. doi:10.3390/plants12132565 PMID:37447126

Anvar, A. A., Ahari, H., & Ataee, M. (2021). Antimicrobial properties of food nanopackaging: A new focus on foodborne pathogens. *Frontiers in Microbiology*, *12*, 690706. doi:10.3389/fmicb.2021.690706 PMID:34322104

Anwar, S., & Prasad, R. (2018). Framework for future telemedicine planning and infrastructure using 5G technology. *Wireless Personal Communications, 100*(1), 193–208. doi:10.100711277-018-5622-8

Aouafi, R., Zaidi, A., Kouachi, S., & Parshad, R. D. (2023). A remark on "Dynamical behavior of a fractional three-species food chain model". *Nonlinear Dynamics, 111*(14), 13641–13651. doi:10.100711071-023-08546-z

Arendt, E., & Dal Bello, F. (2011). *Gluten-free cereal products and beverages*. Elsevier.

Arora, A., Choudhary, V., & Sharma, D. K. (2011). Effect of clay content and clay/surfactant on the mechanical, thermal and barrier properties of polystyrene/organoclay nanocomposites. *Journal of Polymer Research, 18*(4), 843–857. doi:10.100710965-010-9481-6

Arrubla-Hoyos, W., Ojeda-Beltrán, A., Solano-Barliza, A., Rambauth-Ibarra, G., Barrios-Ulloa, A., Cama-Pinto, D., Arrabal-Campos, F. M., Martínez-Lao, J. A., Cama-Pinto, A., & Manzano-Agugliaro, F. (2022). Precision Agriculture and Sensor Systems Applications in Colombia through 5G Networks. *Sensors (Basel), 22*(19), 7295. doi:10.339022197295 PMID:36236394

Arshad, Q. K. U. D., Kashif, A. U., & Quershi, I. M. (2019, January). A review on the evolution of cellular technologies. In *2019 16th International Bhurban Conference on Applied Sciences and Technology (IBCAST)* (pp. 989-993). IEEE.

Arshad, J., Rehman, A. U., Othman, M. T. B., Ahmad, M., Tariq, H. B., Khalid, M. A., Moosa, M. A. R., Shafiq, M., & Hamam, H. (2022). Deployment of wireless sensor network and iot platform to implement an intelligent animal monitoring system. *Sustainability (Basel), 14*(10), 6249. doi:10.3390u14106249

Artawinata, P. C., Lorraine, S., & Waturangi, D. E. (2023). Isolation and characterization of bacteriophages from soil against food spoilage and foodborne pathogenic bacteria. *Scientific Reports, 13*(1), 9282. doi:10.103841598-023-36591-6 PMID:37286897

Arthur, T. M., Kalchayanand, N., Agga, G. E., Wheeler, T. L., & Koohmaraie, M. (2017). Evaluation of bacteriophage application to cattle in lairage at beef processing plants to reduce Escherichia coli O157: H7 prevalence on hides and carcasses. *Foodborne Pathogens and Disease, 14*(1), 17–22. doi:10.1089/fpd.2016.2189 PMID:27870549

Aruldoss, M., Lakshmi Travis, M., & Prasanna Venkatesan, V. (2014). A survey on recent research in business intelligence. *Journal of Enterprise Information Management, 27*(6), 831–866. doi:10.1108/JEIM-06-2013-0029

Arun, K. B., Madhavan, A., Sindhu, R., Binod, P., Pandey, A., Reshmy, R., & Sirohi, R. (2020). Remodeling agro-industrial and food wastes into value-added bioactives and biopolymers. *Industrial Crops and Products, 154*, 112621. doi:10.1016/j.indcrop.2020.112621

Arunvisut, S., Phummanee, S., & Somwangthanaroj, A. (2007). Effect of clay on mechanical and gas barrier properties of blown film LDPE/clay nanocomposites. *Journal of Applied Polymer Science, 106*(4), 2210–2217. doi:10.1002/app.26839

Aslam, F., Aimin, W., Li, M., Ur Rehman, K. (2020). Innovation in the Era of IoT and Industry 5.0: Absolute Innovation Management (AIM) Framework. *Information, 11*(2), 124. doi:10.3390/info11020124

Astm, D. (2004). *04 Standard Specification for Compostable Plastics*. J ASTM Int.

Ateş, E., & Uyanık, O. (2019). Jeofizik yöntemler ile yer ve yapı etkileşimi. *SDÜ Fen Bilimleri Enstitüsü Dergisi, 23*, 46–60.

Ateş, E., & Uyanık, O. (2023). (in press). Comparison of acceleration records measured at different depths. *Turkish Journal of Earthquake Research.*

Atieno, L., Owino, W., Ateka, E. M., & Ambuko, J. (2019). Influence of coating application methods on the postharvest quality of cassava. *International Journal of Food Sciences, 2019*, 1–16. doi:10.1155/2019/2148914 PMID:30863776

Auta, M., Musa, U., Tsado, D. G., Faruq, A. A., Isah, A. G., Raji, S., & Nwanisobi, C. (2018). Optimization of citrus peels D-limonene extraction using solvent-free microwave green technology. *Chemical Engineering Communications, 205*(6), 789–796. doi:10.1080/00986445.2017.1419206

Authority, E. F. S. (2009). The use and mode of action of bacteriophages in food production-Endorsed for public consultation 22 January 2009-Public consultation 30 January–6 March 2009. *EFSA Journal, 7*(5), 1076. doi:10.2903/j.efsa.2009.1076

Awad, W. H., Beyer, G., Benderly, D., Ijdo, W. L., Songtipya, P., del Mar Jimenez-Gasco, M., Manias, E., & Wilkie, C. A. (2009). Material properties of nanoclay PVC composites. *Polymer, 50*(8), 1857–1867. doi:10.1016/j.polymer.2009.02.007

Awasthi, S. K., Sarsaiya, S., Awasthi, M. K., Liu, T., Zhao, J., Kumar, S., & Zhang, Z. (2020). Changes in global trends in food waste composting: Research challenges and opportunities. *Bioresource Technology, 299*, 122555. doi:10.1016/j.biortech.2019.122555 PMID:31866141

Axel, C., Röcker, B., Brosnan, B., Zannini, E., Furey, A., Coffey, A., & Arendt, E. K. (2015). Application of Lactobacillus amylovorus DSM19280 in gluten-free sourdough bread to improve the microbial shelf life. *Food Microbiology, 47*, 36–44. doi:10.1016/j.fm.2014.10.005 PMID:25583336

Ayana, B., Suin, S., & Khatua, B. B. (2014). Highly exfoliated eco-friendly thermoplastic starch (TPS)/poly (lactic acid)(PLA)/clay nanocomposites using unmodified nanoclay. *Carbohydrate Polymers, 110*, 430–439. doi:10.1016/j.carbpol.2014.04.024 PMID:24906776

Aydar, A. Y., Akgün, A., & Dengiz, İ. T. (2019). Glutensiz diyetde kullanilan alternatif hammaddeler. *Geleceğin Dünyasinda*, 20.

Aziez, M., Benharzallah, S., & Bennoui, H. (2019). A full comparison study of service discovery approaches for internet of things. *International Journal of Pervasive Computing and Communications, 15*(1), 30–56. doi:10.1108/IJPCC-04-2019-0038

Azman, N., Khairul, W. M., & Sarbon, N. (2022). A comprehensive review on biocompatible film sensor containing natural extract: Active/intelligent food packaging. *Food Control, 141*, 109189. doi:10.1016/j.foodcont.2022.109189

Babu, S. Z. D., Pandey, D., Naidu, G. T., Sumathi, S., Gupta, A., Bader Alazzam, M., & Pandey, B. K. (2022, July). Analysation of Big Data in Smart Healthcare. In *Artificial Intelligence on Medical Data: Proceedings of International Symposium, ISCMM 2021* (pp. 243-251). Singapore: Springer Nature Singapore.

Bai, C., Dallasega, P., Orzes, G., & Sarkis, J. (2020). Industry 4.0 technologies assessment: A sustainability perspective. *International Journal of Production Economics, 229*, 107776. doi:10.1016/j.ijpe.2020.107776

Bai, X., Nakatsu, C. H., & Bhunia, A. K. (2021). Bacterial biofilms and their implications in pathogenesis and food safety. *Foods, 10*(9), 2117. doi:10.3390/foods10092117 PMID:34574227

Bajpai, V. K., Kamle, M., Shukla, S., Mahato, D. K., Chandra, P., Hwang, S. K., ... Han, Y. K. (2018). Prospects of using nanotechnology for food preservation, safety, and security. *Yao Wu Shi Pin Fen Xi, 26*(4), 1201–1214. PMID:30249319

Baldwin, E. A., Hagenmaier, R., & Bai, J. (2012). *Edible Coatings and Films to Improve Food Quality* (2nd ed.). CRC Press.

Balieiro, A., Santos, R., Pereira, M., Figueiredo, R., Freitas, L., De Alsina, O., & ... . (2016). Adsorption process of molecularly imprinted silica for extraction of lactose from milk. *Brazilian Journal of Chemical Engineering, 33*(2), 361–372. doi:10.1590/0104-6632.20160332s20140089

Ballabio, C., Uberti, F., Di Lorenzo, C., Brandolini, A., Penas, E., & Restani, P. (2011). Biochemical and immunochemical characterization of different varieties of amaranth (Amaranthus L. ssp.) as a safe ingredient for gluten-free products. *Journal of Agricultural and Food Chemistry, 59*(24), 12969–12974. doi:10.1021/jf2041824 PMID:22073907

Ballesteros-Vivas, D., Alvarez-Rivera, G., & Medina, S. J. M. (2019). An integrated approach for the valorization of mango seed kernel: Efficient extraction solvent selection, phytochemical profiling and antiproliferative activity assessment. *Food Research International, 126*, 108616. doi:10.1016/j.foodres.2019.108616 PMID:31732074

Banafaa, M., Shayea, I., Din, J., Azmi, M. H., Alashbi, A., Daradkeh, Y. I., & Alhammadi, A. (2022). 6G mobile communication technology: Requirements, targets, applications, challenges, advantages, and opportunities. *Alexandria Engineering Journal*.

Bandini, L. G., Anderson, S. E., Curtin, C., Cermak, S., Evans, E. W., Scampini, R., Maslin, M., & Must, A. (2010). Food selectivity in children with autism spectrum disorders and typically developing children. *The Journal of Pediatrics, 157*(2), 259–264. doi:10.1016/j.jpeds.2010.02.013 PMID:20362301

Baranwal, A., Srivastava, A., Kumar, P., Bajpai, V. K., Maurya, P. K., & Chandra, P. (2018). Prospects of nanostructure materials and their composites as antimicrobial agents. *Frontiers in Microbiology*, *9*, 422. doi:10.3389/fmicb.2018.00422 PMID:29593676

Barba, F. J., Zhu, Z., Koubaa, M., Sant'ana, A. S., & Orlien, V. (2016). Green alternative methods for the extraction of antioxidant bioactive compounds from winery wastes and by products: A review. *Trends in Food Science & Technology*, *49*, 96–109. doi:10.1016/j.tifs.2016.01.006

Baruah, S., & Dutta, J. (2009). Nanotechnology applications in pollution sensing and degradation in agriculture: A Review. *Environmental Chemistry Letters*, *7*(3), 191–204. doi:10.100710311-009-0228-8

Bas-Bellver, C., Barrera, C., Betoret, N., & Seguí, L. (2020). Turning agri-food cooperative vegetable residues into functional powdered ingredients for the food industry. *Sustainability (Basel)*, *12*(4), 1284. doi:10.3390u12041284

Bastidas, E., Roura, R., Rizzolo, D., Massanés, T., & Gomis, R. (2016). Quinoa (Chenopodium quinoa Willd), from nutritional value to potential health benefits: An integrative review. *Journal of Nutrition & Food Sciences*, *6*(3).

Bastos, G. M., Júnior, M. S. S., Caliari, M., de Araujo Pereira, A. L., de Morais, C. C., & Campos, M. R. H. (2016). Physical and sensory quality of gluten-free spaghetti processed from amaranth flour and potato pulp. *Lebensmittel-Wissenschaft + Technologie*, *65*, 128–136. doi:10.1016/j.lwt.2015.07.067

Baum, S. E., Machalaba, C., Daszak, P., Salerno, R. H., & Karesh, W. B. (2017). Evaluating one health: Are we demonstrating effectiveness? *One Health*, *3*, 5–10. doi:10.1016/j.onehlt.2016.10.004 PMID:28616496

Bautista-Espinoza, P. I., Di Pierro, P., Amaya-Llano, S. L., García-Almendarez, B. E., Mares-Mares, E., Escamilla-García, M., Granados-Arvizu, J. A., Velázquez de la Cruz, G., & Regalado-González, C. (2023). Impact of edible coatings reinforced with mesoporous silica nanoparticles added with cinnamon and lemongrass essential oils applied on sourdough bread. *Food Bioscience*, *55*, 102992. doi:10.1016/j.fbio.2023.102992

Baykut, E. (2021). Bazı tahıl benzeri ürünlerin besin içeriği ve gıda endüstrisinde kullanımı. *Avrupa Bilim ve Teknoloji Dergisi*, *23*, 89–98.

Benbettaïeb, N., Debeaufort, F., & Karbowiak, T. (2019). Bioactive edible films for food applications: Mechanisms of antimicrobial and antioxidant activity. *Critical Reviews in Food Science and Nutrition*, *59*(21), 3431–3455. doi:10.1080/10408398.2018.1494132 PMID:29999405

Benítez, J. J., Castillo, P. M., Del Río, J. C., León-Camacho, M., Domínguez, E., Heredia, A., & Heredia-Guerrero, J. A. (2018). Valorization of tomato processing by-products: Fatty acid extraction and production of bio-based materials. *Materials (Basel)*, *11*(11), 2211. doi:10.3390/ma11112211 PMID:30405081

Benito, A., Ventoura, G., Casadei, M., Robinson, T., & Mackey, B. (1999). Variation in resistance of natural isolates of *escherichia coli* O157 to high hydrostatic pressure, mild heat, and other stresses. *Applied and Environmental Microbiology*, *65*(4), 1564–1569. doi:10.1128/AEM.65.4.1564-1569.1999 PMID:10103251

Ben-Othman, S., Jõudu, I., & Bhat, R. (2020). Bioactives from agrifood wastes: Present insights and future challenges. *Molecules (Basel, Switzerland)*, *25*(3), 510. doi:10.3390/molecules25030510 PMID:31991658

Bensadoun, Kchit, Billotte, Trochu, & Ruiz. (2011). A comparative study of dispersion techniques for nanocomposite made with nanoclays and an unsaturated polyester resin. *Journal of Nanomaterials*.

Berekaa, M. M. (2015). Nanotechnology in food industry; advances in food processing, packaging and food safety. *International Journal of Current Microbiology and Applied Sciences*, *4*(5), 345–357.

Berezina, K., Ciftci, O., & Cobanoglu, C. (2019). *Robots, artificial intelligence, and service automation in travel, tourism and hospitality*. Emerald Publishing Limited.

Bethune, M. T., & Khosla, C. (2008). Parallels between pathogens and gluten peptides in celiac sprue. *PLoS Pathogens*, *4*(2), e34. doi:10.1371/journal.ppat.0040034 PMID:18425213

Bharti, S. K., Pathak, V., Alam, T., Arya, A., Basak, G., & Awasthi, M. G. (2020). Materiality of edible film packaging in muscle foods: A worthwhile conception. *Journal of Packaging Technology and Research*, *4*(1), 117–132. doi:10.100741783-020-00087-9

Bhavani, B., Reddy, D. K., & Waghray, K. (2018). Energy bars made with popped with amranth. *International Journal of Food and Fermentation Technology*, *8*(2), 223–228. doi:10.30954/2277-9396.02.2018.13

Biesiekierski, J. R. (2017). What is gluten? *Journal of Gastroenterology and Hepatology*, *32*(S1), 78–81. doi:10.1111/jgh.13703 PMID:28244676

Biesiekierski, J. R., Muir, J. G., & Gibson, P. R. (2013). Is gluten a cause of gastrointestinal symptoms in people without celiac disease? *Current Allergy and Asthma Reports*, *13*(6), 631–638. doi:10.100711882-013-0386-4 PMID:24026574

Biesiekierski, J. R., Peters, S. L., Newnham, E. D., Rosella, O., Muir, J. G., & Gibson, P. R. (2013). No effects of gluten in patients with self-reported non-celiac gluten sensitivity after dietary reduction of fermentable, poorly absorbed, short-chain carbohydrates. *Gastroenterology*, *145*(2), 320–328. doi:10.1053/j.gastro.2013.04.051 PMID:23648697

Bikiaris, D. N., & Triantafyllidis, K. S. (2013). HDPE/Cu-nanofiber nanocomposites with enhanced antibacterial and oxygen barrier properties appropriate for food packaging applications. *Materials Letters*, *93*, 1–4. doi:10.1016/j.matlet.2012.10.128

Bilgicli, N. (2008). *Utilization of buckwheat flour in gluten-free egg noodle production*. Academic Press.

Bilgiçli, N. (2020). Tahıl benzeri ürünler: Bileşimi, beslenme-sağlık üzerine etkileri ve tahıl ürünlerinde kullanımı. *Food and Health*, *6*(1), 41–56.

Bintsis, T. (2017). Foodborne pathogens. *AIMS Microbiology*, *3*(3), 529–563. doi:10.3934/microbiol.2017.3.529 PMID:31294175

Blagoveshchenskiy, I., Blagoveshchenskiy, V., Besfamilnaya, E., & Sumerin, V. (2020). *Development of databases of intelligent expert systems for automatic control of product quality indicators.* Paper presented at the Journal of physics: conference series. 10.1088/1742-6596/1705/1/012019

Blasi, A. (2018). Scheduling food industry system using fuzzy logic. *Journal of Theoretical and Applied Information Technology*, *96*(19), 6463–6473.

Boettcher, E., & Crowe, S. E. (2013). Celiac Disease: Dispelling Misconceptions and Myths. In Celiac Disease (pp. 255–273). Springer.

Bonaterra, A., Badosa, E., Daranas, N., Francés, J., Roselló, G., & Montesinos, E. (2022). Bacteria as biological control agents of plant diseases. *Microorganisms*, *10*(9), 1759. doi:10.3390/microorganisms10091759 PMID:36144361

Boopathi, S., Pandey, B. K., & Pandey, D. (2023). Advances in Artificial Intelligence for Image Processing: Techniques, Applications, and Optimization. In Handbook of Research on Thrust Technologies' Effect on Image Processing (pp. 73-95). IGI Global.

Boukid, F., Folloni, S., Sforza, S., Vittadini, E., & Prandi, B. (2018). Current trends in ancient grains-based foodstuffs: Insights into nutritional aspects and technological applications. *Comprehensive Reviews in Food Science and Food Safety*, *17*(1), 123–136. doi:10.1111/1541-4337.12315 PMID:33350067

Boussetta, N., Soichi, E., Lanoisellé, J.-L., & Vorobiev, E. (2014). Valorisation of oilseed residues: Extraction of polyphenols from flaxseed hulls by pulsed electric fields. *Industrial Crops and Products*, *52*, 347–353. doi:10.1016/j.indcrop.2013.10.048

Boyacioglu, O., Sulakvelidze, A., Sharma, M., & Goktepe, I. (2016). Effect of a bacteriophage cocktail in combination with modified atmosphere packaging in controlling Listeria monocytogenes on fresh-cut spinach. *Irish Journal of Agricultural and Food Research*, *55*(1), 74–79. doi:10.1515/ijafr-2016-0007

Bozcu, M., Uyanık, O., Çakmak, O., & Türker, A. E. (2007). Geotechnical properties of Esen I HEPP project field. *SDU Journal of Natural and Applied Sciences*, *11*(1), 75–83.

Brabban, A., Hite, E., & Callaway, T. (2005). Evolution of foodborne pathogens via temperate bacteriophage-mediated gene transfer. *Foodborne Pathogens and Disease*, *2*(4), 287–303. doi:10.1089/fpd.2005.2.287 PMID:16366852

Braeken, A., & Liyanage, M. (2021). Highly efficient key agreement for remote patient monitoring in MEC-enabled 5G networks. *The Journal of Supercomputing*, *77*(6), 5562–5585. doi:10.100711227-020-03472-y

Branton, D., Deamer, D. W., Marziali, A., Bayley, H., Benner, S. A., Butler, T., Di Ventra, M., Garaj, S., Hibbs, A., Huang, X., Jovanovich, S. B., Krstic, P. S., Lindsay, S., Ling, X. S., Mastrangelo, C. H., Meller, A., Oliver, J. S., Pershin, Y. V., Ramsey, J. M., ... Schloss, J. A. (2008). The potential and challenges of nanopore sequencing. *Nature Biotechnology*, *26*(10), 1146–1153. doi:10.1038/nbt.1495 PMID:18846088

Breque, M., De Nul, L., & Petridis, A. (2021). *Industry 5.0: towards a sustainable, human-centric and resilient European industry*. European Commission, Directorate-General for Research and Innovation.

Briano, R., Giuggioli, N. R., Girgenti, V., & Peano, C. (2015). Biodegradable and compostable film and modified atmosphere packaging in postharvest supply chain of Raspberry Fruits (cv. Grandeur). *Journal of Food Processing and Preservation*, *39*(6), 2061–2073. doi:10.1111/jfpp.12449

Brito, I. L., de Souza, E. L., Felex, S. S. S., Madruga, M. S., Yamashita, F., & Magnani, M. (2015). Nutritional and sensory characteristics of gluten-free quinoa (Chenopodium quinoa Willd)-based cookies development using an experimental mixture design. *Journal of Food Science and Technology*, *52*(9), 5866–5873. doi:10.100713197-014-1659-1 PMID:26345002

Brown, G., Analyst, P., & Reading, H. (2018). Ultra-reliable low-latency 5G for industrial automation. *Technol. Rep. Qualcomm*, *2*, 52065394.

Brown, S. R., Kozak, S. M., & D'Amico, D. J. (2018). Applications of edible coatings formulated with antimicrobials inhibit Listeria monocytogenes growth on Queso Fresco. *Frontiers in Sustainable Food Systems*, *2*, 1. doi:10.3389/fsufs.2018.00001

BSSC (Building Seismic Safety Council). (1997). *NEHRP Recommended Provisions for Seismic Regulations for New Buildings and Other Structures. Part I, Provisions (FEMA 302), 334*. Author.

Bulka, C. M., Davis, M. A., Karagas, M. R., Ahsan, H., & Argos, M. (2017). The unintended consequences of a gluten-free diet. *Epidemiology (Cambridge, Mass.)*, *28*(3), e24–e25. doi:10.1097/EDE.0000000000000640 PMID:28166100

Bullen, K. E., & Bolt, B. A. (1985). An Introduction to the Theory of Seismology. Cambridge Univ. Press.

Bumbudsanpharoke, N., & Ko, S. (2015). Nano-food packaging: An overview of market, migration research, and safety regulations. *Journal of Food Science*, *80*(5), R910–R923. doi:10.1111/1750-3841.12861 PMID:25881665

Bumbudsanpharoke, N., & Ko, S. (2019). Nanoclays in Food and Beverage Packaging. *Journal of Nanomaterials*, *2019*, 8927167. doi:10.1155/2019/8927167

Bumbudsanpharoke, N., Lee, W., Choi, J. C., Park, S. J., Kim, M., & Ko, S. (2017). Influence of montmorillonite nanoclay content on the optical, thermal, mechanical, and barrier properties of low-density polyethylene. *Clays and Clay Minerals*, *65*(6), 387–397. doi:10.1346/CCMN.2017.064071

Bumunang, E. W., Ateba, C. N., Stanford, K., Niu, Y. D., Wang, Y., & McAllister, T. A. (2020). Activity of bacteriophage and complex tannins against biofilm-forming shiga toxin-producing Escherichia coli from Canada and South Africa. *Antibiotics (Basel, Switzerland)*, *9*(5), 257. doi:10.3390/antibiotics9050257 PMID:32429187

Bumunang, E. W., Zaheer, R., Niu, D., Narvaez-Bravo, C., Alexander, T., McAllister, T. A., & Stanford, K. (2023). Bacteriophages for the Targeted Control of Foodborne Pathogens. *Foods*, *12*(14), 2734. doi:10.3390/foods12142734 PMID:37509826

Cabrera-Chávez, F., de la Barca, A. M. C., Islas-Rubio, A. R., Marti, A., Marengo, M., Pagani, M. A., Bonomi, F., & Iametti, S. (2012). Molecular rearrangements in extrusion processes for the production of amaranth-enriched, gluten-free rice pasta. *Lebensmittel-Wissenschaft + Technologie*, *47*(2), 421–426. doi:10.1016/j.lwt.2012.01.040

Cady, R. K., Farmer, K., Dexter, J. K., & Hall, J. (2012). The bowel and migraine: Update on celiac disease and irritable bowel syndrome. *Current Pain and Headache Reports*, *16*(3), 278–286. doi:10.100711916-012-0258-y PMID:22447132

Calva-Estrada, S. J., Jim'enez-Fernandez, ´. M., & Lugo-Cervantes, E. (2019). Protein-based films: Advances in the development of biomaterials applicable to food packaging. *Food Engineering Reviews*, *11*(2), 78–92. doi:10.100712393-019-09189-w

Campbell, A. (2003). The future of bacteriophage biology. *Nature Reviews. Genetics*, *4*(6), 471–477. doi:10.1038/nrg1089 PMID:12776216

Cao, S., Wang, S., Wang, Q., Lin, G., Niu, B., Guo, R., Yan, H., & Wang, H. (2023). Sodium alginate/chitosan-based intelligent bilayer film with antimicrobial activity for pork preservation and freshness monitoring. *Food Control*, *148*, 109615. doi:10.1016/j.foodcont.2023.109615

Capanoglu, E., Nemli, E., & Tomas-Barberan, F. (2021). Novel approaches in the valorization of agricultural wastes and their applications. *Journal of Agricultural and Food Chemistry*, *2022*, 6787–6804. PMID:35195402

Caperuto, L. C., Amaya-Farfan, J., & Camargo, C. R. O. (2001). Performance of quinoa (Chenopodium quinoa Willd) flour in the manufacture of gluten-free spaghetti. *Journal of the Science of Food and Agriculture*, *81*(1), 95–101. doi:10.1002/1097-0010(20010101)81:1<95::AID-JSFA786>3.0.CO;2-T

Carlucci, D., Nocella, G., De Devitiis, B., Viscecchia, R., Bimbo, F., & Nardone, G. (2015). Consumer purchasing behaviour towards fish and seafood products. Patterns and insights from a sample of international studies. *Appetite*, *84*, 212–227. doi:10.1016/j.appet.2014.10.008 PMID:25453592

Carmody, L. A., Gill, J. J., Summer, E. J., Sajjan, U. S., Gonzalez, C. F., Young, R. F., & LiPuma, J. J. (2010). Efficacy of bacteriophage therapy in a model of Burkholderia cenocepacia pulmonary infection. *The Journal of Infectious Diseases*, *201*(2), 264–271. doi:10.1086/649227 PMID:20001604

Carpin'e, D., Dagostin, J. L. A., Bertan, L. C., & Mafra, M. R. (2015). Development and characterization of soy protein isolate emulsion-based edible films with added coconut oil for olive oil packaging: Barrier, mechanical, and thermal properties. *Food and Bioprocess Technology*, *8*(8), 1811–1823. doi:10.100711947-015-1538-4

Caselato-Sousa, V. M., & Amaya-Farfán, J. (2012). State of knowledge on amaranth grain: A comprehensive review. *Journal of Food Science*, *77*(4), R93–R104. doi:10.1111/j.1750-3841.2012.02645.x PMID:22515252

Castrica, M., Rebucci, R., Giromini, C., Tretola, M., Cattaneo, D., & Baldi, A. (2019). Total phenolic content and antioxidant capacity of agri-food waste and by-products. *Italian Journal of Animal Science*, *18*(1), 336–341. doi:10.1080/1828051X.2018.1529544

Catassi, C., Elli, L., Bonaz, B., Bouma, G., Carroccio, A., Castillejo, G., Cellier, C., Cristofori, F., De Magistris, L., Dolinsek, J., Dieterich, W., Francavilla, R., Hadjivassiliou, M., Holtmeier, W., Körner, U., Leffler, D., Lundin, K., Mazzarella, G., Mulder, C., ... Fasano, A. (2015). Diagnosis of non-celiac gluten sensitivity (NCGS): The Salerno experts' criteria. *Nutrients*, *7*(6), 4966–4977. doi:10.3390/nu7064966 PMID:26096570

Cazón, P., Velazquez, G., Ramírez, J. A., & Vázquez, M. (2017). Polysaccharide-Based Films and Coatings for Food Packaging: A Review. *Food Hydrocolloids*, *68*, 136–148. doi:10.1016/j.foodhyd.2016.09.009

Cela, N., Galgano, F., Perretti, G., Di Cairano, M., Tolve, R., & Condelli, N. (2022). Assessment of brewing attitude of unmalted cereals and pseudocereals for gluten free beer production. *Food Chemistry*, *384*, 132621. doi:10.1016/j.foodchem.2022.132621 PMID:35257999

CEN. (2004). Eurocode 8-Design of Structure for Earthquake Resistance, Part I: General Rules. Seismic Actions and Rules for Buildings. European Standard EN 1998-1, European Committee for Standardization, Brussels.

Chakraborty, P. K., Adhikari, J., & Saha, P. (2018). Facile fabrication of electrospun regenerated cellulose nanofiber scaffold for potential bone-tissue engineering application. *International Journal of Biological Macromolecules*, *122*, 644–652. doi:10.1016/j.ijbiomac.2018.10.216 PMID:30391587

Chakravartula, S. S. N., Cevoli, C., & Balestra, F. (2019). Evaluation of drying of edible coating on bread using NIR spectroscopy. *Journal of Food Engineering*, *240*, 29–37. doi:10.1016/j.jfoodeng.2018.07.009

Chaljub, E., Celorio, M., Cornou, C., Martin, F. D., Haber, E. E., Margerin, L., Marti, J., & Zentner, I. (2021). Numerical simulation of wave propagation in heterogeneous and random media for site effects assessment in the Grenoble Valley. *The 6th IASPEI / IAEE International Symposium: Effects of Surface Geology on Seismic Motion, August 2021*.

Chan, B. K., Sistrom, M., Wertz, J. E., Kortright, K. E., Narayan, D., & Turner, P. E. (2016). Phage selection restores antibiotic sensitivity in MDR Pseudomonas aeruginosa. *Scientific Reports*, *6*(1), 26717. doi:10.1038rep26717 PMID:27225966

Chan, B. K., Turner, P. E., Kim, S., Mojibian, H. R., Elefteriades, J. A., & Narayan, D. (2018). Phage treatment of an aortic graft infected with Pseudomonas aeruginosa. *Evolution, Medicine, and Public Health, 2018*(1), 60–66. doi:10.1093/emph/eoy005 PMID:29588855

Chanishvili, N. (2012). Phage therapy—History from Twort and d'Herelle through Soviet experience to current approaches. *Advances in Virus Research, 83*, 3–40. doi:10.1016/B978-0-12-394438-2.00001-3 PMID:22748807

Chan, M. L., Lau, K. T., Wong, T. T., & Cardona, F. (2011). Interfacial bonding characteristic of nanoclay/polymer composites. *Applied Surface Science, 258*(2), 860–864. doi:10.1016/j.apsusc.2011.09.016

Chan, M. L., Lau, K. T., Wong, T. T., Ho, M. P., & Hui, D. (2011). Mechanism of reinforcement in a nanoclay/polymer composite. *Composites. Part B, Engineering, 42*(6), 1708–1712. doi:10.1016/j.compositesb.2011.03.011

Chaturvedi, R., & Verma, S. (2022). *Artificial Intelligence-Driven Customer Experience: Overcoming The Challenges*. California Management Review Insights.

Chaudhary, A., Gustafson, D., & Mathys, A. (2018). Multi-indicator sustainability assessment of global food systems. *Nature Communications, 9*(1), 848. doi:10.103841467-018-03308-7 PMID:29487286

Chaudhary, P., Fatima, F., & Kumar, A. (2020). Relevance of nanomaterials in food packaging and its advanced future prospects. *Journal of Inorganic and Organometallic Polymers and Materials, 30*(12), 5180–5192. doi:10.100710904-020-01674-8 PMID:32837459

Chaudhry, W. N., Concepcion-Acevedo, J., Park, T., Andleeb, S., Bull, J. J., & Levin, B. R. (2017). Synergy and order effects of antibiotics and phages in killing Pseudomonas aeruginosa biofilms. *PLoS One, 12*(1), e0168615. doi:10.1371/journal.pone.0168615 PMID:28076361

Chauhan, O. P., Nanjappa, C., Ashok, N., Ravi, N., Roopa, N., & Raju, P. S. (2015). Shellac and Aloe vera gel-based surface coating for shelf life extension of tomatoes. *Journal of Food Science and Technology, 52*(2), 1200–1205. doi:10.100713197-013-1035-6 PMID:25694740

Chemat, F., & Strube, J. (2015). *Green extraction of natural products theory and Practice*. Wiley-VCH-Verl. doi:10.1002/9783527676828

Chen, C., Zhang, B., Huang, Q., Fu, X., & Liu, R. (2017). Microwave-assisted extraction of polysaccharides from Moringa oleifera Lam. leaves: Characterization and hypoglycemic activity. *Industrial Crops and Products, 100*, 1–11. doi:10.1016/j.indcrop.2017.01.042

Cheng, J., Lin, X., Wu, X., Liu, Q., Wan, S., & Zhang, Y. (2021). Preparation of a multifunctional silver nanoparticles polylactic acid food packaging film using mango peel extract. *International Journal of Biological Macromolecules, 188*, 678–688. doi:10.1016/j.ijbiomac.2021.07.161 PMID:34343590

Chen, H., Wang, J., Cheng, Y., Wang, C., Liu, H., Bian, H., Pan, Y., Sun, J., & Han, W. (2019). Application of protein-based films and coatings for food packaging: A Review. *Polymers*, *11*(12), 2039. doi:10.3390/polym11122039 PMID:31835317

Chen, X., Zhang, M., Teng, X., & Mujumdar, A. S. (2022). Recent progress in modeling 3D/4D printing of foods. *Food Engineering Reviews*, *14*(1), 1–14. doi:10.100712393-021-09297-6

Chen, Y., Zhang, M., Sun, Y., & Phuhongsung, P. (2022). Improving 3D/4D printing characteristics of natural food gels by novel additives: A review. *Food Hydrocolloids*, *123*, 107160. doi:10.1016/j.foodhyd.2021.107160

Chen, Z., & Chan, I. C. (2022). Smart cities and quality of life: A quantitative analysis of citizens' support for smart city development. *Information Technology & People*, *36*(1), 263–285. doi:10.1108/ITP-07-2021-0577

Cheong, A., Lau, M., Foo, E., Hedley, J., & Bo, J. W. (2016). Development of a robotic waiter system. *IFAC-PapersOnLine*, *49*(21), 681–686. doi:10.1016/j.ifacol.2016.10.679

Chiou, B. S., Yee, E., Wood, D., Shey, J., Glenn, G., & Orts, W. (2006). Effects of processing conditions on nanoclay dispersion in starch-clay nanocomposites. *Cereal Chemistry*, *83*(3), 300–305. doi:10.1094/CC-83-0300

Chitravathi, K., Chauhan, O.P., & Raju, P.S. (2014). *Postharvest biology and technology postharvest shelf-life extension of green chillies (Capsicum annuum L.) using shellac-based edible surface coatings*. Academic Press.

Chiumarelli, M., & Hubinger, M. D. (2014). Evaluation of edible films and coatings formulated with cassava starch, glycerol, carnauba wax and stearic acid. *Food Hydrocolloids*, *38*, 20–27. doi:10.1016/j.foodhyd.2013.11.013

Choi, I., Chang, Y., Kim, S. Y., & Han, J. (2021). Polycaprolactone film functionalized with bacteriophage T4 promotes antibacterial activity of food packaging toward Escherichia coli. *Food Chemistry*, *346*, 128883. doi:10.1016/j.foodchem.2020.128883 PMID:33434864

Ciclitira, P. J., Ellis, H. J., & Lundin, K. E. A. (2005). Gluten-free diet—What is toxic? *Best Practice & Research. Clinical Gastroenterology*, *19*(3), 359–371. doi:10.1016/j.bpg.2005.01.003 PMID:15925842

Cilsalar, H., & Temiz, U. (2020). Probabilistic seismic hazard assessment of city center of Kayseri, Erciyes University. *Journal of Institue of Science and Technology*, *36*(2), 286–298.

Çimen, Ö., Uyanık, O., Elmasdere, E., Korkmaz, K. A., & Keskin, S. N. (2010). Mavikent-Isparta örneğinde mikrobölgeleme çalışmaları için yerel zemin koşullarının belirlenmesi. *SDÜ Fen Bilimleri Enstitüsü Dergisi*, *14*(1), 46–54.

Cimino, A., Elbasheer, M., Longo, F., Nicoletti, L., & Padovano, A. (2023). Empowering Field Operators in Manufacturing: A Prospective Towards Industry 5.0. *Procedia Computer Science*, *217*, 1948–1953. doi:10.1016/j.procs.2022.12.395

Clavijo, V., Baquero, D., Hernandez, S., Farfan, J., Arias, J., Arévalo, A., Donado-Godoy, P., & Vives-Flores, M. (2019). Phage cocktail SalmoFREE® reduces Salmonella on a commercial broiler farm. *Poultry Science*, *98*(10), 5054–5063. doi:10.3382/ps/pez251 PMID:31073590

Clokie, M. R., Millard, A. D., Letarov, A. V., & Heaphy, S. (2011). Phages in nature. *Bacteriophage*, *1*(1), 31–45. Advance online publication. doi:10.4161/bact.1.1.14942 PMID:21687533

Codex Alimentarius Commission. (2009). *Commission regulation (EC) No 41/2009 of 20 January 2009 concerning the composition and labelling of foodstuffs suitable for people intolerant to gluten.* Author.

Coles, D., & Frewer, L. J. (2013). Nanotechnology applied to European food production–A review of ethical and regulatory issues. *Trends in Food Science & Technology*, *34*(1), 32–43. doi:10.1016/j.tifs.2013.08.006

Coles, R., McDowell, D., & Kirwan, M. J. (2003). *Food Packaging Technology*. Crc Press.

Cooper, I. R. (2016). A review of current methods using bacteriophages in live animals, food and animal products intended for human consumption. *Journal of Microbiological Methods*, *130*, 38–47. doi:10.1016/j.mimet.2016.07.027 PMID:27485708

Costa, M. J., Pastrana, L. M., Teixeira, J. A., Sillankorva, S. M., & Cerqueira, M. A. (2023). Bacteriophage delivery systems for food applications: Opportunities and perspectives. *Viruses*, *15*(6), 1271. doi:10.3390/v15061271 PMID:37376571

Costantini, L., Lukšič, L., Molinari, R., Kreft, I., Bonafaccia, G., Manzi, L., & Merendino, N. (2014). Development of gluten-free bread using tartary buckwheat and chia flour rich in flavonoids and omega-3 fatty acids as ingredients. *Food Chemistry*, *165*, 232–240. doi:10.1016/j.foodchem.2014.05.095 PMID:25038671

Coulter, L. B., McLean, R. J., Rohde, R. E., & Aron, G. M. (2014). Effect of bacteriophage infection in combination with tobramycin on the emergence of resistance in Escherichia coli and Pseudomonas aeruginosa biofilms. *Viruses*, *6*(10), 3778–3786. doi:10.3390/v6103778 PMID:25285538

Crossan, M. M., & Apaydin, M. (2010). A multi-dimensional framework of organizational innovation: A systematic review of the literature. *Journal of Management Studies*, *47*(6), 1154–1191. doi:10.1111/j.1467-6486.2009.00880.x

Cugurullo, F. (2018). Exposing smart cities and eco-cities: Frankenstein urbanism and the sustainability challenges of the experimental city. *Environment & Planning A*, *50*(1), 73–92. doi:10.1177/0308518X17738535

Cui, H., Yuan, L., Li, W., & Lin, L. (2017). Edible film incorporated with chitosan and Artemisia annua oil nanoliposomes for inactivation of Escherichia coli O157: H7 on cherry tomato. *International Journal of Food Science & Technology*, *52*(3), 687–698. doi:10.1111/ijfs.13322

Dąbrowska, K., Miernikiewicz, P., Piotrowicz, A., Hodyra, K., Owczarek, B., Lecion, D., Kaźmierczak, Z., Letarov, A., & Górski, A. (2014). Immunogenicity studies of proteins forming the T4 phage head surface. *Journal of Virology*, *88*(21), 12551–12557. doi:10.1128/JVI.02043-14 PMID:25142581

Dadbin, S., Noferesti, M., & Frounchi, M. (2008). Oxygen barrier LDPE/LLDPE/organoclay nano-composite films for food packaging. *Macromolecular Symposia*, *274*(1), 22–27. doi:10.1002/masy.200851404

Dağ, Ş. R. O., & Özkan, A. M. G. (2019). Kinoa (Chenopodium quinoa Willd.) üzerine bir derleme. *Journal of Faculty of Pharmacy of Ankara University*, *43*(3), 309–333.

Dakal, T. C., Kumar, A., Majumdar, R. S., & Yadav, V. (2016). Mechanistic basis of antimicrobial actions of silver nanoparticles. *Frontiers in Microbiology*, *7*, 1831. doi:10.3389/fmicb.2016.01831 PMID:27899918

Dasgupta, N., Ranjan, S., Mundekkad, D., Ramalingam, C., Shanker, R., & Kumar, A. (2015). Nanotechnology in agro-food: From field to plate. *Food Research International*, *69*, 381–400. doi:10.1016/j.foodres.2015.01.005

David, S., Duraipandian, K., Chandrasekaran, D., Pandey, D., Sindhwani, N., & Pandey, B. K. (2023). Impact of blockchain in healthcare system. In *Unleashing the Potentials of Blockchain Technology for Healthcare Industries* (pp. 37–57). Academic Press. doi:10.1016/B978-0-323-99481-1.00004-3

De Arcangelis, E., Cuomo, F., Trivisonno, M. C., Marconi, E., & Messia, M. C. (2020). Gelatinization and pasta making conditions for buckwheat gluten-free pasta. *Journal of Cereal Science*, *95*, 103073. doi:10.1016/j.jcs.2020.103073

De Frutos, M. F., Fotschki, B., Musoles, R. F., & Llopis, J. M. L. (2018). *Gluten-free cereals and pseudocereals: nutrition and health. In Bioactive Molecules in Food.* Springer.

De Jonge, P. A., Nobrega, F. L., Brouns, S. J., & Dutilh, B. E. (2019). Molecular and evolutionary determinants of bacteriophage host range. *Trends in Microbiology*, *27*(1), 51–63. doi:10.1016/j.tim.2018.08.006 PMID:30181062

de la Barca, A. M. C., Rojas-Martínez, M. E., Islas-Rubio, A. R., & Cabrera-Chávez, F. (2010). Gluten-free breads and cookies of raw and popped amaranth flours with attractive technological and nutritional qualities. *Plant Foods for Human Nutrition (Dordrecht, Netherlands)*, *65*(3), 241–246. doi:10.100711130-010-0187-z PMID:20734143

de Oliveira, J. P., Bruni, G. P., El Halal, S. L. M., Bertoldi, F. C., Dias, A. R. G., & da Rosa Zavareze, E. (2019). Cellulose nanocrystals from rice and oat husks and their application in aerogels for food packaging. *International Journal of Biological Macromolecules*, *124*, 175–184. doi:10.1016/j.ijbiomac.2018.11.205 PMID:30471399 de Villiers, A., Venter, P., & Pasch, H. (2016). Recent advances and trends in the liquid-chromatography–mass spectrometry analysis of flavonoids. *Journal of Chromatography. A, 1430*, 16–78. doi:10.1016/j.chroma.2015.11.077 PMID:26718188

De Vriendt, J., Lainé, P., Lerouge, C., & Xu, X. (2002). Mobile network evolution: A revolution on the move. *IEEE Communications Magazine, 40*(4), 104–111. doi:10.1109/35.995858

Debiagi, F., Kobayashi, R. K. T., Nakazato, G., Panagio, L. A., & Mali, S. (2014). Biodegradable active packaging based on cassava bagasse, polyvinyl alcohol and essential oils. *Industrial Crops and Products, 52*, 664–670. doi:10.1016/j.indcrop.2013.11.032

Dębski, B., Gralak, M. A., Bertrandt, J., & Kłos, A. (2013). Minerals and polyphenols content of quinoa (Chenopodium quinoa Willd.) plant. *Probl. Hig. Epidemiol, 94*(2), 300–304.

Dedrick, R. M., Guerrero-Bustamante, C. A., Garlena, R. A., Russell, D. A., Ford, K., Harris, K., ... Schooley, R. T. (2019). Engineered bacteriophages for treatment of a patient with a disseminated drug-resistant Mycobacterium abscessus. *Nature Medicine, 25*(5), 730–733. doi:10.103841591-019-0437-z PMID:31068712

Dehghan-Dehnavi, S., Fotuhi-Firuzabad, M., Moeini-Aghtaie, M., Dehghanian, P., & Wang, F. (2020). Estimating participation abilities of industrial customers in demand response programs: A two-level decision-making tree analysis. *2020 IEEE/IAS 56th Industrial and Commercial Power Systems Technical Conference*. 10.1109/ICPS48389.2020.9176817

Dermeche, S., Nadour, M., Larroche, C., Moulti-Mati, F., & Michaud, P. (2013). Olive mill wastes: Biochemical characterizations and valorization strategies. *Process Biochemistry (Barking, London, England), 48*(10), 1532–1552. doi:10.1016/j.procbio.2013.07.010

Devani, B. M., Jani, B. L., Balani, P. C., & Akbari, S. H. (2019). Optimization of supercritical CO2 extraction process for oleoresin from rotten onion waste. *Food and Bioproducts Processing, 119*, 287–295. doi:10.1016/j.fbp.2019.11.014

Dhall, R. K. (2013). Advances in edible coatings for fresh fruits and vegetables: A review. *Critical Reviews in Food Science and Nutrition, 53*(5), 435–450. doi:10.1080/10408398.2010.541568 PMID:23391012

Dhama, K., Karthik, K., Tiwari, R., Shabbir, M. Z., Barbuddhe, S., Malik, S. V. S., & Singh, R. K. (2015). Listeriosis in animals, its public health significance (food-borne zoonosis) and advances in diagnosis and control: A comprehensive review. *The Veterinary Quarterly, 35*(4), 211–235. doi:10.1080/01652176.2015.1063023 PMID:26073265

Dhumal, C. V., & Sarkar, P. (2018). Composite edible films and coatings from food-grade biopolymers. *Journal of Food Science and Technology, 55*(11), 4369–4383. doi:10.100713197-018-3402-9 PMID:30333633

Dias, M. V., Nilda de Fátima, F. S., Borges, S. V., de Sousa, M. M., Nunes, C. A., de Oliveira, I. R. N., & Medeiros, E. A. A. (2013). Use of allyl isothiocyanate and carbon nanotubes in an antimicrobial film to package shredded, cooked chicken meat. *Food Chemistry*, *141*(3), 3160–3166. doi:10.1016/j.foodchem.2013.05.148 PMID:23871073

Díaz-Pascual, F., Hartmann, R., Lempp, M., Vidakovic, L., Song, B., Jeckel, H., ... Link, H. (2019). Breakdown of Vibrio cholerae biofilm architecture induced by antibiotics disrupts community barrier function. *Nature Microbiology*, *4*(12), 2136–2145. doi:10.103841564-019-0579-2 PMID:31659297

Dickey, J., & Perrot, V. (2019). Adjunct phage treatment enhances the effectiveness of low antibiotic concentration against Staphylococcus aureus biofilms in vitro. *PLoS One*, *14*(1), e0209390. doi:10.1371/journal.pone.0209390 PMID:30650088

Dilucia, F., Lacivita, V., Conte, A., & Nobile, M. A. D. (2020). Sustainable use of fruit and vegetable by-products to enhance food packaging performance. *Foods*, *9*(7), 857. doi:10.3390/foods9070857 PMID:32630106

Domingo-Calap, P., & Delgado-Martínez, J. (2018). Bacteriophages: Protagonists of a post-antibiotic era. [basel]. *Antibiotics (Basel, Switzerland)*, *7*(3), 66. doi:10.3390/antibiotics7030066 PMID:30060506

Dong, C., Wang, B., Li, F., Zhong, Q., Xia, X., & Kong, B. (2020). Effects of edible chitosan coating on Harbin red sausage storage stability at room temperature. *Meat Science*, *159*, 1–9. doi:10.1016/j.meatsci.2019.107919 PMID:31472934

Dong, W., Wang, D., Hu, R., Long, Y., & Lv, L. (2020). Chemical composition, structural and functional properties of soluble dietary fiber obtained from coffee peel using different extraction methods. *Food Research International*, *136*, 109497. doi:10.1016/j.foodres.2020.109497 PMID:32846578

Dorscht, J., Klumpp, J., Bielmann, R., Schmelcher, M., Born, Y., Zimmer, M., Calendar, R., & Loessner, M. J. (2009). Comparative genome analysis of Listeria bacteriophages reveals extensive mosaicism, programmed translational frameshifting, and a novel prophage insertion site. *Journal of Bacteriology*, *191*(23), 7206–7215. doi:10.1128/JB.01041-09 PMID:19783628

Drulis-Kawa, Z., Majkowska-Skrobek, G., Maciejewska, B., Delattre, A.-S., & Lavigne, R. (2012). Learning from bacteriophages-advantages and limitations of phage and phage-encoded protein applications. *Current Protein & Peptide Science*, *13*(8), 699–722. doi:10.2174/138920312804871193 PMID:23305359

Du Toit, A. (2017). The language of phages. *Nature Reviews. Microbiology*, *15*(3), 135–135. PMID:28190882

Dulger & Özpinar. (2021). Use of bacteriophages to improve food safety. *İstanbul Gelişim Üniversitesi Sağlık Bilimleri Dergisi*, (15), 705-712.

Duncan, T. V. (2011). Applications of nanotechnology in food packaging and Food Safety: Barrier Materials, antimicrobials and sensors. *Journal of Colloid and Interface Science*, *363*(1), 1–24. doi:10.1016/j.jcis.2011.07.017 PMID:21824625

Dutta, P. K., Tripathi, S., Mehrotra, G. K., & Dutta, J. (2009). Perspectives for chitosan based antimicrobial films in food applications. *Food Chemistry*, *114*(4), 1173–1182. doi:10.1016/j.foodchem.2008.11.047

Dwivedi, S., Prajapati, P., Vyas, N., Malviya, S., & Kharia, A. (2017). A review on food preservation: Methods, harmful effects and better alternatives. *Asian Journal of Pharmacy and Pharmacology*, *3*, 193–199.

Echegoyen, Y., Rodriguez, S., & Nerin, C. (2016). Nanoclay migration from food packaging materials. *Food Additives & Contaminants. Part A, Chemistry, Analysis, Control, Exposure & Risk Assessment*, *33*(3), 530–539. doi:10.1080/19440049.2015.1136844 PMID:26751017

Echeverría, I., Lopez-Caballero, ´. M. E., Gomez-Guill, ´., ´en, M. C., Mauri, A. N., & Montero, M. P. (2018). Active nanocomposite films based on soy proteins montmorillonite- clove essential oil for the preservation of refrigerated bluefin tuna (Thunnus thynnus) fillets. *International Journal of Food Microbiology*, *266*, 142–149. doi:10.1016/j.ijfoodmicro.2017.10.003 PMID:29216554

Edelman, D. C., & Abraham, M. (2022). Customer Experience in the Age of AI. *Harvard Business Review*, *100*(3-4), 116–125.

Ekici, L., İnanır, C., & Albayrak, S. (2019). Karabuğdayın fitokimyası, farmakolojisi ve biyofonksiyonel özellikleri. *Avrupa Bilim ve Teknoloji Dergisi*, *16*, 713–722.

Ekin, N., & Uyanık, O. (2021a). Comparison of static and dynamic elastic moduli in concrete: Effects of compressive strength, curing conditions and reinforcement. *Civil Engineering (Shiraz)*, *45*(4), 2327–2343. doi:10.100740996-020-00513-7

Ekin, N., & Uyanık, O. (2021b). Anisotropy of reinforced concrete from geophysical methods. *Science of Sintering*, *53*(3), 323–345. doi:10.2298/SOS2103323E

Elangovan, U. (2021). *Industry 5.0: The future of the industrial economy*. CRC Press. doi:10.1201/9781003190677

Elder, J. H., Shankar, M., Shuster, J., Theriaque, D., Burns, S., & Sherrill, L. (2006). The gluten-free, casein-free diet in autism: Results of a preliminary double blind clinical trial. *Journal of Autism and Developmental Disorders*, *36*(3), 413–420. doi:10.100710803-006-0079-0 PMID:16555138

Eleftheriadou, M., Pyrgiotakis, G., & Demokritou, P. (2017). Nanotechnology to the rescue: Using nano-enabled approaches in microbiological food safety and quality. *Current Opinion in Biotechnology*, *44*, 87–93. doi:10.1016/j.copbio.2016.11.012 PMID:27992831

Elik, A., Yanik, D. K., & Göğüş, F. (2020). Microwave-assisted extraction of carotenoids from carrot juice processing waste using flaxseed oil as a solvent. *Lebensmittel-Wissenschaft + Technologie*, *123*, 109100. doi:10.1016/j.lwt.2020.109100

Elli, L., Branchi, F., Tomba, C., Villalta, D., Norsa, L., Ferretti, F., Roncoroni, L., & Bardella, M. T. (2015). Diagnosis of gluten related disorders: Celiac disease, wheat allergy and non-celiac gluten sensitivity. *World Journal of Gastroenterology*, *21*(23), 7110–7119. doi:10.3748/wjg.v21.i23.7110 PMID:26109797

Elsayed, M. M., Elkenany, R. M., Zakari, A. I., & Badawy, B. M. (2023). Isolation and characterization of bacteriophages for combating multidrug-resistant Listeria monocytogenes from dairy cattle farms in conjugation with silver nanoparticles. *BMC Microbiology*, *23*(1), 146. doi:10.118612866-023-02893-y PMID:37217869

Enault, F., Briet, A., Bouteille, L., Roux, S., Sullivan, M. B., & Petit, M.-A. (2017). Phages rarely encode antibiotic resistance genes: A cautionary tale for virome analyses. *The ISME Journal*, *11*(1), 237–247. doi:10.1038/ismej.2016.90 PMID:27326545

Endersen, L., & Coffey, A. (2020). The use of bacteriophages for food safety. *Current Opinion in Food Science*, *36*, 1–8. doi:10.1016/j.cofs.2020.10.006

Erceg, M., Kovačić, T., & Perinović, S. (2010). Isothermal degradation of poly(3- hydroxybutyrate)/organically modified montmorillonite nanocomposites. *Polymer Composites*, *31*(2), 272–278. doi:10.1002/pc.20798

Erken, M., Lutz, C., & McDougald, D. (2013). The rise of pathogens: Predation as a factor driving the evolution of human pathogens in the environment. *Microbial Ecology*, *65*(4), 860–868. doi:10.100700248-013-0189-0 PMID:23354181

Ersan, S., Üstündag, ˘. O. G., Carle, R., & Schweiggert, R. M. (2018). Subcritical water extraction of phenolic and antioxidant constituents from pistachio (Pistacia vera L.) hulls. *Food Chemistry*, *253*, 46–54. doi:10.1016/j.foodchem.2018.01.116 PMID:29502842

Eyiz, V., Tontul, ˙. I., & Türker, S. (2020). The effect of edible coatings on physical and chemical characteristics of fruit bars. *Journal of Food Measurement and Characterization*, *2020*(14), 1775–1783. doi:10.100711694-020-00425-0

Fabra, M. J., Lopez-Rubio, A., & Lagaron, J. M. (2013). High barrier polyhydroxyalcanoate food packaging film by means of nanostructured electrospun interlayers of zein. *Food Hydrocolloids*, *32*(1), 106–114. doi:10.1016/j.foodhyd.2012.12.007

Fabra, M. J., Pardo, P., Martínez-Sanz, M., Lopez-Rubio, A., & Lagar6n, J. M. (2016). Combining polyhydroxyalkanoates with nanokeratin to develop novel biopackaging structures. *Journal of Applied Polymer Science*, *133*(2), app.42695. doi:10.1002/app.42695

Fagerberg, J., Martin, B. R., & Andersen, E. S. (Eds.). (2013). *Innovation studies: Evolution and future challenges*. Oxford University Press. doi:10.1093/acprof:oso/9780199686346.001.0001

Fair & Tor. (2014). Antibiotics and bacterial resistance in the 21st century. *Perspectives in Medicinal Chemistry, 6*.

Falkinham, J. O. III, Pruden, A., & Edwards, M. (2015). Opportunistic premise plumbing pathogens: Increasingly important pathogens in drinking water. *Pathogens (Basel, Switzerland)*, *4*(2), 373–386. doi:10.3390/pathogens4020373 PMID:26066311

Fancello, L., Desnues, C., Raoult, D., & Rolain, J. M. (2011). Bacteriophages and diffusion of genes encoding antimicrobial resistance in cystic fibrosis sputum microbiota. *The Journal of Antimicrobial Chemotherapy*, *66*(11), 2448–2454. doi:10.1093/jac/dkr315 PMID:21816767

Fancello, L., Gandini, S., Pelicci, P. G., & Mazzarella, L. (2019). Tumor mutational burden quantification from targeted gene panels: Major advancements and challenges. *Journal for Immunotherapy of Cancer*, *7*(1), 1–13. doi:10.118640425-019-0647-4 PMID:31307554

FAO. (2012). *Towards the Future We Want: End hunger and make the transition to sustainable agricultural and food systems*. Food and Agriculture Organization of the United Nations Rome.

FAO. (2014). *Food wastage footprint full-cost accounting*. Food Wastage Footprint.

Farhoodi, M. (2016). Nanocomposite materials for food packaging applications: Characterization and safety evaluation. *Food Engineering Reviews*, *8*(1), 35–51. doi:10.100712393-015-9114-2

Farooq, Ali, Shoukat, Paray, & Ali. (2019). *Bacteriophages as biocontrol agents for foodborne diseases*. Academic Press.

Fasihnia, S. H., Peighambardoust, S. H., & Peighambardoust, S. J. (2018). Nanocomposite films containing organoclay nanoparticles as an antimicrobial (active) packaging for potential food application. *Journal of Food Processing and Preservation*, *42*(2), e13488. doi:10.1111/jfpp.13488

Federici, F., Fava, F., Kalogerakis, N., & Mantzavinos, D. (2009). Valorisation of agro-industrial by-products, effluents and waste: concept, opportunities and the case of olive mill wastewaters. *Journal of Chemical Technology & Biotechnology: International Research in Process, Environmental & Clean Technology*, *84*(6), 895–900.

Fellows, P. J. (2017). *Food processing technologies: Princilpes and practices. (4th* (418th ed.). Woodhead Publishing, Elsevier.

Fermoso, F. G., Serrano, A., Alonso-Fariñas, B., Fernández-Bolaños, J., Borja, R., & Rodríguez-Gutiérrez, G. (2018). Valuable compound extraction, anaerobic digestion, and composting: A leading biorefinery approach for agricultural wastes. *Journal of Agricultural and Food Chemistry*, *66*(32), 8451–8468. doi:10.1021/acs.jafc.8b02667 PMID:30010339

Fernández, L., Gutiérrez, D., Rodríguez, A., & García, P. (2018). Application of bacteriophages in the agro-food sector: A long way toward approval. *Frontiers in Cellular and Infection Microbiology*, *8*, 296. doi:10.3389/fcimb.2018.00296 PMID:30186776

Filho, A. M. M., Pirozi, M. R., Borges, J. T. D. S., Pinheiro Sant'Ana, H. M., Chaves, J. B. P., & Coimbra, J. S. D. R. (2017). Quinoa: Nutritional, functional, and antinutritional aspects. *Critical Reviews in Food Science and Nutrition*, *57*(8), 1618–1630. doi:10.1080/10408398.2014.10018 11 PMID:26114306

Flamini, R. (2003). Mass spectrometry in grape and wine chemistry. Part I: Polyphenols. *Mass Spectrometry Reviews*, *22*(4), 218–250. doi:10.1002/mas.10052 PMID:12884388

Flaris, V., & Singh, G. (2009). Recent developments in biopolymers. *Journal of Vinyl and Additive Technology*, *15*(1), 1–11. doi:10.1002/vnl.20171

Fombonne, E. (2003). Epidemiological surveys of autism and other pervasive developmental disorders: An update. *Journal of Autism and Developmental Disorders*, *33*(4), 365–382. doi:10.1023/A:1025054610557 PMID:12959416

Fontana, A. R., Antoniolli, A., & Bottini, R. (2013). Grape pomace as a sustainable source of bioactive compounds: Extraction, characterization, and biotechnological applications of phenolics. *Journal of Agricultural and Food Chemistry*, *61*(38), 8987–9003. doi:10.1021/jf402586f PMID:24004377

Foti, S., Hollender, F., Garofalo, F., Albarello, D., Asten, M., Bard, P. Y., Comina, C., Cornou, C., Cox, B., Giulio, G. D., Forbriger, T., Hayashi, K., Lunedei, E., Martin, A., Mercerat, D., Ohrnberger, M., Poggi, V., Renalier, F., & Sicilia, D. (2018). Guidelines for the good practice of surface wave analysis: A product of the InterPACIFIC project. *Bulletin of Earthquake Engineering*, *16*(6), 2367–2420. doi:10.100710518-017-0206-7

Freitas, L. C., Barbosa, J. R., da Costa, A. L. C., Bezerra, F. W. F., Pinto, R. H. H., & de Carvalho, R. N. Jr. (2021). From waste to sustainable industry: How can agro-industrial wastes help in the development of new products? *Resources, Conservation and Recycling*, *169*, 105466. doi:10.1016/j.resconrec.2021.105466

Frounchi, M., & Dourbash, A. (2009). Oxygen barrier properties of poly (ethylene terephthalate) nanocomposite films. *Macromolecular Materials and Engineering*, *294*(1), 68–74. doi:10.1002/mame.200800238

Fu, X., Belwal, T., Cravotto, G., & Luo, Z. (2020). Sono-physical and sono-chemical effects of ultrasound: Primary applications in extraction and freezing operations and influence on food components. *Ultrasonics Sonochemistry*, *60*, 104726. doi:10.1016/j.ultsonch.2019.104726 PMID:31541966

Gabrielli, M., Cremonini, F., Fiore, G., Addolorato, G., Padalino, C., Candelli, M., De Leo, M. E., Santarelli, L., Giacovazzo, M., & Gasbarrini, A. (2003). Association between migraine and celiac disease: Results from a preliminary case-control and therapeutic study. *The American Journal of Gastroenterology*, *98*(3), 625–629. doi:10.1111/j.1572-0241.2003.07300.x PMID:12650798

Gaikwad, K. K., Singh, S., & Negi, Y. S. (2020). Ethylene scavengers for active packaging of fresh food produce. *Environmental Chemistry Letters*, *18*(2), 269–284. doi:10.100710311-019-00938-1

Gallagher, E., Gormley, T. R., & Arendt, E. K. (2003). Crust and crumb characteristics of gluten free breads. *Journal of Food Engineering*, *56*(2–3), 153–161. doi:10.1016/S0260-8774(02)00244-3

Gallocchio, F., Cibin, V., Biancotto, G., Roccato, A., Muzzolon, O., Carmen, L., Simone, B., Manodori, L., Fabrizi, A., Patuzzi, I., & Ricci, A. (2016). Testing nano-silver food packaging to evaluate silver migration and food spoilage bacteria on chicken meat. *Food Additives & Contaminants. Part A, Chemistry, Analysis, Control, Exposure & Risk Assessment*, *33*(6), 1063–1071. doi:10.1080/19440049.2016.1179794 PMID:27147130

Gallo, M., Ferrara, L., & Naviglio, D. (2018). Application of ultrasound in Food Science and Technology: A perspective. *Foods*, *7*(10), 164. doi:10.3390/foods7100164 PMID:30287795

Galus, S., & Kadzińska, J. (2015). Food applications of emulsion-based edible films and coatings. *Trends in Food Science & Technology*, *45*(2), 273–283. doi:10.1016/j.tifs.2015.07.011

Galus, S., & Kadzinska, J. (2019). Gas barrier and Wetting properties of whey protein isolates based emulsion films. *Polymer Engineering and Science*, *59*(1), 375–383. doi:10.1002/pen.24992

Gambuś, H., Gambuś, F., Pastuszka, D., Wrona, P., Ziobro, R., Sabat, R., Mickowska, B., Nowotna, A., & Sikora, M. (2009). Quality of gluten-free supplemented cakes and biscuits. *International Journal of Food Sciences and Nutrition*, *60*(sup4), 31–50.

Gambus, H., Pastuszka, D., Gambus, F., Wrona, P., Ziobro, R., Mickowska, B., & Sikora, M. (2010). Nutritional and dietary value of gluten-free rolls enriched in amaranth flour. *Journal of Pre-Clinical and Clinical Research*, *4*(2).

Ganesan, A. R., Shanmugam, M., & Bhat, R. (2018). Producing novel edible films from semi refined carrageenan (SRC) and ulvan polysaccharides for potential food applications. *International Journal of Biological Macromolecules*, *112*, 1164–1170. doi:10.1016/j.ijbiomac.2018.02.089 PMID:29454950

Ganguly, S., Dana, K., Mukhopadhyay, T. K., Parya, T. K., & Ghatak, S. (2011). Organophilic nano clay: A comprehensive review. *Transactions of the Indian Ceramic Society*, *70*(4), 189–206. doi:10.1080/0371750X.2011.10600169

Gangwar, M., Rastogi, S., Singh, D., Shukla, A., Dhameja, N., Kumar, D., Kumar, R., & Nath, G. (2021). Study on the effect of oral administration of bacteriophages in charles foster rats with special reference to immunological and adverse effects. *Frontiers in Pharmacology*, *12*, 615445. doi:10.3389/fphar.2021.615445 PMID:33912038

García-Mahecha, M., Soto-Valdez, H., Carvajal-Millan, E., Madera-Santana, T. J., Lomelí-Ramírez, M. G., & Colín-Chávez, C. (2023). Bioactive compounds in extracts from the agro-industrial waste of mango. *Molecules (Basel, Switzerland)*, *28*(1), 458. doi:10.3390/molecules28010458 PMID:36615647

Garcia, P., Martinez, B., Obeso, J., & Rodriguez, A. (2008). Bacteriophages and their application in food safety. *Letters in Applied Microbiology*, *47*(6), 479–485. doi:10.1111/j.1472-765X.2008.02458.x PMID:19120914

Gavahian, M., Chu, Y.-H., Lorenzo, J. M., Mousavi Khaneghah, A., & Barba, F. J. (2020). Essential oils as natural preservatives for bakery products: Understanding the mechanisms of action, recent findings, and applications. *Critical Reviews in Food Science and Nutrition, 60*(2), 310–321. doi:10.1080/10408398.2018.1525601 PMID:30431327

Gebreil, S. Y., Ali, M. I. K., & Mousa, E. A. M. (2020). Utilization of amaranth flour in preparation of high nutritional value bakery products. *Food and Nutrition Sciences, 10*(05), 336–354. doi:10.4236/fns.2020.105025

Ge, H., Fu, S., Guo, H., Hu, M., Xu, Z., Zhou, X., Chen, X., & Jiao, X. (2022). Application and challenge of bacteriophage in the food protection. *International Journal of Food Microbiology, 380*, 109872. doi:10.1016/j.ijfoodmicro.2022.109872 PMID:35981493

Gencay, Y. E., & Brøndsted, L. (2019). Bacteriophages for Biological Control of Foodborne Pathogens. *Food Microbiology: Fundamentals And Frontiers*, 755-786.

Georgiou, K. E., Georgiou, E., & Satava, R. M. (2021). 5G use in healthcare: The future is present. JSLS. *JSLS : Journal of the Society of Laparoendoscopic Surgeons, 25*(4), e2021.00064. doi:10.4293/JSLS.2021.00064

Gill, J., Sandhu, P. S., & Singh, T. (2014). *A review of automatic fruit classification using soft computing techniques.* Paper presented at the Int. Conf. Comput. Syst. Electron. Eng.

Gmoshinski, I. V., Bagryantseva, O. V., Arnautov, O. V., & Khotimchenko, S. A. (2020). Nanoclays in food products: Benefits and possible risks (literature review). *Health Risk Analysis*, (1), 142–164. doi:10.21668/health.risk/2020.1.16.eng

Golebiewski, J., Rozanski, A., Dzwonkowski, J., & Galeski, A. (2008). Low density polyethylene–montmorillonite nanocomposites for film blowing. *European Polymer Journal, 44*(2), 270–286. doi:10.1016/j.eurpolymj.2007.11.002

Golovianko, M., Terziyan, V., Branytskyi, V., & Malyk, D. (2023). Industry 4.0 vs. Industry 5.0: Co-existence, Transition, or a Hybrid. *Procedia Computer Science, 217*, 102–113. doi:10.1016/j.procs.2022.12.206

Gomes Silva, C., Juárez, R., Marino, T., Molinari, R., & García, H. (2011). Influence of excitation wavelength (UV or visible light) on the photocatalytic activity of titania containing gold nanoparticles for the generation of hydrogen or oxygen from water. *Journal of the American Chemical Society, 133*(3), 595–602. doi:10.1021/ja1086358 PMID:21142160

Gomez-Guill, ´., ´en, M. C., Turnay, J., Fernandez-Diaz, M. D., Ulmo, N., Lizarbe, M. A., & Montero, P. (2002). Structural and physical properties of gelatin extracted from different marine species: A comparative study. *Food Hydrocolloids, 16*(1), 25–34. doi:10.1016/S0268-005X(01)00035-2

Gonçalves, C., Lopes, M., Ferreira, J. P., & Belo, I. (2009). Biological treatment of olive mill wastewater by non-conventional yeasts. *Bioresource Technology, 100*(15), 3759–3763. doi:10.1016/j.biortech.2009.01.004 PMID:19231162

Göral, V., Yildirim, N., Kaplan, A., Dede, Ş., & Çelik, M. (2007). Gluten enteropatisi sıklığı. *Akademik Gastroenteroloji Dergisi*, *6*(3), 144–148.

Gouvêa, D. M., Mendonça, R. C. S., Soto, M. L., & Cruz, R. S. (2015). Acetate cellulose film with bacteriophages for potential antimicrobial use in food packaging. *Lebensmittel-Wissenschaft + Technologie*, *63*(1), 85–91. doi:10.1016/j.lwt.2015.03.014

Goyache, F., Bahamonde, A., Alonso, J., López, S., Del Coz, J., Quevedo, J., Ranilla, J., Luaces, O., Alvarez, I., Royo, L. J., & Diez, J. (2001). The usefulness of artificial intelligence techniques to assess subjective quality of products in the food industry. *Trends in Food Science & Technology*, *12*(10), 370–381. doi:10.1016/S0924-2244(02)00010-9

Goyal, S., Pandey, D., Singh, H., Singh, J., Kakkar, R., & Srinivasu, P. N. (2022). Mathematical modelling for prediction of spread of corona virus and artificial intelligence/machine learning-based technique to detect COVID-19 via smartphone sensors. *International Journal of Modelling Identification and Control*, *41*(1-2), 43–52. doi:10.1504/IJMIC.2022.127096

Green, P. H. R., & Cellier, C. (2007). Celiac disease. *The New England Journal of Medicine*, *357*(17), 1731–1743. doi:10.1056/NEJMra071600 PMID:17960014

Groh, K. J., Backhaus, T., Carney-Almroth, B., Geueke, B., Inostroza, P. A., Lennquist, A., Leslie, H. A., Maffini, M., Slunge, D., Trasande, L., Warhurst, A. M., & Muncke, J. (2019). Overview of known plastic packaging-associated chemicals and their hazards. *The Science of the Total Environment*, *651*, 3253–3268. doi:10.1016/j.scitotenv.2018.10.015 PMID:30463173

Gu, B., Zhang, X., Bai, X., Fu, B., & Chen, D. (2019). Four steps to food security for swelling cities. *Nature*, *566*(7742), 31–33. doi:10.1038/d41586-019-00407-3 PMID:30718889

Gundewadi, G., Rudra, S. G., Sarkar, D. J., & Singh, D. (2018). Nanoemulsion based alginate organic coating for shelf life extension of okra. *Food Packaging and Shelf Life*, *18*, 1–12. doi:10.1016/j.fpsl.2018.08.002

Guo, F., Aryana, S., Han, Y., & Jiao, Y. (2018). A review of the synthesis and applications of polymer–nanoclay composites. *Applied Sciences (Basel, Switzerland)*, *8*(9), 1696. doi:10.3390/app8091696

Guo, X.-N., Wei, X.-M., & Zhu, K.-X. (2017). The impact of protein cross-linking induced by alkali on the quality of buckwheat noodles. *Food Chemistry*, *221*, 1178–1185. doi:10.1016/j.foodchem.2016.11.041 PMID:27979076

Gwak, Choi, Lee, Oh, & Park. (2018). *Isolation and characterization of a lytic and highly specific phage against Yersinia enterocolitica as a novel biocontrol agent*. Academic Press.

Hadjivassiliou, M. (2003). Gluten ataxia in perspective: Epidemiology, genetic susceptibility and clinical characteristics. *Brain*, *126*(3), 685–691. doi:10.1093/brain/awg050 PMID:12566288

Hadjivassiliou, M., Grünewald, R. A., Chattopadhyay, A. K., Davies-Jones, G. A. B., Gibson, A., Jarratt, J. A., Kandler, R. H., Lobo, A., Powell, T., & Smith, C. M. L. (1998). Clinical, radiological, neurophysiological, and neuropathological characteristics of gluten ataxia. *Lancet, 352*(9140), 1582–1585. doi:10.1016/S0140-6736(98)05342-2 PMID:9843103

Hadjivassiliou, M., Sanders, D. D., & Aeschlimann, D. P. (2015). Gluten-related disorders: Gluten ataxia. *Digestive Diseases (Basel, Switzerland), 33*(2), 264–268. doi:10.1159/000369509 PMID:25925933

Hafiz, R., Haque, M. R., Rakshit, A., & Uddin, M. S. (2022). Image-based soft drink type classification and dietary assessment system using deep convolutional neural network with transfer learning. *Journal of King Saud University. Computer and Information Sciences, 34*(5), 1775–1784. doi:10.1016/j.jksuci.2020.08.015

Halonen, T., Romero, J., & Melero, J. (Eds.). (2004). GSM, GPRS and EDGE performance: evolution towards 3G/UMTS. John Wiley & Sons.

Hamed, I., Jakobsen, A. N., & Lerfall, J. (2022). Sustainable edible packaging systems based on active compounds from food processing byproducts: A review. *Comprehensive Reviews in Food Science and Food Safety, 21*(1), 198–226. doi:10.1111/1541-4337.12870 PMID:34907649

Hameed, F., Bandral, J. D., Gupta, N., Nayik, G. A., Sood, M., & Rahman, R. (2022). Use of Bacteriophages as a Target Specific Therapy against Food-Borne Pathogens in Food Industry-a Review: Bacteriophage. *Journal of Microbiology, Biotechnology and Food Sciences, 11*(4), e2949–e2949. doi:10.55251/jmbfs.2949

Hanani, N. Z. A., Roos, Y. H., & Kerry, J. P. (2014). Use and application of gelatin as potential biodegradable packaging materials for food products. *International Journal of Biological Macromolecules, 71*, 94–102. doi:10.1016/j.ijbiomac.2014.04.027 PMID:24769086

Han, J. H. (2014). *Innovations in Food Packaging* (2nd ed.). Elsevier Ltd.

Han, J. W., Ruiz-Garcia, L., Qian, J. P., & Yang, X. T. (2018). Foodpackaging: A comprehensive review and future trends. *Comprehensive Reviews in Food Science and Food Safety, 17*(4), 860–877. doi:10.1111/1541-4337.12343 PMID:33350114

Hansen, T. L., Bhander, G. S., Christensen, T. H., Bruun, S., & Jensen, L. S. (2006). Life cycle modelling of environmental impacts of application of processed organic municipal solid waste on agricultural land (EASEWASTE). *Waste Management & Research, 24*(2), 153–166. doi:10.1177/0734242X06063053 PMID:16634230

Haq, I. U., Chaudhry, W. N., Akhtar, M. N., Andleeb, S., & Qadri, I. (2012). Bacteriophages and their implications on future biotechnology: A review. *Virology Journal, 9*(1), 1–8. doi:10.1186/1743-422X-9-9 PMID:22234269

Harada, L. K., Silva, E. C., Campos, W. F., Del Fiol, F. S., Vila, M., Dąbrowska, K., Krylov, V. N., & Balcão, V. M. (2018). Biotechnological applications of bacteriophages: State of the art. *Microbiological Research, 212*, 38–58. doi:10.1016/j.micres.2018.04.007 PMID:29853167

Harrer, A., Bücker, R., Boehm, M., Zarzecka, U., Tegtmeyer, N., Sticht, H., Schulzke, J. D., & Backert, S. (2019). Campylobacter jejuni enters gut epithelial cells and impairs intestinal barrier function through cleavage of occludin by serine protease HtrA. *Gut Pathogens*, *11*(1), 1–16. doi:10.118613099-019-0283-z PMID:30805031

Harrington, J. W., Patrick, P. A., Edwards, K. S., & Brand, D. A. (2006). Parental beliefs about autism: Implications for the treating physician. *Autism*, *10*(5), 452–462. doi:10.1177/1362361306066609 PMID:16940312

Hasan, M., & Ahn, J. (2022). Evolutionary dynamics between phages and bacteria as a possible approach for designing effective phage therapies against antibiotic-resistant bacteria. *Antibiotics (Basel, Switzerland)*, *11*(7), 915. doi:10.3390/antibiotics11070915 PMID:35884169

Hayıt, F., & Gül, H. (2017). *Tam karabuğday unu ve transglutaminaz ilavesinin kısmi pişirilerek dondurulmuş ekşi mayalı ekmeklerin fiziksel ve tekstürel özellikleri üzerine etkisi*. Academic Press.

He, C. X., He, Z. G., & Gao, J. Q. (2010). Microemulsions as drug delivery systems to improve the solubility and the bioavailability of poorly water-soluble drugs. *Expert Opinion on Drug Delivery*, *7*(4), 445–460. doi:10.1517/17425241003596337 PMID:20201713

Heilpern, A. J., & Waldor, M. K. (2003). pIIICTX, a predicted CTXφ minor coat protein, can expand the host range of coliphage fd to include Vibrio cholerae. *Journal of Bacteriology*, *185*(3), 1037–1044. doi:10.1128/JB.185.3.1037-1044.2003 PMID:12533480

Henkel, M., Müller, M. M., Kügler, J. H., Lovaglio, R. B., Contiero, J., Syldatk, C., & Hausmann, R. (2012). Rhamnolipids as biosurfactants from renewable resources: Concepts for next-generation rhamnolipid production. *Process Biochemistry (Barking, London, England)*, *47*(8), 1207–1219. doi:10.1016/j.procbio.2012.04.018

Henriksen, K., Rørbo, N., Rybtke, M. L., Martinet, M. G., Tolker-Nielsen, T., Høiby, N., Middelboe, M., & Ciofu, O. (2019). P. aeruginosa flow-cell biofilms are enhanced by repeated phage treatments but can be eradicated by phage–ciprofloxacin combination: —monitoring the phage–P. aeruginosa biofilms interactions. *Pathogens and Disease*, *77*(2), ftz011. doi:10.1093/femspd/ftz011 PMID:30821815

Hernando-Amado, S., Coque, T. M., Baquero, F., & Martínez, J. L. (2019). Defining and combating antibiotic resistance from One Health and Global Health perspectives. *Nature Microbiology*, *4*(9), 1432–1442. doi:10.103841564-019-0503-9 PMID:31439928

Hervonen, K., Salmi, T. T., Ilus, T., Paasikivi, K., Vornanen, M., Laurila, K., Lindfors, K., Viiri, K., Saavalainen, P., Collin, P., Kaukinen, K., & Reunala, T. (2016). Dermatitis herpetiformis refractory to gluten-free dietary treatment. *Acta Dermato-Venereologica*, *96*(1), 82–86. doi:10.2340/00015555-2184 PMID:26084552

He, Y., Li, H., Fei, X., & Peng, L. (2021). Carboxymethyl cellulose/cellulose nanocrystals immobilized silver nanoparticles as an effective coating to improve barrier and antibacterial properties of paper for food packaging applications. *Carbohydrate Polymers*, *252*, 117156. doi:10.1016/j.carbpol.2020.117156 PMID:33183607

Honarvar, Z., Hadian, Z., & Mashayekh, M. (2016). Nanocomposites in food packaging applications and their risk assessment for health. *Electronic Physician*, *8*(6), 2531–2538. doi:10.19082/2531 PMID:27504168

Hong, S. I., & Rhim, J. W. (2012). Preparation and properties of melt-intercalated linear low density polyethylene/clay nanocomposite films prepared by blow extrusion. *Lebensmittel-Wissenschaft + Technologie*, *48*(1), 43–51. doi:10.1016/j.lwt.2012.03.009

Hosseini, S. F., Rezaei, M., Zandi, M., & Ghavi, F. F. (2013). Preparation and functional properties of fish gelatin–chitosan blend edible films. *Food Chemistry*, *136*(3), 1490–1495. doi:10.1016/j.foodchem.2012.09.081 PMID:23194553

Hosseini, S. S., Khodaiyan, F., Kazemi, M., & Najari, Z. (2019). Optimization and characterization of pectin extracted from sour orange peel by ultrasound assisted method. *International Journal of Biological Macromolecules*, *125*, 621–629. doi:10.1016/j.ijbiomac.2018.12.096 PMID:30543886

Huang, C., Shi, J., Ma, W., Li, Z., Wang, J., Li, J., & Wang, X. J. (2018). Isolation, characterization, and application of a novel specific Salmonella bacteriophage in different food matrices. *Food Research International*, *111*, 631–641. doi:10.1016/j.foodres.2018.05.071 PMID:30007727

Huang, T., Yang, W., Wu, J., Ma, J., Zhang, X., & Zhang, D. (2019). A survey on green 6G network: Architecture and technologies. *IEEE Access : Practical Innovations, Open Solutions*, *7*, 175758–175768. doi:10.1109/ACCESS.2019.2957648

Hu, B., Liu, X., Zhang, C., & Zeng, X. (2017). Food macromolecule based nanodelivery systems for enhancing the bioavailability of polyphenols. *Yao Wu Shi Pin Fen Xi*, *25*(1), 3–15. PMID:28911541

Hulla, J. E., Sahu, S. C., & Hayes, A. W. (2015). Nanotechnology: History and future. *Human and Experimental Toxicology*, *34*(12), 1318–1321. doi:10.1177/0960327115603588 PMID:26614822

Hwang, J.-Y., Kim, J.-E., Song, Y.-J., & Park, J.-H. (2016). Safety of using Escherichia coli bacteriophages as a sanitizing agent based on inflammatory responses in rats. *Food Science and Biotechnology*, *25*(1), 355–360. doi:10.100710068-016-0050-6 PMID:30263278

Hyla, K., Dusza, I., & Skaradzińska, A. (2022). Recent Advances in the Application of Bacteriophages against Common Foodborne Pathogens. *Antibiotics (Basel, Switzerland)*, *11*(11), 1536. doi:10.3390/antibiotics11111536 PMID:36358191

Ignat, I., Volf, I., & Popa, V. I. (2011). A critical review of methods for characterisation of polyphenolic compounds in fruits and vegetables. *Food Chemistry*, *126*(4), 1821–1835. doi:10.1016/j.foodchem.2010.12.026 PMID:25213963

İlkay, H. Ö., & Mutlu, A. A. (2020). Fonksiyonel bir besin: Kinoa. *Beslenme ve Diyet Dergisi*, *48*(3), 92–101.

Ishaq, A., Ebner, P. D., Syed, Q. A., & Ur Rahman, H. U. (2020). Employing list-shield bacteriophage as a bio-control intervention for Listeria monocytogenes from raw beef surface and maintain meat quality during refrigeration storage. *Lebensmittel-Wissenschaft + Technologie*, *132*, 109784. doi:10.1016/j.lwt.2020.109784

Isik, E., Sagir, C., Tozlu, Z., & Ustaoglu, U. S. (2019). Determination of urban earthquake risk for Kırşehir, Türkiye. *Earth Sciences Research Journal*, *23*(3), 237–247. doi:10.15446/esrj.v23n3.60255

Islam, M. S., Yang, X., Euler, C. W., Han, X., Liu, J., Hossen, M. I., & Li, J. (2021). Application of a novel phage ZPAH7 for controlling multidrug-resistant Aeromonas hydrophila on lettuce and reducing biofilms. *Food Control*, *122*, 107785. doi:10.1016/j.foodcont.2020.107785

İşleroğlu, H., Dirim, S. N., & Ertekin, F. K. (2009). Gluten içermeyen, hububat esaslı alternatif ürün formülasyonları ve üretim teknolojileri. *Gida*, *34*(1), 29–36.

Ivanov, S., & Webster, C. (2023). *Restaurants and robots: public preferences for robot food and beverage services*. Academic Press.

Iwata, T. (2015). Biodegradable and bio-based polymers: Future prospects of eco-friendly plastics. *Angewandte Chemie International Edition in English*, *54*(11), 3210–3215. doi:10.1002/anie.201410770 PMID:25583677

Jabeen, N., Maqbool, Q., Bibi, T., Nazar, M., Hussain, S. Z., Hussain, T., Jan, T., Ahmad, I., Maaza, M., & Anwaar, S. (2018). Optimised synthesis of ZnO-nano-fertiliser through green chemistry: Boosted growth dynamics of economically important L. esculentum. *IET Nanobiotechnology / IET*, *12*(4), 405–411. doi:10.1049/iet-nbt.2017.0094 PMID:29768221

Jafari, S. M., & McClements, D. J. (2017). Nanotechnology approaches for increasing nutrient bioavailability. *Advances in Food and Nutrition Research*, *81*, 1–30. doi:10.1016/bs.afnr.2016.12.008 PMID:28317602

Jamshed, A. (2018). An Analysis of Trends, Developments and Transparent Issues in the making of Indian Smart Cities. *EAI Endorsed Transactions on Smart Cities*.

Javadi, A., Srithep, Y., Lee, J., Pilla, S., Clemons, C., Gong, S., & Turng, L. S. (2010). Processing and characterization of solid and microcellular PHBV/PBAT blend and its RWF/nanoclay composites. *Composites. Part A, Applied Science and Manufacturing*, *41*(8), 982–990. doi:10.1016/j.compositesa.2010.04.002

Javaid, M., & Haleem, A. (2020). Critical components of Industry 5.0 towards a successful adoption in the field of manufacturing. *Journal of Industrial Integration and Management*, *5*(03), 327–348. doi:10.1142/S2424862220500141

Jawaid, M., Qaiss, A. & Bouhfid, R. (2016). *Nanoclay reinforced polymer composites*. Academic Press.

Jawaid, M., Qaiss, A., & Bouhfid, R. (2016). *Nanoclay Reinforced Polymer Composites*. Springer. doi:10.1007/978-981-10-1953-1

Jeevahan, J., & Chandrasekaran, M. (2019). Nanoedible films for food packaging: A review. *Journal of Materials Science*, *54*(19), 12290–12318. doi:10.100710853-019-03742-y

Jiang, J., Pi, J., & Cai, J. (2018). The advancing of zinc oxide nanoparticles for biomedical applications. *Bioinorganic Chemistry and Applications*. doi:10.1155/2018/1062562 PMID:30073019

Johnson, P. A., Robinson, P. J., & Philpot, S. (2020). Type, tweet, tap, and pass: How smart city technology is creating a transactional citizen. *Government Information Quarterly*, *37*(1), 101414. Advance online publication. doi:10.1016/j.giq.2019.101414

Jończyk, E., Kłak, M., Międzybrodzki, R., & Górski, A. (2011). The influence of external factors on bacteriophages. *Folia Microbiologica*, *56*(3), 191–200. doi:10.100712223-011-0039-8 PMID:21625877

Jonhed, A., Andersson, C., & J¨arnstrom, ¨. L. (2008). Effects of film forming and hydrophobic properties of starches on surface sized packaging paper. *Packaging Technology & Science*, *21*(3), 123–135. doi:10.1002/pts.783

Joo, M., Lewandowski, N., Auras, R., Harte, J., & Almenar, E. (2011). Comparative shelf life study of blackberry fruit in bio-based and petroleum-based containers under retail storage conditions. *Food Chemistry*, *126*(4), 1734–1740. doi:10.1016/j.foodchem.2010.12.071 PMID:25213952

Joshi, D. C., Chaudhari, G. V., Sood, S., Kant, L., Pattanayak, A., Zhang, K., Fan, Y., Janovská, D., Meglič, V., & Zhou, M. (2019). Revisiting the versatile buckwheat: Reinvigorating genetic gains through integrated breeding and genomics approach. *Planta*, *250*(3), 783–801. doi:10.100700425-018-03080-4 PMID:30623242

Jovanovic-Malinovska, R., Kuzmanova, S., & Winkelhausen, E. (2015). Application of ultrasound for enhanced extraction of prebiotic oligosaccharides from selected fruits and vegetables. *Ultrasonics Sonochemistry*, *22*, 446–453. doi:10.1016/j.ultsonch.2014.07.016 PMID:25116595

Jun, J. W., Park, S. C., Wicklund, A., & Skurnik, M. (2018). Bacteriophages reduce Yersinia enterocolitica contamination of food and kitchenware. *International Journal of Food Microbiology*, *271*, 33–47. doi:10.1016/j.ijfoodmicro.2018.02.007 PMID:29477807

Juri'c, S., Ferrari, G., Velikov, K. P., & Donsì, F. (2019). High-pressure homogenization treatment to recover bioactive compounds from tomato peels. *Journal of Food Engineering*, *262*, 170–180. doi:10.1016/j.jfoodeng.2019.06.011

K, R., G, B., Banat, F., Show, P. L., & Cocoletzi, H. H. (2019). Mango leaf extract incorporated chitosan antioxidant film for active food packaging. *International Journal of Biological Macromolecules, 126*, 1234–1243.

Kahlon, T. S., Avena-Bustillos, R. J., Brichta, J. L., & Kahlon, A. K. (2019). High-protein nutritious flatbreads and an option for gluten-sensitive individuals. *Foods*, *8*(11), 591. doi:10.3390/foods8110591 PMID:31752409

Kah, M., Tufenkji, N., & White, J. C. (2019). Nano-enabled strategies to enhance crop nutrition and protection. *Nature Nanotechnology*, *14*(6), 532–540. doi:10.103841565-019-0439-5 PMID:31168071

Kakani, V., Nguyen, V. H., Kumar, B. P., Kim, H., & Pasupuleti, V. R. (2020). A critical review on Computer Vision and artificial intelligence in food industry. *Journal of Agriculture and Food Research, 2*, 100033. doi:10.1016/j.jafr.2020.100033

Kalita, D., &Baruah, S. (2019). The impact of nanotechnology on food. *Nanomaterials Applications for Environmental Matrices*, 369–379. doi:10.1016/B978-0-12-814829-7.00011-2

Kalpana, S., Priyadarshini, S. R., Leena, M. M., Moses, J. A., & Anandharamakrishnan, C. (2019). Intelligent packaging: Trends and applications in food systems. *Trends in Food Science & Technology, 93*, 145–157. doi:10.1016/j.tifs.2019.09.008

Kanai, K., & Tanaka, T. (1954). Measurement of the microtremor. *Bulletin of the Earthquake Research Institute, University of Tokyo, 32*, 192–209.

Kang, S., Pinault, M., Pfefferle, L. D., & Elimelech, M. (2007). Single-walled carbon nanotubes exhibit strong antimicrobial activity. *Langmuir, 23*(17), 8670–8673. doi:10.1021/la701067r PMID:17658863

Kaptchouang Tchatchouang, C.-D., Fri, J., De Santi, M., Brandi, G., Schiavano, G. F., Amagliani, G., & Ateba, C. N. (2020). Listeriosis outbreak in South Africa: A comparative analysis with previously reported cases worldwide. *Microorganisms, 8*(1), 135. doi:10.3390/microorganisms8010135 PMID:31963542

Karabulut, S., Ozcep, F., Korkmaz, B., & Zarif, H. (2011). Urban seismic microzonation studies at small scales: Sisli/Istanbul (Turkey). *Arabian Journal for Science and Engineering, 36*(1), 75–87. doi:10.100713369-010-0012-x

Kasimanickam, V., Kasimanickam, M., & Kasimanickam, R. (2021). Antibiotics use in food animal production: escalation of antimicrobial resistance: where are we now in combating AMR? *Medical Science, 9*(1), 14. PMID:33669981

Kasinathan, P., Pugazhendhi, S., & Alsharif, M. H. (2022). Realization of Sustainable Development Goals with Disruptive Technologies by Integrating Industry 5.0, Society 5.0, Smart Cities and Villages. *Sustainability (Basel), 14*(22), 15258. doi:10.3390u142215258

Kasman, L. M., & Porter, L. D. (2022). Bacteriophages. StatPearls Publishing.

Kasman, L. M., & Porter, L. D. (2023). Bacteriophages. StatPearls Publishing.

Kaur, S., & Kaur, N. (2017). Development and sensory evaluation of gluten free bakery products using quinoa (Chenopodium Quinoa) flour. *Journal of Applied and Natural Science, 9*(4), 2449–2455. doi:10.31018/jans.v9i4.1552

Kausar, A., Ahmad, I., Aldaghri, O., Ibnaouf, K. H., & Eisa, M. H. (2023). Nanoclay-Reinforced Nanocomposite Nanofibers—Fundamentals and State-of-the-Art Developments. *Minerals (Basel), 13*(6), 817. doi:10.3390/min13060817

Kaźmierczak, Z., Piotrowicz, A., Owczarek, B., Hodyra, K., Miernikiewicz, P., Lecion, D., Harhala, M., Górski, A., & Dąbrowska, K. (2014). Molecular imaging of T4 phage in mammalian tissues and cells. *Bacteriophage*, *4*(2), e28364. doi:10.4161/bact.28364 PMID:24653943

Keçeli, A. (2009). Uygulamalı jeofizik. *TMMOB Jeofizik Mühendisleri Odası Eğitim Yayınları, Ekim Ajans Matbaacılık Hizmetleri*, *9*, 479.

Keet, C. A., Matsui, E. C., Dhillon, G., Lenehan, P., Paterakis, M., & Wood, R. A. (2009). The natural history of wheat allergy. *Annals of Allergy, Asthma & Immunology*, *102*(5), 410–415. doi:10.1016/S1081-1206(10)60513-3 PMID:19492663

Kennouche, S., Le Moigne, N., Kaci, M., Quantin, J. C., Caro-Bretelle, A. S., Delaite, C., & Lopez-Cuesta, J. M. (2016). Morphological characterization and thermal properties of compatibilized poly (3-hydroxybutyrate-co-3-hydroxyvalerate)(PHBV)/poly (butylene succinate)(PBS)/halloysite ternary nanocomposites. *European Polymer Journal*, *75*, 142–162. doi:10.1016/j.eurpolymj.2015.12.009

Keskin, Ş., & Evlice, A. K. (2015). Fırın ürünlerinde kinoa kullanımı. *Tarla Bitkileri Merkez Araştırma Enstitüsü Dergisi*, *24*(2), 150–156.

Khalaj, M. J., Ahmadi, H., Lesankhosh, R., & Khalaj, G. (2016). Study of physical and mechanical properties of polypropylene nanocomposites for food packaging application: Nano-clay modified with iron nanoparticles. *Trends in Food Science & Technology*, *51*, 41–48. doi:10.1016/j.tifs.2016.03.007

Khan, S. T., Al-Khedhairy, A. A., & Musarrat, J. (2015). ZnO and TiO 2 nanoparticles as novel antimicrobial agents for oral hygiene: A review. *Journal of Nanoparticle Research*, *17*(6), 1–16. doi:10.100711051-015-3074-6

Khare, S., Tonk, A., & Rawat, A. (2018). Foodborne diseases outbreak in India: A Review. *International Journal of Food Sciences and Nutrition*, *3*, 9–10.

Khoddami, A., Wilkes, M. A., & Roberts, T. H. (2013). Techniques for analysis of plant phenolic compounds. *Molecules (Basel, Switzerland)*, *18*(2), 2328–2375. doi:10.3390/molecules18022328 PMID:23429347

Kilcher, S., Studer, P., Muessner, C., Klumpp, J., & Loessner, M. J. (2018). Cross-genus rebooting of custom-made, synthetic bacteriophage genomes in L-form bacteria. *Proceedings of the National Academy of Sciences of the United States of America*, *115*(3), 567–572. doi:10.1073/pnas.1714658115 PMID:29298913

Kilincceker, O., & Büyük, G. (2019). Amarant Amaranthus spp'ın Bazı Özellikleri ve Et Ürünlerinde Kullanımı. *Adyutayam Dergisi*, *7*(2), 36–42.

Kim, I., Viswanathan, K., Kasi, G., Thanakkasaranee, S., Sadeghi, K., & Seo, J. (2022). ZnO nanostructures in active antibacterial food packaging: Preparation methods, antimicrobial mechanisms, safety issues, future prospects, and challenges. *Food Reviews International*, *38*(4), 537–565. doi:10.1080/87559129.2020.1737709

Kim, K. M., Son, J. H., Kim, S.-K., Weller, C. L., & Hanna, M. A. (2006). Properties of chitosan films as a function of pH and solvent type. *Journal of Food Science, 71*(3), E119–E124. doi:10.1111/j.1365-2621.2006.tb15624.x

Kim, S. G., Lofgren, E. A., & Jabarin, S. A. (2013). Dispersion of nanoclays with poly (ethylene terephthalate) by melt blending and solid state polymerization. *Journal of Applied Polymer Science, 127*(3), 2201–2212. doi:10.1002/app.37796

Kim, S. W., & Cha, S. H. (2014). Thermal, mechanical, and gas barrier properties of ethylene–vinyl alcohol copolymer-based nanocomposites for food packaging films: Effects of nanoclay loading. *Journal of Applied Polymer Science, 131*(11), 40281–40288. doi:10.1002/app.40289

Kıraç. Altınşehir). Kentleşme ve Jeoloji Sempozyum Kitabı, 19-20 Kasım 1998, 189-202, İstanbul.

Knoerzer, K., Buckow, R., Trujillo, F. J., & Juliano, P. (2015). Multiphysics Simulation of Innovative Food Processing Technologies. *Food Engineering Reviews, 7*(2), 64–81. doi:10.100712393-014-9098-3

Knoerzer, K., Juliano, P., Roupas, P., & Versteeg, C. (2011). *Innovative food processing technologies: Advances in multiphysica simulation*. Wiley-Blackwell. doi:10.1002/9780470959435

Knorr, D., Froehling, A., Jaeger, H., Reineke, K., Schlueter, O., & Schoessler, K. (2011). Emerging technologies in food processing. *Annual Review of Food Science and Technology, 2*(1), 203–235. doi:10.1146/annurev.food.102308.124129 PMID:22129381

Koch, B. J., Hungate, B. A., & Price, L. B. (2017). Food-animal production and the spread of antibiotic resistance: The role of ecology. *Frontiers in Ecology and the Environment, 15*(6), 309–318. doi:10.1002/fee.1505

Koh, J. J., Zhang, X., & He, C. (2018). Fully biodegradable poly(lactic acid)/starch blends: A review of toughening strategies. *International Journal of Biological Macromolecules, 109*, 99–113. doi:10.1016/j.ijbiomac.2017.12.048 PMID:29248552

Ko, M.-J., Kwon, H.-L., & Chung, M.-S. (2016). Pilot-scale subcritical water extraction of flavonoids from satsuma mandarin (Citrus unshiu Markovich) peel. *Innovative Food Science & Emerging Technologies, 38*, 175–181. doi:10.1016/j.ifset.2016.10.008

Kondakci, T., & Zhou, W. (2017). Recent applications of advanced control techniques in food industry. *Food and Bioprocess Technology, 10*(3), 522–542. doi:10.100711947-016-1831-x

Konfo, T. R. C., Djouhou, F. M. C., Hounhouigan, M. H., Dahouenon-Ahoussi, E., Avlessi, F., & Sohounhloue, C. K. D. (2023). Recent advances in the use of digital technologies in agri-food processing: A short review. *Applied Food Research*, 100329.

Kongkaoroptham, P., Piroonpan, T., & Pasanphan, W. (2021). Chitosan nanoparticles based on their derivatives as antioxidant and antibacterial additives for active bioplastic packaging. *Carbohydrate Polymers, 257*, 117610. doi:10.1016/j.carbpol.2020.117610 PMID:33541641

Korehei, R., & Kadla, J. (2013). Incorporation of T4 bacteriophage in electrospun fibres. *Journal of Applied Microbiology*, *114*(5), 1425–1434. doi:10.1111/jam.12158 PMID:23445225

Koren, D., & Rus, K. (2021). Assessment of a city's performance under different earthquake scenarios. *1st Croatian Conference on Earthquake Engineering 1CroCEE*, 109-120. 10.5592/CO/1CroCEE.2021.155

Koshy, R. R., Mary, S. K., Thomas, S., & Pothan, L. A. (2015). Environment friendly green composites based on soy protein isolate – a review. *Food Hydrocolloids*, *50*, 174–192. doi:10.1016/j.foodhyd.2015.04.023

Kowalczyk, D., Skrzypek, T., & Łupina, K. (2020). Effect of carboxymethyl cellulose/candelilla wax edible coating incorporated with ascorbic acid on the physicochemical and sensory qualities of prepackaged minimally processed carrots (Daucus carota L.) during cold storage. *Journal of Food Processing and Preservation*, *44*(9), e14713. doi:10.1111/jfpp.14713

Kreutz, J. M., Adriaanse, M. P. M., van der Ploeg, E. M. C., & Vreugdenhil, A. C. E. (2020). Narrative review: Nutrient deficiencies in adults and children with treated and untreated celiac disease. *Nutrients*, *12*(2), 500. doi:10.3390/nu12020500 PMID:32075276

Kuai, L., Liu, F., Chiou, B.-S., Avena-Bustillos, R. J., McHugh, T. H., & Zhong, F. (2021). Controlled release of antioxidants from active food packaging: A review. *Food Hydrocolloids*, *120*, 106992. doi:10.1016/j.foodhyd.2021.106992

Kuek, M., McLean, S. K., & Palombo, E. A. (2022). Application of bacteriophages in food production and their potential as biocontrol agents in the organic farming industry. *Biological Control*, *165*, 104817. doi:10.1016/j.biocontrol.2021.104817

Kumar Pandey, B., Pandey, D., Nassa, V. K., Ahmad, T., Singh, C., George, A. S., & Wakchaure, M. A. (2021). Encryption and steganography-based text extraction in IoT using the EWCTS optimizer. *Imaging Science Journal*, *69*(1-4), 38–56. doi:10.1080/13682199.2022.2146885

Kumar, A., Pratush, A., & Bera, S. (2021). Significance of nanoscience in food microbiology: current trend and future prospects. *Nanotechnology for Advances in Medical Microbiology*, 249-267.

Kumar, K., Yadav, A. N., Kumar, V., Vyas, P., & Dhaliwal, H. S. (2017). Food waste: A potential bioresource for extraction of nutraceuticals and bioactive compounds. *Bioresources and Bioprocessing*, *4*(1), 18. doi:10.118640643-017-0148-6

Kumar, V., & Singh, L. (2018). Pectin from fruit peels and its uses as pharmaceutical and food grade: A descriptive review. *European Journal of Biomedical and Pharmaceutical Sciences*, *5*, 185–199.

Kupper, C. (2005). Dietary guidelines and implementation for celiac disease. *Gastroenterology*, *128*(4), S121–S127. doi:10.1053/j.gastro.2005.02.024 PMID:15825119

Kutyauripo, I., Rushambwa, M., & Chiwazi, L. (2023). Artificial intelligence applications in the agrifood sectors. *Journal of Agriculture and Food Research*, *11*, 100502. doi:10.1016/j.jafr.2023.100502

Laborde, D., Martin, W., Swinnen, J., & Vos, R. (2020). COVID-19 risks to global food security. *Science, 369*(6503), 500–502. doi:10.1126cience.abc4765 PMID:32732407

Lakshmi, S., Pandey, A., Ravi, N., Chauhan, O., Gopalan, N., & Sharma, R. (2017). Non-destructive quality monitoring of fresh fruits and vegetables. *Defence Life Science Journal, 2*(2), 103–110. doi:10.14429/dlsj.2.11379

Landry, M. P., & Mitter, N. (2019). How nanocarriers delivering cargos in plants can change the GMO landscape. *Nature Nanotechnology, 14*(6), 512–514. doi:10.103841565-019-0463-5 PMID:31168070

Lavelle, K., McDonnell, B., Fitzgerald, G., van Sinderen, D., & Mahony, J. (2023). Bacteriophage-host interactions in Streptococcus thermophilus and their impact on co-evolutionary processes. *FEMS Microbiology Reviews, 47*(4), fuad032. doi:10.1093/femsre/fuad032 PMID:37339909

Lavilla, M., Domingo-Calap, P., Sevilla-Navarro, S., & Lasagabaster, A. (2023). Natural Killers: Opportunities and Challenges for the Use of Bacteriophages in Microbial Food Safety from the One Health Perspective. *Foods, 12*(3), 552. doi:10.3390/foods12030552 PMID:36766081

Leiman & Shneider. (2012). Contractile tail machines of bacteriophages. *Viral Molecular Machines*, 93-114.

Leiman, P. G., Chipman, P. R., Kostyuchenko, V. A., Mesyanzhinov, V. V., & Rossmann, M. G. (2004). Three-dimensional rearrangement of proteins in the tail of bacteriophage T4 on infection of its host. *Cell, 118*(4), 419–429. doi:10.1016/j.cell.2004.07.022 PMID:15315755

Leite, M. S., Santos, M. A., Costa, E. M., Balieiro, A., Lima, Á. S., Sanchez, O. L., & Soares, C. (2019). Modeling of milk lactose removal by column adsorption using artificial neural networks: Mlp and Rbf. *Chemical Industry & Chemical Engineering Quarterly, 25*(4), 369–382. doi:10.2298/CICEQ180606015L

Leng, J., Sha, W., Wang, B., Zheng, P., Zhuang, C., Liu, Q., Wuest, T., Mourtzis, D., & Wang, L. (2022). Industry 5.0: Prospect and retrospect. *Journal of Manufacturing Systems, 65*, 279–295. doi:10.1016/j.jmsy.2022.09.017

Leonard, M. M., Sapone, A., Catassi, C., & Fasano, A. (2017). Celiac disease and nonceliac gluten sensitivity: A review. *Journal of the American Medical Association, 318*(7), 647–656. doi:10.1001/jama.2017.9730 PMID:28810029

Leon-Velarde, C. G., Jun, J. W., & Skurnik, M. (2019). Yersinia phages and food safety. *Viruses, 11*(12), 1105. doi:10.3390/v11121105 PMID:31795231

Leung, V., Groves, L., Szewczyk, A., Hosseinidoust, Z., & Filipe, C. D. (2018). Long-Term Antimicrobial Activity of Phage–Sugar Glasses is Closely Tied to the Processing Conditions. *ACS Omega, 3*(12), 18295–18303. doi:10.1021/acsomega.8b02679

Levin, B. R., Moineau, S., Bushman, M., & Barrangou, R. (2013). The population and evolutionary dynamics of phage and bacteria with CRISPR–mediated immunity. *PLOS Genetics, 9*(3), 1003312. doi:10.1371/journal.pgen.1003312 PMID:23516369

Liang, K. (2014). A DFA-based functional proxy re-encryption scheme for secure public cloud data sharing. *Information Forensics and Security, IEEE Transactions on, 9*(10), 1667-1680.

Liang, J., Yan, H., Zhang, J., Dai, W., Gao, X., Zhou, Y., Wan, X., & Puligundla, P. (2017). Preparation and Characterization of Antioxidant Edible Chitosan Films Incorporated with Epigallocatechin Gallate Nanocapsules. *Carbohydrate Polymers, 171*, 300–306. doi:10.1016/j.carbpol.2017.04.081 PMID:28578967

Li, C., Shi, T., Sun, Y., & Zhang, Y. (2022). A novel method to create efficient phage cocktails via use of phage-resistant bacteria. *Applied and Environmental Microbiology, 88*(6), e02323–e21. doi:10.1128/aem.02323-21 PMID:35080902

Li, F., Yu, H.-Y., Wang, Y.-Y., Zhou, Y., Zhang, H., Yao, J.-M., Abdalkarim, S. Y. H., & Tam, K. C. (2019). Natural biodegradable poly(3-hydroxybutyrate-co-3-hydroxyvalerate) nanocomposites with multifunctional cellulose nanocrystals/graphene oxide hybrids for high-performance food packaging. *Journal of Agricultural and Food Chemistry, 67*(39), 10954–10967. doi:10.1021/acs.jafc.9b03110 PMID:31365242

Li, J., Pettinato, M., Campardelli, R., De Marco, I., & Perego, P. (2022). High-Pressure Technologies for the Recovery of Bioactive Molecules from Agro-Industrial Waste. *Applied Sciences (Basel, Switzerland), 12*(7), 3642. doi:10.3390/app12073642

Li, J., Zhao, F., Zhan, W., Li, Z., Zou, L., & Zhao, Q. (2022). Challenges for the application of bacteriophages as effective antibacterial agents in the food industry. *Journal of the Science of Food and Agriculture, 102*(2), 461–471. doi:10.1002/jsfa.11505 PMID:34487550

Lim, Y., Edelenbos, J., & Gianoli, A. (2022). Dynamics in the governance of smart cities: Insights from South Korean smart cities. *Int. J. Unity Sci.*, 1–23. doi:10.1080/12265934.2022.2063158

Lin, D. M., Koskella, B., & Lin, H. C. (2017). Phage therapy: An alternative to antibiotics in the age of multi-drug resistance. *World Journal of Gastrointestinal Pharmacology and Therapeutics, 8*(3), 162. doi:10.4292/wjgpt.v8.i3.162 PMID:28828194

Liné, C., Larue, C., & Flahaut, E. (2017). Carbon nanotubes: Impacts and behaviour in the terrestrial ecosystem-A review. *Carbon, 123*, 767–785. doi:10.1016/j.carbon.2017.07.089

Ling, H., Lou, X., Luo, Q., He, Z., Sun, M., & Sun, J. (2022). Recent advances in bacteriophage-based therapeutics: Insight into the post-antibiotic era. *Acta Pharmaceutica Sinica. B, 12*(12), 4348–4364. doi:10.1016/j.apsb.2022.05.007 PMID:36561998

Lin, J., Pan, D., Sun, Y., Ou, C., Wang, Y., & Cao, J. (2019). The modification of gelatin films: Based on various cross-linking mechanism of glutaraldehyde at acidic and alkaline conditions. *Food Sciences and Nutrition, 7*(12), 4140–4146. PMID:31890193

Liu, A., Liu, Y., Peng, L., Cai, X., Shen, L., Duan, M., Ning, Y., Liu, S., Li, C., Liu, Y., Chen, H., Wu, W., Wang, X., Hu, B., & Li, C. (2020). Characterization of the narrow-spectrum bacteriophage LSE7621 towards Salmonella enteritidis and its biocontrol potential on lettuce and tofu. *Lebensmittel-Wissenschaft + Technologie, 118*, 108791. doi:10.1016/j.lwt.2019.108791

Liu, C., Hong, Q., Chang, R. Y. K., Kwok, P. C. L., & Chan, H. K. (2022). Phage–Antibiotic therapy as a promising strategy to combat multidrug-resistant infections and to enhance antimicrobial efficiency. *Antibiotics (Basel, Switzerland)*, *11*(5), 570. doi:10.3390/antibiotics11050570 PMID:35625214

Liu, H. K., Guo, Y., & Chen, L. (2023). Crowdsourcing Smart City: SmartTaipei Project. *Proceedings of the 24th Annual International Conference on Digital Government Research*. 10.1145/3598469.3598474

Liu, P., Xu, Y., Gao, X., Zhu, X., Du, M., Wang, Y., Deng, R., & Gao, J. (2017). Optimization of ultrasonic-assisted extraction of oil from the seed kernels and isolation of monoterpene glycosides from the oil residue of Paeonia lactiflora pall. *Industrial Crops and Products*, *107*, 260–270. doi:10.1016/j.indcrop.2017.04.013

Liu, Q., Tang, G. Y., Zhao, C. N., Feng, X. L., Xu, X. Y., Cao, S. Y., Meng, X., Li, S., Gan, R.-Y., & Li, H. B. (2018). Comparison of antioxidant activities of different grape varieties. *Molecules (Basel, Switzerland)*, *23*(10), 2432. doi:10.3390/molecules23102432 PMID:30249027

Liu, Q., Zhang, M., Bhandari, B., Xu, J., & Yang, C. (2020). Effects of nanoemulsion-based active coatingswith composite mixture of star anise essential oil, polylysine, and nisin on the quality and shelf life of ready-to-eat Yao meat products. *Food Control*, *107*, 106771. doi:10.1016/j.foodcont.2019.106771

Li, Y., Tao, F., Cui, K., Song, Y., Nan, L., Cui, C., Li, Y., Yang, J., Wang, Y., & Jiang, L. (2020). Optimization of ultrasonic extraction of anthocyanin in mulberry residue by response surface methodology. *IOP Conference Series. Earth and Environmental Science*, *559*(1), 012024. doi:10.1088/1755-1315/559/1/012024

Lizcano, S. C., Davila, ´. J. A., & Hernandez, ´. V. (2019). Fruit agroindustrial wastes for preparing beverages for medicinal purposes by supercritical fluid extraction technology: Andes berry (Rubus glaucus benth) case. *Prod. Manage. Bev.*, *1*, 151–177. doi:10.1016/B978-0-12-815260-7.00005-5

Lluís Cabedo, E. (2004). Development of EVOH-kaolinite nanocomposites. *Polymer*, *45*(15), 5233–5238. doi:10.1016/j.polymer.2004.05.018

Lone, A., Anany, H., Hakeem, M., Aguis, L., Avdjian, A.-C., Bouget, M., Atashi, A., Brovko, L., Rochefort, D., & Griffiths, M. W. (2016). Development of prototypes of bioactive packaging materials based on immobilized bacteriophages for control of growth of bacterial pathogens in foods. *International Journal of Food Microbiology*, *217*, 49–58. doi:10.1016/j.ijfoodmicro.2015.10.011 PMID:26490649

Longo, F., Padovano, A., & Umbrello, S. (2020). Value-oriented and ethical technology engineering in industry 5.0: A human-centric perspective for the design of the factory of the future. *Applied Sciences (Basel, Switzerland)*, *10*(12), 4182. doi:10.3390/app10124182

Lukinac, J., Jukić, M., Mastanjević, K., & Lučan, M. (2018). Application of computer vision and image analysis method in cheese-quality evaluation: A review. *Ukrainian Food Journal*, *7*(7, Issue 2), 192–214. doi:10.24263/2304-974X-2018-7-2-4

Luo, Y., Wang, Q., & Zhang, Y. (2020). Biopolymer-based nanotechnology approaches to deliver bioactive compounds for food applications: A perspective on the past, present, and future. *Journal of Agricultural and Food Chemistry*, *68*(46), 12993–13000. doi:10.1021/acs.jafc.0c00277 PMID:32134655

Lu, T. K., & Collins, J. J. (2007). Dispersing biofilms with engineered enzymatic bacteriophage. *Proceedings of the National Academy of Sciences of the United States of America*, *104*(27), 11197–11202. doi:10.1073/pnas.0704624104 PMID:17592147

Lutz, M., & Bascuñán-Godoy, L. (2017). The revival of quinoa: a crop for health. *Superfood and Functional Food—an Overview of Their Processing and Utilization.* InTech.

Lu, Z., Wang, J., Gao, R., Ye, F., & Zhao, G. (2019). Sustainable valorisation of tomato pomace: A comprehensive review. *Trends in Food Science & Technology*, *86*, 172–187. doi:10.1016/j.tifs.2019.02.020

Maftoonazad, N., & Ramaswamy, H. S. (2019). Application and evaluation of a pectin based edible coating process for quality change kinetics and shelf-life extension of lime fruit (Citrus aurantifolium). *Coatings*, *9*(5), 1–14. doi:10.3390/coatings9050285

Magsi, H., Sodhro, A. H., Chachar, F. A., Abro, S. A. K., Sodhro, G. H., & Pirbhulal, S. (2018, March). Evolution of 5G in Internet of medical things. In *2018 international conference on computing, mathematics and engineering technologies (iCoMET)* (pp. 1-7). IEEE.

Mahato, K., Prasad, A., Maurya, P., & Chandra, P. (2016). Nanobiosensors: Next generation point-of-care biomedical devices for personalized diagnosis. *Journal of Analytical & Bioanalytical Techniques*, *7*(2).

Majeed, K., Jawaid, M., Hassan, A. A. B. A. A., Bakar, A. A., Khalil, H. A., Salema, A. A., & Inuwa, I. (2013). Potential materials for food packaging from nanoclay/natural fibres filled hybrid composites. *Materials & Design (1980-2015)*, *46*, 391-410.

Majeed, K., Jawaid, M., Hassan, A., Abu Bakar, A., Abdul Khalil, H. P. S., Salema, A. A., & Inuwa, I. (2013). Potential materials for food packaging from nanoclay/natural fibres filled hybrid composites. *Materials & Design*, *46*, 391–410. doi:10.1016/j.matdes.2012.10.044

Majid, I., Nayik, G. A., Dar, S. M., & Nanda, V. (2018). Novel food packaging technologies: Innovations and future prospective. *Journal of the Saudi Society of Agricultural Sciences*, *17*(4), 454–462. doi:10.1016/j.jssas.2016.11.003

Makdoud, S., & Rosentrater, K. A. (2017). Development and Testing of Gluten-Free Pasta Based on Rice, Quinoa and Amaranth Flours. Iowa State University. *Agricultural and Biosystems Engineering, Food Science and Human Nutrition. Journal of Food Research*, *6*(4), 887–1927. doi:10.5539/jfr.v6n4p91

Makkar, T., & Verma, S., Yogesh, & Dubey, A. K. (2018). *Analysis and detection of fruit defect using neural network.* Paper presented at the Data Science and Analytics: 4th International Conference on Recent Developments in Science, Engineering and Technology, REDSET 2017, Gurgaon, India. 10.1007/978-981-10-8527-7_46

Malekzadeh, R., Sachdev, A., & Ali, A. F. (2005). Coeliac disease in developing countries: Middle East, India and North Africa. *Best Practice & Research. Clinical Gastroenterology, 19*(3), 351–358. doi:10.1016/j.bpg.2005.01.004 PMID:15925841

Mallegni, N., Phuong, T. V., Coltelli, M. B., Cinelli, P., & Lazzeri, A. (2018). Poly (lactic acid) (PLA) based tear resistant and biodegradable flexible films by blown film extrusion. *Materials (Basel), 11*(1), 148. doi:10.3390/ma11010148 PMID:29342099

Malucelli, G., Ronchetti, S., Lak, N., Priola, A., Dintcheva, N. T., & la Mantia, F. P. (2007). Intercalation effects in LDPE/o-montmorillonites nanocomposites. *European Polymer Journal, 43*(2), 328–335. doi:10.1016/j.eurpolymj.2006.11.024

Manzoor, M. F., Zeng, X.-A., Rahaman, A., Siddeeg, A., Aadil, R. M., Ahmed, Z., Li, J., & Niu, D. (2019). Combined impact of pulsed electric field and ultrasound on bioactive compounds and FT-ir analysis of Almond extract. *Journal of Food Science and Technology, 56*(5), 2355–2364. doi:10.100713197-019-03627-7 PMID:31168118

Marchi, J., Zborowsky, S., Debarbieux, L., & Weitz, J. S. (2023). The dynamic interplay of bacteriophage, bacteria and the mammalian host during phage therapy. *iScience, 26*(2), 106004. doi:10.1016/j.isci.2023.106004 PMID:36818291

Marín, M., Artola, A., & Sánchez, A. (2019). Optimization of down-stream for cellulases produced under solid-state fermentation of coffee husk. *Waste and Biomass Valorization, 10*(10), 2761–2772. doi:10.100712649-018-0327-5

Martínez-Villaluenga, C., Peñas, E., & Hernández-Ledesma, B. (2020). Pseudocereal grains: Nutritional value, health benefits and current applications for the development of gluten-free foods. *Food and Chemical Toxicology, 137*, 111178. doi:10.1016/j.fct.2020.111178 PMID:32035214

Martiny, T. R., Pacheco, B. S., Pereira, C. M., Mansilla, A., Astorga-España, M. S., Dotto, G. L., Moraes, C. C., & Rosa, G. S. (2020). A novel biodegradable film based on κ-carrageenan activated with olive leaves extract. *Food Science & Nutrition, 8*, 3147–3156.

Marvin, H. J., Bouzembrak, Y., Van der Fels-Klerx, H., Kempenaar, C., Veerkamp, R., & Chauhan, A. (2022). *Digitalisation and Artificial Intelligence for sustainable food systems*. Academic Press.

Marzari, R., Sblattero, D., Righi, M., & Bradbury, A. (1997). Extending filamentous phage host range by the grafting of a heterologous receptor binding domain. *Gene, 185*(1), 27–33. doi:10.1016/S0378-1119(96)00623-3 PMID:9034309

Ma, T., Zhao, H., Wang, J., & Sun, B. (2019). Effect of processing conditions on the morphology and oxidative stability of lipid microcapsules during complex coacervation. *Food Hydrocolloids, 87*, 637–643. doi:10.1016/j.foodhyd.2018.08.053

Matle, I., Mbatha, K. R., & Madoroba, E. (2020). A review of Listeria monocytogenes from meat and meat products: Epidemiology, virulence factors, antimicrobial resistance and diagnosis. *The Onderstepoort Journal of Veterinary Research, 87*(1), 1–20. doi:10.4102/ojvr.v87i1.1869 PMID:33054262

Matthews, L., Low, J., Gally, D., Pearce, M., Mellor, D., Heesterbeek, J., ... Reid, S. (2006). Heterogeneous shedding of Escherichia coli O157 in cattle and its implications for control. *Proceedings of the National Academy of Sciences of the United States of America, 103*(3), 547–552. doi:10.1073/pnas.0503776103 PMID:16407143

Maurya, N. K., & Arya, P. (2018). Amaranthus grain nutritional benefits: A review. *Journal of Pharmacognosy and Phytochemistry, 7*(2), 2258–2262.

Mavani, N. R., Ali, J. M., Othman, S., Hussain, M., Hashim, H., & Rahman, N. A. (2022). Application of artificial intelligence in food industry—A guideline. *Food Engineering Reviews, 14*(1), 134–175. doi:10.100712393-021-09290-z

Ma, Y., Pacan, J. C., Wang, Q., Sabour, P. M., Huang, X., & Xu, Y. (2012). Enhanced alginate microspheres as means of oral delivery of bacteriophage for reducing Staphylococcus aureus intestinal carriage. *Food Hydrocolloids, 26*(2), 434–440. doi:10.1016/j.foodhyd.2010.11.017

Medina-Torres, N., Ayora-Talavera, T., Espinosa-Andrews, H., Sánchez-Contreras, A., & Pacheco, N. (2017). Ultrasound assisted extraction for the recovery of phenolic compounds from vegetable sources. *Agronomy (Basel), 7*(3), 47. doi:10.3390/agronomy7030047

Mehdizadeh, T., Tajik, H., Langroodi, A. M., Molaei, R., & Mahmoudian, A. (2020). Chitosan-starch film containing pomegranate peel extract and Thymus kotschyanus essential oil can prolong the shelf life of beef. *Meat Science, 163*, 108073.

Mehta, R., Kumar, V., Bhunia, H., & Upadhyay, S. N. (2005). Synthesis of poly(lactic acid): A review. Journal of Macromolecular Science. *Part C, 45*(4), 325–349.

Mekonnen, T., Mussone, P., Khalil, H., & Bressler, D. (2013). Progress in bio-based plastics and plasticizing modifications. *Journal of Materials Chemistry. A, Materials for Energy and Sustainability, 1*(43), 13379–13398. doi:10.1039/c3ta12555f

Melikoglu, M., Lin, C. S. K., & Webb, C. (2013). Analysing global food waste problem: Pinpointing the facts and estimating the energy content. *Central European Journal of Engineering, 3*, 157–164.

Melini, V., & Melini, F. (2019). Gluten-free diet: Gaps and needs for a healthier diet. *Nutrients, 11*(1), 170. doi:10.3390/nu11010170 PMID:30650530

Mellinas, A. C., Jim'enez, A., & Garrigos, ´. M. C. (2020). Optimization of microwave-assisted extraction of cocoa bean shell waste and evaluation of its antioxidant, physicochemical and functional properties. *Lebensmittel-Wissenschaft + Technologie, 127*, 109361. doi:10.1016/j.lwt.2020.109361

Mellinas, C., Solaberrieta, I., Pelegrín, C. J., Jiménez, A., & Garrigós, M. C. (2022). Valorization of Agro-Industrial Wastes by Ultrasound-Assisted Extraction as a Source of Proteins, Antioxidants and Cutin: A Cascade Approach. *Antioxidants*, *11*(9), 1739. doi:10.3390/antiox11091739 PMID:36139813

Mellinas, C., Vald'es, A., Ramos, M., Burgos, N., Garrigos, M. D. C., & Jim'enez, A. (2016). Active edible films: Current state and future trends. *Journal of Applied Polymer Science*, *133*(2), 1–15. doi:10.1002/app.42631

Memisi, N., Moracanin, S. V., Milijasevic, M., Babic, J., & Djukic, D. (2015). CIP cleaning processes in the dairy industry. *Procedia Food Science*, *5*, 184–186. doi:10.1016/j.profoo.2015.09.052

Mendieta, C. M., González, G., Vallejos, M. E., & Area, M. C. (2022). Bio-polyethylene furanoate (Bio-PEF) from lignocellulosic biomass adapted to the circular bioeconomy. *BioResources*, *17*(4), 7313–7337. doi:10.15376/biores.17.4.Mendieta

Menta, P. L. R., Andrade, M. E. R., Leocádio, P. C. L., Fraga, J. R., Dias, M. T. S., Cara, D. C., Cardoso, V. N., Borges, L. F., Capettini, L. S. A., Aguilar, E. C., & Alvarez-Leite, J. I. (2019). Wheat gluten intake increases the severity of experimental colitis and bacterial translocation by weakening of the proteins of the junctional complex. *British Journal of Nutrition*, *121*(4), 361–373. doi:10.1017/S0007114518003422 PMID:30554574

Metin, S. (2016). Çölyak hastalığında nutrisyon. *Güncel Gastroenteroloji*, *20*(3), 259–262.

Metselaar, K. I., den Besten, H. M., Abee, T., Moezelaar, R., & Zwietering, M. H. (2013). Isolation and quantification of highly acid resistant variants of Listeria monocytogenes. *International Journal of Food Microbiology*, *166*(3), 508–514. doi:10.1016/j.ijfoodmicro.2013.08.011 PMID:24051176

Mikula, K., Gersz, A., Witek-Krowiak, A., Skrzypczak, D., Izydorczyk, G., & Chojnacka, K. (2022). Agrochemicals in view of circular economy. In *Smart Agrochemicals for Sustainable Agriculture* (pp. 57–80). Academic Press. doi:10.1016/B978-0-12-817036-6.00004-2

Milani, J. M., & Nemati, A. (2022). Lipid-based edible films and coatings: A review of recent advances and applications. *Journal of Packaging Technology and Research*, *6*(1), 11–22. doi:10.100741783-021-00130-3

Miranda, M., Marilene De Mori, M. R., Spricigo, P. C., Pilon, L., Mitsuyuki, M. C., Correa, D. S., & Ferreira, M. D. (2022). Carnauba wax nanoemulsion applied as an edible coating on fresh tomato for postharvest quality evaluation. *Heliyon*, *8*(7), e09803. doi:10.1016/j.heliyon.2022.e09803 PMID:35800251

Mizan, M. F. R., Cho, H. R., Ashrafudoulla, M., Cho, J., Hossain, M. I., Lee, D.-U., & Ha, S.-D. (2020). The effect of physico-chemical treatment in reducing Listeria monocytogenes biofilms on lettuce leaf surfaces. *Biofouling*, *36*, 1243–1255. PMID:33401969

Mohammad, Z. H., Ahmad, F., Ibrahim, S. A., & Zaidi, S. (2022). Application of nanotechnology in different aspects of the food industry. *Discover Food*, *2*(1), 12. doi:10.100744187-022-00013-9

Mokwena, K. K., & Tang, J. (2012). Ethylene vinyl alcohol: A review of barrier properties for packaging shelf stable foods. *Critical Reviews in Food Science and Nutrition*, *52*(7), 640–650. doi:10.1080/10408398.2010.504903 PMID:22530715

Molinaro, S., Cruz Romero, M., Boaro, M., Sensidoni, A., Lagazio, C., Morris, M., & Kerry, J. (2013). Effect of nanoclay-type and PLA optical purity on the characteristics of PLA-based nanocomposite films. *Journal of Food Engineering*, *117*(1), 113–123. doi:10.1016/j.jfoodeng.2013.01.021

Monteiro, A. L., Freitas Souza, M., Lins, H. A., Teófilo, T. M. S., Barros Júnior, A. P., Silva, D. V., & Mendonça, V. (2021). A new alternative to determine weed control in agricultural systems based on artificial neural networks (ANNs). *Field Crops Research*, *263*, 108075. doi:10.1016/j.fcr.2021.108075

Morales-Jiménez, M., Gouveia, L., Yáñez-Fernández, J., Castro-Muñoz, R., & Barragán-Huerta, B. (2020). Production, preparation and characterization of microalgae-based biopolymer as a potential bioactive film. *Coatings*, *10*(2), 120. doi:10.3390/coatings10020120

Morrar, R., Arman, H., & Mousa, S. (2017). The fourth industrial revolution (Industry 4.0): A social innovation perspective. *Technology Innovation Management Review*, *7*(11), 12–20. doi:10.22215/timreview/1117

Morsy, M. K., Khalaf, H. H., Sharoba, A. M., El-Tanahi, H. H., & Cutter, C. N. (2014). Incorporation of essential oils and nanoparticles in pullulan films to control foodborne pathogens on meat and poultry products. *Journal of Food Science*, *79*(4), M675–M684. doi:10.1111/1750-3841.12400 PMID:24621108

Motosaka, M., & Nagana, M. (1997). Analysis of amplification characteristics of ground motions in the heavily damaged belt zone during the 1995 Hyogo-Ken Nanbu (Kobe) Earthquake. *Earthquake Engineering & Structural Dynamics*, *26*(3), 377–393. doi:10.1002/(SICI)1096-9845(199703)26:3<377::AID-EQE649>3.0.CO;2-0

Moye, Z. D., Woolston, J., & Sulakvelidze, A. (2018). Bacteriophage applications for food production and processing. *Viruses*, *10*(4), 205. doi:10.3390/v10040205 PMID:29671810

Mozafari, M., Johnson, C., Hatziantoniou, S., & Demetzos, C. (2008). Nanoliposomes and their applications in food nanotechnology. *Journal of Liposome Research*, *18*(4), 309–327. doi:10.1080/08982100802465941 PMID:18951288

Mulder, C. J. J., Van Bergeijk, J. D., Jansen, T. L. T. A., & Uil, J. J. (1993). Coeliac Disease. *Scandinavian Journal of Gastroenterology, 28*(sup200), 42–47.

Murrieta-Martínez, C. L., Soto-Valdez, H., Pacheco-Aguilar, R., Torres-Arreola, W., Rodríguez-Felix, F., Marquez, ´., & Ríos, E. (2018). Edible protein films: Sources and behavior. *Packaging Technology & Science*, *31*(3), 113–122. doi:10.1002/pts.2360

Nabeel, M., & Bertino, E. (2012). Privacy-Preserving Fine-Grained Access Control in Public Clouds. *IEEE Data Eng. Bull.*, *35*(4), 21–30.

Nahavandi, S. (2019). Industry 5.0—A human-centric solution. *Sustainability (Basel)*, *11*(16), 4371. doi:10.3390u11164371

Nandhini, M., Kala, K. U., Thangadarshini, M., & Madhusudhana Verma, S. (2022). Deep Learning model of sequential image classifier for crop disease detection in plantain tree cultivation. *Computers and Electronics in Agriculture*, *197*, 106915. doi:10.1016/j.compag.2022.106915

Naqash, F., Gani, A., Gani, A., & Masoodi, F. A. (2017). Gluten-free baking: Combating the challenges-A review. *Trends in Food Science & Technology*, *66*, 98–107. doi:10.1016/j.tifs.2017.06.004

Nasr, N. F. (2015). Applications of nanotechnology in food microbiology. *International Journal of Current Microbiology and Applied Sciences*, *4*(4), 846–853.

Naylor, S. W., Low, J. C., Besser, T. E., Mahajan, A., Gunn, G. J., Pearce, M. C., McKendrick, I. J., Smith, D. G. E., & Gally, D. L. (2003). Lymphoid follicle-dense mucosa at the terminal rectum is the principal site of colonization of enterohemorrhagic Escherichia coli O157: H7 in the bovine host. *Infection and Immunity*, *71*(3), 1505–1512. doi:10.1128/IAI.71.3.1505-1512.2003 PMID:12595469

Nazir, A., Xu, X., Liu, Y., & Chen, Y. (2023). Phage Endolysins: Advances in the World of Food Safety. *Cells*, *12*(17), 2169. doi:10.3390/cells12172169 PMID:37681901

Ncube, L. K., Ude, A. U., Ogunmuyiwa, E. N., Zulkifli, R., & Beas, I. N. (2020). Environmental impact of food packaging materials: A review of contemporary development from conventional plastics to polylactic acid based materials. *Materials (Basel)*, *13*(21), 4994. doi:10.3390/ma13214994 PMID:33171895

Nehra, V., Marietta, E. V, & Murray, J. A. (2014). Celiac disease and its therapy: current approaches and new advances. *Wheat and Rice in Disease Prevention and Health*, 143–155.

Nelson, D. L., & Cox, M. M. (2008). *Lehninger Principles of Biochemistry* (5th ed.). W.H. Freeman and Company.

Nerin, C., Aznar, M., & Carrizo, D. (2016). Food contamination during food process. *Trends in Food Science & Technology*, *48*, 63–68. doi:10.1016/j.tifs.2015.12.004

Newberry, C., McKnight, L., Sarav, M., & Pickett-Blakely, O. (2017). Going gluten free: The history and nutritional implications of today's most popular diet. *Current Gastroenterology Reports*, *19*(11), 1–8. doi:10.100711894-017-0597-2 PMID:28948465

Newell, D., Elvers, K., Dopfer, D., Hansson, I., Jones, P., James, S., ... Connerton, I. (2011). Biosecurity-based interventions and strategies to reduce Campylobacter spp. on poultry farms. *Applied and Environmental Microbiology*, *77*(24), 8605–8614. doi:10.1128/AEM.01090-10 PMID:21984249

Newell, D., & Fearnley, C. (2003). Sources of Campylobacter colonization in broiler chickens. *Applied and Environmental Microbiology*, *69*(8), 4343–4351. doi:10.1128/AEM.69.8.4343-4351.2003 PMID:12902214

Nguyen, M. M., Gil, J., Brown, M., Cesar Tondo, E., Soraya Martins De Aquino, N., Eisenberg, M., & Erickson, S. (2020). Accurate and sensitive detection of Salmonella in foods by engineered bacteriophages. *Scientific Reports*, *10*(1), 17463. doi:10.103841598-020-74587-8 PMID:33060781

Niland, B., & Cash, B. D. (2018). Health benefits and adverse effects of a gluten-free diet in non–celiac disease patients. *Gastroenterology & Hepatology*, *14*(2), 82. PMID:29606920

Nile, S. H., Baskar, V., Selvaraj, D., Nile, A., Xiao, J., & Kai, G. (2020). Nanotechnologies in food science: Applications, recent trends, and future perspectives. *Nano-Micro Letters*, *12*(1), 1–34. doi:10.100740820-020-0383-9 PMID:34138283

Ni, Y., Sun, J., & Wang, J. (2021). Enhanced antimicrobial activity of konjac glucomannan nanocomposite films for food packaging. *Carbohydrate Polymers*, *267*, 118215. doi:10.1016/j.carbpol.2021.118215 PMID:34119169

Nouri, A., Yaraki, M. T., Ghorbanpour, M., & Wang, S. (2018). Biodegradable κ-carrageenan/nanoclay nanocomposite films containing Rosmarinus officinalis L. extract for improved strength and antibacterial performance. *International Journal of Biological Macromolecules*, *115*, 227–235. doi:10.1016/j.ijbiomac.2018.04.051 PMID:29660461

Nourmohammadi, A., Azar, H. H., & Aminzare, M. (2023). The effects of whey protein/nanoclay biocomposite containing Thymus fedtschenkoi Ronneger essential oil and resveratrol on the shelf life of liqvan cheese during refrigerated storage. *Lebensmittel-Wissenschaft + Technologie*, *187*, 115175. doi:10.1016/j.lwt.2023.115175

Nyabako, T., Mvumi, B. M., Stathers, T., Mlambo, S., & Mubayiwa, M. (2020). Predicting Prostephanus truncatus (Horn) (Coleoptera: Bostrichidae) populations and associated grain damage in smallholder farmers' maize stores: A machine learning approach. *Journal of Stored Products Research*, *87*, 101592. doi:10.1016/j.jspr.2020.101592

O'connell, A., Ruegg, P., Jordan, K., O'brien, B., & Gleeson, D. (2016). The effect of storage temperature and duration on the microbial quality of bulk tank milk. *Journal of Dairy Science*, *99*(5), 3367–3374. doi:10.3168/jds.2015-10495 PMID:26947309

O'Shea, N., Greene, D., & Fenelon, M. A. (2023). *Artificial Intelligence in Food Safety Reference Module in Food Science*. Elsevier.

O'Shea, Y. A., & Boyd, E. F. (2002). Mobilization of the Vibrio pathogenicity island between Vibrio cholerae isolates mediated by CP-T1 generalized transduction. *FEMS Microbiology Letters*, *214*(2), 153–157. doi:10.1111/j.1574-6968.2002.tb11339.x PMID:12351223

Okamoto, S. (1973). Introduction to earthquake engineering. University of Tokyo Press.

Olicón-Hernández, D. R., Acosta-Sánchez, Á., Monterrubio-López, R., & Guerra-Sánchez, G. (2019). Chitosan and Opuntia ficus indica mucilage as the base of a polymeric edible film for the protection of tomatoes against Rhizopus stolonifer. *TIP Revista Especializada en Ciencias Químico-Biológicas, 22*(1), 1–9.

Oliveira Filho, J. G. D., Silva, G. D. C., Oldoni, F. C. A., Miranda, M., Florencio, C., Oliveira, R. M. D. D., Gomes, M. P., & Ferreira, M. D. (2022). Edible coating based on carnauba wax nanoemulsion and Cymbopogon martinii essential oil on papaya postharvest preservation. *Coatings*, *12*(11), 1700. doi:10.3390/coatings12111700

Olivo, P. M., da Silva Scapim, M. R., Maia, L. F., Miazaki, J., Rodrigues, B. M., Madrona, G. S., ... dos Santos Pozza, M. S. (2020). Probiotic coating for ripened cheeses with lactobacillus acidophilus and lactobacillus helveticus inclusion. *Journal of Agricultural Studies*, *8*(3), 152–170. doi:10.5296/jas.v8i3.16052

Oluwafemi, J., Ofuyatan, O., Sadiq, O., Oyebisi, S., Abolarin, J., & Babaremu, K. (2018). Review of world earthquakes. *International Journal of Civil Engineering and Technology*, *9*, 440–464.

Opie, R. S., O'Neil, A., Itsiopoulos, C., & Jacka, F. N. (2015). The impact of whole-of-diet interventions on depression and anxiety: A systematic review of randomised controlled trials. *Public Health Nutrition*, *18*(11), 2074–2093. doi:10.1017/S1368980014002614 PMID:25465596

Opperman, C. J., Wojno, J. M., & Brink, A. J. (2022). Treating bacterial infections with bacteriophages in the 21st century. *Southern African Journal of Infectious Diseases*, *37*(1), 346. doi:10.4102ajid.v37i1.346 PMID:35399556

Oruç, B. (2014). Structural interpretation of southern part of western Anatolian using analytic signal of the second order gravity gradients and discrete wavelet transform analysis. *Journal of Applied Geophysics*, *103*, 82–98. doi:10.1016/j.jappgeo.2014.01.008

Otoni, C. G., Avena-Bustillos, R. J., Azeredo, H. M., Lorevice, M. V., Moura, M. R., Mattoso, L. H., & McHugh, T. H. (2017). Recent advances on edible films based on fruits and vegetables—A review. *Comprehensive Reviews in Food Science and Food Safety*, *16*(5), 1151–1169. doi:10.1111/1541-4337.12281 PMID:33371612

Otoni, C. G., Lodi, B. D., Lorevice, M. V., Leitão, R. C., Ferreira, M. D., de Moura, M. R., & Mattoso, L. H. (2018). Optimized and scaled-up production of cellulose-reinforced biodegradable composite films made up of carrot processing waste. *Industrial Crops and Products*, *121*, 66–72. doi:10.1016/j.indcrop.2018.05.003

Öziçer, S., & Uyanık, O. (2017). Beton dayanımının yerinde p dalga hızından belirlenmesi ve İzmir örneği. *SDU International Journal of Technological Sciences*, *9*(1), 1–16.

Özkaya, V., & Özkaya, Ş. Ö. (2018). Çölyak Hastalığına Diyetetik Yaklaşım. *Selcuk University Medical Journal, 34*(4).

Padalino, L., Conte, A., & Del Nobile, M. A. (2016). Overview on the general approaches to improve gluten-free pasta and bread. *Foods*, *5*(4), 87. doi:10.3390/foods5040087 PMID:28231182

Padil, V. V., Kumar, K. A., Murugesan, S., Torres-Mendieta, R., Wacławek, S., Cheong, J. Y., Černík, M., & Varma, R. S. (2022). Sustainable and safer nanoclay composites for multifaceted applications. *Green Chemistry*, *24*(8), 3081–3114. doi:10.1039/D1GC03949K

Palacios, F., Diago, M. P., & Tardaguila, J. (2019). A non-invasive method based on computer vision for grapevine cluster compactness assessment using a mobile sensing platform under field conditions. *Sensors (Basel), 19*(17), 3799. doi:10.339019173799 PMID:31480754

Pamukçu, O., Gönenç, T., Uyanık, O., Sözbilir, H., & Çakmak, O. (2014). A microgravity model for the city of Izmir (western Anatolia) and its tectonic implementations. *Acta Geophysica, 62*(4), 849–871. doi:10.247811600-014-0203-z

Pamuk, E., Gönenç, T., Özdağ, Ö. C., & Akgün, M. (2018). 3D bedrock structure of Bornova Plain and its surroundings (İzmir/Western Turkey). *Pure and Applied Geophysics, 175*(3), 1–16.

Pana, V. S., Babalola, O. P., & Balyan, V. (2022). 5G radio access networks: A survey. *Array (New York, N.Y.), 14*, 100170. doi:10.1016/j.array.2022.100170

Pandey, B. K., Pandey, S. K., & Pandey, D. (2011). A survey of bioinformatics applications on parallel architectures. *International Journal of Computer Applications, 23*(4), 21–25. doi:10.5120/2877-3744

Pandey, J. K. (2022). Investigating Role of IoT in the Development of Smart Application for Security Enhancement. In *IoT Based Smart Applications* (pp. 219–243). Springer International Publishing.

Pandey, S., Zaidib, M. G. H., & Gururani, S. K. (2013). Recent developments in clay-polymer nano composites. *Scientific Journal of Review, 2*(11), 296–328.

Pandey, V. K., Upadhyay, S. N., Niranjan, K., & Mishra, P. K. (2020). Antimicrobial biodegradable chitosan-based composite Nano-layers for food packaging. *International Journal of Biological Macromolecules, 157*, 212–219. doi:10.1016/j.ijbiomac.2020.04.149 PMID:32339572

Pandhi, S., Kumar, A., & Alam, T. (2019). Probiotic edible films and coatings: Concerns, applications and future prospects. *Journal of Packaging Technology and Research, 3*(3), 261–268. doi:10.100741783-019-00069-6

Pandit, M. A., Kumar, J., Gulati, S., Bhandari, N., Mehta, P., Katyal, R., Rawat, C. D., Mishra, V., & Kaur, J. (2022). Major biological control strategies for plant pathogens. *Pathogens (Basel, Switzerland), 11*(2), 273. doi:10.3390/pathogens11020273 PMID:35215215

Panzella, L., Moccia, F., Nasti, R., Marzorati, S., Verotta, L., & Napolitano, A. (2020). Bioactive phenolic compounds from agri-food wastes: An update on green and sustainable extraction methodologies. *Frontiers in Nutrition, 7*, 60. doi:10.3389/fnut.2020.00060 PMID:32457916

Park, J. Y., Moon, B. Y., Park, J. W., Thornton, J. A., Park, Y. H., & Seo, K. S. (2017). Genetic engineering of a temperate phage-based delivery system for CRISPR/Cas9 antimicrobials against Staphylococcus aureus. *Scientific Reports, 7*(1), 44929. doi:10.1038rep44929 PMID:28322317

Partel, V., Charan Kakarla, S., & Ampatzidis, Y. (2019). Development and evaluation of a low-cost and smart technology for precision weed management utilizing artificial intelligence. *Computers and Electronics in Agriculture, 157*, 339–350. doi:10.1016/j.compag.2018.12.048

Parvinzadeh, M., Moradian, S., Rashidi, A., & Yazdanshenas, M. E. (2010). Effect of the addition of modified nanoclays on the surface properties of the resultant polyethylene terephthalate/clay nanocomposites. *Polymer-Plastics Technology and Engineering*, *49*(9), 874–884. doi:10.1080/03602551003664628

Patel, P. (2019). *Edible packaging*. Academic Press.

Patel, N. K., & Lacy, B. E. (2018). Another reason to avoid the gluten-free fad? *Clinical Gastroenterology and Hepatology*, *16*(2), 184–185. doi:10.1016/j.cgh.2017.10.002 PMID:28987505

Pavli, F., Argyri, A. A., Skandamis, P., Nychas, G.-J., Tassou, C., & Chorianopoulos, N. (2019). Antimicrobial activity of oregano essential oil incorporated in sodium alginate edible films: Control of Listeria monocytogenes and spoilage in ham slices treated with high pressure processing. *Materials (Basel)*, *12*(22), 1–26. doi:10.3390/ma12223726 PMID:31718078

Pawar, S. N., & Edgar, K. J. (2012). Alginate derivatization: A review of chemistry, properties and applications. *Biomaterials*, *33*(11), 3279–3305. doi:10.1016/j.biomaterials.2012.01.007 PMID:22281421

Pelissari, F. M., Ferreira, D. C., Louzada, L. B., dos Santos, F., Corr^ea, A. C., Moreira, F. K. V., & Mattoso, L. H. (2019). Starch-based edible films and coatings: An eco-friendly alternative for food packaging. In M. T. P. Silva Clerici & M. Schmiele (Eds.), Starches for food application (pp. 359–420). Academic Press.

Pellecchia, M. T., Scala, R., Perretti, A., De Michele, G., Santoro, L., Filla, A., Ciacci, C., & Barone, P. (1999). Cerebellar ataxia associated with subclinical celiac disease responding to gluten-free diet. *Neurology*, *53*(7), 1606. doi:10.1212/WNL.53.7.1606-a PMID:10534283

Pellegrini, N., & Agostoni, C. (2015). Nutritional aspects of gluten-free products. *Journal of the Science of Food and Agriculture*, *95*(12), 2380–2385. doi:10.1002/jsfa.7101 PMID:25615408

Penagini, F., Dilillo, D., Meneghin, F., Mameli, C., Fabiano, V., & Zuccotti, G. V. (2013). Gluten-free diet in children: An approach to a nutritionally adequate and balanced diet. *Nutrients*, *5*(11), 4553–4565. doi:10.3390/nu5114553 PMID:24253052

Penchalaraju, M., & Shireesha, B. (2013). Preservation of foods by high-pressure processing. *RE:view*.

Pereira de Abreu, D. A., Paseiro Losada, P., Angulo, I., & Cruz, J. M. (2007). Development of new polyolefin films with nanoclays for application in food packaging. *European Polymer Journal*, *43*(6), 2229–2243. doi:10.1016/j.eurpolymj.2007.01.021

Perera, K. Y., Hopkins, M., Jaiswal, A. K., & Jaiswal, S. (2023). Nanoclays-containing bio-based packaging materials: Properties, applications, safety, and regulatory issues. *Journal of Nanostructure in Chemistry*, 1–23. doi:10.100740097-023-00525-5 PMID:36747507

Pérez-Hernández, H., Pérez-Moreno, A., Sarabia-Castillo, C. R., García-Mayagoitia, S., Medina-Pérez, G., López-Valdez, F., Campos-Montiel, R. G., Jayanta-Kumar, P., & Fernández-Luqueño, F. (2021). Ecological drawbacks of nanomaterials produced on an industrial scale: Collateral Effect on Human and Environmental Health. *Water, Air, and Soil Pollution*, *232*(10), 1–33. doi:10.100711270-021-05370-2 PMID:34658457

Petkoska, A. T., Daniloski, D., D'Cunha, N. M., Naumovski, N., & Broach, A. T. (2021). Edible packaging: Sustainable solutions and novel trends in food packaging. *Food Research International*, *140*, 109981. doi:10.1016/j.foodres.2020.109981 PMID:33648216

Phillips, C., & Jiao, J. (2023). Artificial Intelligence & Smart City Ethics: A Systematic Review. *2023 IEEE International Symposium on Ethics in Engineering, Science, and Technology (ETHICS)*, 1-5. 10.1109/ETHICS57328.2023.10154961

Pickard, J. M., Zeng, M. Y., Caruso, R., & Núñez, G. (2017). Gut microbiota: Role in pathogen colonization, immune responses, and inflammatory disease. *Immunological Reviews*, *279*(1), 70–89. doi:10.1111/imr.12567 PMID:28856738

Pinton, M. B., dos Santos, B. A., Lorenzo, J. M., Cichoski, A. J., Boeira, C. P., & Campagnol, P. C. (2021). Green Technologies as a strategy to reduce NaCl and phosphate in meat products: An overview. *Current Opinion in Food Science*, *40*, 1–5. doi:10.1016/j.cofs.2020.03.011

Pinzon, M. I., Sanchez, L. T., Garcia, O. R., Gutierrez, R., Luna, J. C., & Villa, C. C. (2020). Increasing shelf life of strawberries (Fragaria ssp) by using a banana starch-chitosan-Aloe vera gel composite edible coating. *International Journal of Food Science & Technology*, *55*(1), 92–98. doi:10.1111/ijfs.14254

Pires, D. P., Costa, A. R., Pinto, G., Meneses, L., & Azeredo, J. (2020). Current challenges and future opportunities of phage therapy. *FEMS Microbiology Reviews*, *44*(6), 684–700. doi:10.1093/femsre/fuaa017 PMID:32472938

Pirsa, S., & Asadi, S. (2021). Innovative smart and biodegradable packaging for margarine based on a nano composite polylactic acid/lycopene film. *Food Additives & Contaminants. Part A, Chemistry, Analysis, Control, Exposure & Risk Assessment*, *38*(5), 856–869. doi:10.1080/19440049.2021.1891299 PMID:33709879

Pirsa, S., Mohtarami, F., & Kalantari, S. (2020). Preparation of biodegradable composite starch/tragacanth gum/nanoclay film and study of its physicochemical and mechanical properties. *Chemical Review and Letters*, *3*(3), 98–103.

Pizoń, J., & Gola, A. (2023). Human–Machine Relationship—Perspective and Future Roadmap for Industry 5.0 Solutions. *Machines*, *11*(2), 203. doi:10.3390/machines11020203

Pojić, M., Mišan, A., & Tiwari, B. (2018). Eco-innovative technologies for extraction of proteins for human consumption from renewable protein sources of plant origin. *Trends in Food Science & Technology*, *75*, 93–104. doi:10.1016/j.tifs.2018.03.010

Polanía, A. M., Londoño, L., Ramírez, C., Bolivar, G., & Aguilar, C. N. (2023). Valorization of pineapple waste as novel source of nutraceuticals and biofunctional compounds. *Biomass Conversion and Biorefinery*, *13*(5), 3593–3618. doi:10.100713399-022-02811-8

Połaska, M., & Sokołowska, B. (2019). Bacteriophages—A new hope or a huge problem in the food industry. *AIMS Microbiology*, *5*(4), 324–346. doi:10.3934/microbiol.2019.4.324 PMID:31915746

Pooja Saklani, P., nath, S., Kishor Das, S., & Singh, S. M. (2019). A review of edible packaging for foods. *International Journal of Current Microbiology and Applied Sciences*, *8*(7), 2885–2895. doi:10.20546/ijcmas.2019.807.359

Pourmolaie, H., Khosrowshahi Asl, A., Ahmadi, M., Zomorodi, S., & Naghizadeh Raeisi, S. (2018). The effect of Guar and Tragacanth gums as edible coatings in Cheddar cheese during ripening. *Journal of Food Safety*, *38*(6), 1–9. doi:10.1111/jfs.12529

Pradhan, N., Singh, S., Ojha, N., Shrivastava, A., Barla, A., Rai, V., & Bose, S. (2015). Facets of nanotechnology as seen in food processing, packaging, and preservation industry. *BioMed Research International*, *2015*, 2015. doi:10.1155/2015/365672 PMID:26613082

Pramanik, S., Pandey, D., Joardar, S., Niranjanamurthy, M., Pandey, B. K., & Kaur, J. (2023, October). An overview of IoT privacy and security in smart cities. In AIP Conference Proceedings (Vol. 2495, No. 1). AIP Publishing. doi:10.1063/5.0123511

Principi, N., Silvestri, E., & Esposito, S. (2019). Advantages and limitations of bacteriophages for the treatment of bacterial infections. *Frontiers in Pharmacology*, *10*, 513. doi:10.3389/fphar.2019.00513 PMID:31139086

Putra, N. R., Rizkiyah, D. N., Abdul Aziz, A. H., Che Yunus, M. A., Veza, I., Harny, I., & Tirta, A. (2023). Waste to Wealth of Apple Pomace Valorization by Past and Current Extraction Processes: A Review. *Sustainability (Basel)*, *15*(1), 830. doi:10.3390u15010830

Pyra, A., Filik, K., Szermer-Olearnik, B., Czarny, A., & Brzozowska, E. (2020). New insights on the feature and function of tail tubular protein b and tail fiber protein of the lytic bacteriophage φYeO3-12 specific for Yersinia enterocolitica serotype O: 3. *Molecules (Basel, Switzerland)*, *25*(19), 4392. doi:10.3390/molecules25194392 PMID:32987777

Qian, J., Dai, B., Wang, B., Zha, Y., & Song, Q. (2022). Traceability in food processing: Problems, methods, and performance evaluations—A review. *Critical Reviews in Food Science and Nutrition*, *62*(3), 679–692. doi:10.1080/10408398.2020.1825925 PMID:33016094

Quirós, P., Colomer-Lluch, M., Martínez-Castillo, A., Miró, E., Argente, M., Jofre, J., Navarro, F., & Muniesa, M. (2014). Antibiotic resistance genes in the bacteriophage DNA fraction of human fecal samples. *Antimicrobial Agents and Chemotherapy*, *58*(1), 606–609. doi:10.1128/AAC.01684-13 PMID:24165177

Rafiee, Z., Nejatian, M., Daeihamed, M., & Jafari, S. M. (2019). Application of different nanocarriers for encapsulation of curcumin. *Critical Reviews in Food Science and Nutrition*, *59*(21), 3468–3497. doi:10.1080/10408398.2018.1495174 PMID:30001150

Raghav, P. K., & Saini, M. (2018). Development of mint (Mentha viridis L.) herbal edible coating for shelf life enhancement of cucumber (Cucumis sativus). *International Journal of Green and Herbal Chemistry*, *7*(2), 379–391.

Rahimi-Midani, A., Lee, S. W., & Choi, T. J. (2021). Potential solutions using bacteriophages against antimicrobial resistant bacteria. *Antibiotics (Basel, Switzerland)*, *10*(12), 1496. doi:10.3390/antibiotics10121496 PMID:34943708

Rahimi, S., & Mikani, M. (2019). Lycopene green ultrasound-assisted extraction using edible oil accompany with response surface methodology (RSM) optimization performance: Application in tomato processing wastes. *Microchemical Journal*, *146*, 1033–1042. doi:10.1016/j.microc.2019.02.039

Raja Santhi, A., & Muthuswamy, P. (2023). Industry 5.0 or industry 4.0 S? Introduction to industry 4.0 and a peek into the prospective industry 5.0 technologies. *International Journal on Interactive Design and Manufacturing*, *17*(2), 947–979. doi:10.100712008-023-01217-8

Raj, B., Jagadish, R. S., Srinivas, P., & Siddaramaiah. (2005). Crosslinking of low-density polyethylene by diisocyanates for superior barrier properties. *Journal of Applied Polymer Science*, *96*(no. 4), 1193–1199. doi:10.1002/app.21355

Rajumesh, S. (2023). Promoting sustainable and human-centric industry 5.0: a thematic analysis of emerging research topics and opportunities. *Journal of Business and Socio-Economic Development*.

Ramirez-Asis, E., Vilchez-Carcamo, J., Thakar, C. M., Phasinam, K., Kassanuk, T., & Naved, M. (2022). A review on role of artificial intelligence in food processing and manufacturing industry. *Materials Today: Proceedings*, *51*, 2462–2465. doi:10.1016/j.matpr.2021.11.616

Ramos, M., Vald'es, A., Beltr'an, A., & Garrigos, ´. C. M. (2016). Gelatin-based films and coatings for food packaging applications. *Coatings*, *6*(4), 41. doi:10.3390/coatings6040041

Ramos-Vivas, J., Elexpuru-Zabaleta, M., Samano, M. L., Barrera, A. P., Forbes-Hernández, T. Y., Giampieri, F., & Battino, M. (2021). Phages and enzybiotics in food biopreservation. *Molecules (Basel, Switzerland)*, *26*(17), 5138. doi:10.3390/molecules26175138 PMID:34500572

Ranchordás, S. (2019). Nudging citizens through technology in smart cities. *International Review of Law Computers & Technology*, *34*(3), 254–276. doi:10.1080/13600869.2019.1590928

Rangaraj, V. M., Rambabu, K., Banat, F., & Mittal, V. (2021). Effect of date fruit waste extract as an antioxidant additive on the properties of active gelatin films. *Food Chemistry*, *355*, 129631. doi:10.1016/j.foodchem.2021.129631 PMID:33799252

Ranjha, M. M., Amjad, S., Ashraf, S., Khawar, L., Safdar, M. N., Jabbar, S., Nadeem, M., Mahmood, S., & Murtaza, M. A. (2020). Extraction of polyphenols from Apple and pomegranate peels employing different extraction techniques for the development of functional date bars. *International Journal of Fruit Science*, *20*(sup3). doi:10.1080/15538362.2020.1782804

Rary, E., Anderson, S. M., Philbrick, B. D., Suresh, T., & Burton, J. (2020). Smart sanitation—Biosensors as a public health tool in sanitation infrastructure. *International Journal of Environmental Research and Public Health*, *17*(14), 5146. doi:10.3390/ijerph17145146 PMID:32708793

Rashidi, L., & Khosravi-Darani, K. (2011). The applications of nanotechnology in food industry. *Critical Reviews in Food Science and Nutrition*, *51*(8), 723–730. doi:10.1080/10408391003785417 PMID:21838555

Rastogi, R., Vellinga, W. P., Rastogi, S., Schick, C., & Meijer, H. E. H. (2004). The three-phase structure and mechanical properties of poly (ethylene terephthalate). *Journal of Polymer Science. Part B, Polymer Physics*, *42*(11), 2092–2106. doi:10.1002/polb.20096

Rawat, S. (2015). Food spoilage: Microorganisms and their prevention. *Asian J. Plant Sci. Res.*, *5*, 47–56.

Raya, R. R., Varey, P., Oot, R. A., Dyen, M. R., Callaway, T. R., Edrington, T. S., Kutter, E. M., & Brabban, A. D. (2006). Isolation and characterization of a new T-even bacteriophage, CEV1, and determination of its potential to reduce Escherichia coli O157:H7 levels in sheep. *Applied and Environmental Microbiology*, *72*(9), 6405–6410. doi:10.1128/AEM.03011-05 PMID:16957272

Razaak, M., Kerdegari, H., Davies, E., Abozariba, R., Broadbent, M., Mason, K., ... Remagnino, P. (2019). An integrated precision farming application based on 5G, UAV and deep learning technologies. *Computer Analysis of Images and Patterns: CAIP 2019 International Workshops, ViMaBi and DL-UAV, Salerno, Italy, September 6, 2019 Proceedings*, *18*, 109–119.

Reddy, M. M., Vivekanandhan, S., Misra, M., Bhatia, S. K., & Mohanty, A. K. (2013). Biobased plastics and bionanocomposites: Current status and future opportunities. *Progress in Polymer Science*, *38*(10-11), 1653–1689. doi:10.1016/j.progpolymsci.2013.05.006

Reichert, C. L., Bugnicourt, E., Coltelli, M. B., Cinelli, P., Lazzeri, A., Canesi, I., Braca, F., Martínez, B. M., Alonso, R., Agostinis, L., Verstichel, S., Six, L., Mets, S. D., Gómez, E. C., Ißbrücker, C., Geerinck, R., Nettleton, D. F., Campos, I., Sauter, E., ... Schmid, M. (2020). Bio-based packaging: Materials, modifications, industrial applications and sustainability. *Polymers*, *12*(7), 1558. doi:10.3390/polym12071558 PMID:32674366

Restrepo, A. E., Rojas, J. D., García, O. R., Sánchez, L. T., Pinzón, M. I., & Villa, C. C. (2018). Mechanical, barrier, and color properties of banana starch edible films incorporated with nanoemulsions of lemongrass (Cymbopogon citratus) and rosemary (Rosmarinus officinalis) essential oils. *Food Science & Technology International*, *24*(8), 705–712. doi:10.1177/1082013218792133 PMID:30060685

Rhim, J.-W., Hong, S.-I., & Ha, C.-S. (2009). Tensile, water vapor barrier and antimicrobial properties of PLA/nanoclay composite films. *Lebensmittel-Wissenschaft + Technologie*, *42*(2), 612–617. doi:10.1016/j.lwt.2008.02.015

Riley, L. W. (2020). Extraintestinal foodborne pathogens. *Annual Review of Food Science and Technology*, *11*(1), 275–294. doi:10.1146/annurev-food-032519-051618 PMID:32004092

Robledo, N., Vera, P., Lopez, L., Yazdani-Pedram, M., Tapia, C., & Abugoch, L. (2018). Thymol nanoemulsions incorporated in quinoa protein/chitosan edible films; antifungal effect in cherry tomatoes. *Food Chemistry*, *246*, 211–219. doi:10.1016/j.foodchem.2017.11.032 PMID:29291841

Rocha, J. M., Lorite, G. S., & Miilumäki, N. (2016). Novel polylactic acid (PLA)-based active packaging with incorporation of nanoparticles and its performance throughout shelf-life of fresh-cut fruit. In *XIX National Congress of Biochemistry*. Sociedade Portuguesa de Bioquímica, University of Minho.

Rodrigues, D. C., Cunha, A. P., de Brito, E. S., Gallao, M., De Azeredo, H. M. C., & Rodrigues, D. C. (2016). Mesquite seed gum and palm fruit oil emulsion edible films: Influence of oil content and sonication. *Food Hydrocolloids*, *56*, 227–235. doi:10.1016/j.foodhyd.2015.12.018

Rojas-Graü, M. A., Soliva-Fortuny, R., & Martín-Belloso, O. (2009). Edible coatings to incorporate active ingredients to fresh-cut fruits: A review. *Trends in Food Science & Technology*, *20*(10), 438–447. doi:10.1016/j.tifs.2009.05.002

Rojas, L. F., Zapata, P., & Ruiz-Tirado, L. (2022). Agro-industrial waste enzymes: Perspectives in circular economy. *Current Opinion in Green and Sustainable Chemistry*, *34*, 100585. doi:10.1016/j.cogsc.2021.100585

Roopashree, K., Kanmani, Babitha, & Pavanalaxmi. (2022). Smart Farming with IoT: A Case Study. *Internet of Things and Analytics for Agriculture*, *3*, 273–286. doi:10.1007/978-981-16-6210-2_13

Rosenboom, J. G., Hohl, D. K., Fleckenstein, P., Storti, G., & Morbidelli, M. (2018). Bottle-grade polyethylene furanoate from ring-opening polymerisation of cyclic oligomers. *Nature Communications*, *9*(1), 2701. doi:10.103841467-018-05147-y PMID:30042388

Rost, P., Banchs, A., Berberana, I., Breitbach, M., Doll, M., Droste, H., Mannweiler, C., Puente, M. A., Samdanis, K., & Sayadi, B. (2016). Mobile network architecture evolution toward 5G. *IEEE Communications Magazine*, *54*(5), 84–91. doi:10.1109/MCOM.2016.7470940

Rovera, C., Ghaani, M., & Farris, S. (2020). Nano-inspired oxygen barrier coatings for food packaging applications: An overview. *Trends in Food Science & Technology*, *97*, 210–220. doi:10.1016/j.tifs.2020.01.024

Rowat, S. J., Legge, R. L., & Moresoli, C. (2021). Plant protein in material extrusion 3D printing: Formation, plasticization, prospects, and challenges. *Journal of Food Engineering*, *308*, 110623. doi:10.1016/j.jfoodeng.2021.110623

Roy, R. B., Tudu, B., Bandyopadhyay, R., & Bhattacharyya, N. (2019). Application of electronic nose and tongue for beverage quality evaluation. Engineering tools in the beverage industry, 229-254.

Rutz, J. K., Borges, C. D., Zambiazi, R. C., da Rosa, C. G., & da Silva, M. M. (2016). Elaboration of microparticles of carotenoids from natural and synthetic sources for applications in food. *Food Chemistry*, *202*, 324–333. doi:10.1016/j.foodchem.2016.01.140 PMID:26920301

Ryan, E. M., Gorman, S. P., Donnelly, R. F., & Gilmore, B. F. (2011). Recent advances in bacteriophage therapy: How delivery routes, formulation, concentration and timing influence the success of phage therapy. *The Journal of Pharmacy and Pharmacology*, *63*(10), 1253–1264. doi:10.1111/j.2042-7158.2011.01324.x PMID:21899540

Sabaghi, M., Maghsoudlou, Y., Khomeiri, M., & Ziaiifar, A. M. (2015). Active edible coating from chitosan incorporating green tea extract as an antioxidant and antifungal on fresh walnut kernel. *Postharvest Biology and Technology*, *110*, 224–228. doi:10.1016/j.postharvbio.2015.08.025

Sabbağ, N., & Uyanık, O. (2016). Time-dependent change of seismic velocities on low strength concrete. *The Online Journal of Science and Technology*, *6*(4), 49–57.

Sabbağ, N., & Uyanık, O. (2017). Prediction of reinforced concrete strength by ultrasonic velocities. *Journal of Applied Geophysics*, *170*(103832), 13–23. doi:10.1016/j.jappgeo.2017.04.005

Sabbağ, N., & Uyanık, O. (2018a). Determination of the reinforced concrete strength by apparent resistivity depending on the curing conditions. *Journal of Applied Geophysics*, *155*, 13–25. doi:10.1016/j.jappgeo.2018.03.007

Sabbağ, N., & Uyanık, O. (2018b). Determination of reinforced dry concrete strength by electrical resistivity method. *Pamukkale Univ Journal of Engineering Sciences*, *24*(6), 1230–1236. doi:10.5505/pajes.2017.76702

Sabbağ, N., & Uyanık, O. (2020). Determination of anisotropy of reinforced concretes by using resistivity method. *Pamukkale Univ Journal of Engineering Sciences*, *26*(3), 570–578. doi:10.5505/pajes.2019.00008

Sabbah, M., Di Pierro, P., Dell'Olmo, E., Arciello, A., & Porta, R. (2019). Improved shelf-life of Nabulsi cheese wrapped with hydrocolloid films. *Food Hydrocolloids*, *96*, 29–35. doi:10.1016/j.foodhyd.2019.05.010

Safari, J., & Zarnegar, Z. (2014). Advanced Drug Delivery Systems: Nanotechnology of health design a review. *Journal of Saudi Chemical Society*, *18*(2), 85–99. doi:10.1016/j.jscs.2012.12.009

Safdar, M. N., Kausar, T., & Nadeem, M. (2017). Comparison of ultrasound and maceration techniques for the extraction of polyphenols from the mango peel. *Journal of Food Processing and Preservation*, *41*(4), e13028. doi:10.1111/jfpp.13028

Saha, D., & Bhattacharya, S. (2010). Hydrocolloids as thickening and gelling agents in food: A critical review. *Journal of Food Science and Technology*, *47*(6), 587–597. doi:10.100713197-010-0162-6 PMID:23572691

Saha, K. K., & Zude-Sasse, M. (2022). Estimation of chlorophyll content in banana during shelf life using LiDAR laser scanner. *Postharvest Biology and Technology*, *192*, 112011. doi:10.1016/j.postharvbio.2022.112011

Saha, R. K., Saengudomlert, P., & Aswakul, C. (2016). Evolution toward 5G mobile networks-A survey on enabling technologies. *Engineering Journal (New York)*, *20*(1), 87–119. doi:10.4186/ej.2016.20.1.87

Salem, H. S. (2000). Poisson's ratio and the porosity of surface soils and shallow sediments, determined from seismic compressional and shear wave velocities. *Geotechnique*, *50*(4), 461–463. doi:10.1680/geot.2000.50.4.461

Sandey, K., Qureshi, M., Meshram, B., Agrawal, A., & Uprit, S. (2017). Robotics—An Emerging Technology in Dairy Industry. *Int. J. Eng. Trends Technol*, *43*, 58–62. doi:10.14445/22315381/IJETT-V43P210

Sandri, L. T. B., Santos, F. G., Fratelli, C., & Capriles, V. D. (2017). Development of gluten-free bread formulations containing whole chia flour with acceptable sensory properties. *Food Science & Nutrition*, *5*(5), 1021–1028. doi:10.1002/fsn3.495 PMID:28948020

Sangroniz, A., Zhu, J.-B., Tang, X., Etxeberria, A., Chen, E. Y. X., & Sardon, H. (2019). Packaging materials with desired mechanical and barrier properties and full chemical recyclability. *Nature Communications*, *10*(1), 3559. doi:10.103841467-019-11525-x PMID:31395871

Santana, J. C., Gardim, R. B., Almeida, P. F., Borini, G. B., Quispe, A. P., Llanos, S. A., Heredia, J. A., Zamuner, S., Gamarra, F. M. C., Farias, T. M. B., Ho, L. L., & Berssaneti, F. T. (2020). Valorization of chicken feet by-product of the poultry industry: High qualities of gelatin and biofilm from extraction of collagen. *Polymers*, *12*(3), 529. doi:10.3390/polym12030529 PMID:32121646

Sapone, A., Bai, J. C., Ciacci, C., Dolinsek, J., Green, P. H. R., Hadjivassiliou, M., Kaukinen, K., Rostami, K., Sanders, D. S., Schumann, M., Ullrich, R., Villalta, D., Volta, U., Catassi, C., & Fasano, A. (2012). Spectrum of gluten-related disorders: Consensus on new nomenclature and classification. *BMC Medicine*, *10*(1), 1–12. doi:10.1186/1741-7015-10-13 PMID:22313950

Sasidharan, A., & Venugopal, V. (2020). Proteins and co-products from seafood processing discards: Their recovery, functional properties and applications. *Waste and Biomass Valorization*, *11*(11), 5647–5663. doi:10.100712649-019-00812-9

Saturni, L., Ferretti, G., & Bacchetti, T. (2010). The gluten-free diet: Safety and nutritional quality. *Nutrients*, *2*(1), 16–34. doi:10.3390/nu2010016 PMID:22253989

Saucedo-Pompa, S., Rojas-Molina, R., Aguilera-Carbó, A. F., Sáenz-Galindo, A., & de La Garza, H. (2009). Edible film based on candelilla wax to improve the shelf life and quality of avocado. *Food Research International*, *42*(4), 511–515. doi:10.1016/j.foodres.2009.02.017

Sayed, H. S., Sakr, A. M., & Hassan, N. M. M. (2016). Effect of pseudo cereal flours on technological, chemical and sensory properties of pan bread. *World J. Dairy Food Sci*, *11*(1), 10–17.

Scarfato, P., Di Maio, L., Milana, M. R., Giamberardini, S., Denaro, M., & Incarnato, L. (2017). Performance properties, lactic acid specific migration and swelling by simulant of biodegradable poly (lactic acid)/nanoclay multilayer films for food packaging. *Food Additives & Contaminants. Part A, Chemistry, Analysis, Control, Exposure & Risk Assessment*, *34*(10), 1730–1742. doi:10.1080/19440049.2017.1321786 PMID:28434378

Scarfato, P., Incarnato, L., Di Maio, L., Dittrich, B., & Schartel, B. (2016). Influence of a novel organo-silylated clay on the morphology, thermal and burning behavior of low density polyethylene composites. *Composites. Part B, Engineering*, *98*(1), 444–452. doi:10.1016/j.compositesb.2016.05.053

Schieber, A. (2017). Side streams of plant food processing as a source of valuable compounds: Selected examples. *Annual Review of Food Science and Technology*, *8*(1), 97–112. doi:10.1146/annurev-food-030216-030135 PMID:28068488

Schirone, M., Visciano, P., Tofalo, R., & Suzzi, G. (2019). Foodborne pathogens: Hygiene and safety. Frontiers Media SA.

Schober, T. J. (2009). Manufacture of gluten-free specialty breads and confectionery products. *Gluten-free. Food Science and Technology (Campinas)*, 130–180.

Schoenlechner, R., Drausinger, J., Ottenschlaeger, V., Jurackova, K., & Berghofer, E. (2010). Functional properties of gluten-free pasta produced from amaranth, quinoa and buckwheat. *Plant Foods for Human Nutrition (Dordrecht, Netherlands)*, *65*(4), 339–349. doi:10.100711130-010-0194-0 PMID:20972627

Schotten, H. D., Sattiraju, R., Serrano, D. G., Ren, Z., & Fertl, P. (2014, June). Availability indication as key enabler for ultra-reliable communication in 5G. In *2014 European Conference on Networks and Communications (EuCNC)* (pp. 1-5). IEEE. 10.1109/EuCNC.2014.6882630

SDG. (2023). *The Sustainable Development Goals Report 2023: Special edition*. https://unstats.un.org/sdgs/report/2023/The-Sustainable-Development-Goals-Report-2023.pdf

Seed, K. D. (2015). Battling phages: How bacteria defend against viral attack. *PLoS Pathogens*, *11*(6), e1004847. doi:10.1371/journal.ppat.1004847 PMID:26066799

See, J., & Murray, J. A. (2006). Gluten-free diet: The medical and nutrition management of celiac disease. *Nutrition in Clinical Practice*, *21*(1), 1–15. doi:10.1177/011542650602100101 PMID:16439765

Segatto, M. L., Stahl, A. M., Zanotti, K., & Zuin, V. G. (2022). Green and sustainable extraction of proteins from agro-industrial waste: An overview and a closer look to Latin America. *Current Opinion in Green and Sustainable Chemistry*, *37*, 100661. doi:10.1016/j.cogsc.2022.100661

Senkaya, G. V., Şenkaya, M., Karslı, H., & Gueney, R. (2020). Integrated shallow seismic imaging of a settlement located in a historical landslide area. *Bulletin of Engineering Geology and the Environment*, *79*(4), 1781–1796. doi:10.100710064-019-01612-0

Senturk Parreidt, T., Müller, K., & Schmid, M. (2018). Alginate-based edible films and coatings for food packaging applications. *Foods*, *7*(10), 170. doi:10.3390/foods7100170 PMID:30336642

Serin, Y., & Akbulut, G. (2017). Çölyak hastalığı ve glutensiz diyet tedavisine güncel yaklaşım. *Journal of the Turkish Society of Intensive Care*, *2*(3).

Shafiee-Jood, M., & Cai, X. (2016). Reducing food loss and waste to enhance food security and environmental sustainability. *Environmental Science & Technology*, *50*(16), 8432–8443. doi:10.1021/acs.est.6b01993 PMID:27428555

Shahbazi, M., Ahmadi, S. J., Seif, A., & Rajabzadeh, G. (2016). Carboxymethyl cellulose film modification through surface photo-crosslinking and chemical crosslinking for food packaging applications. *Food Hydrocolloids*, *61*, 378–389. doi:10.1016/j.foodhyd.2016.04.021

Shahidi, F., & Hossain, A. (2020). Preservation of aquatic food using edible films and coatings containing essential oils: A review. *Critical Reviews in Food Science and Nutrition*, 1–40. PMID:32847401

Sharma, C., Dhiman, R., Rokana, N., & Panwar, H. (2017). Nanotechnology: An untapped resource for food packaging. *Frontiers in Microbiology*, *8*, 1735. doi:10.3389/fmicb.2017.01735 PMID:28955314

Sharma, S., Chatterjee, S., Datta, S., Prasad, R., Dubey, D., Prasad, R. K., & Vairale, M. G. (2017). Bacteriophages and its applications: An overview. *Folia Microbiologica*, *62*(1), 17–55. doi:10.100712223-016-0471-x PMID:27718043

Sharma, S., Cheng, S. F., Bhattacharya, B., & Chakkaravarthi, S. (2019). Efficacy of free and encapsulated natural antioxidants in oxidative stability of edible oil: Special emphasis on nanoemulsion-based encapsulation. *Trends in Food Science & Technology*, *91*, 305–318. doi:10.1016/j.tifs.2019.07.030

Sharma, S., Sanpui, P., Chattopadhyay, A., & Ghosh, S. S. (2012). Fabrication of antibacterial silver nanoparticle—Sodium alginate–chitosan composite films. *RSC Advances*, *2*(13), 5837–5843. doi:10.1039/c2ra00006g

Sheikh, H., Mitchell, P., & Foth, M. (2023). Foth More-than-human smart urban governance: A research agenda Digit. *Geogr. Soc.*, *4*, 100045. Advance online publication. doi:10.1016/j.diggeo.2022.100045

Shi, C., Zhao, Z., Jia, Z., Hou, M., Yang, X., Ying, X., & Ji, Z. (2023). Artificial neural network-based shelf life prediction approach in the food storage process: A review. *Critical Reviews in Food Science and Nutrition*, 1–16. doi:10.1080/10408398.2023.2245899 PMID:37688408

Shin, Y. J., Jang, S.-A., & Song, K. B. (2011). Preparation and mechanical properties of rice bran protein composite films containing gelatin or red algae. *Food Science and Biotechnology*, *20*(3), 703–707. doi:10.100710068-011-0099-1

Shit, S. C., & Shah, P. M. (2014). Edible polymers: Challenges and opportunities. *JournalofPolymers*, *2014*, 1–13.

Shivamathi, C. S., Moorthy, I. G., Kumar, R. V., Soosai, M. R., Maran, J. P., Kumar, R. S., & Varalakshmi, P. (2019). Optimization of ultrasound assisted extraction of pectin from custard apple peel: Potential and new source. *Carbohydrate Polymers*, *225*, 115240. doi:10.1016/j.carbpol.2019.115240 PMID:31521287

Shkoporov, A. N., & Hill, C. (2019). Bacteriophages of the human gut: The "known unknown" of the microbiome. *Cell Host & Microbe*, *25*(2), 195–209. doi:10.1016/j.chom.2019.01.017 PMID:30763534

Sikorski, Z. Z. E., & Kolakowska, A. (2011). *Chemical, Biological, and Functional Aspects of Food Lipids* (2nd ed.). CRC Press.

Sillankorva, S. M., Oliveira, H., & Azeredo, J. (2012). Bacteriophages and their role in food safety. *International Journal of Microbiology*, *2012*, 863945. doi:10.1155/2012/863945 PMID:23316235

Silva, F. A., Finkler, L., & Finkler, C. L. L. (2018). Effect of edible coatings based on alginate/pectin on quality preservation of minimally processed 'Espada' mangoes. *Journal of Food Science and Technology*, *55*(12), 5055–5506. doi:10.100713197-018-3444-z PMID:30483001

Singh, P., Arora, A., Strand, T. A., Leffler, D. A., Catassi, C., Green, P. H., Kelly, C. P., Ahuja, V., & Makharia, G. K. (2018). Global prevalence of celiac disease: Systematic review and meta-analysis. *Clinical Gastroenterology and Hepatology*, *16*(6), 823–836. doi:10.1016/j.cgh.2017.06.037 PMID:29551598

Siracusa, V., & Lotti, N. (2018). Biobased Plastics for Food Packaging. In *Reference Module in Food Science* (pp. 1–19). Elsevier. doi:10.1016/B978-0-08-100596-5.22413-X

Siracusa, V., Lotti, N., Munari, A., & Dalla Rosa, M. (2015). Poly (butylene succinate) and poly (butylene succinate-co-adipate) for food packaging applications: Gas barrier properties after stressed treatments. *Polymer Degradation & Stability*, *119*, 35–45. doi:10.1016/j.polymdegradstab.2015.04.026

Siracusa, V., Rocculi, P., Romani, S., & Rosa, M. D. (2008). Biodegradable polymers for food packaging: A review. *Trends in Food Science & Technology*, *19*(12), 634–643. doi:10.1016/j.tifs.2008.07.003

Sivaramakrishnan, H. P., Senge, B., & Chattopadhyay, P. K. (2004). Rheological properties of rice dough for making rice bread. *Journal of Food Engineering*, *62*(1), 37–45. doi:10.1016/S0260-8774(03)00169-9

Smith, H. W., Huggins, M. B., & Shaw, K. M. (1987). Factors influencing the survival and multiplication of bacteriophages in calves and in their environment. *Microbiology*, *133*(5), 1127–1135. doi:10.1099/00221287-133-5-1127 PMID:3309178

Soares, F. L. P., de Oliveira Matoso, R., Teixeira, L. G., Menezes, Z., Pereira, S. S., Alves, A. C., Batista, N. V., de Faria, A. M. C., Cara, D. C., Ferreira, A. V. M., & Alvarez-Leite, J. I. (2013). Gluten-free diet reduces adiposity, inflammation and insulin resistance associated with the induction of PPAR-alpha and PPAR-gamma expression. *The Journal of Nutritional Biochemistry*, *24*(6), 1105–1111. doi:10.1016/j.jnutbio.2012.08.009 PMID:23253599

Sohail, H. A., Coffey, A., Debrowska, K., Meyer, I. M., Middelboe, M., Sohail, M., & Clokie, M. R. (2020). Bacteriophages: Emerging applications in medicine, food, and biotechnology. *PHAGE (New Rochelle, N.Y.)*, *1*(2), 75–82. doi:10.1089/phage.2020.29004.has PMID:36147894

Solíz-Guerrero, J. B., de Rodriguez, D. J., Rodríguez-García, R., Angulo-Sánchez, J. L., & Méndez-Padilla, G. (2002). Quinoa saponins: concentration and composition analysis. International Journal of Trends in New Crops and New Uses.

Souri, J., Almasi, H., Hamishehkar, H., & Amjadi, S. (2021). Sodium caseinate-coated and β-cyclodextrin/vitamin E inclusion complex-loaded nanoliposomes: A novel stabilized nanocarrier. *Lebensmittel-Wissenschaft + Technologie*, *151*, 112174. doi:10.1016/j.lwt.2021.112174

Srichuwong, S., Curti, D., Austin, S., King, R., Lamothe, L., & Gloria-Hernandez, H. (2017). Physicochemical properties and starch digestibility of whole grain sorghums, millet, quinoa and amaranth flours, as affected by starch and non-starch constituents. *Food Chemistry*, *233*, 1–10. doi:10.1016/j.foodchem.2017.04.019 PMID:28530552

Stejskal, N., Miranda, J. M., Martucci, J. F., Ruseckaite, R. A., Barros-Velázquez, J., & Aubourg, S. P. (2020). Quality enhancement of refrigerated hake muscle by active packaging with a protein concentrate fromSpirulina platensis. *Food and Bioprocess Technology*, *13*(7), 1110–1118. doi:10.100711947-020-02468-z

Stevens, J. R., Newton, R. W., Tlusty, M., & Little, D. C. (2018). The rise of aquaculture by-products: Increasing food production, value, and sustainability through strategic utilisation. *Marine Policy*, *90*, 115–124. doi:10.1016/j.marpol.2017.12.027

Stoitsis, G., Papakonstantinou, M., Karvounis, M., & Manouselis, N. (2023). The role of Big Data and Artificial Intelligence in food risk assessment and prediction. In M. E. Knowles, L. E. Anelich, A. R. Boobis, & B. Popping (Eds.), *Present Knowledge in Food Safety* (pp. 1032–1044). Academic Press. doi:10.1016/B978-0-12-819470-6.00041-X

Strange, J. E., Leekitcharoenphon, P., Møller, F. D., & Aarestrup, F. M. (2021). Metagenomics analysis of bacteriophages and antimicrobial resistance from global urban sewage. *Scientific Reports*, *11*(1), 1600. doi:10.103841598-021-80990-6 PMID:33452346

Strathdee, S. A., Hatfull, G. F., Mutalik, V. K., & Schooley, R. T. (2023). Phage therapy: From biological mechanisms to future directions. *Cell*, *186*(1), 17–31. doi:10.1016/j.cell.2022.11.017 PMID:36608652

Strydom, A. (2015). *Control of Listeria monocytogenes in an avocado processing facility*. University of the Free State.

Stuckmann, P. (2002). *The GSM evolution: mobile packet data services*. John Wiley & Sons.

Subramaniyan, A. K., & Sun, C. T. (2006). Enhancing compressive strength of unidirectional polymeric composites using nanoclay. *Composites. Part A, Applied Science and Manufacturing*, *37*(12), 2257–2268. doi:10.1016/j.compositesa.2005.12.027

Sudesh, K., & Iwata, T. (2008). Sustainability of biobased and biodegradable plastics. *Clean (Weinheim)*, *36*(5-6), 433–442. doi:10.1002/clen.200700183

Suhag, R., Kumar, N., Petkoska, A. T., & Upadhyay, A. (2020). Film formation and deposition methods of edible coating on food products: A review. *Food Research International*, *136*, 109582. doi:10.1016/j.foodres.2020.109582 PMID:32846613

Sukumaran, A. T., Nannapaneni, R., Kiess, A., & Sharma, C. S. (2015). Reduction of Salmonella on chicken meat and chicken skin by combined or sequential application of lytic bacteriophage with chemical antimicrobials. *International Journal of Food Microbiology*, *207*, 8–15. doi:10.1016/j.ijfoodmicro.2015.04.025 PMID:25950852

Summers, W. C. (2012). The strange history of phage therapy. *Bacteriophage*, *2*(2), 130–133. doi:10.4161/bact.20757 PMID:23050223

Summers, W. C. (2021). The discovery of bacteriophages and the historical context. Bacteriophages: Biology, Technology. *Therapy*, 387–400.

Suresh, P. V., Kudre, T. G., & Johny, L. C. (2018). Sustainable valorization of seafood processing by-product/discard. In R. R. Singhania, R. A. Agarwal, R. P. Kumar, & R. K. Sukumaran (Eds.), *Waste to wealth* (pp. 111–139). Springer. doi:10.1007/978-981-10-7431-8_7

Svircev, A., Roach, D., & Castle, A. (2018). Framing the future with bacteriophages in agriculture. *Viruses*, *10*(5), 218. doi:10.3390/v10050218 PMID:29693561

Swaminathan, B., & Gerner-Smidt, P. (2007). The epidemiology of human listeriosis. *Microbes and Infection*, *9*(10), 1236–1243. doi:10.1016/j.micinf.2007.05.011 PMID:17720602

Tan, G. K. (2021). Citizens go digital: A discursive examination of digital payments in Singapore's Smart Nation project. *Urban Studies (Edinburgh, Scotland)*, *59*(12), 2582–2598. doi:10.1177/00420980211039407

Tang, Y., Dananjayan, S., Hou, C., Guo, Q., Luo, S., & He, Y. (2021). A survey on the 5G network and its impact on agriculture: Challenges and opportunities. *Computers and Electronics in Agriculture*, *180*, 105895. doi:10.1016/j.compag.2020.105895

Tang, Y., & Tsao, R. (2017). Phytochemicals in quinoa and amaranth grains and their antioxidant, anti-inflammatory, and potential health beneficial effects: A review. *Molecular Nutrition & Food Research*, *61*(7), 1600767. doi:10.1002/mnfr.201600767 PMID:28239982

Tavassoli-Kafrani, E., Shekarchizadeh, H., & Masoudpour-Behabadi, M. (2016). Development of edible films and coatings from alginates and carrageenans. *Carbohydrate Polymers*, *137*, 360–374. doi:10.1016/j.carbpol.2015.10.074 PMID:26686140

Tawakkal, I. S. M. A., Cran, M. J., Miltz, J., & Bigger, S. W. (2014). A review of poly(lactic acid)-based materials for antimicrobial packaging. *Journal of Food Science*, *79*(8), R1477–R1490. doi:10.1111/1750-3841.12534 PMID:25039867

TBDY (Türk Bina Deprem Yönetmeliği). (2018). Türkiye Bina Deprem Yönetmeliği. Afet ve Acil Durum Yönetim Başkanlığı, 30364, 416.

Tegmark, M. (2016). Benefits and risks of artificial intelligence. *Future of Life*, 29-31.

Teixeira-Costa, B. E., & Andrade, C. T. (2021). Natural polymers used in edible food packaging—History, function and application trends as a sustainable alternative to synthetic plastic. *Polysaccharides*, *3*(1), 32–58. doi:10.3390/polysaccharides3010002

Terfansky, M., Thangavelu, B., Fritz, B., & Khoshnevis, B. (2013) 3D printing of food for space missions. *Proceedings of AIAA SPACE 2013 Conference and Exposition*. 10.2514/6.2013-5346

Thakur, R., Pristijono, P., Bowyer, M., Singh, S. P., Scarlett, C. J., Stathopoulos, C. E., & Vuong, Q. V. (2019). A starch edible surface coating delays banana fruit ripening. *Lebensmittel-Wissenschaft + Technologie*, *100*, 341–347. doi:10.1016/j.lwt.2018.10.055

Thakur, R., Pristijono, P., Scarlett, C. J., Bowyer, M., Singh, S. P., & Vuong, Q. V. (2019). Starch-based films: Major factors affecting their properties. *International Journal of Biological Macromolecules*, *132*, 1079–1089. doi:10.1016/j.ijbiomac.2019.03.190 PMID:30926503

Theethira, T. G., & Dennis, M. (2015). Celiac disease and the gluten-free diet: Consequences and recommendations for improvement. *Digestive Diseases (Basel, Switzerland)*, *33*(2), 175–182. doi:10.1159/000369504 PMID:25925920

Thung, T. Y., Lee, E., Mahyudin, N. A., Anuradha, K., Mazlan, N., Kuan, C. H., Pui, C. F., Ghazali, F. M., Ab Rashid, N.-K. M., & Rollon, W. D. (2019). Evaluation of a lytic bacteriophage for bio-control of Salmonella Typhimurium in different food matrices. *Lebensmittel-Wissenschaft + Technologie*, *105*, 211–214. doi:10.1016/j.lwt.2019.02.033

Thung, T. Y., Lee, E., Mahyudin, N. A., Wan Mohamed Radzi, C. W. J., Mazlan, N., Tan, C. W., & Radu, S. (2020). Partial characterization and in vitro evaluation of a lytic bacteriophage for biocontrol of Campylobacter jejuni in mutton and chicken meat. *Journal of Food Safety*, *40*(2), e12770. doi:10.1111/jfs.12770

Tiwari, S., Bahuguna, P. C., & Walker, J. (2022). Industry 5.0: A macroperspective approach. In Handbook of Research on Innovative Management Using AI in Industry 5.0 (pp. 59-73). IGI Global.

Tiwari, A., & Khawas, R. (2021). *Food Waste and Agro By-Products: A Step towards Food Sustainability*. IntechOpen. doi:10.5772/intechopen.96177

Toepfl, S., Heinz, V., & Knorr, D. (2006). Applications of pulsed electric fields technology for the food industry. *Pulsed Electric Fields Technology for the Food Industry*, 197–221. doi:10.1007/978-0-387-31122-7_7

Tokatlı, K., & Demirdöven, A. (2020). Effects of chitosan edible filmcoatings on the physicochemical and microbiological qualities of sweet cherry (Prunus avium L.). *Scientia Horticulturae*, *259*, 108656. doi:10.1016/j.scienta.2019.108656

Tokusoglu, O. (2015). Introduction to innovative food processing. *Natural Science and Discovery*, *1*(4), 85–87. doi:10.20863/nsd.89151

Tomadoni, B., Moreira, M. D. R., Pereda, M., & Ponce, A. G. (2018). Gellan-based coatings incorporated with natural antimicrobials in fresh-cut strawberries: Microbiological and sensory evaluation through refrigerated storage. *Lebensmittel-Wissenschaft + Technologie*, *97*, 384–389. doi:10.1016/j.lwt.2018.07.029

Torres-Barceló, C. (2018). Phage therapy faces evolutionary challenges. *Viruses*, *10*(6), 323. doi:10.3390/v10060323 PMID:29895791

Torres-Barceló, C., & Hochberg, M. E. (2016). Evolutionary rationale for phages as complements of antibiotics. *Trends in Microbiology*, *24*(4), 249–256. doi:10.1016/j.tim.2015.12.011 PMID:26786863

Toseef, M., & Khan, M. J. (2018). An intelligent mobile application for diagnosis of crop diseases in Pakistan using fuzzy inference system. *Computers and Electronics in Agriculture*, *153*, 1–11. doi:10.1016/j.compag.2018.07.034

Trabelsi, M., Casprini, E., Fiorini, N., & Zanni, L. (2023). Unleashing the value of artificial intelligence in the agri-food sector: Where are we? *British Food Journal*, *125*(13), 482–515. doi:10.1108/BFJ-11-2022-1014

Trbojevich, R. A., Khare, S., Lim, J. H., Watanabe, F., Gokulan, K., Krohmaly, K., & Williams, K. (2020). Assessment of silver release and biocidal capacity from silver nanocomposite food packaging materials. *Food and Chemical Toxicology*, *145*, 111728. doi:10.1016/j.fct.2020.111728 PMID:32898598

Trifol, J., Plackett, D., Sillard, C., Szabo, P., Bras, J., & Daugaard, A. E. (2016). Hybrid poly (lactic acid)/nanocellulose/nanoclay composites with synergistically enhanced barrier properties and improved thermomechanical resistance. *Polymer International*, *65*(8), 988–995. doi:10.1002/pi.5154

Türker, E., Kaya, M. A., Kamacı, Z., Uyanık, O., Büyükköse, N., Mutlutürk, M., Yalcın, A., & Özkan, F. (1996). Dinar Afet Bölgesi Geoteknik Raporu. *Afet İşleri Genel Müdürlüğü-SDÜ*. (Unpublished).

Türker, E., Uyanık, O., Işık, H., & Özalpaydın, C. (1999). *Gölcük Donanma Üssü Yerleşim Alanının Jeolojik ve Jeoteknik Etüdü Projesi.* MTA-SDU (Unpublished).

Türksoy, S., & Özkaya, B. (2006). Gluten ve çölyak hastalığı. Türkiye, 9, 24–26.

Uçak, I. (2019). Physicochemical and antimicrobial effects of gelatin based edible films incorporated with garlic peel extract on the rainbow trout fillets. *Progress in Nutrition*, *21*(1), 232–240.

Uddin, F. (2008). Clays, Nanoclays, and Montmorillonite Minerals. *Metallurgical and Materials Transactions. A, Physical Metallurgy and Materials Science*, *39*(12), 2804–2814. doi:10.100711661-008-9603-5

Ulugergerli, E. U., Seyitoğlu, G., Başokur, A. T., Kaya, C., Dikmen, Ü., & Candansayar, M. E. (2007). The geoelectrical structure of northwestern Anatolia, Turkey. *Pure and Applied Geophysics*, *164*(5), 999–1026. doi:10.100700024-007-0200-0

Ulugergerli, E. U., & Uyanık, O. (2007). Statistical correlations between seismic wave velocities and SPT blow counts and the relative density of soils. *Journal of Testing and Evaluation*, *35*(2), 187–191. doi:10.1520/JTE100159

Umaraw, P., & Verma, A. K. (2017). Comprehensive review on application of edible film on meat and meat products: An eco-friendly approach. *Critical Reviews in Food Science and Nutrition*, *57*(6), 1270–1279. doi:10.1080/10408398.2014.986563 PMID:26090546

USDA. (2011). *National Nutrient Database for Standard Reference. Food Composition Database United States Department of Agriculture*. USDA.

USGS. (2000). The local effects of strong ground motion shaking: *USGS Responds Report to an Earthquake.*

Usman, M., Farooq, M., Wakeel, A., Nawaz, A., Cheema, S. A., Rehman, H., Ashraf, I., & Sanaullah, M. (2020). Nanotechnology in agriculture: Current status, challenges and future opportunities. *The Science of the Total Environment*, *721*, 137778. doi:10.1016/j.scitotenv.2020.137778 PMID:32179352

Uyanık, O. (2014). Klasik ve jeofizik yöntemlerle yapı incelemeleri (Yapı Jeofiziği). TMMOB Jeofizik Mühendisleri Odası, JFMO Eğitim Yayınları.

Uyanık, O., Çekmen V. (2009). Zeminlerin basınç ve kesme mukavemetlerinin sismik hızlardan tahmini. *İstanbul Üniversitesi Yerbilimleri Dergisi*, *22*(2), 95-104.

Uyanık, N. A. (2022). Determination of the radiological risk and the cancer effect caused by geological units and samples from Afyon, Turkey. *Pure and Applied Geophysics*, *179*(4), 1295–1308. doi:10.100700024-022-02978-1

Uyanik, N. A., Akkurt, İ., & Uyanik, O. (2010). A ground radiometric study of uranium, thorium and potassium in Isparta, Turkey. *Annals of Geophysics*, *53*(5-6), 25–30.

Uyanik, N. A., Öncü, Z., Uyanik, O., & Akkurt, İ. (2015a). Determination of natural radioactivity from 232th with gamma-ray spectrometer in Dereköy-Yazır (Southwestern Anatolia). *Acta Physica Polonica A*, *128*(2B, No 2-B), B441–B442. doi:10.12693/APhysPolA.128.B-441

Uyanik, N. A., Oncu, Z., Uyanik, O., & Bozcu, M. (2022). Determination of alteration zones and geological unit limits using natural radioactivity properties of Sandikli-Suhut areas. *Journal of Applied Geophysics*, *196*, 104525. doi:10.1016/j.jappgeo.2021.104525

Uyanik, N. A., Öncü, Z., Uyanik, O., Bozcu, M., Akkurt, I., Günoglu, K., & Yagmurlu, F. (2015b). Distribution of natural radioactivity from 40k radioelement in volcanics of Sandıklı-Suhut (Afyon) Area. *Acta Physica Polonica A*, *128*(2B, No 2-B), B438–B440. doi:10.12693/APhysPolA.128.B-438

Uyanık, N. A., Uyanık, O., & Akkurt, İ. (2013c). Micro-zoning of natural radioactivity and seismic velocities of potential residential areas in volcanic fields: The case of Isparta (Turkey). *Journal of Applied Geophysics*, *98*, 191–204. doi:10.1016/j.jappgeo.2013.08.020

Uyanık, N. A., Uyanık, O., Gür, F., & Aydın, İ. (2013d). Natural radioactivity of brick sand brick material in the Salihli-Turgutlu area of Turkey. *Environmental Earth Sciences*, *68*(2), 499–506. doi:10.100712665-012-1754-4

Uyanık, O. (2002). *Kayma dalga hızına bağlı potansiyel sıvılaşma analiz yöntemi*. Dokuz Eylül University Natural and Applied Sciences, Doktora Tezi.

Uyanık, O. (2006). Sıvılaşır ya da sıvılaşmaz zeminlerin yinelemeli gerilme oranına bir seçenek. *DEU. Engineering Faculty Journal of Science and Engineering*, *8*(2), 79–91.

Uyanık, O. (2011). The porosity of saturated shallow sediments from seismic compressional and shear wave velocities. *Journal of Applied Geophysics*, *73*(1), 16–24. doi:10.1016/j.jappgeo.2010.11.001

Uyanık, O. (2012). Sismik hızlardan beton dayanımının belirlenmesi. *Jeofizik Bülteni*, *23*(70), 25–30.

Uyanık, O. (2015). Deprem ağır hasar alanlarının önceden belirlenmesi ve şehir planlaması için makro ve mikro bölgelendirmelerin önemi. *SDU Fen Bilimleri Enstitüsü Dergisi*, *19*(2), 24–38.

Uyanık, O. (2019). Estimation of the porosity of clay soils using seismic P- and S-wave velocities. *Journal of Applied Geophysics*, *170*(103832), 103832. doi:10.1016/j.jappgeo.2019.103832

Uyanık, O. (2020). Soil liquefaction analysis based on soil and earthquake parameters. *Journal of Applied Geophysics*, *176*, 104004. doi:10.1016/j.jappgeo.2020.104004

Uyanık, O., & Çatlıoğlu, B. (2014). Elektrik özdirenç ve sismik kırılma yöntemlerinden heyelan geometrisinin belirlenmesi. *SDU Fen Bilimleri Enst. Dergisi*, *18*(3), 22–29.

Uyanık, O., & Çatlıoğlu, B. (2015). Sismik hızlardan yoğunluğun belirlenmesi. *Jeofizik*, *17*, 3–15.

Uyanık, O., Ekinci, B., & Uyanık, N. A. (2013a). Liquefaction analysis from seismic velocities and determination of lagoon limits Kumluca/Antalya example. *Journal of Applied Geophysics*, *95*, 90–103. doi:10.1016/j.jappgeo.2013.05.008

Uyanık, O., Ekin, N., & Çoşkun, O. (2021). Zemin ve deprem parametrelerinden ivme azalım ilişkisi. *DEÜ Fen ve Mühendislik Dergisi*, *23*(68), 575–593. doi:10.21205/deufmd.2021236820

Uyanık, O., Kaptan, K., Gülay, F. G., & Tezcan, S. (2011). Beton dayanımının tahribatsız ultrasonik yöntemle tayini. *Yapı Dünyası*, *184*, 55–58.

Uyanık, O., & Sabbağ, N. (2013). Determination of landslide geometry by geophysical methods. *4th International Geosciences Student Conference Proceeding Book*.

Uyanık, O., Sabbağ, N., Uyanık, N. A., & Öncü, Z. (2019). Prediction of mechanical and physical properties of some sedimentary rocks from ultrasonic velocities. *Bulletin of Engineering Geology and the Environment*, *78*(8), 6003–6016. doi:10.100710064-019-01501-6

Uyanık, O., Şenli, G., & Çatlıoğlu, B. (2013b). Determination from non-destructive geophysics methods of concrete quality of buildings. *SDU Int. Technologic Science*, *5*(2), 156–165.

Uyanık, O., & Taktak, A. G. (2009). Kayma dalga hızı ve etkin titreşim periyodundan sıvılaşma çözümlemesi için yeni bir yöntem, *SDU Fen Bilimleri Enst. Dergisi*, *13*(1), 74–81.

Uyanık, O., & Türker, E. (2007). Fethiye Eşen II HES şalt ve santral sahasındaki potansiyel heyelanının yerteknik özellikleri ve yorumu. *SDU Fen Bilimleri Enst. Dergisi*, *11*(1), 84–90.

Uyanık, O., Türker, E., & İsmailov, T. (2006). *Sığ sismik mikro-bölgeleme ve Burdur/Türkiye örneği. Ekologiya və Su Təsərrüfatı Elmi-Texniki və istehsalat Jurnalı.*

Uyanık, O., & Ulugergerli, E. U. (2008). Quality control of compacted grounds using seismic velocities. *Near Surface Geophysics*, *6*(5), 299–306. doi:10.3997/1873-0604.2008004

Valdés, A., Ramos, M., Beltrán, A., Jiménez, A., & Garrigós, M. C. (2017). State of the art of antimicrobial edible coatings for food packaging applications. *Coatings*, *7*(4), 56. doi:10.3390/coatings7040056

van Hemert, S., Breedveld, A. C., Rovers, J. M. P., Vermeiden, J. P. W., Witteman, B. J. M., Smits, M. G., & de Roos, N. M. (2014). Migraine associated with gastrointestinal disorders: Review of the literature and clinical implications. *Frontiers in Neurology*, *5*, 241. doi:10.3389/fneur.2014.00241 PMID:25484876

Van Mierlo, B., Leeuwis, C., Smits, R., & Woolthuis, R. K. (2010). Learning towards system innovation: Evaluating a systemic instrument. *Technological Forecasting and Social Change*, *77*(2), 318–334. doi:10.1016/j.techfore.2009.08.004

Vandamme, E. J., & Mortelmans, K. (2019). A century of bacteriophage research and applications: Impacts on biotechnology, health, ecology and the economy! *Journal of Chemical Technology and Biotechnology*, *94*(2), 323–342. doi:10.1002/jctb.5810

Varolgüneş, F. K. (2021). Success factors for post-disaster permanent housing: Example of Turkish earthquakes. *The Turkish Online Journal of Design, Art and Communication*, *11*(1), 115-130.

Vassileva, S., & Mileva, S. (2010). AI-based software tools for beer brewing monitoring and control. *Biotechnology, Biotechnological Equipment*, *24*(3), 1936–1939. doi:10.2478/V10133-010-0060-0

Velickova, E., Winkelhausen, E., Kuzmanova, S., Moldao-Martins, M., & Alves, V. D. (2015). Characterization of multilayered and composite edible films from chitosan and beeswax. *Food Science & Technology International*, *21*(2), 83–93. doi:10.1177/1082013213511807 PMID:24285830

Vengarai Jagannathan, B., Kitchens, S., Priyesh Vijayakumar, P., Price, S., & Morgan, M. (2021). Efficacy of bacteriophage cocktail to control E. coli O157: H7 contamination on baby spinach leaves in the presence or absence of organic load. *Microorganisms*, *9*(3), 544. doi:10.3390/microorganisms9030544 PMID:33800760

Venskutonis, P. R., & Kraujalis, P. (2013). Nutritional components of amaranth seeds and vegetables: A review on composition, properties, and uses. *Comprehensive Reviews in Food Science and Food Safety*, *12*(4), 381–412. doi:10.1111/1541-4337.12021 PMID:33412681

Verbeek, C. J., & van den Berg, L. E. (2010). Extrusion processing and properties of protein-based thermoplastics. *Macromolecular Materials and Engineering*, *295*(1), 10–21. doi:10.1002/mame.200900167

Vierk, K. A., Koehler, K. M., Fein, S. B., & Street, D. A. (2007). Prevalence of self-reported food allergy in American adults and use of food labels. *The Journal of Allergy and Clinical Immunology*, *119*(6), 1504–1510. doi:10.1016/j.jaci.2007.03.011 PMID:17451802

Vikram, A., Tokman, J. I., Woolston, J., & Sulakvelidze, A. (2020). Phage Biocontrol Improves Food Safety by Significantly Reducing the Level and Prevalence of Escherichia coli O157:H7 in Various Foods. *Journal of Food Protection*, *83*(4), 668–676. doi:10.4315/0362-028X.JFP-19-433 PMID:32221572

Vikram, A., Woolston, J., & Sulakvelidze, A. (2021). Phage biocontrol applications in food production and processing. *Current Issues in Molecular Biology*, *40*, 267–302. doi:10.21775/cimb.040.267 PMID:32644048

Vilcacundo, R., & Hernández-Ledesma, B. (2017). Nutritional and biological value of quinoa (Chenopodium quinoa Willd.). *Current Opinion in Food Science*, *14*, 1–6. doi:10.1016/j.cofs.2016.11.007

Vilela, C., Kurek, M., Hayouka, Z., Röcker, B., Yildirim, S., Antunes, M. D. C., Nilsen-Nygaard, J., Pettersen, M. K., & Freire, C. S. (2018). A concise guide to active agents for active food packaging. *Trends in Food Science & Technology*, *80*, 212–222. doi:10.1016/j.tifs.2018.08.006

Villalpando-Aguilar, J. L., Matos-Pech, G., López-Rosas, I., Castelán-Sánchez, H. G., & Alatorre-Cobos, F. (2022). Phage therapy for crops: Concepts, experimental and bioinformatics approaches to direct its application. *International Journal of Molecular Sciences*, *24*(1), 325. doi:10.3390/ijms24010325 PMID:36613768

Volta, U., Caio, G., De Giorgio, R., Henriksen, C., Skodje, G., & Lundin, K. E. (2015). Non-celiac gluten sensitivity: A work-in-progress entity in the spectrum of wheat-related disorders. *Best Practice & Research. Clinical Gastroenterology*, *29*(3), 477–491. doi:10.1016/j.bpg.2015.04.006 PMID:26060112

Wada, A., Towhata, I., Tamura, K., & Qu, Z. (2018). Moving toward cities where earthquakes will not cause a grievous disaster. *Japan Architectural Review*, *1*(4), 10. doi:10.1002/2475-8876.12061

Wagenaar, J. A., Van Bergen, M. A., Mueller, M. A., Wassenaar, T. M., & Carlton, R. M. (2005). Phage therapy reduces Campylobacter jejuni colonization in broilers. *Veterinary Microbiology*, *109*(3–4), 275–283. doi:10.1016/j.vetmic.2005.06.002 PMID:16024187

Wagh, R. V., Priyadarshi, R., & Rhim, J.-W. (2023). Novel bacteriophage-based food packaging: An innovative food safety approach. *Coatings*, *13*(3), 609. doi:10.3390/coatings13030609

Waheed, M., Butt, M. S., Shehzad, A., Adzahan, N. M., Shabbir, M. A., Suleria, H. A. R., & Aadil, R. M. (2019). Eggshell calcium: A cheap alternative to expensive supplements. *Trends in Food Science & Technology*, *91*, 219–230. doi:10.1016/j.tifs.2019.07.021

Walia, J. S., Hämmäinen, H., Kilkki, K., & Yrjölä, S. (2019). 5G network slicing strategies for a smart factory. *Computers in Industry, 111*, 108–120. doi:10.1016/j.compind.2019.07.006

Wan, Liu, & Deng. (2012). HASBE: a hierarchical attributebased solution for flexible and scalable access control in cloud computing. *Information Forensics and Security, IEEE Transactions on, 7*(2), 743-754.

Wang, X., Puri, V. M., & Demirci, A. (2020). Equipment Cleaning, sanitation, and maintenance. *Food Engineering Series*, 333–353. doi:10.1007/978-3-030-42660-6_13

Wang, H., Ding, F., Ma, L., & Zhang, Y. (2021). Edible films from chitosan-gelatin: Physical properties and food packaging application. *Food Bioscience, 40*, 100871. doi:10.1016/j.fbio.2020.100871

Wang, J., & Hu, X. (2021). Convolutional neural networks with gated recurrent connections. *IEEE Transactions on Pattern Analysis and Machine Intelligence, 44*(7), 3421–3435. doi:10.1109/TPAMI.2021.3054614 PMID:33497326

Wang, S., Lu, A., & Zhang, L. (2016). Recent advances in regenerated cellulose materials. *Progress in Polymer Science, 53*, 169–206. doi:10.1016/j.progpolymsci.2015.07.003

Wang, X., & Gao, L. (2020). *When 5G Meets Industry 4.0*. Springer. doi:10.1007/978-981-15-6732-2

Wei, L. T., & Yazdanifard, R. (2013). Edible food packaging as an eco-friendly technology using green marketing strategy. *Global Journal of Commerce & Management Perspective, 2*(6), 8–11.

Whichard, J., Sriranganathan, M. N., & Pierson, F. W. (2003). Suppression of Salmonella growth by wild-type and large-plaque variants of bacteriophage Felix O1 in liquid culture and on chicken frankfurters. *Journal of Food Protection, 66*(2), 220–225. doi:10.4315/0362-028X-66.2.220 PMID:12597480

White, J. C., & Gardea-Torresdey, J. (2018). Achieving food security through the very small. *Nature Nanotechnology, 13*(8), 627–629. doi:10.103841565-018-0223-y PMID:30082813

Whitley, L. (2008). Food Processing Handbook - edited by Brennan J C, Wiley-VCH. International Journal of Dairy Technology, 61(3), 309–310. doi:10.1111/j.1471-0307.2008.00386.x

Williams, K., Ward, M., Dhungyel, O., Hall, E., & Van Breda, L. (2014). A longitudinal study of the prevalence and super-shedding of Escherichia coli O157 in dairy heifers. *Veterinary Microbiology, 173*(1-2), 101–109. doi:10.1016/j.vetmic.2014.07.001 PMID:25064268

Wirtz, B. W., Becker, M., & Schmidt, F. W. (2021). Smart city services: An empirical analysis of citizen preferences. *Public Organization Review, 22*(4), 1063–1080. doi:10.100711115-021-00562-0

Wójcicki, M., Błażejak, S., Gientka, I., & Brzezicka, K. (2019). The concept of using bacteriophages to improve the microbiological quality of minimally processed foods. *Acta Scientiarum Polonorum. Technologia Alimentaria, 18*, 373–383. PMID:31930789

Wong, C. W., Delaquis, P., Goodridge, L., Lévesque, R. C., Fong, K., & Wang, S. (2020). Inactivation of Salmonella enterica on post-harvest cantaloupe and lettuce by a lytic bacteriophage cocktail. *Current Research in Food Science, 2*, 25–32.

Woo, J., & Ahn, J. (2014). Assessment of synergistic combination potential of probiotic and bacteriophage against antibiotic-resistant Staphylococcus aureus exposed to simulated intestinal conditions. *Archives of Microbiology, 196*(10), 719–727. doi:10.100700203-014-1013-z PMID:25015717

Wu, J. H. Y., Neal, B., Trevena, H., Crino, M., Stuart-Smith, W., Faulkner-Hogg, K., Louie, J. C. Y., & Dunford, E. (2015). Are gluten-free foods healthier than non-gluten-free foods? An evaluation of supermarket products in Australia. *British Journal of Nutrition, 114*(3), 448–454. doi:10.1017/S0007114515002056 PMID:26119206

Xia, C., Wang, W., Wang, L., Liu, H., & Xiao, J. (2019). Multilayer zein/gelatin films with tunable water barrier property and prolonged antioxidant activity. *Food Packaging and Shelf Life, 19*, 76–85. doi:10.1016/j.fpsl.2018.12.004

Xiong, Y., Chen, M., Warner, R. D., & Fang, Z. (2020a). Incorporating nisin and grape seed extract in chitosan-gelatine edible coating and its effect on cold storage of fresh pork. *Food Control, 110*, 1–8. doi:10.1016/j.foodcont.2019.107018

Xue & Hong. (2014). A Dynamic Secure Group Sharing Framework in Public Cloud Computing. *Cloud Computing, IEEE Transactions on, 2*(4), 459-470.

Xu, P., Zeng, Q., Cao, Y., Ma, P., Dong, W., & Chen, M. (2017). Interfacial modification on polyhydroxyalkanoates/starch blend by grafting in-situ. *Carbohydrate Polymers, 174*, 716–772. doi:10.1016/j.carbpol.2017.06.048 PMID:28821123

Yadav, S., Mehrotra, G. K., Bhartiya, P., Singh, A., & Dutta, P. K. (2020). Preparation, physicochemical and biological evaluation of quercetin-based chitosan-gelatin film for food packaging. *Carbohydrate Polymers, 227*, 115348. doi:10.1016/j.carbpol.2019.115348 PMID:31590881

Yadav, S., Mehrotra, G. K., & Dutta, P. K. (2021). Chitosan based ZnO nanoparticles loaded gallic-acid films for active food packaging. *Food Chemistry, 334*, 127605. doi:10.1016/j.foodchem.2020.127605 PMID:32738726

Yahiaoui, F., Benhacine, F., Ferfera-Harrar, H., Habi, A., Hadj-Hamou, A. S., & Grohens, Y. (2015). Development of antimicrobial PCL/nanoclay nanocomposite films with enhanced mechanical and water vapor barrier properties for packaging applications. *Polymer Bulletin, 72*(2), 235–254. doi:10.100700289-014-1269-0

Yang, S. C., Lin, C. H., Aljuffali, I. A., & Fang, J. Y. (2017). Current pathogenic *Escherichia coli* foodborne outbreak cases and therapy development. *Archives of Microbiology, 199*(6), 811–825. doi:10.100700203-017-1393-y PMID:28597303

Yan, T., Liang, L., Yin, P., Zhou, Y., Mahdy Sharoba, A., Lu, Q., Dong, X., Liu, K., Connerton, I. F., & Li, J. (2020). Application of a novel phage LPSEYT for biological control of Salmonella in foods. *Microorganisms*, *8*(3), 400. doi:10.3390/microorganisms8030400 PMID:32178465

Yap, X. Y., Gew, L. T., Khalid, M., & Yow, Y. Y. (2023). Algae-Based Bioplastic for Packaging: A Decade of Development and Challenges (2010–2020). *Journal of Polymers and the Environment*, *31*(3), 833–851. doi:10.100710924-022-02620-0

Yavari, F., & Pilevari, N. (2020). Industry revolutions development from Industry 1.0 to Industry 5.0 in manufacturing. *Journal of Industrial Strategic Management*, *5*(2), 44–63.

Yaver, E., & Bilgiçli, N. (2020). Pseudocereals: Composition, effect on nutrition-health and usage in cereal products. *Food and Health*, *6*(1), 41–56. doi:10.3153/FH20006

Yeşil, Y., & Uyanık, O. (2012). Zemin büyütmesinin sismik hızlardan belirlenmesi. SDÜ Fen Bilimleri Enstitüsü, Yüksek Lisans Tezi.

Yildirim, E. (2020). Çölyak hastalığı ve glutensiz besleme. *Genel Sağlık Bilimleri Dergisi*, *2*(3), 175–187.

Yildirim, S., Röcker, B., Pettersen, M. K., Nilsen-Nygaard, J., Ayhan, Z., Rutkaite, R., Radusin, T., Suminska, P., Marcos, B., & Coma, V. (2018). Active packaging applications for Food. *Comprehensive Reviews in Food Science and Food Safety*, *17*(1), 165–199. doi:10.1111/1541-4337.12322 PMID:33350066

Yin, I. X., Zhang, J., Zhao, I. S., Mei, M. L., Li, Q., & Chu, C. H. (2020). The antibacterial mechanism of silver nanoparticles and its application in dentistry. *International Journal of Nanomedicine*, *15*, 2555–2562. doi:10.2147/IJN.S246764 PMID:32368040

Yin, S., & Yu, Y. (2022). An adoption-implementation framework of digital green knowledge to improve the performance of digital green innovation practices for industry 5.0. *Journal of Cleaner Production*, *363*, 132608. doi:10.1016/j.jclepro.2022.132608

Yin, Y., & He, G. (2008). A fast high-intensity pulsed electric fields (pef)-assisted extraction of dissoluble calcium from bone. *Separation and Purification Technology*, *61*(2), 148–152. doi:10.1016/j.seppur.2007.10.005

Yoon, S.-D. (2014). Cross-linked potato starch-based blend films using ascorbic acid as a plasticizer. *Journal of Agricultural and Food Chemistry*, *62*(8), 1755–1764. doi:10.1021/jf4024855 PMID:23909738

Yost, E., & Cheng, Y. (2021). Customers' risk perception and dine-out motivation during a pandemic: Insight for the restaurant industry. *International Journal of Hospitality Management*, *95*, 102889. doi:10.1016/j.ijhm.2021.102889 PMID:36540682

Yotova, L., Yaneva, S., & Marinkova, D. (2013). Biomimetic nanosensors for determination of toxic compounds in food and agricultural products. *Journal of Chemical Technology & Metallurgy*, *48*(3).

Yu, Wang, Ren, & Lou. (2010). Attribute based data sharing with attribute revocation. *Proceedings of the 5th ACM Symposium on Information, Computer and Communications Security,* 261-270.

Yuksel, A. S., Uyanik, O., & Er, K. (2020). Development of a fuzzy logic based online visualization application for 2D geotechnical cross-section modeling. *Earth Science Informatics, 13*(4), 1523–1538. doi:10.100712145-020-00500-6

Yu, Y., Jin, T. Z., Fan, X., & Wu, J. (2018). Biochemical degradation and physical migration of polyphenolic compounds in osmotic dehydrated blueberries with pulsed electric field and thermal pretreatments. *Food Chemistry, 239,* 1219–1225. doi:10.1016/j.foodchem.2017.07.071 PMID:28873543

Yu, Y., Jin, T. Z., Fan, X., & Xu, Y. (2017). Osmotic dehydration of blueberries pretreated with pulsed electric fields: Effects on dehydration kinetics, and microbiological and nutritional qualities. *Drying Technology, 35*(13), 1543–1551. doi:10.1080/07373937.2016.1260583

Zabihollahi, N., Alizadeh, A., Almasi, H., Hanifian, S., & Hamishekar, H. (2020). Development and characterization of carboxymethyl cellulose based probiotic nanocomposite film containing cellulose nanofiber and inulin for chicken fillet shelf-life extension. *International Journal of Biological Macromolecules, 160,* 409–417. doi:10.1016/j.ijbiomac.2020.05.066 PMID:32416305

Zaïdi, A. (2019). Necessary and Sufficient Conditions for the Existence of Robust Whitening Matrices. *IEEE Signal Processing Letters, 26*(6), 863–867. doi:10.1109/LSP.2019.2909651

Zarski, M., Wojcik, B., Miszczak, J., Blachowski, B., & Ostrowski, M. (2022). Computer vision based inspection on post-earthquake with UAV synthetic dataset. *IEEE Access : Practical Innovations, Open Solutions, 10,* 108134–108144. doi:10.1109/ACCESS.2022.3212918

Żbikowska, K., Michalczuk, M., & Dolka, B. (2020). The use of bacteriophages in the poultry industry. *Animals (Basel), 10*(5), 872. doi:10.3390/ani10050872 PMID:32443410

Zehetmeyer, G., Soares, R. M. D., Brandelli, A., Mauler, R. S., & Oliveira, R. V. B. (2012). Evaluation of polypropylene/montmorillonite nanocomposites as food packaging material. *Polymer Bulletin, 68*(8), 2199–2217. doi:10.100700289-012-0722-1

Zhang, H. Q., Barbosa-Cãnovas, G. V., Bala, B. V. M., Dunne, C. P., Farkas, D. F., & C., Y. J. T. (2011). Nonthermal Processing Technologies for Food. Wiley.

Zhang, H., Liang, Y., Li, X., & Kang, H. (2020). Effect of chitosan-gelatin coating containing nano-encapsulated tarragon essential oil on the preservation of pork slices. *Meat Science, 166,* 108137. doi:10.1016/j.meatsci.2020.108137 PMID:32272381

Zhang, H., Li, X., Kang, H., & Peng, X. (2021). Antimicrobial and antioxidant effects of edible nanoemulsion coating based on chitosan and Schizonepeta tenuifolia essential oil in fresh pork. *Journal of Food Processing and Preservation, 45*(11), e15909. doi:10.1111/jfpp.15909

Zhang, J., Bates, J., & Abbott, P. (2022). State-steered smartmentality in Chinese smart urbanism. *Urban Studies (Edinburgh, Scotland), 59*(14), 2933–2950. doi:10.1177/00420980211062888

Zhang, X., Niu, Y. D., Nan, Y., Stanford, K., Holley, R., McAllister, T., & Narváez-Bravo, C. (2019). SalmoFresh effectiveness in controlling salmonella on romaine lettuce, mung bean sprouts and seeds. *International Journal of Food Microbiology, 305*, 108250.

Zhao, R., Torley, P., & Halley, P. J. (2008). Emerging biodegradable materials: Starch-and protein-based bio-nanocomposites. *Journal of Materials Science, 43*(9), 3058–3071. doi:10.100710853-007-2434-8

Zhong, M., Yang, Y., Yao, H., Fu, X., Dobre, O. A., & Postolache, O. (2019). 5G and IoT: Towards a new era of communications and measurements. *IEEE Instrumentation & Measurement Magazine, 22*(6), 18–26. doi:10.1109/MIM.2019.8917899

Zhou, Y. H., Mujumdar, A. S., Vidyarthi, S. K., Zielinska, M., Liu, H., Deng, L. Z., & Xiao, H. W. (2021). Nanotechnology for food safety and security: A Comprehensive Review. *Food Reviews International*, 1–21.

Zhu, L., & Xanthos, M. (2004). Effects of process conditions and mixing protocols on structure of extruded polypropylene nanocomposites. *Journal of Applied Polymer Science, 93*(4), 1891–1899. doi:10.1002/app.20658

Zia, S., & Alkheraije, K. A. (2023). Recent trends in the use of bacteriophages as replacement of antimicrobials against food-animal pathogens. *Frontiers in Veterinary Science, 10*, 1162465. doi:10.3389/fvets.2023.1162465 PMID:37303721

Zimmer, K.-P. (2011). Nutrition and celiac disease. *Current Problems in Pediatric and Adolescent Health Care, 41*(9), 244–247. doi:10.1016/j.cppeds.2011.04.004 PMID:21939908

Žlabur, J., Voća, S., Brnčić, M., & Rimac-Brnčić, S. (2018). New trends in food technology for green recovery of bioactive compounds from plant materials. *Role of Materials Science in Food Bioengineering*, 1–36. doi:10.1016/B978-0-12-811448-3.00001-2

Zook, M. (2017). Crowd-sourcing the smart city: Using big geosocial media metrics in urban governance. *Big Data & Society, 4*(1), 4. doi:10.1177/2053951717694384

Zuidmeer, L., Goldhahn, K., Rona, R. J., Gislason, D., Madsen, C., Summers, C., Sodergren, E., Dahlstrom, J., Lindner, T., Sigurdardottir, S. T., McBride, D., & Keil, T. (2008). The prevalence of plant food allergies: A systematic review. *The Journal of Allergy and Clinical Immunology, 121*(5), 1210–1218. doi:10.1016/j.jaci.2008.02.019 PMID:18378288

Zulfiqar, F., Navarro, M., Ashraf, M., Akram, N. A., & Munné-Bosch, S. (2019). Nanofertilizer use for sustainable agriculture: Advantages and limitations. *Plant Science, 289*, 110270. doi:10.1016/j.plantsci.2019.110270 PMID:31623775

About the Contributors

Shilpa Mehta has completed her PhD degree from Auckland University of Technology, New Zealand. She has five years of teaching experience from different reputed colleges and universities of India. Currently, she is working as a Teaching Assistant at Auckland University of Technology (AUT), Auckland, New Zealand. She was awarded Summer Doctoral Research Scholarship for her PhD research. She has worked on various interdisciplinary research projects and her research interests include Radio Frequency Integrated Circuits, RF front ends, Optimization, Internet of Things, Wireless Communication, Artificial Intelligence, Healthcare, Radars, and Software-defined Radios, Smart Cities. She is working as a Series Editor for one of the Bentham book series. As an editor, she has edited two books with IGI Global and one with Nova Science Publishers. Additionally, she is currently working on different edited books to be published by IGI Global, Apple Academic Press, Scrivener-Wiley, and Springer Nature, respectively.

Fakhar Islam is working as lecturer in the Department of Food Science and Technology at NUR International University Lahore. Moreover, he has published more than 45 international publications having impact factor of 120 and 3 international book chapters.

Ali Imran is working as associate professor in the Department of Food Science and Technology at Government college university Faisalabad. Moreover, he has published more than 120 international publications having impact factor of 350 and 10 international chapters and 5 books published as editor. He is also working as deputy director of industrial linkages at Government College University Faisalabad.

* * *

Arisha Ahmad has completed her Bachelor's degree Doctor of Dietetics and Nutritional Sciences from The University of Lahore (UOL), Lahore and currently completing her MS Food, Nutrition and Dietetics degree from Nur International University (NIU), Lahore.

Shabbir Ahmad is working as Assistant professor in the Department of Food Science and Technology, Muhammad Nawaz Shareef University of Agriculture, Multan. Moreover, he has published 25 articles in National and international journals having impact factor of 35 and 4 international book chapters.

Darakhshan Ahmed completed her Masters in Food Sciences and technology from UET Lahore with the research focused on fortification of buttermilk with beetroot juice as a source of Folic Acid. Recently she is working as a Quality Assurance officer in Nishat Hotel, Lahore

Tugba Aktar is an Associate Professor of Food Engineering Deparment at Alanya Alaaddin Keykubat University. She majored in Food Engineering at the Erciyes University and has a degree of MSc and Ph.D. in Food Science and Nutrition from the University of Leeds in 2012 and 2015, respectively. Her research interests include sensory studies, texture analysis, rheology analysis, modeling, consumer behavior, as well as human sensation, perception mechanisms, and mechanoreceptors. She is involved in several infrastructural and research projects and currently works on experimental research in the abovementioned areas.

Zunaira Basharat is PhD Scholar at School of Food & Biology Engineering, Jiangsu University, Zhenjiang, China. She has 07 International Publications with an impact factor of 20.2.

Aliza Batool currently enrolled in PhD Food Science and Technology in the Department of Food science and Technology Muhammad Nawaz Sharif University of Agriculture, Multan, Pakistan. I am a PhD Scholar and hold MSc hons degree in Food Science and Technology with research title 'Nutraceutical potential of flaxseed oil with special reference to cancer'. And now in PhD am working on the topic 'Hypolipidemic potential of olive oil and its food application'. I have great interest in the management of different diseases through healthy diet and functional food. I have write multiple research paper, peer reviewed paper and book chapters. I present my research work in national and international level and my area of research is, food nutrition, product development, food technology, food safety and quality.

İpek Bayrakcı, is graduated from Acıbadem University with double major at the Department of Nutrition and Dietetics and Department of Psychology in 2019 and 2020, respectively. In 2023, she started her MSc in Food Engineering Department at Alanya Alaaddin Keykubat University. She continues to work as a dietitian and her research interests include gluten-free product studies, nutrition models in diseases and alternative foods.

Korkmaz Bellitürk is Full Professor of Soil Science and Plant Nutrition Department of Agriculture Faculty at the Tekirdağ Namık Kemal University, in Tekirdağ, Türkiye. He served as project head and researcher in 29 international projects. He has 155 articles, 1 patent, 15 book editors, 21 book chapters and 3 books.

Pankaj Dadheech is currently working as a Professor & Deputy Head in the Department of Computer Science & Engineering (NBA Accredited), Swami Keshvanand Institute of Technology, Management & Gramothan (SKIT), Jaipur, Rajasthan, India (Accredited by NAAC A++ Grade). He has more than 18 years of experience in teaching. He is currently working a Professor & Dy. HOD in the Department of Computer Science & Engineering (NBA Accredited), Swami Keshvanand Institute of Technology, Management & Gramothan (SKIT), Jaipur, Rajasthan, India. He has published 25 Patents at Intellectual Property India, Office of the Controller General of Patents, Design and Trade Marks, Department of Industrial Policy and Promotion, Ministry of Commerce and Industry, Government of India. He has published 8 International Patents (USA, South African, Australian, Germany) & 2 Copyrights. He has 73 publications in various International & National Journals, 63 papers in various National & International conferences. He has published 9 Books & 35 Book Chapters.

Nevbahar Ekin currently works at the Department of Geophysical Engineering, Süleyman Demirel University in Isparta-Turkey. She completed her PhD in "Süleyman Demirel Univesity (Turkey)" under supervision of Assoc. Prof. Dr. Osman Uyanık. Her researches are in Counstruction Geophysics, Electrical Prospecting, Soil dynamic, soil mechanic, Geotechnics. She is involved to a number of different national project.

Umar Farooq is currently working as Professor in the Department of Food Science and Technology Muhammad Nawaz Sharif University of Agriculture, Multan, Pakistan. By profession I am a researcher and hold PhD degree in Food Science and Technology. I have 17 funded projects as s principal or co-principal investigator, which amounted to 110.51 million Pakistani rupees. I have 69 international peer-reviewed publications with an approximate accumulated impact factor of 150 and also authored two books and contributed 11 chapters to the edited books. In parent University I also working as Dean of the Faculty of Food and Home Sciences, Chairman Department of Human Nutrition, Additional Director of ORIC, Principal Officer of Academics, and Chairman Admission Committee. My area of research is food microbiology, food nutrition, food safety and quality and I also present my research work at various national and international level.

Nida Firdous is working as Assistant Professor in Department of Food Science and Technology, FF&HS, MNS-University of Agriculture, Multan. Experience of working as Assistant Food Safety Officer, Scientific officer and master trainer in Punjab Food Authority for 5 years. Worked as Visiting Scholar at University of Minnesota, USA. Published 12 peer reviewed research papers in national and international journals. Authored one book and worked on various research projects. Received letter of appreciation and multiple honoraria rewards from Govt. of Pakistan on multiple occasions for working on food safety related issues in Punjab. Acquired multiple trainings on HACCP Level-IV Highfield Level 4 Award in HACCP for Management (CODEX Principles) (RQF), International Food Standards, Global market programs: A pathway towards GFSI-Recognized certification, Quality Management System and Six Sigma, E-Codes and their critical halal aspects in Food industries, Traceability: A concept in halal food chain, and "Supercritical fluid extraction technique: Isolation of Functional and Nutraceutical food components".

Predrag Ilić currently works at Institute for Protection and Ecology of the Republic of Srpska. Predrag does research in Environmental Engineering and Ecological Engineering. He has a long experience in the fields of environmental protection, wastewater, chemicals, waste management, air protection, soil protection, energy and greenhouse gas, environmental impact assessment. He used to work as an expert in the fields of chemicals, environmental protection and waste management plans.

Zafar Iqbal is working as Food Safety Officer in Research and Development section, Punjab Food Authority. He has experience of working as research associate in many research projects as well has strong background of working in various food processing industries. He has authored six research article in reputed journals and worked as visiting scholar in Michigan State University, USA. He worked as a team member to develop various analytical method for adulteration detection.

Maliha Islam is working as an intern at Department of research and development at Vital Agri Nutrients Pvt Ltd. Moreover she has published one international book chapter.

Zulqurnain Khan is working as Assistant professor in the Institute of Plant breeding and genetics at Muhammad Nawaz Shareef University of Agriculture Multan Pakistan. Moreover he has published more than 50 international publications having impact factor of more 100 and 4 international books.

Samreen Latif currently serves as a Lecturer in the Department of Clinical Nutrition at Nur International University. She holds a master's degree in Food Nutrition and Dietetics from Nur International University Lahore and is actively pursuing a Ph.D. in Human Nutrition and Dietetics. Ms. Latif specializes in qualitative studies within the realm of research, showcasing a dedicated focus on advancing knowledge in her field.

Paavanalaxmi is an Assistant Professor at Sahyadri College of Engineering & Management, Mangaluru. She obtained her B.E. (Electronics & Communication Engineering) from NMAMIT, Nitte and M. Tech. (Digital Communication and Networking) from GIT, Belgaum, Karnataka, India. She has published 10 research papers in various international journals. She has authored three Book Chapters which is published by CRC Press, Springer, IoP Publications. Her research interest includes Embedded System, Control System, Communication and Networking.

Praveen Kumar M. is working as an Assistant Professor in the Department of Electronics and Communication Engineering in Sahyadri College of Engineering and Management, Mangaluru, Affiliated to Visvesvaraya Technological University. He is having more than 8 years of experience in the teaching field.

Zain Mushtaq is working as an Assistant Professor in the Department of Soil Science at University of the Punjab Lahore. Moreover he has published more than 20 international publications having impact factor of 51 and 2 book chapters.

Roopashree Nayak completed BE in Electronics and Communication Engineering and Masters in Digital Electronics and Communication, have been into teaching for more than 15 years and research interest in Artificial Neural Networks. She has published many research papers in various international journals. She has authored Four Book Chapters which is published by CRC Press, Springer, IoP Publications.

Digvijay Pandey is a lecturer Department of technical Education, Research Scholar IET Lucknow, India.

Amir Parnian is affiliated as an Assistant Professor of environmental soil chemistry and faculty staff at the National Salinity Research Center, Yazd, Iran. His major experiences are environmental remediation and composting, but he has some experience with crop science, landscape management, and preservation. His current research efforts are environmental remediation in saline conditions, phosphorus uptake, and behavior in saline environments.

Muhammad Usama Pervaiz is M.Phil. Scholar at Department of Clinical Sciences & Nutrition, Chester University, UK. He has 01 International Publications with an impact factor of 2.9.

Salwa Saeed is working as an intern at Department of research and development at Vital Agri Nutrients Pvt Ltd. Moreover she has published one international book chapter.

Nitin N. Sakhare received B. E. Computer degree in Computer Engineering from Vishwakarma Institute of Information Technology, Pune affiliated to Savitribai Phule Pune University, formerly known as Pune University, in 2012 and Master's degree in Computer Engineering from Sinhgad College of Engineering, Pune affiliated to Savitribai Phule Pune University, formerly known as Pune University, in 2012. He is currently working toward the Ph.D. degree at the Department of Computer Science and Engineering, KL University, Vijaywada, Andhra Pradesh. He is currently working as an assistant professor in the department of Computer Engineering of Vishwakarma Institute of Information Technology, Pune. His research interests include Data Science, Machine Learning, Computer Networking and Internet of Things.

Vetrivel S. C. is a faculty member in the Department of Management Studies, Kongu Engineering College (Autonomous), Perundurai, Erode Dt. Having experience in Industry 20 years and Teaching 16 years. Awarded with Doctoral Degree in Management Sciences in Anna University, Chennai. He has organized various workshops and Faculty Development Programmes. He is actively involved in research and consultancy works. He acted as a resource person to FDPs & MDPs to various industries like, SPB ltd, Tamilnadu Police, DIET, Rotary school and many. His areas of interest include Entrepreneurship, Business Law, Marketing and Case writing. Articles published more than 100 International and National Journals. Presented papers in more than 30 National and International conferences including IIM Bangalore, IIM Kozhikode, IIM Kashipur and IIM Indore. He was a Chief Co-ordinator of Entrepreneurship and Management Development Centre (EMDC) of Kongu Engineering College, he was instrumental in organizing various Awareness Camps, FDP, and TEDPs to aspiring entrepreneurs which was funded by NSTEDB – DST/GoI.

Afshan Shafi working as an assistant professor in Department of Human Nutrition and Dietetics, faculty of Food and Home sciences, MNS-University of Agriculture, Multan. She has completed PhD in 2018 from Sargodha University. She has published 41 research articles with total impact factor of 102 and 13 international book chapters.

Muhammad Sibt-e-Abbas is working as Assistant Professor in the Department of Food Science and Technology at MNS University of Agriculture Multan. Moreover, he has published more than 15 international publications having impact factor of 20 and 1 international book chapter along with participation in several national & International conferences/seminars as Guest Speaker. Dr. Sibt-e-Abbas also worked as Visiting Research Scholar at Texas A&M University, Texas USA.

Tahira Siddique is currently working as Deputy Director, Islamabad Food Authority. Previously she has worked as Assistant Food Safety officer in many districts of Punjab in Punjab Food Authority. She has a grasp on food safety related issues and authored multiple research papers as well in reputed peer reviewed journals.

Muhammad Asim Ijaz Sidhu is currently working as Deputy Director in Punjab Food Authority having extensive experience related o food safety and regulations. He has also worked as Deputy Director in Research and Development section of Punjab Food Authority and plays his part in developing adulteration detection techniques. He authored various publication in reputed journals as well.

Hafiz Suleria is an Australian Professor. He is working as senior Lecturer at The University of Melbourne, Australia.

Alghayia Tahir is a Consultant Clinical Nutritionist/Dietitian who completed her bachelor's in Food Science & Nutrition from GC University Faisalabad and has currently completing her MS Food, Nutrition and Dietetics degree from Nur International University (NIU), Lahore. She is working as Teaching Fellow in the Department of Clinical Nutrition at NIU since 2020. She has had a practice experience for 6 years displaying a vast diversity of clientele seeking her expertise.

Muhammad Rizwan Tariq is Assistant Professor at Department of Food Sciences, University of the Punjab, Lahore, Pakistan. He has 47 International Publications with 10 h-index and i-10 index.

Mohanasundaram Thangamuthu has completed his Doctoral degree in Management from Anna University. He has done his Master of Business Administration (MBA) from Bharathiar University. He has 20 years of total experience in Industry and Academics. He has published 40 research articles in reputed Journals (some are indexed in Scopus, ABDC, IEEE, UGC Care) and presented many research papers at International Conferences organized by IIMs, IITs, IISC and other reputed institutions. He is actively involved in research and consultancy works for Government, public and private institutions. He has received grants and financial assistance from various agencies like AICTE, ICSSR, AIMS, ICWA and NHRC for research projects and for organizing research programmes. He is a recognized Research Supervisor in Visvesvaraya Technological University, Belagavi. He has organised many FDPs/Workshops and MDPs. He has acted as a resource person for more than 60 programmes related to Management research that are organised across the country (including BITS Pilani, IMS Gaziabad, VESIM Mumbai, Osmania University, Hyderabad). His favourite subjects of teaching are; Predictive Analytics, International Finance, Financial Management, Bank Management, Financial and Management Accounting, Security Analysis & Portfolio Management and Business Research Methodology. He is a certified financial education Resource Person of the Securities and Exchange Board of India (SEBI). He has received the national level 'Best Young Teacher Award-2018' from the Association of Indian Management Schools (AIMS) and Best Young Faculty award from Nehru Group of Institutions, Coimbatore.

Muhammad Usman working as lecturer in Department of Human Nutrition and Dietetics Faculty of Food and Home Sciences MNS University of Agriculture Multan. He has completed Masters in 2020 from National Institute of Food science and Technology and medalist of his bachelor and master degrees. He has published 3 research articles and 3 Book chapters.

Nurten Ayten Uyanık currently works at the Vocational School of Tech. Sciences, Isparta University of Applied Sciences in Isparta-Turkey. She completed her PhD in "Süleyman Demirel Univesity (Turkey)" under supervision of Prof. Dr. İskender Akkurt. Her researches are in Natural Radioactivity, Magnetic Prospecting, Soil dynamic, Geotechnics. She is involved to a number of different national project.

Osman Uyanık currently works at the Department of Geophysical Engineering, Süleyman Demirel University in Isparta-Turkey. He completed his PhD in "Dokuz Eylül Univesity (Turkey)" under supervision of Prof. Dr. A. Güngör Taktak. His researches are in Seismic Refraction, Soil dynamic, Soil mechanic, Geotechnics, Counstruction Geophysics, Natural Radioactivity. He is involved to a number of different national project.

Wisha Waheed has completed her MS Food, Nutrition and Dietetics degree from Nur International University (NIU), Lahore and currently completing her PhD from University of Management and Technology, Lahore. She is working as lecturer in the Department of Clinical Nutrition at NIU since 2021. Her research interests include qualitative and cross-sectional researches in the field of nutrition.

Index

H

I

L

M

N

P

Q

R

S

T

U

W

www.ingramcontent.com/pod-product-compliance
Lightning Source LLC
LaVergne TN
LVHW080345011026
849161LV00009B/2197
* 9 7 9 8 3 6 9 3 0 8 1 9 6 *